BUILDING ECONOMICS: THEORY AND PRACTICE

BUILDING ECONOMICS: THEORY AND PRACTICE

Rosalie T. Ruegg
and
Harold E. Marshall

VNR Springer Science+Business Media, LLC

Library of Congress Catalog Card Number 90-33143
Originally published by Van Nostrand Reinhold in 1990.
Softcover reprint of the hardcover 1st edition 1990.

ISBN 978-1-4757-4690-7 ISBN 978-1-4757-4688-4 (eBook)
DOI 10.1007/978-1-4757-4688-4

16 15 14 13 12 11 10 9 8 7 6 5 4 3 2 1

Library of Congress Cataloging-in-Publication Data

Ruegg, Rosalie T.
Building economics: theory and practice / Rosalie T. Ruegg and Harold E. Marshall.
p. cm.
Includes bibliographical references.

1. Building—Economic aspects. I. Marshall, Harold E. II. Title.
TH437.R88 1990
332.63'243—dc20 90-33143
CIP

To
Bill Ruegg
and
Janet, Casey, and Brian Marshall

Contents

PART 4

APPLICATIONS OF ECONOMIC EVALUATION METHODS AND RISK ANALYSIS TECHNIQUES

APPENDIXES

INDEX

Foreword

We no longer build buildings like we used to nor do we pay for them in the same way. Buildings today are no longer only shelter but are also life support systems, communication terminals, data manufacturing centers, and much more. Buildings are incredibly expensive tools that must be constantly adjusted to function efficiently. The economics of building has become as complex as its design.

When buildings were shelter they lasted longer than their builders. The average gothic master mason lived 35 or 40 years. Cathedrals took 3 or 4 hundred years to build. Cost estimates were verified by great great grandchildren of the original designer.

Today, creative economics has become as important as creative design and creative building. The client brings builder, contractor, architect, and facilities manager to account in their life time. The cost of building can therefore no longer be left to chance or act of god.

Solutions are no longer as ingeniously simple as those proposed by a Florentine builder early in the 15th century. He proposed to center the dome of S. Maria del Fiore on a great mound of earth mixed with pennies. When the job was done street urchins would carry away the dirt in their search for the pennies. This was a serious suggestion offered by an early construction manager before Brunelleschi solved the problem more sensibly.

When OPEC increased oil prices fourfold in the early 1970s and then doubled them again in 1979, dirt and penny economics went out the window. Building costs became serious stuff. The cost of energy did more to improve the basic design and cost procedures for architecture than eclecticism, modern architecture, post modern architecture, and deconstructivism combined.

During the troubled years following the oil embargo architects, designers, builders, and students were aided in their search for understanding of the new economic conditions by a series of small books, research reports, and journal articles written by Ms. Rosalie Ruegg and Dr. Harold Marshall.

I was privileged to work with these two—economists so thoroughly familiar with their mysterious science that they could make it comprehensible to builders, engineers, and architects. They made it not only understandable, but interesting and often fascinating.

You will find in this book the distillation of two decades of national and international experience in the field of building economics. They have carried the crusade of understandable economics to architects, builders, engineers, real estate developers, facilities planners, managers, and building economists throughout the U.S. and Europe.

This book is based on real world problems and solid experience. You do not have to be an economist or an accountant to understand it. The approach is systematic, practical problem solving. All design and building decisions are weighed against cost and value. All the tools you might need are here.

Anyone, from economists to those who never bothered their head about the matter before, can use the book and will find it fascinating. A knowledge of building is helpful but not essential. The techniques apply to any investment decision even though the focus is on building and building systems. The book will add to your knowledge if you have some to begin with or can be a good beginning.

Try it—you will like it.

Forrest Wilson

Preface

WHY BUILDING ECONOMICS?

This book will show you how to apply the concepts and methods of economics to decisions about the location, design, engineering, construction, management, operation, rehabilitation, and disposition of buildings. Use of these tools can lower costs and increase profits from buildings.

WHO NEEDS THIS BOOK?

This book is for people concerned with the economic efficiency of buildings:

- For practicing architects, engineers, real estate developers, facilities planners and managers, building economists, and other professionals who make decisions about buildings, it is a comprehensive, readable reference guide.
- For professors and students of advanced undergraduate and graduate courses in schools of architecture, engineering, economics, quantity surveying, and business administration, it is an educational textbook with theory demonstrated in many examples.
- For those designing and attending short training courses and workshops in building economics and life-cycle cost analysis, it is a source book.
- For engineers and architects preparing for professional exam boards, it is a valuable study aid.
- For researchers and analysts who evaluate the potential of new building technologies and provide analyses in support of public policies affecting buildings, it is a concise, practical guide.

This book assumes neither that you are an economist nor that you aspire to become an economist. It does assume that you are interested in improving the economy of your buildings.

No specialized background in mathematics or statistics is required. Basic algebra and elementary statistics are sufficient for understanding the concepts and for applying most of the methods and techniques.

WHY THIS BOOK?

There are few books which combine economic concepts with building technology, and most of them tend to be theoretical and macro-oriented; i.e., directed towards general analysis rather than particular building decisions. Most do not describe economic methods to evaluate building investments or provide help in finding appropriate data for evaluations. This book fills the void with a practical guide grounded in theory.

WHY WE WROTE IT

Extensive first-hand experience with day-to-day real-world problems enabled us to develop practical approaches to building evaluation, which we bring to this book. Our active involvement with professional groups gave us a broad awareness of current practices and needs in the field both in the United States and throughout the world, and enabled us to provide you with state-of-the-art methods and techniques which are sound and well accepted. As a result of years of working at the intersection of theory and practice, we have attempted to bring to you a unique book, solid in theory and rich in practical application.

Over the past 10 years, we have taught building economics in short courses to more than 1,000 building professionals in the United States, Europe, and Asia. Their enthusiastic response and favorable comments convinced us of the need for a book like this which meets the special needs of the building community. Our consulting and research experience helped us focus on those special needs.

We worked extensively with state and federal government agencies to develop economic guidelines for making building decisions. For example, we developed guidelines for the U.S. Department of Energy for use by all federal agencies in evaluating energy conservation decisions in more than 400,000 federal buildings. We also developed training courses and audiovisual training films in building economics for the U.S. Department of Energy, as well as courses for the U.S. General Services Administration and the Army Corps of Engineers.

We worked for many years with standards-writing groups and professional societies to formulate building economics methods and standards. These groups include the American Society for Testing and Materials (ASTM), with its more than 30,000 members from industry, government, and academia; the American Association of Cost Engineers (AACE); the American Society of Heating, Refrigerating and Air-Conditioning Engineers (ASHRAE); the American Consulting Engineers Council (ACEC); the International Energy Agency (IEA); the International Council for Building Research, Studies, and Documentation (CIB), which sponsors a Building Economics Working Commission with members from 50 countries; and the Architectural Research Centers Consortium (ARCC) Panel on Building Economics and Industry Studies.

We also provided economic consulting for engineers and architects on individual building projects, and for researchers and analysts investigating advanced building technologies.

ABOUT THE CONTENTS

The book is well balanced between the theoretical and practical issues of building economics. It has four major parts: Part 1 presents six methods of economic evaluation and shows how and how not to use them. The methods covered are life-cycle costing, net benefits (net savings), benefit-to-cost (savings-to-investment) ratio, internal rate of return, overall rate of return, and payback. Simple problems illustrate the methods.

Part 2 explains what you need to know to apply the methods. For those who do not understand the time value of money and discounting, it provides a tutorial. It also explains how to perform evaluations in the face of inflation, how to adjust for tax effects, and how to find and use necessary data.

Part 3 presents techniques for dealing with uncertainty and risk. Techniques covered are conservative benefit and cost estimating, breakeven analyses, sensitivity analysis, risk-adjusted discount rate, certainty equivalent technique, input estimation using expected values, mean-variance criterion and coefficient of variation, decision analysis, simulation, mathematical/analytical technique, portfolio analysis, and the capital asset pricing model.

Part 4 demonstrates how to apply the methods of economic evaluation and techniques for dealing with uncertainty and risk to solve real-world problems. Quick reference guides direct you to case studies which use particular methods and techniques and show you where to find cases applicable to special topics, such as energy conservation, equipment replacement, and leasing versus buying. The case studies are in a separate part rather than in Parts 1 and 3 for two reasons. First, the case studies illustrate combinations of economic methods and risk analysis techniques. Second, the case studies are sufficiently comprehensive and complex that they deserve a focused presentation.

The case studies breathe life into the theory, helping the novice understand fundamentals and the seasoned professional see new problem-solving capabilities. They are taken from business, government, nonprofit organizations, and homeowners. The cases demonstrate how to use the methods presented in Part 1 and the techniques presented in Part 3 to make decisions such as the following:

- Which alternative building design and size to select? Which mechanical system? Which furnishings?
- Where to locate a building?
- Whether to lease or buy equipment and buildings?
- What size building investment to make?
- Which projects to select when funds are limited?
- What combination of interdependent building systems to use?
- Whether to invest in a new building technology?

We designed the case studies for instructional use, presenting them in a consistent, step-by-step format. They are more comprehensive and realistic than typical textbook problems, and are logically and systematically formulated and solved. The essential logic is first given in words, followed by quantitative analysis.

By walking through the problem formulations and solutions, practitioners can gain confidence in their ability to solve their own problems. Professors will find the cases appropriate for take-home exam questions and for special projects.

Seven appendixes provide a glossary, data and data sources, an annotated list of computer programs, worksheets, and other supporting information.

HOW TO PROCEED

If you come to the book with no prior experience or training in building economics, you may want to read Parts 1 through 3 completely before going to the examples in Part 4. Alternatively, as you read about each method in Part 1 and each technique in Part 3, you may wish to have a look at the specific case illustrations of that method or technique in Part 4. The best way to learn how to use the methods and techniques is first to study them, next master the material in Part 2, and then work through the sample problems.

If you are already knowledgeable about the economic methods and risk analysis techniques, you may wish to proceed directly to those case studies in Part 4 which hold special interest for you.

We welcome your comments and suggestions for improvements.

ACKNOWLEDGMENTS

A number of people influenced, supported, and encouraged the preparation of the book. Deserving special thanks for their valuable technical reviews are Dr.

Tung Au, professor in the Department of Civil Engineering of Carnegie Mellon University; Dr. Ranko Bon, associate professor of building economics and technology in the School of Architecture and Planning of the Massachusetts Institute of Technology; Dr. Carol Rawie, economist at the U.S. Environmental Protection Agency; and Dr. Amrik Rakhra, economist at the National Research Council of Canada. In addition, we thank Ms. Ida Wooten Tripp, writer, who provided helpful editorial suggestions.

Part 1

METHODS OF ECONOMIC EVALUATION

1

Building Economics at a Glance

1.1 BUILDING DECISIONS AND THE NEED FOR ECONOMIC TOOLS

Building owners generally wish to lower costs or increase profits. To accomplish this, buildings must be located, designed, engineered, constructed, managed, and operated with an eye to the economic consequences of these decisions. Owners need to select locations which enhance revenue opportunities, or lower costs, or both. Architects need to consider the owning and operating costs of alternative building designs. Mechanical and structural engineers need to take into account the economy of alternative designs and sizes of building systems and components. Architects and engineers need to work together to make economic tradeoffs between the building envelope and mechanical systems. Construction companies and builders need to select cost-effective materials, equipment, and construction techniques. Building managers and operators need to establish cost-effective maintenance, repair, and replacement policies, and to decide when and what to renovate. In short, those who design, engineer, construct, manage, operate, and own buildings are faced with numerous decisions which affect the economics of buildings. Typical examples of decisions that affect the economics of buildings are presented in Chart 1-1.

Those who use economics tools to improve their decision making are practicing building economics. Building professionals and owners, however, are not always well trained or experienced in selecting and applying tools of economic analysis to make economic building decisions. Moreover, there are few books and training courses which treat the application of economics to building decisions.[1] The unfortunate result is that many opportunities for improving the economic performance of buildings are lost. This book supplies both the conceptual base and practical tools to help the building community make better economic decisions about buildings.

[1]Books that do combine economic concepts with building technology have been written by Bon (1989), Hillebrandt (1974), and Stone (1975).

Chart 1-1. Decisions which affect the economics of buildings

Deciding to accept or reject a project
Deciding where to locate a building
Choosing between leasing and buying
Selecting among alternative building and system designs and sizes (envelope, mechanical system, furnishings, layout)
Choosing among alternative building investments
Establishing operation, maintenance, and repair policies
Choosing a replacement schedule
Combining interdependent building projects (envelope design and HVAC system, windows and lighting systems, fire sprinkler systems and use of fire-resistant building materials)

1.2 CONCEPTS OF ECONOMIC EFFICIENCY

Improving the economic efficiency or cost effectiveness of buildings is at the heart of this book. Before going further, we want you to understand basic concepts of economic efficiency which will help you understand the book, diagnose economic problems, and select and apply appropriate economic methods and techniques.

1.2.1 Economic Efficiency and Cost Effectiveness

A building decision is "economically efficient" from the standpoint of the decision maker if net benefits (profits) are maximized as a result of the decision.[2] For example, if investing in an office building is both (1) profitable and (2) more profitable than other available options for an investor, it is the economically efficient investment choice for that investor. If a building location will yield greater revenue but will cost the same or less than alternative locations,

[2]We use the term "economic efficiency" in a more narrow sense than it is generally used in economic theory. There the term generally refers to the maximum improvement in social welfare under the prevailing income distribution. Resource use is said to be economically efficient when the allocation among competing uses yields maximum total benefits to members of society. Market imperfections can cause the interests of the individual to diverge from the interest of society. Sometimes a decision which maximizes an individual's net benefits may have external effects ("externalities") which adversely affect other members of society and cause the results to be economically inefficient for society. For example, constructing a tall building in location A may maximize the owner's profits and, hence, be economically efficient from the standpoint of the owner, but by blocking the view of neighboring buildings or increasing traffic congestion, it may not be economically efficient for society. Reflecting our concern in this book with decisions at many different levels, we use the term "economic efficiency" in a limited sense to mean that resources available to the party in question—which may be a building owner, a group of owners, a community, or society—are used so as to yield the maximum return to that party.

it is the economically efficient choice. If a building design enhances productivity while costing no more than competing designs, it is the economically efficient choice. If an essential building system of one design provides the same level of performance as its alternatives and costs less, it is the economically efficient choice.

We can also say that the choice in the latter example is "cost effective." A building decision is cost effective if it results in benefits equal to those of alternative decisions and has lower costs. Cost effectiveness is a subset of economic efficiency. For example, a heating, ventilating, and air conditioning (HVAC) system which satisfies the heating and cooling requirements of a building at minimum life-cycle costs, including both fuel and system costs, is the cost-effective HVAC system.

Note that if a building system is optional, that is, the building can meet performance requirements without the system, it is necessary to compare alternative system designs not only against one another, but also against the option of not having the system, in order to determine whether it is cost effective. For example, in considering whether to install a central vacuuming system for routine cleaning, alternative vacuuming systems must compete not only against one another, but also against the option of the existing method of cleaning. To decide whether to add a solar energy heating system, compare alternative solar energy system designs with not only one another, but also with the option of using a conventional system alone. If an optional system reduces the life-cycle costs of a building, it is cost effective.

Sometimes the terms "economically efficient" and "cost effective" are used loosely to describe individual investments whose benefits or savings exceed their costs, but which may not be the *most* economically efficient or *most* cost effective alternatives. For example, triple-glazed windows may be cost effective relative to single-glazed windows in that savings are estimated to exceed costs. Double glazing, however, might offer additional net savings and hence be the more cost-effective choice for windows. (See Section 1.2.2, "Economic Optimization.")

Also note that economic efficiency is not the same as engineering efficiency. One furnace, for example, may be more "efficient" than another in the sense that it delivers more heat for a given quantity of fuel. Yet, it may not be economically efficient if higher first costs outweigh savings from reduced fuel consumption.

1.2.2 Economic Optimization

"Economic optimization" is the process in which economic analysis is applied to determine the most economically efficient or cost-effective choice among alternatives. The optimal design or size of a building or building system maximizes net benefits or minimizes life-cycle costs for a desired level of performance.[3]

The related graphs in Figures 1-1a–c demonstrate the concept of optimization for a sample building problem. The problem is to decide what level of energy conservation (e.g., resistance value of insulation) is economically efficient (or cost effective in this case) in a given building. The figures demonstrate that two methods—maximizing net savings and minimizing life-cycle costs—can be used to arrive at the same optimal decision. In each of the figures, the physical quantity of inputs used to conserve energy is measured on the horizontal axis, and dollar costs are measured on the vertical axis.

Figure 1-1a illustrates optimization using the "net benefits or net savings method." With this method, savings in energy costs are the benefits. The objective is to find the level of conservation for which the difference between benefits (or savings) and costs of conservation is greatest.

Figure 1-1a shows the total costs of conservation rising slowly at first but then rising more sharply as it becomes more difficult to add additional inputs of conservation. The total cost curve for a building investment will typically bend upward.

In contrast, total conservation benefits (or energy savings) rise at a decreasing rate as more and more inputs of conservation are added to the building. As long as the benefits (or savings) curve lies above the costs curve, energy conservation is profitable. The conservation level, Q_c, at which the curves are most distant, with the benefits (or savings) curve above the costs curve, is the most profitable. The level at which the curves intersect indicates a breakeven investment, i.e., benefits (or savings) fully offset costs. On the other hand, the energy conservation investment loses money if the cost curve rises above the benefits or savings curve.

Figure 1-1b is an alternative graph which uses "marginal analysis" to illustrate the optimal level of conservation based on the net benefits (net savings) method. This figure depicts changes in the total benefits (or savings) curve and in the total costs curve of Figure 1-1a as inputs of energy conservation are increased; that is, in terms of calculus, Figure 1-1b is the derivative of Figure 1-1a. The point of intersection of the two marginal curves in Figure 1-1b in-

---

[3]Optimization is sometimes used to describe the process of determining the most desired choice where noneconomic and sometimes nonquantifiable benefits and costs are included in the objective function. This type of optimization requires weighting the relative importance of multiple objectives to determine the optimum design or size of the building. We focus in this book on the more narrow definition of the term, and all our examples involve benefits and costs that are measurable in monetary terms.

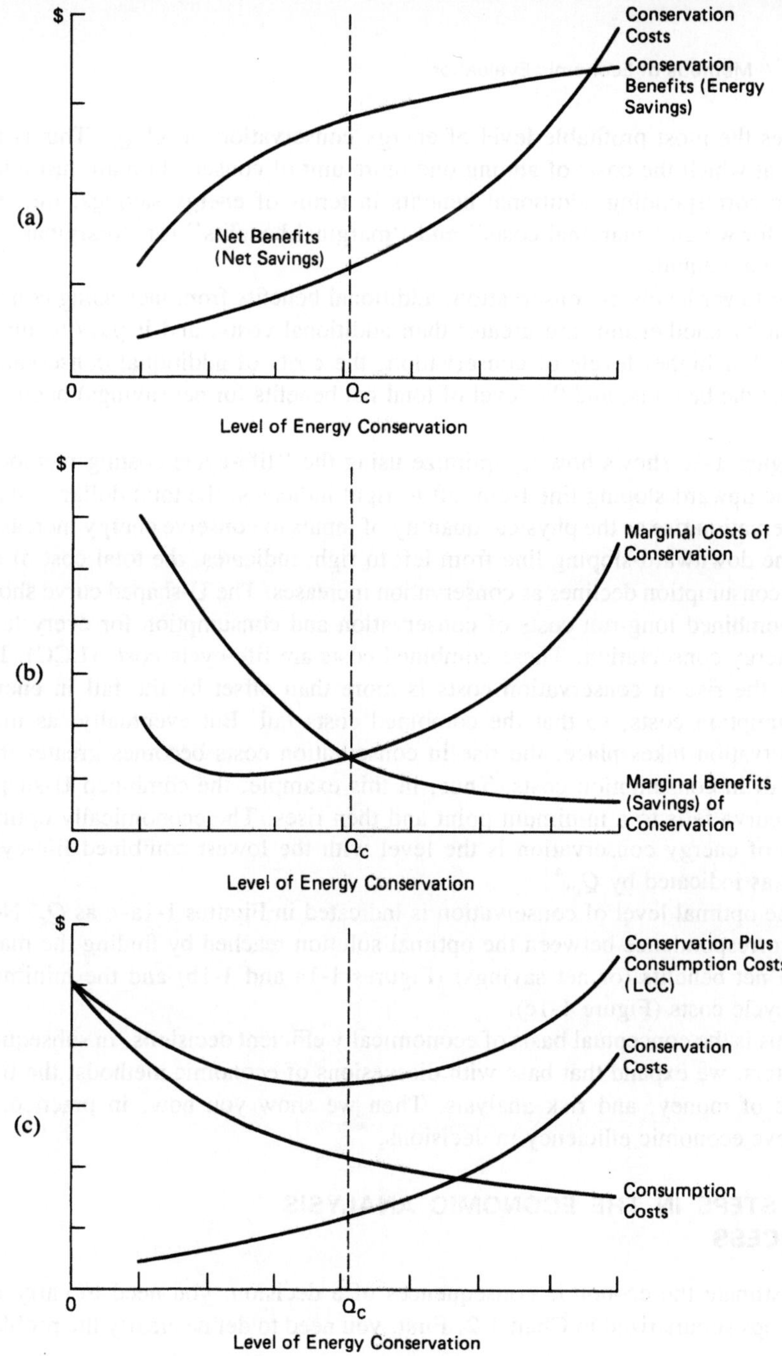

Figure 1-1. (a) Level of energy conservation that maximizes net benefits or net savings. (b) Levels of energy conservation that maximizes net benefits or net savings. (c) Level of energy conservation that minimizes life-cycle costs.

dicates the most profitable level of energy conservation, level Q_c. This is the level at which the costs of adding one more unit of conservation are just equal to the corresponding additional benefits in terms of energy savings, i.e., the level for which "marginal costs" and "marginal benefits" (or "marginal savings") are equal.

For lower levels of conservation, additional benefits from increasing conservation by another unit are greater than additional costs, and it pays to invest more. For higher levels of conservation, the costs of additional conservation exceed the benefits, and the level of total net benefits (or net savings) begins to fall.

Figure 1-1c shows how to optimize using the "life-cycle costing method." As the upward-sloping line from left to right indicates, the total dollar costs of conservation rise as the physical quantity of inputs to conserve energy increases. As the downward-sloping line from left to right indicates, the total cost of energy consumption declines as conservation increases. The U-shaped curve shows the combined long-run costs of conservation and consumption for every level of energy conservation. These combined costs are life-cycle costs (LCC). Initially the rise in conservation costs is more than offset by the fall in energy consumption costs, so that the combined costs fall. But eventually, as more conservation takes place, the rise in conservation costs becomes greater than the fall in consumption costs. Thus, in this example, the combined U-shaped cost curve falls to a minimum point and then rises. The economically optimal level of energy conservation is the level with the lowest combined life-cycle cost, as indicated by Q_c.[4]

The optimal level of conservation is indicated in Figures 1-1a–c as Q_c. Note the correspondence between the optimal solution reached by finding the maximum net benefits (or net savings) (Figures 1-1a and 1-1b) and the minimum life-cycle costs (Figure 1-1c).

This is the conceptual basis of economically efficient decisions. In subsequent chapters, we expand that base with discussions of economic methods, the time value of money, and risk analysis. Then we show you how, in practice, to achieve economic efficiency in decisions.

1.3 STEPS IN THE ECONOMIC ANALYSIS PROCESS

To estimate the economic consequences of a decision, you need to carry out the steps summarized in Chart 1-2. First, you need to define clearly the problem

---

[4]For a monotonically rising life-cycle cost curve, see Ruegg, *et al.* (1982, p. 12).

Chart 1-2. Steps in the economic analysis process

1. Define the problem and your objective.
2. Identify feasible alternatives for accomplishing your objective, taking into account any constraints.
3. Determine whether an economic analysis is necessary, and if so, the level of effort which is warranted.
4. Select a method or methods of economic evaluation.
5. Select a technique that accounts for uncertainty and/or risk if the data to be used with the economic method are uncertain.
6. Compile data and make assumptions called for by the economic method and risk analysis technique.
7. Compute a measure of economic performance.
8. Compare the economic consequences of alternatives and make a decision, taking into account any nonquantified effects and the risk attitude of the decision maker.

and your objective. For example, the problem might be the need to relocate your business. Your objective might be to choose the most profitable location from among various building sites.

Second, you need to identify feasible alternatives for accomplishing your objective, taking into account any constraints. It is imperative to identify good alternatives because, no matter how good your economic evaluation is, the outcome can be no better than the best alternative considered. It is important to take into account constraints in order to avoid unnecessary economic evaluation of alternatives which are not acceptable on other grounds. Constraints may arise from technical, physical, functional, statutory, building code, personal, or budgetary considerations. For example, constraints on location might narrow potential candidate building sites to those costing under $500,000, lying within a particular geographical area, and being served by a major transportation artery.

Third, you should determine whether an economic analysis is necessary and, if so, the level of effort which is warranted. If an alternative has clear cost and revenue advantages over its competitors, additional analysis probably is not needed. For example, if location A costs no more than locations B and C and is expected to result in higher sales, the choice is straightforward.[5] But if none of the alternatives has clear cost and revenue advantages, an economic analysis probably is needed to make the choice. For example, if location A will yield higher revenues than location B, but B will cost less than A, the choice is not obvious.

[5]We make this point, although it seems self-evident, because in our years of experience, we frequently have seen economic studies of alternatives conducted without a preliminary screening.

An analysis can be anything from a "back of the envelope" assessment, costing little or nothing, to an elaborate, carefully documented, and expensive study. The level of effort that goes into an analysis is usually a judgment call based on the importance of the decision and the magnitude of potential gains and losses from making a poor decision.

Fourth, you need to select a method or methods of economic evaluation from among the six methods we present in this book. For example, to evaluate the economics of locations A, B, and C, the net benefits method would be appropriate.

Fifth, you should select a technique that accounts for uncertainty and/or risk if the data to be used with the economic method are uncertain. For example, if expected sales to be generated at the three locations are highly uncertain, you might use sensitivity analysis to make a better decision.

Sixth, you must compile the data and make the assumptions called for by the economic method and risk analysis technique. Only items of cost and revenue on which the choice has an impact are relevant. For example, you would need to compile land costs, construction costs, utilities and other operating costs, quantity of sales, unit prices, real estate taxes, and transportation costs if these are the items affected by the choice of locations A, B, or C.

Seventh, use the data and assumptions to measure economic performance. This may be done manually or with a computer program.

Eighth, compare the economic consequences of alternatives and make a decision, taking into account any nonquantified effects and the risk attitude of the decision maker. For example, if location A offers greater prestige for the company than location B, but this benefit of A is not fully reflected in the data, it should nevertheless be considered in the selection decision. Location B may have higher estimated net benefits than C, but B may entail the risk of incurring higher than estimated costs. A risk-averse decision maker may not be willing to take a chance on B.

1.4 QUICK REFERENCE TO PRACTICAL TOOLS FOR EVALUATING THE ECONOMIC CONSEQUENCES OF DECISIONS

In Chapters 2 through 7, we describe six methods for measuring the economic consequences of alternatives: life-cycle cost, net benefits, benefit-to-cost ratio, internal rate of return, overall rate of return, and payback methods. We present the methods as they are used in deterministic analyses, without regard to uncertainty, reserving that discussion for later.

For the most part, the evaluation methods as usually formulated share the following distinguishing features:

Chart 1-3. Methods at a glance: directory of quick-reference charts for methods

Method	Chart	Page
Life-cycle cost (LCC)	2-1	
Net benefits (NB)	3-1	
Net savings (NS)	3-1	
Benefit-to-cost ratio (BCR)	4-1	
Savings-to-investment Ratio (SIR)	4-1	
Internal rate of return (IRR)	5-1	
Overall rate of return (ORR)	6-1	
Discounted payback (DPB)	7-1	
Simple payback (SPB)	7-1	

1. They look beyond first costs to include projections of significant future dollar amounts over the relevant study period.[6]
2. They adjust dollar amounts occurring at different times to time-equivalent values to permit valid comparisons.[7]
3. They account only for those effects which are expressible in dollars or other monetary units. Effects not measured in dollars must be uniform across alternatives or taken into account in some other way by the decision maker.
4. They are geared to comparing alternatives in order to make choices among them.
5. They are suitable for making decisions about buildings.

The methods are closely related but differ in formulation, unit of measure, and interpretation. The differences can be important, although there often is flexibility in the choice of methods for use for a given decision.

We provide technically correct formulations that overcome the shortcomings found in typical formulations. We also have made our formulations consistent with the standard economic methods on building economics published by ASTM. We highlight the salient points about each method in a series of six charts for convenient reference. Chart 1-3 is a directory to the methods charts,

---

[6]The payback method as usually formulated is an exception, because it takes into account future amounts only up to the point where payback is achieved, rather than over the investor's entire study period.

[7]The simple version of the payback method is an exception because it does not adjust dollar values to time-equivalent values.

showing where they are located. If you are thoroughly familiar with the methods, you may only wish to refer to the charts for a quick overview. If not, you may wish to read the chapters which cover each method in depth and give many capsule examples to amplify concepts.

Chart 1-4 is a quick-reference guide for those who are familiar with the economic methods but are unsure about their appropriate uses. It shows which economic methods to use for frequently made building decisions. Chart 1-5 is a quick reference guide for those whose interest is in evaluating a particular type of building decision or using a particular method. It is a directory showing where in Part 4 to find case illustrations of correct application of the economic methods to day-to-day building decisions.

Integral to evaluation methods which take a long-run perspective is the concept of the time value of money. In brief, the concept is that a dollar[8] received or paid today is not equivalent to a dollar received or paid at a future time. Amounts must be adjusted to a common time before they can be meaningfully combined. Although any common time will do, the common-time convention most often selected is the *present*. This means expressing all amounts as though they occur in one lump sum at the beginning of the study period. Another common-time convention which is used is *annual*, which means expressing all amounts as though they occur in uniform, annually recurring amounts. The time-adjustment procedure, usually called "discounting," is reflected in the formulas for the evaluation methods in Chapters 2 through 7, but the detailed hows and whys of discounting are left to Chapter 8.

To apply the six methods of economic evaluation, you need to know what to do about inflation, how to select a study period and discount rate, how to estimate benefits and costs in dollars, and how to adjust for the effects of taxes and financing. Chapters 9 through 13 treat these topics.

It is naive to apply the economic methods of evaluation deterministically when economic evaluations are carried out under conditions of uncertainty. Part 3 describes techniques for taking into account uncertainty and risk in conjunction with economic evaluation. Chapter 14 discusses risk exposure and risk attitude in theory. Chapters 15 to 18 describe sensitivity analysis, the risk-adjusted discount rate, the certainty equivalent technique, decision analysis, simulation, and other techniques that account for uncertainty and risk.

Chart 1-6 is a quick reference guide for those who wish to assess the risk of a particular building decision. It is a directory to case illustrations in Part 4

[8]For convenience, "dollar" is used in place of the more general term "monetary unit," but other monetary units, such as yen, pounds, or marks, could be substituted for dollars without changing the meaning.

Chart 1-4. Which methods to use for typical building decisions

	Method					
Decisions	Life-Cycle Cost (LCC)[a]	Net Benefits (NB) and Net Savings (NS)	Benefit-to-Cost Ratio (BCR) and Savings-to-Investment Ratio (SIR)	Internal Rate of Return (IRR)	Overall Rate of Return (ORR)	Discounted Payback (DPB) and Simple Payback (SPB)
Acceptance/rejection	○	Use	Use	†	Use	†
Design	○	Use	*	Don't use	*	Don't use
Size	○	Use	*	Don't use	*	Don't use
Combination	**, ‡	‡	Use	Don't use	Use	Don't use
Ranking	Don't use	Don't use	Use	Don't use	Use	Don't use
Location	○	Use	*	Don't use	*	Don't use
Lease/Buy	○	Use	*	Don't use	*	Don't use
Replacement	○	Use	Don't use	Don't use	Don't use	Don't use

[a]Note that two or more LCC measures are required to make comparisons.

Symbols:

○ = Use if costs predominate.

† = Recommended only under limited circumstances, as described in the section.

* = Recommended only if incremental benefits and costs are used and each design/size is considered.

** = Recommended only if all projects are within a given building or facility.

‡ = Recommended if overall life-cycle costs or aggregate net benefits for projects in combination are computed; not recommended if life-cycle costs or net benefits for individual projects are compared.

Chart 1-5. Directory of economic methods applied to building decisions

Decision	Chapter	Method	Section
Which design to choose	19	LCC	19.1
		NB	19.2
		NS	19.1
Whether to accept or reject	20	LCC	20.1 and 20.3
		NB	20.4 and 20.5
		NS	20.1 and 20.3
		BCR	20.2
		SIR	20.1 and 20.3
		ORR (adjusted rate of return)	20.4
		DPB	20.3
Where to locate	21	LCC (annual cost)	21.1
		NB (annual value)	21.2
Lease or buy	22	LCC	22
What size	23	NS	23.1
		LCC	23.2
Which project to choose when budget is limited	24	BCR	24.1
		BCR	24.2
When to replace	25	LCC (annual cost)	25.1
		LCC (present value)	25.2
Which combination of interdependent systems to choose	26	Aggregate LCC	26.2
		Sequential SIR	26.2

Chart 1-6. Directory of techniques for treating uncertainty and risk applied to building decisions

Decision	Chapter	Technique	Section
Which design to choose	19	Sensitivity analysis	19.1
		Decision analysis	19.3
Whether to accept or reject	20	Breakeven analysis	20.1
		Simulation	20.4
		Mathematical/analytical	20.5
Where to locate	21	Sensitivity analysis	21
Lease or buy	22	Risk-Adjusted Discount Rate	22
		Sensitivity analysis	22
What size	23	Sensitivity analysis	23.1

which show how to apply techniques for treating uncertainty and risk to specific building decisions.

REFERENCES

Bon, Ranko. 1989. *Building as an Economic Process: An Introduction to Building Economics*. Englewood Cliffs, New Jersey: Prentice-Hall, Inc.

Hillebrandt, Patricia M. 1974. *Economic Theory and the Construction Industry*. London: Macmillan Press, Ltd.

Ruegg, Rosalie T., G. Thomas Sav, Jeanne W. Powell, and E. Thomas Pierce. 1982. *Economic Evaluation of Solar Energy Systems in Commercial Buildings*. NBSIR 82-2540. Gaithersburg, MD: National Bureau of Standards.

Stone, P. A. 1975. *Building Design Evaluation: Costs in Use*. London: E. & F. N. Spon, Ltd.

2

Life-Cycle Cost (LCC)

The LCC method is used to make cost-effective choices for a given building, facility, or system. Emphasis is on cost effectiveness because the LCC method is used to evaluate alternatives which compete primarily on the basis of costs. Thus target performance standards are defined, and only candidates which meet the standards are considered. The LCC associated with each candidate is computed, and the candidate with the lowest LCC is the cost-effective solution. If revenue and other benefits predominate, one of the other evaluation methods, such as net benefits, is usually more appropriate than LCC.[1]

Chart 2-1 summarizes key features of the LCC method for quick reference. The remainder of this chapter provides an in-depth treatment. If you are not already acquainted with the time-value of money concept, we recommend that you first read Chapter 2 for perspective and then read Chapter 8 to understand the time-value concept before continuing with Part 1.

2.1 FORMULATION OF LCC

The LCC method is simple. First, compute the LCC for a particular course of action by summing all significant, time-adjusted costs associated with it over the relevant period of time.[2] Include initial and future costs that are affected by the decision; exclude others. For example, in evaluating the choice of exterior wall construction, include the building's energy cost if it is affected by the

[1]Positive cash flows can be subtracted from costs in computing LCC. But if positive cash flows are sufficiently large to cause the LCC value to become negative, a different method should be used.

[2]Because of its name, the LCC method is often described as a method for assessing the costs of an item over its service life. But we recommend that it be used to assess costs over the relevant period for making the decision, which may be longer or shorter than the service life of the asset. We call the relevant time for computing LCCs the "study period." See Chapter 10.

Chart 2-1. Quick reference for the LCC method

Abbreviation	LCC Unit of Measure $
Description	Sums all significant, time-adjusted costs relevant to a given course of action over a study period.
Formula (PV)	$LCC_{A1} = \sum_{t=0}^{N} \frac{C_t}{(1+d)^t}$
Applications	**Approach**
1. Accept/reject	Compare the LCC of a building with and without a given system or modification and accept it if it lowers the LCC.
2. Design/size	Compare LCCs of design or size alternatives and choose the one resulting in the lowest LCC.
3. Locate cost-driven businesses	Compare LCCs associated with alternative locations and choose the location resulting in the lowest LCC.
4. Replacements	For replacement in kind, identify the replacement period for which the system LCC (annual cost) is minimized. For replacement with an unlike system, identify the changeover time for which the system LCC (present value) is minimized.
5. Lease/buy	Compare LCCs of alternative means of securing the use of an asset over a given time and choose the one with the lowest LCC.
6. Combine interdependent systems	Compare overall LCCs based on selection of alternative combinations of interdependent systems and choose the combination that minimizes overall LCC, taking into account interdependencies.
7. Allocate a limited budget within a given building or facility	Compare overall LCCs based on selection of alternative sets of candidate projects, each set fitting within the available budget, and choose the set that minimizes overall LCC.

Special Considerations

- Suitable when focus is on costs rather than benefits. Benefits must be held constant among alternatives or subtracted from costs.
- Effects not measured in dollars must be uniform across alternatives or should be taken into account in some other way.
- LCCs for two or more mutually exclusive project alternatives if measured in present value dollars must always be computed over the same study period and often must be if measured in annual value dollars; a single LCC measure is not adequate for decision making.

Notation: LCC_{A1} = LCC associated with choosing alternative A1; C_t = sum of all relevant costs occurring in period t; d = discount rate; and PV = present value.

choice; otherwise, omit it.[3] Subtract from costs any positive cash flows, such as salvage or resale values, that are to be taken into account in making the decision.[4]

Repeat the LCC computation for each mutually exclusive alternative from which a choice is to be made.[5] Then compare the LCCs associated with each alternative to see which is lowest. If other factors are about the same, select the alternative with the minimum LCC.

LCC analysis is usually performed in either present value or annual value dollars. These are alternative, but time-equivalent, ways of expressing streams of costs over time. A present value LCC model expresses LCCs as a lump-sum amount at the present. An annual value LCC model expresses LCCs as a uniformly recurring annual amount over the study period. We give each formulation.

In most cases, the choice of using a present value or an annual value cost model is simply a matter of convenience or preference on the part of an analyst. We use present value in most LCC applications in Part 4. One reason is that calculating the annual value usually requires first calculating the present value and then amortizing the result over the study period, which adds an extra calculation. Another reason is that many people find present values more comprehensible than annual values.

Despite our preference, we include an annual value cost model because certain types of problems can be solved more easily using it rather than using a present value cost model. For example, when the problem to be solved is how

---

[3]It may be argued that it does not matter if costs which are unaffected by the decision are included in LCCs, provided that they are included equally in the LCCs of all alternatives compared. They will simply offset one another. We caution you, however, against a pitfall of this practice: the inclusion of large sums which are not affected by the decision may cause a reduction in LCC to appear trivial in amount in comparison with the total, even though it may represent an attractive return on investment. See Section 19.1, for an example.

[4]Difficult-to-quantify values, such as comfort and productivity benefits in excess of target performance standards, are often omitted.

[5]Alternatives are mutually exclusive if accepting one means not accepting the others. For example, locating a building on site A is mutually exclusive to locating it on site B. Insulating a house is mutually exclusive to not insulating it. Insulating it with R-30 insulation is mutually exclusive to insulating it with R-11. Allocating a budget of $10,000 to fund projects 1, 2, and 3, which together cost $10,000, is mutually exclusive to using the budget to fund projects 4 and 5, which also cost $10,000.

When evaluating a building system which is discretionary (such as storm doors), it is necessary to compare the LCC with the building system against the LCC for the "baseline" alternative of not having the system. This comparison is necessary to determine if the investment is economic. When evaluating a building system which is mandatory, such as a roof, the baseline is an alternative type of the system, such as a different type of roof.

long to hold equipment until replacing it in kind or where to locate a building—and the decision hinges on costs—an annual value cost model is often useful.

Annual value cost models are particularly helpful for evaluating alternatives that have different lives. To compare alternatives on the basis of present value costs requires that all alternatives be evaluated over the same study period.[6] But an annual value cost model allows different evaluation periods *provided* that none of the annual costs would change if alternatives were instead evaluated for the same study period. For example, suppose that you are comparing three types of solar window films, with lives of 5, 8, and 10 years, for a building with a holding period of 8 years. Ask yourself if the annual value LCC for each film is likely to be the same if you evaluated them using 5, 8, and 10 years, respectively, as if you used a common study period of 8 years. This condition is often ignored by users of annual value models.[7] The result is distorted economic decisions.

2.1.1 Present Value Cost Model

A simple formulation of LCC using a present value model is the following:[8]

$$PV\,LCC_{A1} = \sum_{t=0}^{N} \frac{C_t}{(1 + d)^t} \tag{2-1}$$

where

$PV\,LCC_{A1}$ = LCC in present value dollars of a building or system which is estimated to result if the decision is made to choose a given alternative, $A1$,

C_t = sum of all relevant[9] costs, including initial and future costs, arising from the selection of alternative $A1$, less any allowable positive cash flows, such as those from resale, occurring in period t,

---

[6]See Chapter 10 for guidelines on selecting a study period.

[7]See Chapter 10 for further discussion and examples of the appropriate use of annual value models for comparing alternatives.

[8]This formulation has been adopted by the American Society for Testing and Materials (ASTM 1989). The initial ASTM standard practice on LCC was based in part on Ruegg, Petersen, and Marshall (1980).

[9]Relevant costs (and benefits) are those which are affected by the decision.

N = number of periods comprising the study period, and
d = discount rate used to adjust cash flows to the present value.

Another present value formulation of LCC itemizes principal constituents:

$$PV\,LCC_{A1} = I_p + E_p + M_p + R_p - S_p \tag{2-2}$$

where

I_p = present value investment costs of alternative $A1$,
E_p = present value energy costs associated with alternative $A1$,
M_p = present value nonfuel operating and maintenance costs associated with alternative $A1$,
R_p = present value repair and replacement costs associated with alternative $A1$, and
S_p = present value resale value (or scrap or salvage value) less disposal costs associated with alternative $A1$.

Table 2-1 shows an "anatomy" of an LCC analysis. The first column lists the principal elements. The remainder of the table is divided into two parts to show the comparison of two mutually exclusive alternatives. The alternatives are to accept or reject a discretionary system or project for a building, such as a waste-heat recovery system. The first part, denoted by bars over symbols, gives expected costs if the system is *not* accepted. The second part, denoted by the absence of bars over symbols, gives expected costs if the system *is* accepted. Each main part, in turn, is subdivided vertically to denote time differences. The column headed "Present Time" shows a summation of amounts as of the present. The remaining columns show future amounts, where 1, 2, 3, . . . , N indicate the number of elapsed years from the present until each future amount occurs. N, the study period, is the total number of years over which the building owner, operator, or other decision maker is concerned about the costs associated with the decision.

The example highlights the point that, when deciding whether or not to invest in a discretionary building system, the baseline alternative is that of not making the investment. This brings us to a question which is sometimes a source of confusion. What costs (and benefits) go into computing the LCC for the "do-nothing" alternative? The answer is to include those relevant items upon which the decision turns. In the case of the waste-heat recovery system, the LCC without the system includes energy costs, maintenance and repair costs if these change with the addition of the system, and an investment cost of zero.

Table 2-1 displays investment costs as though they occur at the present time.

Table 2-1. Anatomy of an LCC analysis

Cost Elements	Compute $\overline{LCC}$ Without the Project						Compute LCC With the Project					
		Future Years						Future Years				
	Present Time	1	2	3	. . .	N	Present Time	1	2	3	. . .	N
Investment (I)	$\bar{I}_p$						I_p					
Energy (E)	$\bar{E}_p = \sum_{j=1}^{N} \frac{(\bar{E}_j)}{(1+d)^j}$	$\bar{E}_1$	$\bar{E}_2$	$\bar{E}_3$	. . .	$\bar{E}_N$	$E_p = \sum_{j=1}^{N} \frac{(E_j)}{(1+d)^j}$	E_1	E_2	E_3	. . .	E_N
Nonfuel Operation and Maintenance (M)	$\bar{M}_p = \sum_{j=1}^{N} \frac{(\bar{M}_j)}{(1+d)^j}$	$\bar{M}_1$	$\bar{M}_2$	$\bar{M}_3$	. . .	$\bar{M}_N$	$M_p = \sum_{j=1}^{N} \frac{(M_j)}{(1+d)^j}$	M_1	M_2	M_3	. . .	M_N
Repairs and Replacements (R)	$\bar{R}_p = \sum_{j=1}^{N} \frac{(\bar{R}_j)}{(1+d)^j}$	$\bar{R}_1$	$\bar{R}_2$	$\bar{R}_3$	. . .	$\bar{R}_N$	$R_p = \sum_{j=1}^{N} \frac{(R_j)}{(1+d)^j}$	R_1	R_2	R_3	. . .	R_N
Resale/Scrap[a] (S)	$(\bar{S}_p) = \frac{\bar{S}_N}{(1+d)^N}$					$\bar{S}_N$	$(S_p) = \frac{S_N}{(1+d)^N}$					S_N
Total Present Value Life Cycle Cost (LCC)	$\overline{LCC} = \bar{I}_p + \bar{E}_p + \bar{M}_p + \bar{R}_p - \bar{S}_p$						$LCC = I_p + E_p + M_p + R_p - S_p$					
	Compare LCC with $\overline{LCC}$ and accept Project if LCC is lower											

[a]Usually a positive amount that is subtracted from costs.

Nomenclature: d = discount rate, $1/(1+d)^j$ = single present value discount formula for discount rate d and year j (explained in Chapter 8); N = the number of years in the study period (explained in Chapter 10); subscript p indicates present values (explained in Chapter 8); and the bar over a symbol signifies a cost incurred if the project is not accepted.

Source: Ruegg and Petersen (1987, p. 9).

Often in economic evaluations, analysts make the simplifying assumption that a project is put into place instantaneously, and they ignore details of project scheduling. This approach is useful and appropriate in many cases, particularly when choosing among alternatives which have similar scheduling periods. But when acquisition or construction periods differ among alternatives, it may be necessary to account for the timing of investment costs. This can be done by expressing construction or acquisition costs as a series of phased-in costs over time, or as a lump-sum future amount, instead of as a lump-sum amount at the beginning of the study period. When investment costs are phased in over a construction period or delayed to a future year, the onset of operating costs normally is also delayed. (A planning and construction period with delayed investment is demonstrated in Section 23.2.)

Because we assume that investment costs occur instantly in the illustration of Table 2-1, we subscript them with "p" to signify that they are in "present value dollars." We assume that all of the other amounts occur in future years, and therefore are in "future dollars," as signified by the subscripts $1 \ldots N$, which denote future years of occurrence.

We adjust future amounts to present value amounts (e.g., $\overline{E}_p$) to make them time equivalent with investment costs. Then we combine them (as shown in the lower part of Table 2-1) to find LCC without the project ($\overline{\text{LCC}}$) and LCC with the project (LCC). The last line of Table 2-1 compares $\overline{\text{LCC}}$ and LCC. Having LCC less than $\overline{\text{LCC}}$ means that the building's LCC is lower with the project than without it. Thus, the project is cost effective. Having LCC greater than $\overline{\text{LCC}}$ means that the project is not cost effective.

2.1.2 Annual Value Cost Model

The LCC can be computed in annual value dollars as follows:

$$AV\,LCC_{A1} = \left(\sum_{t=0}^{N} \frac{C_t}{(1+d)^t}\right)\left(\frac{d(1+d)^N}{(1+d)^N - 1}\right) \tag{2-3}$$

which is the same as

$$AV\,LCC_{A1} = (PV\,LCC_{A1})\left(\frac{d(1+d)^N}{(1+d)^N - 1}\right) \tag{2-3a}$$

where

$AV\,LCC_{A1}$ = LCC, in annual value dollars, of a building or system which is estimated to result if the decision is to choose a given alternative, $A1$, and

$\dfrac{d(1+d)^N}{(1+d)^N - 1}$ = formula for the uniform capital recovery factor used to convert present value amounts to annual value equivalents.

2.2 APPLICATIONS OF THE LCC METHOD

The LCC method can be applied to many building decisions for which the determining factor is cost effectiveness. It is useful for deciding whether to accept or reject a given investment aimed at lowering building costs; identifying cost-effective designs and sizes of building and building systems, building locations, and replacement times; deciding whether to lease or buy buildings, furnishings, and equipment; finding the cost-minimizing combination of interdependent building systems; and, when the budget is limited, finding the set of independent or interdependent projects of variable design and size within the budget which will minimize the long-run costs of a given building or facility while meeting performance requirements.

We discuss each of these applications, outlining the approach, stating the decision rule, giving simple examples to illustrate the concepts, and referring you to detailed case studies of the applications in Part 4. We treat the problem of budget constraint separately, because it is necessary to make accept/reject, design and sizing, and other decisions whether or not the budget is constrained. We conclude the chapter with examples of both correct and incorrect use of the LCC method. We include examples of incorrect but, unfortunately, not uncommon, use because we think they may help steer practitioners away from typical pitfalls in application.

2.2.1 Accept/Reject Decisions

You can use the LCC method to determine if an investment in a given system or modification is worthwhile. The approach is to compute and compare the LCC of the building with and without the investment. The decision rule is to accept the investment if it lowers the LCC.

Table 2-1 demonstrates the concept of using LCC for accept/reject decisions. Table 2-2 provides a numerical example. Of three potential projects, those in buildings A and C are accepted because they lower the LCC. The project in building B is rejected because it raises the LCC.

Chapter 20 demonstrates in detail with two case studies how to use LCC for an accept/reject decision. The first case is whether to retrofit a computer facility for waste-heat recovery. The second case is whether to add a solar hot water system to a launderette on a military base.

Table 2-2. LCC for accept/reject decisions: illustration

(1) Project Description	(2) LCC with Project (PV $)	(3) LCC without Project (PV $)	(4) Decision
Insulate building A	20,000	35,000	Accept
Replace windows in building B	15,000	12,000	Reject
Install waste-heat recovery system in building C	75,000	100,000	Accept

Nomenclature: PV $ = present value dollars. Costs are expressed as a single lump-sum amount occurring at the present.

2.2.2 Design and Sizing Decisions

You can use the LCC method to select the cost-effective design or size of a building or system provided alternative designs or sizes all satisfy performance standards. Compute the LCC for each alternative design or size, making sure to include the same cost elements in the evaluation of each design or size. Choose the design or size that minimizes the LCC for the building.[10]

Table 2-3 illustrates the use of LCC for designing and sizing a building system. It shows the cost-effective solar energy system to be design B in size 2.

Other examples of design and size decisions which might be assisted by the LCC method are whether to pay extra to have a stone exterior wall with low maintenance costs instead of a stucco wall with higher maintenance costs; whether to choose single, double, or triple glazing for windows; and whether to have small, medium-sized, or large windows on the north, east, west, and south sides of a building.

Chapter 19 demonstrates, with two detailed case studies, how to use LCC to select between alternative building designs. Chapter 23 demonstrates, with two additional detailed case studies, how to use LCC to size building systems.

2.2.3 Location Decisions

The LCC method can assist in selecting a building's location when the decision hinges on costs, rather than on revenue or other benefits. Compute and compare the LCCs associated with alternative locations. Choose the location with the minimum LCC, other factors being the same.

[10] If the budget is limited, it may be necessary to forego cost-effective building systems or increments to systems. See Section 2.2.7 for a discussion of system design and sizing in the context of a budget limitation.

Table 2-3. LCC for design/size decisions: illustration

(1) Competing Designs/Sizes of a Solar Energy System	(2) LCC (PV $)
Design A	
Size 1	65,000
Size 2	58,000
* Design B	
Size 1	50,000
* Size 2	45,000
Size 3	52,000

Nomenclature: PV $ = present value dollars. Costs are expressed as a single lump-sum amount occurring at the present.

* designates least-cost design and size.

Table 2-4 illustrates the use of LCC to locate a plant which produces an item distributed and marketed nationally. Site 2 is selected because it has the lowest LCC. Chapter 21 presents a detailed case study of how to use an annual value LCC model in locating a manufacturing plant which has no strong market ties.

2.2.4 Replacement Decisions

The LCC method can help decide when to replace building systems and components whose owning and operating costs change over time. When an asset is to be replaced in kind, we recommend an annual cost model.

As illustrated by Table 2-5, the approach is to find the holding period with the minimum annual cost. Since the annual cost is lowest if the system is re-

Table 2-4. LCC for location decisions: illustration

(1) Location	(2) Plant Cost (PV $)	(3) Labor Cost (PV $)	(4) Materials Cost (PV $)	(5) Transportation Cost (PV $)	(6) (= 1 + 2 + 3 + 4 + 5) LCC (PV $)
Site 1	800,000	2,000,000	500,000	90,000	3,390,000
Site 2	500,000	1,500,000	900,000	200,000	3,100,000*
Site 3	1,000,000	2,200,000	400,000	50,000	3,650,000

Nomenclature: PV $ = present value dollars. Costs are expressed as a single lump-sum amount occurring at the present.

* designates minimum LCC value.

Table 2-5. LCC for replacement decisions: illustration

(1) Elapsed Time Before Replacing Building System (Years)	(2) LCC of Owning and Operating System (AV $)
10	5,000
11	4,500
*12	4,000
13	5,000
14	6,500

Nomenclature: AV $ = annual value dollars. Costs are expressed as a uniform annual amount recurring over the study period.

* designates the optimal replacement period and economic system life.

placed every 12 years, that period is its optimal replacement period. It is also its "economic life."

When replacement is with an unlike system and the decision is when to change over to the new system, we recommend a present value cost model. Compare the present value costs associated with alternative changeover times and choose the changeover time which offers the lowest present value cost.

Chapter 25 presents two detailed case studies of replacement decisions. The first case study uses an annual value LCC model to decide when to replace with like equipment. The second one uses a present value LCC model to determine when to replace existing equipment with improved equipment.

2.2.5 Lease/Buy Decisions

You can use LCC to help decide whether to lease or buy a building, furnishings, or equipment. A comparison of the LCCs of alternative means of obtaining an asset over a given time indicates which is most cost effective from the standpoint of direct costs.[11]

Table 2-6 illustrates the use of LCC to decide whether to lease or buy furniture. In this example, leasing is the cost-effective choice. Chapter 22 provides a detailed case study using LCC to decide whether to lease or buy a building for an expanding business. The case study uses a present value LCC model, but annual value models can also be used for lease or buy decisions.

---

[11]Chapter 22 points out that there are often economic effects not captured in an LCC analysis which are important to the lease/buy decision.

Table 2-6. LCC for lease/buy decisions: illustration

(1) Decision	(2) Acquisition (PV $)	(3) Ending Value (Current $)	(4) Ending Value (PV $)	(5) (= 2 − 4) LCC (PV $)
* Lease furniture	10,000	0	0	10,000
Buy furniture	15,000	7,000	3,000	12,000

Nomenclature: PV $ = present value dollars. Costs are expressed as a single lump-sum amount occurring at the present.

* designates the cost-effective choice.

2.2.6 Decisions About Interdependent Systems

You can use LCC to determine the cost-effective combination of interdependent systems for a new or existing building. First, compute and compare overall LCCs for the building[12] based on selections of alternative combinations of systems. Then choose the combination that minimizes the overall LCC for the building, taking into account the interdependencies.

An example of system interdependence is illustrated in Table 2-7. The problem is to decide how much insulation in the walls and what level of efficiency in the HVAC system will be cost effective in combination. The overall LCC for the building is minimized by combining insulation to level 1 with replacement of the HVAC system. Without replacement of the HVAC system, insulation to level 3 is cost effective. But with HVAC replacement, it does not pay to increase insulation beyond level 1.

A case study in Chapter 26 shows two approaches to determining the cost-effective combination of energy conservation investments in a house. Attic insulation, storm windows, caulking, and heating system efficiency are evaluated jointly.

2.2.7 Budget Allocation Decisions

You can use LCC to allocate a limited budget in order to achieve the largest reduction in long-run costs for a given building or facility. Compute and compare the overall LCC of the building or facility with alternative sets of candidate

[12]It is important to define the measure of LCC broadly enough to encompass costs affected by all of the combinations of the systems being compared. For example, in comparing combination A, which includes labor-saving equipment, with combination B, which includes energy-saving equipment, the measure of LCC used for both should include both labor and energy costs.

Table 2-7. LCC for interdependent system decisions: illustration

(1) Alternative Energy Conservation Investments	(2) LCC of Building A (PV $)
Do nothing	100,000
Insulate walls of building A to level 1	75,000
Insulate walls of building A to level 2	65,000
Insulate walls of building A to level 3	63,000
Replace HVAC of building A	55,000
* Insulate walls to level 1 and replace HVAC	50,000
Insulate walls to level 2 and replace HVAC	52,000
Insulate walls to level 3 and replace HVAC	53,000

Nomenclature: PV $ = present value dollars. Costs are expressed as a single lump-sum amount occurring at the present.

* designates the cost-effective combination.

projects, each set fitting within the available budget. Identify the set that minimizes the overall LCC.[13]

The sets considered may contain alternative designs and sizes of given building systems, as well as different systems. Building systems may be independent or interdependent. If they are interdependent, adjustments must be made to LCCs to reflect interdependencies.

An example of a budget allocation problem which can be solved by the LCC method is illustrated in Table 2-8. Given a budget of $50,000 and five independent, cost-effective modifications to a building totaling $100,000 in investment costs, which modifications should be selected for funding? Table 2-8 shows the five projects configured into three sets that are affordable under the $50,000 budget constraint. Choosing the second set minimizes the overall LCC for the building.

Whether project increments are considered in the budget allocation process depends on the approach taken to designing and sizing projects. In principle, designing and sizing individual projects should be performed simultaneously with project selection. This is necessary to achieve the condition for economic

[13]It is necessary to define overall LCC broadly in order to encompass all costs affected by projects within alternative sets of candidate projects and to measure overall LCC consistently for the alternative sets.

Table 2-8. LCC for budget allocation decisions

(1) Alternative Sets of Projects for Building A	(2) Investment Cost ($)	(3) Overall LCC of Building A (PV $)
Without investment	0	500,000
Set 1: projects A, C, D	50,000	300,000
* Set 2: project B	50,000	275,000
Set 3: project A, C, E	50,000	325,000

Nomenclature: PV $ = present value dollars. Costs are expressed as a single lump-sum amount occurring at the present.

* designates the cost-effective set.

efficiency that the last dollar spent on each project yields equal benefits or savings. Otherwise, overall LCC for the building can be reduced merely by shifting funds among competing projects.

The principle of requiring that increments to projects within a given building compete for funding clearly applies when allocating a single budget. For example, if you hire a contractor to make a building as cost effective as possible with a one-time investment of a specified amount, the contractor must follow the principle of expanding each project only if the expansion saves more than an investment in another project in that building. Otherwise, he or she will miss potential cost reductions. (See Section 24.1 for a detailed case study treating allocation of a single budget among competing projects and increments to a project. Although it is solved using a method different from LCC, it illustrates the principle discussed here.)

In practice, however, a simultaneous solution is often infeasible. The design and sizing of projects may be undertaken by different organizational units at different times. Furthermore, funds to be allocated may come in a series of budgets over time, rather than in a single budget.

When there is a series of related budgets to be allocated, optimizing the return to each budget, as though it were the only one to come, leads to underdesigning and undersizing individual projects. On the other hand, designing and sizing projects as though there were no budget constraint leads to overdesigning and oversizing projects. In this case, a practical rule of thumb is to design and size each project to minimize the LCC prior to the budget allocation, despite the tendency toward overdesigning and oversizing.

For example, suppose that a building manager's current budget is sufficient either to add insulation in the walls and attic to R-30 (the level which has been found to minimize the LCC) or to add a solar water heating system with 100

ft^2 of collector area (the size which has been found to minimize the LCC), but not both. It would likely be better to do one or the other fully with the funds on hand, and delay the other until the next budget is received, rather than to do both partially.

This is true because the cost of returning to complete a project at a later time tends to be higher than the additional cost of completing it as part of the initial effort. The recurrence of "fixed costs," which are incurred regardless of the size of the job, drives up the costs of a multistage completion process. Higher costs offset potential savings, thereby reducing a project's cost-effective size. As a result, potential long-run savings are foregone. A quantitative analysis of tradeoffs between funding projects at lower levels and doing them sooner, versus funding them at higher levels and delaying them or not doing them at all, is required to improve on the rule of thumb for project design, size, and selection. (See Section 24.2 for a case study which evaluates the requirements for project increments to compete for funding versus predesigning and presizing projects when there are a series of budgets to allocate. Although it uses a method different from LCC to allocate the budget, it illustrates the principle discussed here.)

2.2.8 LCC Method Not for All Decisions

We do not recommend the LCC method for evaluating investments which generate significant revenues or other benefits; ranking individual projects competing for a limited budget; or allocating budgets among projects in separate facilities.

2.2.9 Summary of Points to Note in Applying the LCC Method

You can choose among mutually exclusive alternatives on the basis of their LCCs if the following conditions are met. All alternatives must meet performance requirements. Positive cash flows, such as those from resale, must be subtracted from costs. And effects not measured in dollars must be either insignificant, uniform across alternatives, or taken into account in some other way (outside the LCC analysis). LCCs must be provided for at least two alternatives. For discretionary investment decisions, a baseline LCC must be provided for the alternative of not making the investment.

If an analysis is performed in present value dollars, competing alternatives must be evaluated over the same period of time. If an analysis is performed in annual value dollars, competing alternatives may be evaluated over different periods of time, but only under limited conditions. (For further discussion, see Chapter 10.)

It pays to accept any investment in a building which will lower its LCC as long as the performance objectives of the building are met. But if the budget is limited, it may be necessary to forego cost-effective projects or increments to projects. To minimize the LCC of a building or facility for a given level of performance and a fixed budget, it is necessary to shift funds among cost-reducing projects until the last dollar spent on each project results in equal cost reductions.

2.3 EXAMPLES OF CORRECT USE

Example 1. A new hospital is to be built. Design A of the hospital has an estimated LCC over the 25-year study period of $80 million; design B has an estimated LCC over the same period of $60 million. Both meet performance standards. Design B is the cost-effective design choice.

Example 2. A public school gymnasium is to be renovated. A decision must be made on whether or not to install an automatic light-sensing device (with an override switch) to dim or switch off the electric lighting system automatically when daylight is adequate to meet lighting needs. The device has an estimated life of 10 years.

Taking into account only the relevant costs affected by the decision, the LCC of the gym over the 10-year study period is estimated at $25,000 without the device and $18,000 with it. Installing the device is therefore estimated to be cost effective.

Example 3. An architect must decide what type of exterior doors to put in the main entrances to an office building. Choices which meet performance requirements equally well are automatic sliding doors, with an estimated LCC of $10,000, and revolving doors, with an estimated LCC of $7,000, both taken over a 30-year life. The revolving doors are estimated to be the cost-effective choice.

Example 4. A public housing bureau is retrofitting a complex of houses for energy conservation. Two modifications are being considered: adding attic insulation [up to resistance (R) level R38] and replacing the old oil furnace with a high-efficiency unit. Funding is available for any combination of modifications determined to be the cost-effective retrofit package. Taking into account affected costs over a 15-year study period, the LCC of a typical house in the complex without either modification is $15,000; with R38 alone, $11,000; with the new furnace system alone, $10,000; with a combination of the new furnace system and R19, $9,500; and with other combinations, the LCC is higher than $9,500. The cost-effective retrofit package is estimated to be R19 in combination with the new furnace.

Example 5. A building manager is given a budget of $50,000 to reduce the long-run costs of a pharmaceutical plant. Two project combinations are possible within the budget: Projects 1 and 2 together or Project 3 alone. Without any of the projects, the plant's LCC (based only on relevant costs over the designated study period of 10 years) is $1,200,000. Projects 1 and 2 together are estimated to reduce the LCC over the 10-year study period to $1,000,000. Project 3 alone is estimated to reduce the LCC over the 10-year study period to $900,000. The estimated cost-effective allocation of the budget is to choose project 3.

2.4 EXAMPLES OF INCORRECT USE

Example 1. A real estate investment company must decide if it will invest in either an apartment complex or a luxury hotel, both of which have similar first costs. The LCC of the apartment complex over its anticipated holding period is $5 million, and the LCC of the luxury hotel over its anticipated holding period is $3 million. The company invests in the hotel because it has the lower LCC.

This is *not* a valid conclusion. First, the LCCs for the two buildings do not provide any indication of the profitability of one building relative to another. Second, the LCCs provide no indication of profitability relative to other building or nonbuilding investment opportunities. Furthermore, if the holding periods are different and the costs are expressed as present value equivalents, the LCCs do not even indicate the comparative costs of the two buildings.

Example 2. A building manager has identified five independent projects in a given building which are cost effective but has insufficient funds to undertake all of them. In order to allocate the limited budget among candidate projects, he ranks them in ascending order of their LCCs, with the lowest LCC project receiving highest funding priority.

This is *not* a valid method of allocating a limited budget. Ranking projects according to their LCCs will not reliably select the set of projects which, for the budget, minimizes overall LCC for the building.

REFERENCES

American Society for Testing and Materials (ASTM). 1989. *Standard Practice for Measuring Life-Cycle Costs of Building and Building Systems.* E 917-89. Philadelphia: ASTM.

Ruegg, Rosalie T. and Stephen R. Petersen. 1987. *Comprehensive Guide for Least-Cost Energy Decisions.* SP 709. Gaithersburg, MD: National Bureau of Standards.

Ruegg, Rosalie T., Stephen R. Petersen, and Harold E. Marshall. 1980. *Recommended Practice for Measuring Life-Cycle Costs of Buildings and Building Systems.* NBSIR 80-2040. Gaithersburg, MD: National Bureau of Standards.

3

Net Benefits (NB) and Net Savings (NS)

The NB method is reliable, straightforward, and widely applicable for finding the economically efficient choice among building alternatives. It measures the amount of net benefits from investing in a candidate project instead of investing in the foregone opportunity. NB is computed by subtracting the time-adjusted costs of an investment from its time-adjusted benefits. If NB is positive, the investment is economic; if it is zero, the investment is as good as the next best investment opportunity;[1] if it is negative, the investment is uneconomic.[2] Emphasis is on economic efficiency rather than cost effectiveness because the method is appropriate for evaluating alternatives which compete on benefits, such as revenue and other advantages which are measured in dollars, in addition to costs.

The NS method is the NB method recast to fit the situation where there are no important benefits in terms of revenue or the like, but there are reductions in future costs. By treating savings just like revenue benefits, we reformulate the NB method as the NS method.

Chart 3-1 summarizes key features of the NB (NS) method for quick reference.

3.1 FORMULATION OF NB

As with the LCC method, there is the option of expressing NB (or NS) in either present value or annual value terms.

---

[1]This condition is imposed by discounting costs and benefits with a rate based on the opportunity cost of capital.

[2]We use the generic term "economic" as a synonym for both "economic efficiency" and "cost effective," where "economic efficiency" is used in connection with a benefits-producing investment and "cost effective" is used with a cost-reducing investment. "Uneconomic" thus means economically inefficient or cost ineffective.

Chart 3-1. Quick reference for net benefits or net savings method

Abbreviation	NB and NS	Unit of Measure $
Description	Finds the difference between time-adjusted benefits (or savings) and costs of a course of action relative to a mutually exclusive alternative course of action.	

Formulas

(PV)

$$NB_{A1:A2} = \sum_{t=0}^{N} \frac{B_t - \overline{C}_t}{(1 + d)^t}$$

$$NS_{A1:A2} = \sum_{t=0}^{N} \frac{S_t - \overline{I}_t}{(1 + d)^t}$$

$$= LCC_{A2} - LCC_{A1}$$

Aggregate NB:

$$\text{Max} \sum_{p=1}^{n} NB_p X_p, \quad \text{where } X_p = 0 \text{ or } 1$$

$$\text{Subject to} \sum_{p=1}^{n} I_p X_p \le L$$

Applications	Approach
1. Accept/reject	Compute NB or NS for an investment relative to its base-case alternative, and find the investment acceptable if NB or NS is positive.
2. Design/size	Compute NB or NS for each alternative relative to a fixed baseline, and select the alternative design or size which maximizes NB or NS. or Compute incremental NBs or NSs in successive pairwise comparisons, changing the baseline each time to the current best and continuing to expand the investment as long as incremental NB or NS remains positive.
3. Location	Compute NB or NS for each alternative location relative to a fixed baseline alternative, and choose the location with maximum NB or NS, other factors being the same.
4. Lease/buy	Compute NB for leasing relative to buying and choose leasing if NB is positive, other factors being the same.
5. Combine interdependent systems	Compute aggregate NB or NS for each combination of interdependent systems and choose the combination that maximizes aggregate NB or NS, adjusting for interdependencies.

Chart 3-1. (*Continued*)

Applications	Approach
6. Allocate a limited budget	Compute aggregate NB or NS for each alternative set of candidate projects, each set fitting within the available budget, and choose the set that maximizes aggregate NB or NS.

Special Considerations

- Effects not measured in dollars must be uniform across alternatives or should be taken into account in some other way.
- NB or NS expresses in a single number a comparison of two mutually exclusive project alternatives over the same study period (one alternative may be that of doing nothing).

Notation: PV = present value; $NB_{A1:A2}$ = NB associated with choosing alternative A1 as compared with a mutually exclusive alternative, A2; N = number of years in study period; B_t = benefits in year t of A1 less those of A2; $\overline{C}_t$ = costs in year t of A1 less those of A2; d = discount rate; S_t = savings in year t of A1 as compared with A2; $\overline{I}_t$ = investment costs in year t of A1 less those of A2; NS = net savings; LCC = life-cycle cost; n = number of candidate projects which may be combined in different combinations; NB_p = net benefits of project p; X_p = 0 if project p is not included in a given combination or 1 if it is; I_p = initial investment cost for project p at the base time; L = budget constraint.

3.1.1 Present Value Model

A present value model expresses NB as a lump-sum amount received at the present:

$$PV\,NB_{A1:A2} = \sum_{t=0}^{N} \frac{(B_t - \overline{C}_t)}{(1 + d)^t} \tag{3-1}$$

where

$PV\,NB_{A1:A2}$ = NB, (i.e., benefits net of costs), in present value dollars, attributed to a given alternative, A1, compared with those of a mutually exclusive alternative, A2 (which may be the alternative of doing nothing),

B_t = relevant benefits (i.e., positive cash flows such as revenues or other advantages which are assigned a dollar value) associated with a given alternative, A1, less relevant benefits for a mutually exclusive alternative, A2, in period t, and

$\overline{C}_t$ = relevant costs (i.e., negative cash flows) associated with a given alternative, A1, less relevant costs for a mutually exclusive alternative, A2, in period t.

Note that costs are not defined in equation 3-1 exactly as they are in equation 2-1. In the LCC method, the baseline case and the alternatives are evaluated separately. But in the NB method—as well as in the remaining methods—the baseline case receives neither a separate evaluation nor an individual measure. Inherent in the NB measure of worth is a comparison of an alternative against the baseline or another alternative.

The present value version of the NB (or NS) method is also called the "net present worth" or "net present value" method, or, for short, simply the "present worth" or "present value" method. The term "net present value" is an abbreviation of "present value of net benefits."

A simplistic formulation of NB is given in equation 3-2. It does not emphasize the distinction between benefits and costs, as equation 3-1 does, but simply focuses on net cash flow. Furthermore, it may not make explicit the comparison of alternatives. However, it gives results identical to those of equation 3-1, provided that a_t is as defined below.

$$NPW = \sum_{t=0}^{N} \frac{a_t}{(1 + d)^t} \tag{3-2}$$

where

NPW = net present worth (or value), and
a_t = net cash flow in period t resulting from B_t less $\overline{C}_t$.

3.1.2 Annual Value Model

NB may be expressed in annual value dollars, i.e., as a uniformly recurring annual receipt over a designated period of time, as given by equation 3-3:

$$AV\,NB_{A1:A2} = \left(\sum_{t=0}^{N} \frac{B_t - \overline{C}_t}{(1 + d)^t}\right)\left(\frac{d(1 + d)^N}{(1 + d)^N - 1}\right) \tag{3-3}$$

where

$AV\,NB_{A1:A2}$ = NB (i.e., benefits net of costs), in annual value dollars, attributed to a given alternative, A1, as compared with those of a mutually exclusive alternative, A2, and

$\frac{d(1 + d)^N}{(1 + d)^N - 1}$ = formula for the capital recovery discount factor (see Chapter 8).

3.1.3 Aggregate NB or Aggregate NS Model

You will need a special version of the formula for combining interdependent projects or for allocating a limited budget among competing projects. It is the aggregate NB or NS formula for finding the combination of projects which will maximize aggregate net benefits or net savings.[3]

$$\text{Max} \sum_{p=1}^{n} NB_p \, X_p \tag{3-4}$$

$$\text{Subject to} \sum_{p=1}^{n} I_p X_p \leq L$$

where

n = number of candidate projects and separable increments to projects,
NB_p = net benefits or net savings for project p,
X_p = 0 if project p is not among these chosen for a given combination and 1 if it is, and
I_p = initial investment cost for project p at the base time, and
L = budget constraint.

Using equation 3-4 to find the combination of interdependent projects which maximizes aggregate NB or NS requires that the summation be adjusted to reflect interdependence each time another project or increment to a project is taken. The budget may or may not be limited in this application. (See Section 3.3.5.)

Using equation 3-4 to find the combination of projects which maximizes aggregate NB or NS for a limited budget requires that the investment costs of the projects in each combination fit within the budget constraint. (See Section 3.3.6.)

3.2 FORMULATION OF NS

Equation 3-5 reformulates the NB method to allow it to be used when the advantages of an investment are cost reductions, i.e., savings instead of receipts.

$$PV\,NS_{A1:A2} = \sum_{t=0}^{N} \frac{S_t - \bar{I}_t}{(1 + d)^t} \tag{3-5}$$

[3]Only the aggregate NB formula is given. The aggregate NS formula is identical, except that NS is substituted for NB.

where

$PV\,NS_{A1:A2}$ = NS, in present value dollars, attributed to a given alternative, A1, as compared with those of a mutually exclusive alternative, A2, which may be the alternative of doing nothing,

S_t = cost savings plus any benefits associated with a given alternative, A1, as compared with a mutually exclusive alternative, A2, in period t, where

$$\sum_{t=0}^{N} S_t = \sum_{t=0}^{N} (B_t - \tilde{C}_t), \tag{3-6}$$

$$\left| \sum_{t=0}^{N} \tilde{C}_t \right| >> \sum_{t=0}^{N} B_t \quad \text{and} \quad \sum_{t=0}^{N} \tilde{C}_t < 0 \tag{3-7}$$

and

$$\tilde{C}_t = \overline{C}_t - \overline{I}_t \tag{3-8}$$

$\overline{I}_t$ = investment costs associated with a given alternative, A1, less those associated with a mutually exclusive alternative, A2, in period t, and

$$\overline{I}_t = \overline{C}_t - \tilde{C}_t \tag{3-9}$$

Equation 3-10, which takes the difference between the LCCs of two alternatives, gives the same result as equation 3-5:

$$NS_{A1:A2} = LCC_{A2} - LCC_{A1} \tag{3-10}$$

Appendix B provides worksheets for calculating NS. Instructions which accompany the worksheets explain, step by step, how to use them.

3.3 APPLICATIONS OF THE NB OR NS METHOD

The NB method is applicable to a wider array of decisions than the NS and LCC methods because it accounts for variations in benefits as well as costs among alternatives. In addition, the NS method is applicable to a wider array of decisions than the LCC method, despite the fact that it is usually calculated by taking the difference in LCCs, and often appears only as a supplementary measure in LCC studies.

NB or NS is more versatile than the LCC method for budget allocation decisions. The LCC method is limited to finding the set of candidate projects with the largest impact on the costs of a single building or facility. (This is done by

Table 3-1. NB for accept/reject decisions: illustration

(1) Investment	(2) NB (PV $)	(3) Decision
Apartment building	600,000	Accept
Amusement park	−1,000,000	Reject

Nomenclature: PV $ = present value dollars. NB are expressed as a single lump-sum amount occurring at the present.

computing the overall LCC for the building or facility, based on different combinations of projects, and determining for which combination overall LCC reaches a minimum.) But NB or NS can be used not only to find the set of candidate projects with the largest impact on a single building or facility, but also to find the set of projects with the largest impact *across* buildings and facilities. (This is done by computing aggregate NB or NS for alternative sets of projects that are affordable within a given budget—in one or many buildings—and determining the set which maximizes aggregate NB or NS.)

3.3.1 Accept/Reject Decisions

When used in the context of a for-profit investment, net benefits and net savings are synonymous with long-run, time-adjusted profits. The NB (NS) method is therefore widely applicable to decisions which are made on the basis of profitability.

Table 3-1 illustrates the use of NB to decide whether to accept or reject two projects. The decision rule is to accept the investment if NB is positive and reject it if NB is negative, i.e., if net losses result. Hence, the apartment building investment is accepted, and the amusement park investment is rejected.

Chapter 20 illustrates the use of NB in two detailed case studies to evaluate the profitability of investments in public storage facilities and shopping centers. It also shows the use of NS in two additional case studies to evaluate the cost effectiveness of investment in waste-heat recovery and solar hot water heating.

The NB (NS) method is also applicable to evaluating investments of public and other nonprofit organizations. For example, a government agency might use NB to decide whether construction of a new recreation center will generate net benefits to the public. The decision rule is to accept the investment if net social benefits are positive and if it meets budgetary and other constraints.[4]

[4]Benefit-cost analysis in support of certain public decisions tends to encompass broader concerns, more dimensions, and greater complexity than benefit-cost analysis applied to many private sector

Table 3-2. NS for design/size decisions: illustration of fixed baseline approach

(1) Alternative Window Designs	(2) NS (PV $)
Single glazing (base case)[a]	
Double glazing	800
* Low-emissivity double glazing	1,000
Triple glazing	300

Nomenclature: PV $ = present value dollars. NS is expressed as a single lump-sum amount occurring at the present.
[a]Single glazing is the fixed baseline against which each of the other window designs is compared.
* designates the cost-effective design decision.

3.3.2 Design and Sizing Decisions

The NB or NS method can be used to design and size a building or system. There are two approaches: (1) the total approach using a fixed baseline and (2) the incremental approach using a shifting baseline.[5]

The fixed baseline approach entails comparing all alternative designs or sizes against the same baseline. The decision rule is to choose the alternative with the largest NB or NS or, if none has a positive NB or NS, to choose the baseline alternative. Table 3-2 illustrates the use of NS with the fixed baseline approach to compare the cost effectiveness of alternative window designs. Double-glazed windows, low-emissivity doubled-glazed windows, and triple-glazed windows are each compared against single-glazed windows. Low-emissivity double-glazed windows are indicated to have higher NS relative to single-glazed windows than the alternatives and, therefore, are the economic choice.

The incremental approach entails using a shifting baseline. Incremental NB or NS values are computed by performing pairwise comparisons of design or size alternatives. One usually orders the set of alternatives from lowest to highest investment cost. If the first alternative yields a positive NB or NS relative to the baseline, it is the current best. The alternative with the next highest in-

decisions. The straightforward profitability approach used with private sector evaluations and some public sector decisions is often inadequate to address broader issues of social welfare. For a treatment of benefit-cost analysis for evaluating social welfare issues, see Mishan (1976).

[5]The decision to accept or reject a discretionary project is often combined with the design and sizing decisions. In this case, the fixed baseline, or the initial value of the shifting baseline, should be the best alternative to having the project. But in the case of a mandatory system, or when the decision to accept has already been made, the fixed baseline or initial value of the shifting baseline should be the project design or size with the lowest investment cost.

Table 3-3. NS for design/size decisions: illustration of incremental approach

(1)	(2)	(3)	(4)
From Carpet Quality Grade	To Carpet Quality Grade		
	Grade 2	Grade 3*	Grade 4
		Incremental NS in PV $	
Grade 1	5,000		
Grade 2		2,000	
Grade 3			−1,000

Nomenclature: PV $ = present value dollars. NS are expressed as a single lump-sum amount occurring at the present.
* designates quality grade that maximizes NS.

vestment cost is then compared with the current best to see if it yields a positive NB or NS. If it does, it becomes the baseline for the next pairwise comparison. If it does not, the baseline remains the same and yet another alternative is compared. The investment is expanded until there are no remaining alternatives that generate a positive incremental NB or NS.

Table 3-3 illustrates the incremental approach to designing and sizing with NS. Four levels of carpet quality are compared. The first comparison is between grade 2 and the baseline, grade 1. Grade 2 yields a positive incremental NS and, therefore, becomes the new baseline for comparison with grade 3. Incremental NS are positive for grade 3, making it the new baseline for comparison with grade 4. Net losses result from grade 4. Therefore, the expansion of investment in higher-quality carpeting stops with grade 3.

3.3.3 Location Decisions

Although many factors go into the choice of locations, the NB or NS method may provide useful information for the decision. Compute and compare the NB or NS associated with alternative locations and, if using a fixed baseline alternative, favor the location with the maximum NB. (See Section 3.3.2 for a discussion of the baseline.)

Businesses that sell a service rather than a product typically have sales and revenue which are highly sensitive to store location. Public service organizations, such as libraries, also typically have benefits which are highly sensitive to location. Their location decisions may best be assisted by use of the NB method. Organizations whose location decision hinges primarily on costs can use the NS method as a substitute for the LCC method to assist the decision.

Chapter 21 shows, in a detailed case study, how to use NB to evaluate the comparative profitability of alternative locations for a highly market-oriented service.

3.3.4 Lease/Buy Decisions

Follow essentially the same procedures to apply the NB or NS method to lease or buy decisions as for design and sizing applications. Compute NB or NS for leasing relative to buying and favor leasing if NB is positive and other factors are the same. But, as explained in Chapter 22, there may be other factors not captured in the benefit-cost analysis which strongly influence the decision.

3.3.5 Combining Interdependent Systems

The NB or NS method can be used to select the most profitable combination of interdependent investments, but only if a special version of the method is applied: the aggregate NB or aggregate NS method. The approach is to compute aggregate NB or NS for alternative combinations of interdependent projects, adjusting for project interdependencies. Select the combination having the highest aggregate NB or aggregate NS. (See section 3.1.3.)

3.3.6 Budget Allocation Decisions

The aggregate NB or NS approach can also be used to allocate a limited budget among competing projects either within a given building or facility or across buildings or facilities. The approach is to compute the aggregate NB or NS for each combination of candidate projects fitting within the budget and select the combination which maximizes aggregate NB or NS.[6] (Projects within alternative combinations may be independent or interdependent.) Note that the aggregate approach does *not* indicate the order in which projects within the selected group should receive funding priority.

3.3.7 NB or NS Method Not for All Decisions

A project's individual NB or NS is *not* appropriate for assigning it priority for budget allocation. Table 3-4 shows how choosing projects with the highest *individual* NB or NS may *not* maximize the *aggregate* NB or NS for the budget.

[6]If there are funds remaining after allocation to a given combination of projects, the residual is assumed to yield a return equal to that on the next best available investment (as indicated by the discount rate). This means that NB for any residual funds is zero.

Table 3-4. NB and NS not suited for direct ranking of projects for budget allocation

Projects	Investment Costs (PV $)	NB (PV $)	Ranking According to Individual NB
A	5,000	10,000	1
B	2,000	6,000	2
C	2,000	5,000	3
D	1,000	2,000	4

Budget constraint: $5,000
Project selection according to individual NB ranking: Project A
I = $5,000
NB = $10,000

Project selection according to aggregate NB approach: Projects B, C, and D
I = $5,000
NB = $13,000

Nomenclature: PV $ = present value dollars. NB are expressed as a single lump-sum amount occurring at the present.

Four independent projects, A through D, compete for a limited budget of $5,000. Project A appears most economic on the basis of its individual NB ranking. Yet spending the $5,000 budget on projects B, C, and D, instead of on A, will yield a higher NB in the aggregate than project A.

3.3.8 Summary of Points to Note in Applying the NB or NS Method

As with the LCC method, evaluate competing alternatives over the same period of time if the analysis is performed in present value dollars. You can relax this requirement if the analysis is performed in annual value dollars, but only under limited conditions. (For further discussion, see Chapter 10.)

It pays to accept any investment or increment to the investment which has a positive NB or NS. But if the budget is limited, it may be necessary to forego investments and increments to investments even though their NB or NS is positive. A condition for maximizing NB or NS with a limited budget is that the return on the last dollar spent on each investment be equal.

Effects *not* measured in dollars are left out of the NB or NS analysis. Therefore, you should ascertain that these effects are either insignificant or uniform across alternatives. If they are neither insignificant nor uniform, you will need to take them into account in some other way.[7]

3.4 EXAMPLES OF CORRECT USE OF THE NB OR NS METHOD

Example 1. A corporation must decide whether to upgrade its manufacturing plant to improve the quality of its product (which will be sold at a higher price than the existing product). Taking into account downtime, renovation costs, changes in future operating costs, and changes in future revenue, the NB of upgrading the plant is positive. The upgrade is approved.

Example 2. An investment specialist is looking for real estate investments for clients. Shares are available in a limited partnership which invests in small shopping centers. The specialist uses the NB method to evaluate the profitability of the investment for a client. The specialist does not recommend the investment because the NB, after taxes, is negative.

Example 3. The owner of a movie theater is considering relocating the theater to increase profits. Two new potential locations have been identified. Location A, the more costly of the two, is on a major bus line, which is expected to increase attendance and revenues. Location B, in a lower-rent district, will reduce costs but will also reduce revenues because it is difficult to reach by public transportation and has little parking space available. The owner uses the NB method to compare the economics of the two sites and chooses location A on the basis of its higher NB.

Example 4. A county board must decide whether to dispose of increasing quantities of trash by building an incinerator and disposing of the ash or by dumping the trash directly into a landfill. The incinerator/ash disposal approach entails millions of dollars of capital investment and a transportation system to move trash to the incinerator and ash to the depository. The incinerator will be

---

[7]The analytical hierarchy process model, developed by Saaty (1988), is an example of a method which allows users to use qualitative criteria in evaluating investment alternatives. The user specifies evaluation criteria, which may include difficult-to-quantify criteria such as beauty, as well as economic criteria; specifies the investment alternatives to be evaluated, such as alternatives A and B; makes pairwise comparisons between criteria to establish weights for them (e.g., beauty is twice as important as costs); rates each investment alternative with respect to each criterion (e.g., alternative A is three times more beautiful than alternative B); and then computes a final score for each alternative.

used to fire a power generator owned by the municipality. The sale of electricity will generate revenue. The landfill approach requires a large area of land and a transportation system to move trash to the site. The landfill has lower initial costs. Both approaches adversely affect surrounding property values and the tax base. The board must present to the public a detailed economic evaluation of all options considered. The NB method is used to compare the costs and benefits of the incinerator relative to those of the landfill.

Example 5. The U.S. Army Corps of Engineers is considering a multipurpose water resources project which entails constructing a dam and reservoir. Costs include, in addition to construction costs, opportunities foregone to grow crops on flooded land and the loss of houses to the reservoir pool. Benefits are expected from reduced flood damage to crops and property in the adjacent floodplain, as well as from irrigation, recreation, and power generation. The Corps of Engineers uses the NB method to evaluate the economic merits of the project.

Example 6. Designing a new building with windows which are easier to wash than windows of conventional design is estimated to a raise the first cost but reduce maintenance costs. The easy-to-wash windows are evaluated using the NS method and approved on the grounds that they result in a positive NS.

3.5 EXAMPLES OF INCORRECT USE OF THE NB AND NS METHODS

Example 1. A plant manager of a research facility is looking for ways to obtain the largest, long-run savings in the operation costs of the facility by spending a renovation budget of $100,000. More than a dozen projects have been identified which are estimated to save more than they cost, but their first costs total $200,000, twice the budget. Since the intent is to maximize savings from the fixed budget of $100,000, the plant manager accepts individual projects in order of their NS, accepting first the project with the greatest NS. But this approach is faulty. It may not result in the selection of projects which produce the greatest aggregate NS.

Example 2. The energy manager of a large chemical plant computes the NS for an on-site cogeneration plant as an alternative to purchasing electricity and steam from municipal suppliers. The NS is estimated at $2 million, but the manager rejects the project on the grounds that project costs are $2.5 million, that is, \$0.5 million in excess of NS. The conclusion to reject the project stems from an invalid interpretation of NS as savings rather than *net* savings. An NS of \$2 million means that the project earns \$2 million over and above the \$2.5 million in costs. Thus the project is estimated to be cost effective.[8]

REFERENCES

Mishan, E.J. 1976. *Cost-Benefit Analysis, An Introduction.* New York: Praeger Publishers.

Saaty, Thomas L. 1988. *Multicriteria Decision Making: The Analytic Hierarchy Process*. Pittsburgh: University of Pittsburgh Press.

[8]It may appear that this error would not be made, but it is a common error to net out costs twice, implicitly if not explicitly. Another common error is to ignore the fact that a project with a zero or small NB or NS nevertheless earns at a rate equal to or greater than the investor's minimum acceptable rate of return.

4

Benefit-to-Cost Ratio (BCR) and Savings-to-Investment Ratio (SIR)

BCR and SIR are numerical ratios whose size indicates the economic performance of an investment. The BCR is computed as benefits, net of future non-investment costs, divided by investment costs.[1] The SIR is savings divided by investment costs. The SIR is the BCR method recast to fit the situation where an investment's primary advantage is lower costs. SIR is to BCR what NS is to NB.

A ratio less than 1.0 indicates an uneconomic investment; a ratio of 1.0 indicates an investment whose benefits or savings just equal its costs; and a ratio greater than 1.0 indicates an economic project. A ratio of, say, 6.0 means that the investor can expect to receive on average $6 for every $1 invested, over and above the required rate of return imposed by the discount rate.

Chart 4-1 summarizes for quick reference key features of the BCR and SIR methods.

4.1 FORMULATION OF THE BCR AND SIR

In formulating the ratio, give attention to placement of costs and benefits in the numerator and denominator. We recommend that you place in the denominator only those costs on which you wish to maximize the return. For example, to maximize the return on equity funds, place in the denominator only the part of the investment budget representing equity funds. To maximize the return on the combined total of equity and borrowed investment funds, place their sum in the denominator of the ratio.

[1]Benefit-cost ratio and cost-benefit ratio, both represented by B/C, are common expressions for the BCR. We prefer "benefit-to-cost" ratio over "benefit-cost" ratio because it reinforces the notion that we have a ratio. We prefer "benefit-to-cost" ratio to "cost-benefit" ratio for the additional reason that analysts sometimes mistakenly divide costs by benefits when using a "cost-benefit" ratio.

Chart 4-1. Quick Reference Guide for the BCR and SIR methods.

Abbreviations	BCR, SIR	Unit of Measure	Pure number

Description	Find the ratio of benefits (less future costs) to investment costs or the ratio of savings to investment costs, where all amounts are time-equivalent values.

Formulas

(*PV*)

$$BCR_{A1:A2} = \sum_{t=0}^{N} \frac{(B_t - \ddot{C}_t)/(1 + d)^t}{\ddot{I}_t/(1 + d)^t}$$

$$SIR_{A1:A2} = \sum_{t=0}^{N} \frac{(\ddot{S}_t)/(1 + d)^t}{\ddot{I}_t/(1 + d)^t}$$

Applications	Approach
1. Accept/reject	Compute the ratio based on benefits (or savings) and costs of having the investment versus not having it, and find the investment acceptable if the BCR or SIR is greater than 1.0.
2. Design/size	Compute incremental BCRs and SIRs by comparing design/size alternatives in pairwise comparisons, changing the baseline each time to the current best, and expanding the investment until there are no remaining alternatives with incremental ratios greater than 1.0.
3. Combine interdependent systems	An approximation approach is to compute the BCR or SIR for each interdependent building system in its least expensive size, without regard to interdependencies; select the system with the highest ratio; and recalculate ratios on all interdependent systems and increments to systems based on acceptance of the first. Select another system or increment to the first-selected system, recalculating ratios on remaining options based on acceptance of the second. Continue this process until there are no more options having SIRs greater than 1.0.
4. Allocate a budget	Rank each project and project increment according to its ratio, with top priority for funding assigned to the project with the highest ratio. Allocate the budget in accordance with priority rankings. If the budget is not mostly exhausted by taking projects in BCR or SIR order, test alternative project combinations within budget to see which has the highest aggregate NB or NS.

Special Considerations

- Effects not measured in dollars must be insignificant or uniform across alternatives or should be taken into account in some other way.
- BCR and SIR express in a single number a comparison of two mutually exclusive project alternatives over the same study period. (One alternative may be to do nothing.)

Chart 4-1. *(Continued)*

Abbreviations	BCR, SIR	Unit of Measure	Pure number

- How the ratio is formulated, with placement of items in the numerator and denominator, can affect decisions. It is important to use a formulation which fits the investor's objective.

Notation: PV = present value; $BCR_{A1:A2}$ = benefits-to-cost ratio for alternative A1 as compared with a mutually exclusive alternative, A2; N = number of years in study period; B_t = benefits in year t of A1 less those of A2; $\check{C}_t$ = costs in year t of A1 less those of A2, excluding investment costs shifted to the denominator: $\check{I}_t$ = that part of investment costs in time t on which the return is to be maximized; d = discount rate; SIR = savings-to-investment ratio; and $\check{S}_t$ = savings from alternative A1 as compared with alternative A2.

How you formulate the ratio is important because varying the placement of cost and benefit items between the numerator and denominator can affect its value. Changing the placement of items will *not* cause a project which appears economic by one formulation of the ratio to appear uneconomic by a different version, and it will *not* cause a project which appears uneconomic by one formulation to appear economic by another.[2] But placement can affect the numerical value of a ratio for a given investment. Thus, by one formulation of the BCR or SIR, a project may appear more or less economically worthwhile than by another. The result is that the relative ranking of competing independent investments can be affected by the placement of items in the numerator or denominator, and this can bias selection away from the most efficient projects.

The following examples show how biasing effects, detrimental to economic efficiency, can result from certain formulations. Subtracting maintenance costs from the numerator, but adding energy costs to the denominator, will bias selection toward projects higher in maintenance costs and lower in energy costs. This is because subtracting a given amount from the numerator reduces the ratio less than adding the same amount to the denominator. Similarly, subtracting resale value from investment costs in the denominator biases selection of projects toward those higher in resale value and lower in other benefits, because adding a given amount to the numerator increases the ratio less than subtracting an identical amount from the denominator.[3]

Table 4-1 illustrates how formulation of the ratio can affect its value, as well as the investment decision. Using the formulation we recommend in Chart 4-1 and Section 4.1.1 results in the project ranking shown in column 6. In contrast, using a formulation of the BCR which places maintenance costs in the denom-

[2]See Marshall and Ruegg (1981, p. 8) for a proof showing that the placement of items between the numerator and denominator will not make an economic project appear uneconomic or vice versa.

[3]It is possible that an investor will prefer a formulation of the ratio that has a bias as defined here. For instance, the investor may wish to maximize the return on a particular type of funds such as working capital, which might be the constraining factor.

Table 4-1. Effects of BCR and SIR formulation: illustration

(1) Competing Projects to Rank for Budget Allocation	(2) Investment ($)	(3) Maintenance Cost (PV $)	(4) Revenue (PV $)	(5) BCR (Chart 4-1)[a]	(6) Ranking According to BCR in Col. 5	(7) BCR (Different Formulation)[b]	(8) Ranking According to BCR in Col. 7	(9) NB (PV $)
Project A	2,000	300	5,000	2.35	3	2.17	2	2,700
Project B	2,000	0	5,000	2.50	2	2.50	1	3,000
Project C	2,000	1,000	6,200	2.60	1	2.07	3	3,200

Nomenclature: PV $ = present value dollars. NB is expressed as a single lump-sum amount occurring at the present.

[a]BCR computed using equation 4-1 as given in Section 4.1.1 and Chart 4-1.

[b]Alternative formulation used to compute BCR: revenue/(investment + maintenance costs). Note that we use this formulation only for purposes of illustration and do not recommend it.

inator results in the project ranking in column 8. From column 9 we can compute the aggregate NB for a given budget. For a budget of $2,000, the column 6 ranking results in an aggregate NB of $3,200, compared with $3,000 for the column 8 ranking. For a budget of $4,000, the column 6 ranking results in an aggregate NB of $6,200 compared with $5,700 for the column 8 ranking. This shows that failing to follow the recommended placement of costs may bias selection toward an inefficient project.

4.1.1 Formulation of the BCR

Equation 4-1 is a formulation of the BCR which avoids biasing effects.[4] It places future benefits and costs in the numerator, except for that part of the investment expenditure on which the investor wishes to maximize the return. Equation 4-1 is consistent with the economic efficiency objective of maximizing NB for the available investment budget. This formulation has two features that alternative formulations often lack. First, investment costs can be specified for the current year or for a multiyear planning period. Second, investment costs can be specified as equity funds only, total investment costs, or any part of investment costs you choose.

$$BCR_{A1:A2} = \sum_{t=0}^{N} \frac{(B_t - \ddot{C}_t)/(1+d)^t}{\ddot{I}_t/(1+d)^t} \tag{4-1}$$

where

$BCR_{A1:A2}$ = Benefit-to-Cost ratio computed from present value benefits and costs[5] for a given alternative, A1, as compared with that of a mutually exclusive alternative, A2,

B_t = relevant benefits associated with a given alternative, A1, less relevant benefits for a mutually exclusive alternative, A2, in period t,

$\ddot{C}_t$ = relevant costs associated with a given alternative, A1, less relevant costs of a mutually exclusive alternative, A2, in period t, where both exclude all or part of investment costs, $\ddot{I}_t$,

[4]This formulation has been adopted by the American Society for Testing and Materials (ASTM 1989).

[5]The ratio also can be computed using annual value benefits and costs by converting the present value numerator and denominator to annual values. How to convert present values to annual values is shown in equation 2-3 and explained in Chapter 8.

shifted to the denominator of the BCR ratio:

$$\ddot{C}_t = \overline{C}_t - \ddot{I}_t, \text{ and} \tag{4-2}$$

$\ddot{I}_t$ = that part of investment costs in time t (shifted from $\overline{C}_t$) on which the return is to be maximized, and $\ddot{I}_t \leq \overline{I}_t$.

Note that $\ddot{I}_t$ consists solely of investment costs on which the investor wishes to maximize returns, while $\overline{I}_t$, used in Chapter 3's NB and NS formulations, consists of total investment costs. Therefore, $\ddot{I}_t$ may be equal to or less than $\overline{I}_t$. For example, when part of investment costs are borrowed and the investor wishes to maximize the return on equity, only the part of investment paid out of equity funds is included in $\ddot{I}_t$.

4.1.2 Formulation of the SIR

Equation 4-3 formulates the SIR to avoid the biasing effects described earlier.

$$SIR_{A1:A2} = \sum_{j=0}^{N} \frac{\ddot{S}_t/(1 + d)^t}{\ddot{I}_t/(1 + d)^t} \tag{4-3}$$

where

$SIR_{A1:A2}$ = Savings-to-investment ratio computed from present value savings and costs[6] for a given alternative, A1, as compared with that of a mutually exclusive alternative, A2, and

$\ddot{S}_t$ = cost savings, plus any benefits associated with a given alternative, A1, as compared with those of a mutually exclusive alternative, A2, in period t, computed as the present value of benefits combined with reductions in costs, $\ddot{C}_t$:

$$\sum_{t=0}^{N} \ddot{S}_t = \sum_{t=0}^{N} (B_t - \ddot{C}_t) \tag{4-4}$$

where

$$\left|\sum_{t=0}^{N} \ddot{C}_t\right| >> \sum_{t=0}^{N} B_t \quad \text{and} \quad \sum_{t=0}^{N} \ddot{C}_t < 0 \tag{4-5}$$

[6]The ratio also can be computed from annual value savings and costs by converting the present value numerator and denominator to annual values.

4.2 APPLICATIONS OF THE RATIO METHOD

You can use BCR or SIR to determine if a project is acceptable on economic grounds. If you compute the ratio on the basis of incremental rather than total benefits and costs, you can use it to design and size projects and to make other choices among mutually exclusive alternatives, such as alternative locations.

You can use BCR or SIR to combine interdependent building systems whether or not there is a budget constraint. But a primary application of BCR or SIR is to set funding priorities among projects when there *is* a limited budget. This tends to be its primary application.

4.2.1 Accept/Reject Decisions

To use the ratio to decide whether to accept or reject a given investment, calculate the ratio based on the benefits (or savings) and costs of making the investment versus not making it. Accept the investment if its BCR or SIR is greater than 1.0. The higher the ratio, the more desirable the investment on economic grounds.

Table 4-2 illustrates using the BCR for accept/reject decisions. The dam construction project in region 1 is rejected because its BCR is less than 1.0. The levee and the dredging projects are accepted (provided that there is a sufficient budget) because their BCRs are more than 1.0. Chapter 20 gives several detailed case examples of using BCR for accept/reject decisions.

4.2.2 Design and Sizing Decisions

To design or size buildings or building systems, compare alternatives on the basis of their *incremental* BCR or SIR. To compute incremental ratios, arrange design/size alternatives in increasing order of their investment costs for pairwise comparisons.[7] In the first pairwise comparison, compare the additional benefits or savings of the alternative just above the baseline against its additional investment costs. If the ratio is greater than 1.0, that alternative becomes the new baseline. Expand the investment as long as the budget allows, until there are no remaining alternatives that generate incremental ratios greater than 1.0.

This approach is demonstrated in Table 4-3 with data presented in Table 4-4. Look first at Table 4-4, which shows five size alternatives for a project arranged in increasing order of their investment costs, beginning with the zero size op-

[7]For a different ordering of alternatives for pairwise comparisons, see Au and Au (1983). They arrange mutually exclusive alternatives in ascending order of the present values of their total costs.

Table 4-2. BCR for accept/reject decisions: illustration

(1) Candidate Projects	(2) Investment Costs (PV $)	(3) Benefits (PV $)	(4) Noninvestment Costs (PV $)	(5) BCR $(5) = \frac{(3) - (4)}{(2)}$	(6) Decision
Construct dam in region 1	4,000,000	10,000,000	8,000,000	0.50	Reject
Build levee in region 2	25,000,000	100,000,000	5,000,000	3.80	Accept
Dredge channel in region 3	10,000,000	60,000,000	6,000,000	5.40	Accept

Nomenclature: PV $ = present value dollars. NB is expressed as a single lump-sum amount occurring at the present.

Table 4-3. Incremental BCRs for sizing decision: illustration[a]

(1)	(2)	(3)	(4)	(5)
	Size Compared Against Baseline (to This Size)			
Baseline Size (from This Size)	A	B	C*	D
0	5.0	4.6	4.1	3.9
A		3.0	2.2	1.9
B			1.3	1.0
C				0.5

[a]Based on data in Table 4-4.

Note: Only incremental BCRS on the lower diagonal are appropriate for sizing. We show BCRs in the shaded region to emphasize that BCRs may not reveal the economic size unless computed on the smallest possible increments.

*Designates the economically efficient size.

tion. For each size, column 2 gives total investment costs; column 3, total benefits.

Table 4-3 shows incremental BCRs from pairwise comparisons. The incremental BCRs are computed from the data in Table 4-4 by subtracting total benefits of the baseline from total benefits of the alternative (column 3), subtracting total investment costs of the baseline from total investment costs of the alternative (column 2), and then dividing the former by the latter.

Table 4-4. Project data for sizing decision: illustration

(1) Project Size Alternatives	(2) Total Investment (PV $)	(3) Total Benefits (PV $)	(4) Total NB (4) = (3) − (2) (PV $)
0	0	0	0
A	200,000	1,000,000	800,000
B	250,000	1,150,000	900,000
*C	290,000	1,200,000	910,000
D	310,000	1,210,000	900,000

Nomenclature: PV $ = present value dollars. NB is expressed as a single lump-sum amount occurring at the present.

*Designates the economically efficient size.

The BCR ratios in the unshaded portion of Table 4-3, below the diagonal, are computed on the smallest size increments indicated in Table 4-4. Size A compared against size 0 gives a BCR of 5.0 (i.e., \$1,000,000 ÷ \$200,000). Since it is larger than 1.0, size A becomes the new baseline in the second pairwise comparison. Size B compared against size A gives a BCR of 3.0 (i.e., \$150,000 ÷ \$50,000), making size B the new baseline. In the third comparison, size C compared against size B gives a BCR of 1.3. Size D compared against size C gives a BCR of 0.5. Since the ratio is less than 1.0, size D does not become the new baseline. In the absence of a budget constraint, the project is expanded to size C. You can verify that size C is the size which maximizes NB by looking at column 4 of Table 4-4. Total NB from investment in size C are \$910,000, which is greater than NB from other sizes.

Entries in the shaded area show BCRs which are not computed on the smallest available increments. Using them can lead to selection of an uneconomic size.

4.2.2.1 Alternative Approach to Pairwise Comparisons

In addition to the biasing effect, another problem may arise if the ratio is formulated to place in the denominator amounts other than investment costs. If you place noninvestment amounts in the denominator, you may get negative values in the denominator of the ratio when you array alternatives in increasing order of their investment costs for pairwise comparisons. If this happens, the preceding decision rule (i.e., to expand the investment as long as the budget allows until there are no remaining alternatives that generate incremental ratios greater than 1.0) may fail. You can avoid both problems by using equations 4-1 and 4-3.

To address the problem of having negative values appear in the denominator of the ratio, Au (1988) proposes a new approach to performing incremental BCR analysis using pairwise comparisons. This approach does not require ordering alternatives in any specific way, and it applies a new set of decision rules. While it solves the problem of negative values in the denominator, it does not acknowledge the biasing effect that results from placing noninvestment costs in the denominator.

The approach is to list all available alternatives in any order. Beginning with the first one listed, compute the incremental ratio for successive pairwise comparisons. With each comparison, apply decision rules to select the better of the two for the next comparison, continuing until all alternatives have been compared. The last accepted alternative is the best of all the alternatives. The decision rules can be summarized as follows:[8]

[8]The symbols used by Au (1988) are changed to correspond to those used here.

Where A2 is the baseline, i.e., the current best,

1. Prefer alternative A1 over A2 if the incremental investment cost of A1 relative to A2 is positive and the incremental BCR exceeds 1.0; otherwise, prefer A2.
2. Prefer alternative A1 over A2 if the incremental investment cost of A1 relative to A2 is negative and the incremental BCR is less than 1.0; otherwise, prefer A2.
3. Be indifferent between the alternatives if the incremental BCR equals 1.0.

4.2.2.2 BCR and SIR Prone to Misuse in Designing and Sizing

A common mistake is to calculate a BCR (SIR) for each design or size alternative relative to a fixed baseline and to choose the one with the highest BCR or SIR. But the design or size with the highest BCR or SIR is unlikely to be the one with the highest NB or NS. Table 4-3, for example, shows size A to have the highest BCR at 5.0. But we find that A is not the economically efficient size.

Why may the ratio computed on totals incorrectly signal the cost-effective design or size? It is because the ratio measures the average return per investment dollar (over and above the required rate of return). Since the average begins to fall before marginal benefits (marginal savings) drop below marginal costs, basing design and size decisions on total ratios will lead to underinvestment.

Suppose, for example, that spending $100 yields benefits of $1,000, an average return per dollar invested of $10. Suppose that spending $200 instead of $100 yields benefits of $1,400, an average return per dollar invested of $7. The BCR is lower on the larger investment but, at the margin, spending the additional $100 yields additional benefits of $400, an average return per dollar invested of $4. This makes it an economic investment. From spending the first $100, NB is $900, and from spending a total of $200, NB rises to $1,200.

Recall from Chapter 1 that NB (NS) is maximized at the design/size for which marginal benefits (marginal savings) just equal marginal costs (i.e., where the incremental BCR equals 1.0). The mathematical relationship between average and marginal returns makes it imperative to use incremental ratios in making choices among mutually exclusive alternatives such as design and sizing.[9]

---

[9]If the BCR or SIR computed on total benefits and costs rises as investment size increases, it is possible to choose the economic design or size on the basis of total BCR or SIR without incremental analysis. This is because we know that the ratio computed on totals can rise only if the incremental ratio is greater than the ratio on the preceding investment amount. For example, if the ratio on size 2 relative to size 1, the smallest, is 2.0, and the ratio on size 3 relative to size 1 is 4.0, we know intuitively that the incremental ratio on size 3 relative to size 2 must be greater than 1.0. (In fact, we know that it is greater than 4.0.) We know, therefore, to prefer size 3 over size 2.

4.2.3 Combining Interdependent Systems

The problem of finding the cost-effective combination of building systems may exist even when funds are ample. This is true when two or more systems are interdependent. You can use incremental BCRs and SIRs in a sequential approach to approximate the combination of interdependent building systems which will maximize aggregate NB.

The sequential approach to applying incremental ratios is practical when there are few alternatives. But the procedure becomes cumbersome as the number of alternatives increases. It is necessary to recompute ratios each time a selection is made.

The first step is to compute BCRs or SIRs for each interdependent building system in its least expensive size, without regard to the interdependencies. Select the system with the highest ratio. Then recalculate ratios on all interdependent systems and increments to systems based on acceptance of the first. In a second selection round, tentatively select the option with the next highest ratio—which could be another system or an increment to the first-selected system. Recalculate ratios on the remaining options, based on acceptance of the second. In addition, recalculate the ratio on the first-selected system to make sure that it does not fall below the acceptance point as the second option is added. (If the first option is no longer acceptable, the ratios on the remaining options should be based only on interdependence with the second selection.) Continue this process until there are no more options having SIRs greater than 1.0. Chapter 26 illustrates this approach in detail.

4.2.4 Budget Allocation Decisions

BCRs or SIRs can guide the allocation of limited funds among competing building investments. The investments may be either independent or interdependent[10] and of fixed or varying design and size. Set priorities among competing projects according to their ratios, and give funding priority to projects yielding the greatest return per dollar invested. The procedure begins by computing incremental BCRs or SIRs—whichever is appropriate—for each project in its smallest investment size (''base-size project'') and for each project increment. Eliminate at the outset base-size projects and increments to projects having ratios less than 1.0. Rank each remaining base-size project and project increment according to its ratio, with top priority for funding assigned to the base-size

---

[10] If projects are interdependent, care must be taken to check project rankings each time an additional selection is made.

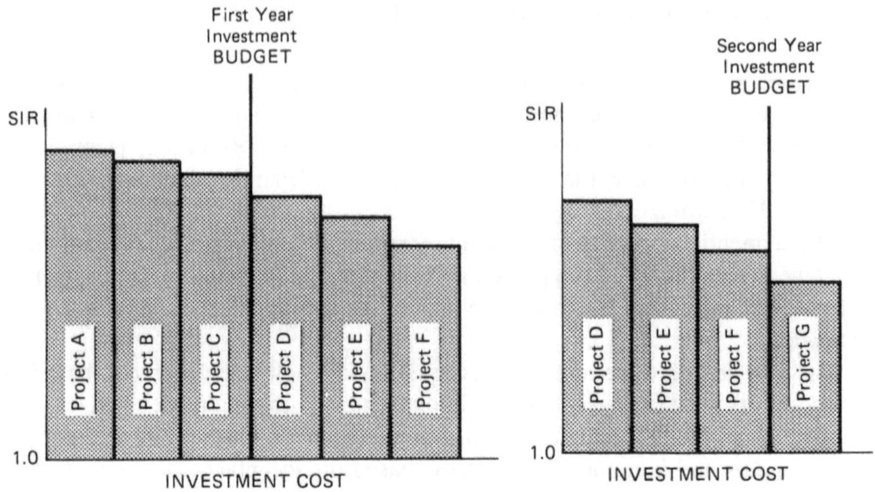

Figure 4-1. Budget allocation by SIR.

project with the highest ratio. Allocate the budget in accordance with the priority rankings.[11]

Figure 4-1 illustrates the use of SIRs to select among projects for funding over two budget cycles. For simplicity, the illustration shows independent projects without competing design and size options. Six candidate projects are cost effective (i.e., their SIRs are greater than 1.0) in the first year's budget cycle. However, the budget in that year allows only the first three to be done. In the second year, the budget allows the remaining three projects to be accepted. (Reevaluate them if you think that the SIRs may have changed significantly.) A new candidate project (G) is omitted because the budget is insufficient to fund all available cost-effective projects, and Project G has a lower SIR than the other candidates.

---

[11]An exception to this ranking procedure is necessary when the incremental ratio increases as investment in a given project is expanded. Suppose, for example, that the incremental BCR on project A in size 1 is 5.0, the incremental BCR on size 2 of project A relative to size 1 is 10.0, and the total BCR on project A in size 2 is 7.0. Suppose that project A in size 2 requires the same investment cost as another competing project, B, which has a BCR of 6.0. The described ranking procedure would rank project B (BCR = 6.0) higher than project A in size 1 (BCR = 5.0). (The size 2 increment is then not chosen because size 1 is not chosen.) But NB would be maximized by choosing project A in size 2 (BCR = 7.0) instead of project B (BCR = 6.0) for the same investment cost. In this exceptional case, compute total BCR or SIR for project A in both sizes, 1 and 2. Prefer project A in size 2 over project B if funding is sufficient. Otherwise, prefer project A in size 1.

The ratio ranking procedure is a reliable way to maximize NB (NS) from the available budget, provided that the budget can be totally expended while taking projects exactly in descending order of their ratios. This is demonstrated in Table 4-5. The example shows eight independent projects ranked in descending order of their SIRs. With a budget of $15,000, selecting projects A, B, and C is estimated to maximize aggregate NS. And with a budget of $40,000, giving priority to projects A, B, C, D, E, and F is estimated to maximize aggregate NS. This can be verified by computing aggregate NS for each combination of projects that fits within the budget.

But if it is necessary to depart from the rankings due to "lumpiness" in project costs, the ratio procedure may not identify the set of projects which will maximize aggregate NB for the budget. It may be possible to increase aggregate NB by skipping over higher-ranked projects and increments to projects and selecting lower-ranked projects or increments. Therefore, when the budget is not fully exhausted following the ratio-ranking procedure, it is advisable to test alternative sets of projects using aggregate NB or aggregate NS[12] to see which set maximizes aggregate NB or aggregate NS for the budget.

Although the ratio-ranking procedure is not fail-safe in allocating a budget, it indicates which sets of projects to test with the aggregate NB or aggregate NS method. This is demonstrated in Table 4-6. It shows how to select projects for funding when project costs are lumpy and the budget cannot be fully allocated by strict adherence to BCR or SIR ranking.

Table 4-6 lists seven independent projects. (In order to focus on the problem of lumpy project costs, the illustration is shown for independent projects without competing design and size options.) Investment costs, in column 2, total almost $20 million. Savings are in column 3 and NS (i.e., savings minus costs) in column 4.

If the available budget is as much as $20 million, it pays to accept all the projects because they each have a positive NS (and an SIR greater than 1.0). But suppose that a lesser budget is available—say, only $10 million—and the objective is to maximize aggregate NS for the $10 million expenditure. It is necessary to select from among the seven candidates those for funding. First, compute the SIR for each candidate, as shown in column 5. Next, rank the projects in descending order of SIRs, as shown in column 6.

Project E is first (SIR = 12.5), using $2 million of the budget. Project F is second (SIR = 12.0), using another $3 million, for a total budget allocation for E and F of $5 million. Project G is third (SIR = 9.0), for a total allocation

[12]See Section 3.1.3, for the aggregate NB formula and Sections 3.3.6 and 3.3.7 for discussion of using the aggregate NB or aggregate NS method for budget allocation.

Table 4-5. **Budget allocation by SIR: illustration**[a]

(1) Projects	(2) Investment Costs (PV $)	(3) Cumulative Investment Costs (PV $)	(4) Total Savings (PV $)	(5) NS	(6) SIR	(7) Ranking
A	4,000	4,000	50,000	46,000	12.5	1
B	2,500	6,000	16,000	13,500	6.4	2
C	8,000	14,500	32,000	24,000	4.0	3
D	2,500	17,000	6,000	3,500	2.4	4
E	15,000	32,000	33,000	18,000	2.2	5
F	8,000	40,000	12,000	4,000	1.5	6
G	2,000	42,000	2,500	500	1.3	7
H	25,000	67,000	27,500	2,500	1.1	8

Nomenclature: PV $ = present value dollars. Net benefits are expressed as a single lump-sum amount occurring at the present.
[a]Present values based on a discount rate of 8%.

Table 4-6. Budget allocation by SIR when project costs are lumpy: illustration

(1) Projects	(2) Investment Costs ($1 Million)	(3) Savings ($1 Million)	(4) NS ($1 Million)	(5) SIR	(6) Ranking (No Budget Constraint)
A	0.2	0.9	0.7	4.5	7
B	2.0	10.0	8.0	5.0	6
C	1.6	12.0	10.4	7.5	5
D	10.0	80.0	70.0	8.0	4
E	2.0	25.0	23.0	12.5	1
F	3.0	36.0	33.0	12.0	2
G	1.0	9.0	8.0	9.0	3
Totals	$19.8	$172.9	$153.1		

Option within Budget of $10 Million	(1) Select All Projects Except D	(2) Select Only Project D
Expenditure (million):	$ 9.8	$10.0
Aggregate NS (million):	$83.1	$70.0

Source: Ruegg (1987, p. 137).

of $6 million. Now look at the fourth-ranked project, Project D (SIR = 8.0). Its cost of $10 million, when added to the costs of the projects already tentatively accepted, exceeds the budget. The SIR ranking procedure breaks down due to lumpy costs. At this point, there are two ways to allocate all or most of the budget:[13]

1. Skip project D and continue selecting projects according to SIR rankings until the $10 million budget is allocated ($0.2 million remain unallocated).
2. Drop preceding projects tentatively selected and instead spend the entire budget on project D.

Keeping in mind the objective—which is to select the set of projects which maximizes aggregate NS for the available budget—compare aggregate NS for the two options. Option 1 (take all projects except D) yields NS of $83.1 mil-

[13]Recall that unspent funds will earn at the opportunity rate, which is equivalent to the discount rate. Thus NS from unallocated funds are zero.

lion. Option 2 (take only Project D) yields NS of $70.0 million. Select option 1 to maximize NS for the limited budget.

Chapter 24 shows, in two detailed case studies, how to use the ratio-ranking technique to guide budget allocation. One of the case studies shows how to allocate a single-period, limited budget among competing projects and project increments. The other shows how to allocate a multiperiod, limited budget.

4.2.5 Ratio Method Not for All Decisions

We recommend against using ratios alone for budget allocation unless there is a fairly close match between project costs and the budget. If the match is not close enough to allow most of the funds to be spent while adhering to ratio rankings, we suggest using aggregate NB or NS as a supplementary measure. We also recommend against using ratios computed on total costs and benefits for selecting designs and sizes.

4.2.6 Summary of Points to Note in Applying the Ratio Method

How the ratio is formulated, with placement of items in the numerator and denominator, can affect decisions. It is important to use a formulation which fits the intended use.

Effects not measured in dollars are not reflected in BCR and SIR ratios. Unless incommensurable effects are either insignificant or uniform across alternatives, the BCR or SIR alone may not provide reliable direction. Remember to take important effects omitted from the analysis into account in some other way.[14]

4.3 EXAMPLES OF CORRECT USE OF THE RATIO METHOD

Example 1. A county park and recreation authority must decide whether to build a community aquatic center. A BCR of 5.8 is estimated for the center, and it is recommended for approval.

Example 2. A plant manager for a research facility is looking for ways to obtain the largest long-run savings in operational costs of the facility from a

[14]Incommensurable effects are often taken into account informally, but they may also be formally modeled. For an example of a model which includes quantified and unquantified variables, see Saaty (1988).

budget of $100,000 which is earmarked for plant renovation. More than a dozen projects have been identified which are estimated to save more than they cost, but their total initial investment cost, $200,000, is twice the budget. The plant manager accepts individual projects in descending order of their SIRs, starting with the project with the highest SIR, and allocates the entire $100,000 without deviating from SIR rankings.

Example 3. An engineer is considering three alternative mechanical systems for a new retail store: (1) a single constant volume (CV) system, (2) a single variable air volume (VAV) system, and (3) a high-efficiency VAV system with an economizer cycle. The budget is sufficient to purchase the system which will be most economical in the long run. The first system has the lowest first costs and highest energy costs; the second has higher first costs and lower energy costs than the first; and the third has the highest first costs and lowest energy costs of the three. The SIR of the second system, using the first as the baseline, is 2.6. The SIR of the third system, using the second as the baseline, is 1.4. The engineer concludes that the third system, the high-efficiency VAV system with an economizer cycle, is the cost-effective choice, because its additional first cost is more than compensated for by its additional energy savings.

4.4 EXAMPLES OF INCORRECT USE OF THE RATIO METHOD

Example 1. A public housing analyst is using SIRs to determine cost-effective levels of insulation [measured in resistance (R) values] for uninsulated attics of public housing. Installing R-11 has an SIR of 15.0. Installing R-19 has an SIR of 12.0. Installing R-30 has an SIR of 7.0. The analyst incorrectly concludes that R-11 is the most cost-effective level of insulation. This conclusion is unjustified because the marginal savings of additional insulation may exceed its marginal costs. (See Section 4.2.2.2.)

Example 2. The BCR on providing a public recreation center without a swimming pool is 3.5. The incremental BCR on adding a swimming pool to the center is 2.0. The swimming pool addition is vetoed by the city government because its BCR is less than that for the center without the pool. But it is incorrect to conclude that the swimming pool addition is uneconomic. As long as the incremental BCR exceeds 1.0, incremental benefits exceed incremental costs and the addition is economic.

REFERENCES

American Society for Testing and Materials (ASTM). 1989. *Standard Practice for Measuring Benefit-to-Cost and Savings-to-Investment Ratios for Buildings and Building Systems*. E 964. Philadelphia: ASTM.

Au, Tung, and Thomas P. Au. 1983. *Engineering Economics for Capital Investment Analysis*. Newton, Massachusetts: Allyn and Bacon, Inc.

Au, Tung. 1988. Profit measures and methods of economic analysis for Capital Project Selection. *Journal of Management in Engineering*. 4(3): 217–228.

Marshall, Harold E., and Rosalie J. Ruegg. 1981. *Recommended Practice for Measuring Benefit/Cost and Savings-to-Investment Ratios for Buildings and Building Systems*. NBSIR 81-2397. Gaithersburg, MD: National Bureau of Standards.

Ruegg, Rosalie T. 1987. *Life-Cycle Costing Manual for the Federal Energy Management Program*. NBS Handbook 135. Gaithersburg, MD: National Bureau of Standards.

Saaty, Thomas L. 1988. *Multicriteria Decision Making: The Analytic Hierarchy Process*. Pittsburgh: University of Pittsburgh Press.

5

Internal Rate-of-Return (IRR)

The IRR is a measure of the percentage yield on investment. The IRR is compared against the investor's minimum acceptable rate of return (MARR)[1] to ascertain the economic attractiveness of the investment. If the IRR exceeds the MARR, the investment is economic. If it is less than the MARR, the investment is uneconomic. If the IRR equals the MARR, the investment's benefits or savings just equal its costs.

The IRR method has shortcomings which limit its usefulness.[2] (See Section 5-3 for more about shortcomings.) Furthermore, it is often misunderstood and misused. For most applications, one of the other methods is better. If a measure of yield is desired, the overall rate of return method described in Chapter 6 provides a more correct measure of how an alternative will contribute to an investor's economic objective.

Despite its shortcomings, the IRR method continues to be widely used in industrial and financial circles.[3] For that reason we treat it, but we urge caution in its use.

Chart 5-1 summarizes key points of the IRR method for quick reference.

5.1 FORMULATION OF IRR

Compute the IRR by solving for the value of the discount rate which will result in NB (or NS) equal to zero when used to discount benefits (or savings) and

[1]The MARR is based on the opportunity cost of capital and is identical to the discount rate, as discussed in Chapters 8 and 11.

[2]Hirshleifer (1958) and Beaves (1988) are among the authors who have discussed the shortcomings of the IRR.

[3]Awareness of its shortcomings by practitioners is growing. This is indicated by an article in *LOTUS*, a computer magazine for managers and professionals, "What's Wrong with @IRR?" The author, Nelson (1989), a CPA, points out that the fault of the IRR calculation provided by LOTUS 1, 2, 3 or Symphony, two widely used spreadsheet computer programs, is not with the programming, but with the IRR method itself. He explains how to calculate what we call the "overall rate of return (ORR)," using LOTUS or Symphony. EXCEL, another spreadsheet program, includes the ORR as a built-in function, calling it the "modified internal rate of return," or MIRR. Inclusion of the ORR in spreadsheet programs will increase its use by practitioners.

Chart 5-1. Quick reference for the IRR method

Abbreviations	IRR	Unit of Measure %
Description	The percentage yield on the portion of investment funds which remains committed to a project in various periods, provided the project's cash flow has a unique value of IRR.	
Formulas (PV)	Find the value of d^* for which $NB_{A1:A2} = \sum_{t=0}^{N} \frac{B_t - \overline{C}_t}{(1 + d^*)^t} = 0$ Or find the value of d^* for which $NS_{A1:A2} = \sum_{t=0}^{N} \frac{S_t - \overline{I}_t}{(1 + d^*)^t} = 0$	
Applications	Approach	
Accept/reject	Compute the percentage rate of return based on the benefits (or savings) and costs of making an investment versus not making it, and find the investment acceptable if its rate of return exceeds the investor's minimum acceptable rate of return.	

Special Considerations

- Effects not measured in dollars must be insignificant or uniform across alternatives or should be taken into account in some other way.
- A single IRR expresses how a project alternative, A1, performs relative to a mutually exclusive alternative, A2 (which may be doing nothing), over the same study period.
- IRR has limitations as a decision tool.

Notation: PV = present value; $NB_{A1:A2}$ = NB associated with choosing alternative A1 as compared with a mutually exclusive alternative, A2; N = number of years in study period; B_t = benefits in year t of A1 less those of A2; $\overline{C}_t$ = costs in year t of A1 less those of A2; d^* = IRR expressed as a decimal, i.e., a discount rate for which $NB = 0$; NS = net savings; S_t = savings in year t of A1 less those of A2; $\overline{I}_t$ = investment costs in year t of A1 less those of A2.

costs. The solution discount rate converted to a percent is the IRR. Equations 5-1 and 5-2[4] are formed by setting equations 3-1 and 3-5, for NB and NS, equal to zero and specifying the discount rate as the unknown, d^*.

Find the value of d^* for which

$$NB_{A1:A2} = \sum_{t=0}^{N} \frac{B_t - \overline{C}_t}{(1 + d^*)^t} = 0 \quad (5\text{-}1)$$

[4]Equations 5-1 and 5-2 are in present value form. For examples of annual value models of LCC and NB, see equations 2-3 and 3-3.

where

$NB_{A1:A2}$ = Net benefits, i.e., benefits net of costs, in present value dollars, attributed to a given alternative, A1, compared with that of a mutually exclusive alternative, A2 (which may be the alternative of doing nothing),

B_t = relevant benefits (i.e., positive cash flows such as revenues or other advantages which are assigned a dollar value) associated with a given alternative, A1, less relevant benefits for a mutually exclusive alternative, A2, in period t,

$\overline{C}_t$ = relevant costs (i.e., negative cash flows) associated with a given alternative, A1, less relevant costs for a mutually exclusive alternative, A2, in period t,

N = length of the study period, and

d^* = IRR expressed as a decimal.

Or, if emphasis is on cost reductions, find the value of d^* for which

$$NS_{A1:A2} = \sum_{t=0}^{N} \frac{S_t - \overline{I}_t}{(1 + d^*)^t} = 0 \qquad (5\text{-}2)$$

where

$NS_{A1:A2}$ = Net savings, in present value dollars, attributed to a given alternative, A1, as compared with that of a mutually exclusive alternative, A2, which may be the alternative of doing nothing,

S_t = cost savings, plus any benefits, associated with a given alternative, A1, as compared with those of a mutually exclusive alternative, A2, in period t, and

$\overline{I}_t$ = investment costs associated with a given alternative, A1, less those associated with a mutually exclusive alternative, A2, in period t.

5.2 COMPUTATION OF IRR

5.2.1 Trial-and-Error Procedure

Typically, a trial-and-error procedure is used to find the solution discount rate in equations 5-1 and 5-2. The procedure is to compute NB or NS by substituting a trial discount rate, d^*, into the equation. A positive NB or NS means that the solution IRR is greater than the trial rate. A negative NB or NS means that the solution IRR is less than the trial rate. By a series of iterations, the rate at which NB or NS approximates zero can be reached.

When solving for the IRR manually, conventional practice is to estimate the IRR by interpolation once it is closely bracketed, thereby reducing the number of computations.[5] It is often possible to guess approximately what the correct rate will be by inspecting cash flows. A good guess can reduce the number of trials.[6] Approximation procedures include mathematical approaches to determining upper and lower bounds within which the IRR(s) will occur.[7] The availability of computer software reduces the tedious work of a manual search.

Here is an example of how to compute IRR using the trial-and-error procedure with equation 5-1. Suppose that an investment opportunity requires two installment payments consisting of an initial outlay of $1,000 and an additional outlay of $1,000 at the end of the second year. You expect to receive back in benefits $4,000 at the end of the fourth year and another $1,000 at the end of the tenth year, when the building is sold. The IRR is determined as follows:

1. Test the NB outcome for 25%:

$$NB = -\$1{,}000 - \frac{\$1{,}000}{(1 + 0.25)^2} + \frac{\$4{,}000}{(1 + 0.25)^4} + \frac{\$1{,}000}{(1 + 0.25)^{10}}$$

$$= \$106$$

2. Since NB is positive, 25% must be too low. Test the NB outcome for 28%:

$$NB = -\$1{,}000 - \frac{\$1{,}000}{(1 + 0.28)^2} + \frac{\$4{,}000}{(1 + 0.28)^4} + \frac{\$1{,}000}{(1 + 0.28)^{10}}$$

$$= -\$36.$$

3. Since the NB outcome is negative, the 28% trial rate must be too high. Knowing that the solution value lies between 25% and 28%, estimate it

[5]Use of interpolation over large intervals is not recommended because multiple solutions may occur within the interval. This is in addition to the inaccuracies associated with linear interpolation.

[6]Grant and Ireson (1982) illustrate how simple preliminary calculations can help one guess a good trial value.

[7]See, for example, Hajdasinski (1988).

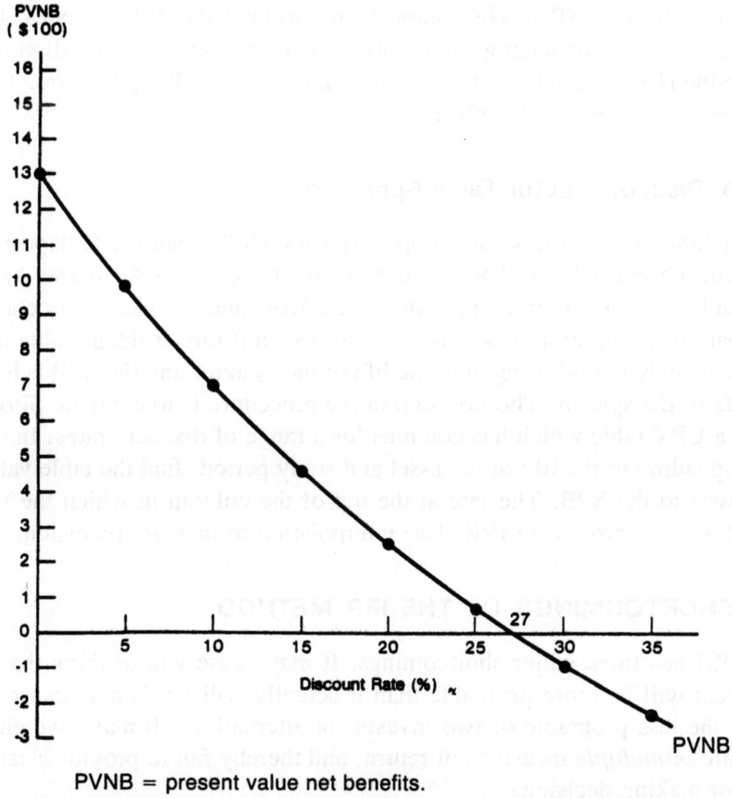

PVNB = present value net benefits.

Figure 5-1. Graphical approach to determining IRR

by interpolation as follows:[8]

$$IRR = 25\% + \left[\frac{\$106}{\$106 + \$36}\right](28\% - 25\%)$$

$$= 25\% + (0.75)(3\%) = 27.25\%$$

5.2.2 Graphical Display

Graphical displays, as shown in Figure 5-1, sometimes help in determining the IRR. With NB or NS measured on the *y*-axis and discount rates on the *x*-axis,

---

[8]The simple proportional interpolation used here for illustration entails the assumption of a linear relationship between function and argument. Since the relationship is not linear, the result is an approximation. For descriptions of other interpolation procedures, see Scarborough (1950).

plot the values of NB or NS obtained for various trial discount rates. Draw a profile curve by connecting the points. The IRR is equal to the discount rate where the NB curve intersects the x-axis, i.e., where NB equals zero. The IRR in Figure 5-1 is approximately 27%.

5.2.3 Discount Factor Table Approach

Under limited conditions, the simple payback (SPB) method (Chapter 7) and uniform present value (UPV) discount factor tables (Chapter 8 and Appendix G) can be used together to approximate the IRR quickly. Necessary conditions are that yearly net cash flows are more or less uniform, residual value is minimal, the study period is equal to the life of the system, and the SPB is less than the life of the system. The first step in the procedure is to compute SPB. Then go to a UPV table which has columns for a range of discount rates. In the row corresponding to the life of the asset and study period, find the table value that is closest to the SPB. The rate at the top of the column in which the value is found is the approximate IRR. Use interpolation to increase precision.

5.3 SHORTCOMINGS OF THE IRR METHOD

The IRR has three major shortcomings. It may cause you to think that an investment will be more profitable than it actually will be. It may cause you to select the less profitable of two investment alternatives. It may give either *no* measure or *multiple* measures of return, and thereby fail to provide clear direction for making decisions.

5.3.1 IRR May Overstate Profitability

The IRR provides only a partial indication of how a project with multiyear cash flows satisfies an investor's economic objectives. For a project whose benefits or savings do not all occur at the end of the investor's study period—and for most building projects they do not—the IRR fails to give a measure of overall yield associated with the project over the study period.[9]

---

[9]If a project entails an initial investment cost and all benefits accrue at the end of the study period, i.e., if there are no multiyear cash flows, the IRR provides a correct measure of overall as well as internal yield, because the two yields are identical. This is because there are no interim receipts to be reinvested at some other rate. An analogy can be made with the yield on a zero coupon bond versus the yield on a coupon bond. A zero coupon bond pays its face value at maturity. The investor knows exactly what the overall yield will be on a zero coupon bond held to maturity. In contrast, a coupon bond's receipts can be reinvested at a rate which may differ from the bond's direct yield. The investor's overall return may be higher or lower than the direct yield. This is the reason

Yet, the IRR is often presented and (mis)understood to mean the compound annual yield which an investor can expect *over the study period*. It is not surprising that the measure would be interpreted this way. After all, most investors expect an economic performance measure to provide a comparison of their wealth at the end of a project with their wealth just before undertaking the project.

But the IRR actually provides this *only* if all interim proceeds from a project are reinvested over the remainder of the study period at the same rate as the yield on the original investment. Hence, when it is presented and understood as a measure of the investor's return over the study period, the IRR may be said to have implicit in it the assumption of reinvestment of interim proceeds at a rate equal to the IRR.[10]

The case for an implicit reinvestment assumption can be supported by showing the results when the assumption is made explicit. Take, for example, the previous example of calculating IRR with the trial-and-error method. The project entailed an initial outlay of \$1,000, an additional outlay of \$1,000 at the end of the second year, benefits in the fourth year of \$4,000, and benefits of \$1,000 at the end of the tenth year. By trial and error, the IRR was calculated to be 27.25%. If we modify equation 5-1 to make explicit reinvestment of the interim \$4,000 receipt at 27.25% (and discounting future investment costs at 27.25%), we find the solution value of d^* as follows:

$$NB = \left[\frac{(\$4{,}000(1 + 0.2725)^6) + \$1{,}000}{(1 + d^*)^{10}}\right] - \left[\$1{,}000 + \frac{\$1{,}000}{(1 + 0.2725)^2}\right] = 0$$

$$= \left[\frac{\$17{,}983}{(1 + d^*)^{10}}\right] - \$1{,}618 = 0$$

$$d^* = -1 + \left[\frac{\$17{,}983}{\$1{,}618}\right]^{1/10} = 0.2723 \text{ or } 27.23\%$$

---

investors prefer zero coupon bonds if they expect rates to fall and coupon bonds if they expect rates to rise.

[10]Beaves (1988) is among those who makes the case that there is an implicit reinvestment rate in the IRR; Lohmann (1988) is among those who argue that there is not. It is our opinion (Ruegg, 1989, in press, and Marshall, 1986) that there is an implicit reinvestment rate assumption when the IRR is put forward as a measure of the yield over the study period on a project with multiyear cash flows. The reinvestment rate assumption is there unless the IRR is clearly stated as a measure of the yield on original investment funds only.

The $4,000 receipt at the end of year 4 is compounded forward for 6 years at a rate of 27.25%. The IRR calculated with explicit reinvestment of interim receipts at 27.25% is identical to that obtained using equation 5-1, which shows no explicit reinvestment.[11]

Interpreting the IRR as the yield over the entire study period, with its attendant implicit assumption about the reinvestment rate, is unrealistic. Ask yourself about the validity of assuming that receipts from different projects can be reinvested at the same rates received on each initial investment. Suppose, for example, that you are evaluating two projects which will take place over the same period of time. One investment is estimated to have an exceptionally high internal yield of 60%, the other an internal yield of 15%. Does it make sense to assume that all net receipts from the one investment will be reinvested to earn 60%, while all net receipts from the other will be reinvested to earn only 15%? If a 60% return is available, why reinvest at 15%? Does it make sense to assume that there are as many reinvestment rates as there are projects to be evaluated?

A more reasonable assumption is that the reinvestment rate is the rate of return on the best available alternative use of funds at the time. This is the rate indicated by the minimum acceptable rate of return or discount rate (Chapter 11). It should be approximately the same rate for reinvestment of receipts from various projects of similar risk.

The IRR tends to overstate the profitability of projects over the study period because the implicit reinvestment rate of an economically efficient project, by definition, exceeds the MARR (which is the correct indicator of the reinvestment rate). Hence, implicit reinvestment is at a rate higher than the correct reinvestment rate. For instance, if project receipts in the preceding example can actually be reinvested at 12% instead of 27%, the overall profitability of the preceding project will be closer to 17% than 27%.[12]

5.3.2 IRR May Cause Selection of Less Profitable Alternatives

If you choose among projects on the basis of their IRRs, you may not choose the most profitable of them. The implicit reinvestment assumption is again the culprit.

Consider the example illustrated in table 5-1. Suppose that you are the manager of a manufacturing plant, and you have a current budget of $100,000 to

[11]The results are identical except for a small error in the originally computed rate caused by linear interpolation.

[12]The overall rate of return, ORR, is found to be 17.3% by applying equation 6-1, i.e., ORR = $[(\$4{,}000(1 + 0.12)^6 + \$1{,}000)/(\$1{,}000 + \$1000(1 + 0.12)^{-2})]^{1/10} - 1 = 17.3\%$. (See Chapter 6.)

Table 5-1. IRR may not indicate the cost-effective project choice for a limited budget

(1)	(2)	(3)	(4)	(5)	(6)	(7)	(8)	(9)
			Cash-Flow Profile ($) Year					
Independent Projects	0	1	2	3	4	5	IRR (%)	NB[a] ($)
A	−100,000	5,000	10,000	10,000	30,000	140,000	16.4	32,970
B	−100,000	60,000	30,000	30,000	10,000	10,000	18.1	17,050

[a]The minimum acceptable rate of return and discount rate are 9%. Net benefits are stated in present value dollars.
Note: Illustration of this IRR shortcoming is not limited to the two-project case. Two are shown for simplicity.
Source: Ruegg (1989, in press, p. 23).

be allocated. You can spend it on either of two projects, A or B, each costing $100,000.

Project A entails investment in a new product line. It will generate a gradually rising revenue stream as the product is brought to market, with a surge of revenue in the fifth year when demand for the product is concentrated. Project B entails a cost-reducing investment in an existing product line. Its savings will start out strong but peter out by the end of the fifth year due to falling demand for the product. The investor's best alternative investment opportunity offers an annual return of 9%.

Table 5-1 shows cash flow profiles for the two competing projects. The IRR on project A is 16.4%; and on project B, 18.1%. Based on IRRs, project B appears the better choice. But in column 9 of the table, compare net benefits from the two projects over the 5-year study period. Project A's present value net benefits are about $33,000, while project B's are only about $17,000. Given that both require an initial investment of $100,000, project A is the more profitable choice. The reason the IRR fails to indicate the economically efficient choice is that the investor expects to earn 9% per annum on reinvested receipts, not 16.4% or 18.1%. Project A will earn 16.4% on the original investment and 9% on interim receipts; project B will earn 18.1% on the original investment and 9% on interim receipts. Project A has the higher net benefits because funds remain invested at 16.4% longer than project B's remain invested at 18.1%. (See the treatment of ORRs for projects A and B in Section 6.4.4.)

5.3.3 Possibility of No Unique Solution

Another problem is that there may not be a unique value of IRR. Table 5-2 demonstrates the problem. The project initially requires a cash outflow of

Table 5-2. Multiple solution values for IRR: illustration

(1) Year	(2) Cash Flow for Project A ($)
0	−2,040
1	+900
2	+900
3	+900
4	+900
5	+900
6	+900
7	+900
8	−4,980

Find d^* for which $NB = -\$2,040 + \$900([(1 + d^*)^7 - 1]/[d^*(1 + d^*)^7]) - \$4,980(1/(1 + d^*)^8) = 0$, where $([1 + d^*)^7 - 1]/[d^*(1 + d^*)^7])$ is the uniform present value discount formula (see Chapter 8):

For trial $d^* = 9\%$, $NB = -\$10$
For trial $d^* = 10\%$, $NB = +\$18$
By interpolation, $d^* = 9.4\%$

For trial $d^* = 20\%$, $NB = +\$46$
For trial $d^* = 25\%$, $NB = -\$31$
By interpolation, $d^* = 23.0\%$

$2,040. A series of cash inflows occurs in years 1 through 7 of $900 per year. An outflow occurs again at the end of the study period of $4,980. This cash flow pattern produces two solution values of the IRR, 9.4% and 23.0%. More than one sign change in a project's cash flow can result in multiple solution values for the IRR. In this case, the method does not provide clear guidance.[13]

5.4 ACCEPT/REJECT DECISIONS

Because of shortcomings, we discourage use of the IRR, except for deciding whether or not to accept a given project. And this application is conditional on there being a unique value of IRR.

The IRR is reliable for accept/reject decisions for the following reason: If the

[13]When a project's cash flow produces multiple values of IRR, you can derive meaning from the values by following Wohl's ordering procedure as described in Au and Au (1983). It is, however, a lengthy and cumbersome process.

IRR is greater than the MARR, the overall yield, taking into account reinvestment of interim receipts at the discount rate,[14] will also be greater than the MARR. For example, suppose that the IRR on a project with multiyear cash flow is 12% and the discount rate is 10%. The overall yield over the study period, taking into account reinvestment of interim receipts, will fall between 10% and 12%. The overall yield has to exceed 10%. The project remains acceptable in terms of having a rate of return greater than the MARR.

5.5 EXAMPLES OF CORRECT USE OF THE IRR METHOD

Example 1. Investment in an apartment building has an estimated IRR of 15%. An investor whose MARR is 10% accepts the project as profitable.

Example 2. A corporation's MARR is 12%. Modernizing its plant is estimated to have an IRR of 20%. The corporation accepts modernization.

5.6 EXAMPLES OF INCORRECT USE OF THE IRR METHOD

Example 1. Project A comes in two sizes. The IRR on the project in size 1 is 14%. The IRR on the project in size 2 is 9%. The investor's MARR is 6%. The investor chooses size 1 because it has the higher IRR. But the incremental IRR on size 2 may, in fact, exceed the investor's MARR, thereby indicating that size 2 is the economic size.

Example 2. A facilities manager must choose between projects X and Y for building 1. She wishes to choose the project with the highest overall yield over the 10-year intended holding period for the building. Project X, which has most of its benefits early in the study period, has an IRR of 16%. Project Y, which has most of its benefits late in the study period, has an IRR of 14%. The opportunity cost is expected to average 10% over the period. The manager chooses project X because its IRR is higher. This decision neglects that the overall rate of return may actually be greater for project Y than for project X.

---

[14]Recall that we set the reinvestment rate equal to the discount rate, which in turn equals the MARR.

REFERENCES

Au, Tung, and Thomas P. Au. 1983. *Engineering Economics for Capital Investment Analysis*. Boston: Allyn and Bacon, Inc.

Beaves, 1988. Net present value and rate of return: implicit and explicit reinvestment rate assumption. *The Engineering Economist* 33(4):275–302.

Grant, Eugene L., and W. Grant Ireson. 1982. *Principles of Engineering Economy*. New York: The Ronald Press Company.

Hajdasinsk, Miroslaw M. 1988. On bounding the internal rates of return of a project. *The Engineering Economist* 33(3):235–271.

Hirshleifer, J. 1958. On the theory of optimal investment decision. *The Journal of Political Economy* 66(4):329–352.

Lohmann, Jack. R. 1988. The IRR, NPV and the fallacy of the reinvestment rate assumptions. *The Engineering Economist* 33(4):303–330.

Marshall, Harold E. 1986. Advantages of the adjusted rate of return. *Cost Engineering* 28(2):32–37.

Marshall, Harold E. 1983. *Recommended Practice for Measuring Net Benefits and Internal Rates of Return for Investments in Buildings and Building Systems*. NBSIR 83-2657. Gaithersburg, MD: National Bureau of Standards.

Nelson, Stephen L. 1989. What's wrong with @IRR? *Lotus*, July, 58–62.

Ruegg, Rosalie T. 1989 (in press). The internal rate of return and the reinvestment rate controversy: an attempt at clarification. *Persistent Problems in Investment Appraisal*. American Institute of Chemical Engineering Symposium Series.

Scarborough, J. B. 1950. *Numerical Mathematical Analysis*. Baltimore: Johns Hopkins University Press.

6

Overall Rate-of-Return (ORR)

The ORR[1] is the annual yield from a project over the study period, taking into account reinvestment of interim receipts. Project earnings and earnings from reinvestment are accumulated to the end of the study period and set equal to the present value of cost to compute the ORR.

The ORR was developed to provide a rate-of-return measure of economic performance without the limitations of the IRR. A growing number of practitioners are adopting it in place of the IRR.[2]

Unlike the IRR, the ORR is a unique solution value and is easy to compute with and without a computer. The reinvestment rate is explicit. The ORR provides a correct measure of performance over the study period provided the reinvestment rate is set correctly. And it gives results consistent with NB (NS) in applications they share.[3] If used, for example, to choose between projects A and B in Table 5-1, the ORR, unlike the IRR, indicates the project with the greater NB. (In Table 5-1, the ORR on project A is 12.5% and that on project B is 10%; hence project A is preferred.)

Chart 6-1 provides a quick-reference summary of the ORR.

---

[1]A variety of names have been given to this measure of the rate of return. For example, Weston and Brigham (1981) call it the "adjusted IRR" or IRR*; White et al. (1984) call it the "external rate of return," or ERR; Au and Au (1983) call it the "overall rate of return," or ORR; and the American Society of Testing and Materials calls it the "adjusted internal rate of return," or AIRR. The name "overall rate of return" is adopted here because it seems most descriptive of the measure.

[2]Jaffe (1977) provides an excellent account of developments over the period from the 1930's to the mid-1970s in economics and finance to identify and address shortcomings of the IRR. This includes recognition of the need to make explicit the reinvestment rate at a rate equal to the discount rate, i.e., to use the ORR instead of the IRR. He observes that "by about 1970, most of the questions and controversies on the IRR had been solved in finance, but in real estate investment analysis, the literature only begins to appear in 1969." The debate over whether the ORR is a necessary substitute for a flawed IRR and recommendations for improving rate-of-return measures continues in the engineering economics literature. See, for example, Beaves (1988) and Lohmann (1988).

[3]A condition for obtaining results from the ORR method consistent with results from the NB and NS methods is that the reinvestment rate in the ORR be set equal to the discount rate. This is because the reinvestment rate implicitly equals the discount rate in the NB (NS) method.

Chart 6-1. Quick reference guide for the ORR method

Abbreviations	ORR	Unit of Measure %
Description	Finds the annual compound percentage yield from a project over the study period, taking into account reinvestment of interim receipts.	

Formula

(PV) $$ORR_{A1:A2} = \left[\frac{\sum_{t=0}^{N} (B_t - \ddot{C}_t)(1 - r)^{N-t}}{\sum_{t=0}^{N} \ddot{I}_t (1 + r)^{-t}} \right]^{1/N} - 1$$

Applications	Approach
1. Accept/reject	Compute ORR based on the benefits (or savings) and costs of accepting an investment versus not accepting it. Find the investment acceptable if the rate of return exceeds the investor's MARR.
2. Design/size	Compute incremental ORRs by comparing design/size alternatives in pairwise comparisons, starting with the alternative with the lowest initial investment cost and changing the baseline each time to the current best. Expand the investment until there are no remaining alternatives with incremental ORRs greater than the MARR.
3. Combine interdependent systems	Compute the ORR for each interdependent building system in its smallest size, without regard to interdependencies. Select the system with the highest ORR. Recalculate ORRs on all interdependent systems and increments to systems based on acceptance of the first. Select the system or increment to the first-selected system with the next highest ORR. Recalculate the ORRs on remaining options and previously selected options, based on acceptance of the second. Continue this approximation process until there are no more options having ORRs greater than the MARR, taking into account interdependencies.
4. Allocate a budget	Rank each project and project increment in descending order of its ORR, and allocate the budget in accordance with priority rankings. If the budget is not all or mostly exhausted by taking projects in ORR order, test alternative project combinations within the budget to see which has the highest aggregate NB or NS.

Special Considerations

- Effects not measured in dollars must be insignificant or uniform across alternatives or should be taken into account in some other way.
- ORR expresses in a single number a comparison of project A1 against a mutually exclusive alternative, A2 (which may be not doing the project), over the same study period.
- ORR is preferable to IRR on economic efficiency grounds.
- The reinvestment rate, r, has the same value as the discount rate or MARR.

Notation: PV = present value; $ORR_{A1:A2}$ = ORR associated with choosing alternative A1 as compared with a mutually exclusive alternative, A2; N = number of years in study period; B_t = benefits in year t of A1 less those of A2; $\ddot{C}_t$ = costs in year t of A1 less those of A2; r = reinvestment rate; and $\ddot{I}_t$ = investment costs in year t of A1 less those of A2.

6.1 FORMULATION OF ORR

As shown in equation 6-1, the ORR can be calculated by taking the Nth root of the ratio of the terminal value of all cash flows (except investment costs) to the present value of investment costs and subtracting 1.

$$ORR_{A1:A2} = \left[\frac{\sum_{t=0}^{N} (B_t - \ddot{C}_t)(1 + r)^{N-t}}{\sum_{t=0}^{N} \ddot{I}_t (1 + r)^{-t}} \right]^{1/N} - 1 \qquad (6\text{-}1)$$

where

$ORR_{A1:A2}$ = Overall rate of return on a given investment alternative, A1, as compared with that of a mutually exclusive alternative, A2, over a designated study period,

B_t = relevant benefits associated with a given alternative, A1, less relevant benefits for a mutually exclusive alternative, A2, in period t,

$\ddot{C}_t$ = relevant costs associated with a given alternative, A1, less relevant costs of a mutually exclusive alternative, A2, in period t, where both exclude all or part of investment costs, $\ddot{I}_t$, shifted to the denominator of the ratio,

$\ddot{I}_t$ = that part of investment costs in time t on which the return is to be maximized,

r = the rate at which available net receipts can be reinvested,[4]

N = length of the study period,

$\sum_{t=0}^{N} (B_t - \ddot{C}_t)(1 + r)^{N-t}$ = terminal value, and

$\sum_{t=0}^{N} \ddot{I}_t (1 + r)^{-t}$ = present value investment costs on which return is to be maximized.

When the reinvestment rate is made explicit, all investment costs are easily expressible as a time-equivalent initial outlay and all noninvestment cash flows as a time-equivalent terminal amount. This allows a straightforward comparison of the amount of money that comes out of a project (its terminal value) with

---

[4]The reinvestment rate, r, should reflect the earnings rate on the next best investment opportunity, i.e., normally it should be equal to the investor's MARR and discount rate.

the amount of money put into a project (its present value investment costs). It also makes possible the use of the closed-form solution given in equation 6-1.

Two convenient formulations of ORR which can reduce computations are equations 6-2 and 6-3. Use equation 6-2 when you have already calculated the BCR. Use equation 6-3 when you have already calculated present value NB.[5]

$$ORR = -1 + (1 + r)\,(BCR)^{1/N} \tag{6-2}$$

where

BCR = benefit-to-cost ratio.

$$ORR = -1 + (1 + r)\left(1 + \frac{PVNB}{\sum_{t=0}^{N} \ddot{I}_t(1 + r)^{-t}}\right)^{1/N} \tag{6-3}$$

where

$PVNB$ = present value of NB (see Chapter 3).

6.1.1 Specifying the Reinvestment Rate

To solve for ORR using equations 6-1, 6-2, or 6-3 requires that you specify a reinvestment rate (r). Even for projects where you do not intend to reinvest (e.g., earnings go to consumption), the opportunity for reinvestment exists and should be taken into account in the analysis.

A reasonable assumption for specifying the value of the reinvestment rate, r, is that interim proceeds can be reinvested at the best earnings rate then current. In concept, the discount rate expresses this rate. Therefore, the projected discount rate and the projected reinvestment rate normally should be the same. They both might vary over time.[6]

Specifying a reinvestment rate equal to the discount rate brings the ORR in line with the BCR and SIR methods and, when the ORR is used correctly, with the NB method. This is because those methods entail the implicit reinvestment assumption that net proceeds are reinvested at a rate equal to the discount rate.

---

[5]For a derivation of equation 6-3, see Marshall (1983, p. 29).

[6]In this book, both the discount rate and the reinvestment rate are represented in equations as constant rates, but it is recognized that variable discount rates and reinvestment rates could be specified.

Beaves (1988) argues that there may be reasons stemming from market imperfections and uncertainty why the reinvestment rate may differ from the discount rate. These reasons include transactions costs, indivisibility of investment opportunities, and differences in the time horizon. But he acknowledges that aside from these factors, the discount rate is the appropriate indicator of the reinvestment rate. He also asserts—and we agree—that the specified reinvestment rate should be used to discount future investment costs in calculating the ORR.

Note that specifying a reinvestment rate other than the discount rate in calculating the ORR causes the ORR, like the IRR, to diverge from other methods.[7] Some who argue for the use of a reinvestment rate different from the discount rate remedy this problem by reformulating the other methods also to contain an explicit reinvestment assumption.[8] This makes it possible to specify consistent reinvestment assumptions when calculating the NB, NS, BCR, SIR, and ORR. Taking our approach of setting the reinvestment rate and discount rate equal and holding both constant over the study period avoids the need for reformulation.

6.1.2 Specifying Investment Costs

In formulating the ratio which appears in equation 6-1, it is important to specify $\bar{I}_t$ as consisting solely of those investment costs on which the investor wishes to maximize the return. This requirement is similar to that for specifying the BCR and SIR ratios. In both cases, it matters whether costs are subtracted from the numerator or added to the denominator, and whether benefits are added to the numerator or subtracted from the denominator. Placement of items can affect the measure of economic performance and the choice among alternatives. (For a discussion of biasing effects which can result from various formulations of a ratio of benefits or savings to costs, see Section 4.1.)

6.2 RELATIONSHIP OF ORR TO IRR

If the explicit reinvestment rate (r) is lower than the IRR, the ORR will be lower than the IRR, and vice versa. The more concentrated investment earnings are early in the study period, the greater will be the difference between the ORR and IRR.

---

[7]If the reinvestment rate were allowed to vary by period in calculating the ORR, it would be necessary also to vary the discount rate by period in calculating the BCR (SIR) or NB (NS) to preserve consistency among the methods.

[8]See, for example, Beaves (1988).

There are two conditions under which the IRR and ORR give identical results. (1) If there is a single initial investment which yields a single lump-sum benefit at the end of the study period, IRR and ORR are identical. This is because there are no interim receipts to reinvest. (2) If the reinvestment rate, *r*, used in calculating the ORR happens to be the same as the calculated IRR, then the IRR and ORR are identical. This requires that all receipts from an investment be reinvested to earn exactly the same return as the original investment.[9]

6.3 ADVANTAGES AND DISADVANTAGES OF THE ORR METHOD

6.3.1 Advantages

The chief advantage of the ORR over the IRR is that it is a more accurate measure of the rate of return which an investor can expect over a designated study period from an investment with multiple cash flows. Other advantages are that the ORR is easier to compute manually, gives a unique solution, and provides investment guidance consistent with the other methods of economic evaluation in applications they share.

6.3.2 Disadvantages

The ORR needs to be recomputed if the discount rate (reinvestment rate) is changed, whereas the IRR does not. This characteristic has won the IRR friends among businesses that frequently change their MARRs. The disadvantage of the ORR seems small, however, when compared with the need for a valid economic measure. Furthermore, the greater simplicity of calculating the ORR, compared with the IRR, lessens the disadvantage.

6.4 APPLICATIONS OF THE ORR METHOD

You can use the ORR for the same applications as the BCR and SIR. But use it when you want to measure economic performance as an annual rate of return instead of as a ratio.

You can use it to decide whether to accept or reject a project, to combine interdependent projects, and to allocate limited funding among competing uses. You can also use it to design and size projects, provided that you compute it on project increments rather than on total benefits and costs.

---

[9]These two conditions are proven and illustrated in Ruegg (1989, in press, pp. 15-16).

Table 6-1. ORR for accept/reject decisions: illustration

(1) Independent Projects	(2) Study Period (Years)	(3) Cash Flow in Year *t* ($1,000) 0	(4) 1	(5) 2	(6) 3	(7) 4	(8) 5	(9) 6	(10) MARR & Discount Rate (%)	(11) ORR (%)	(12) Decision
X	5	−20	5	10	15	20	25		12	34.5	Accept
Y	6	−15	2	4	4	4	4	2	12	10.3	Reject

6.4.1 Accept/Reject Decisions[10]

To decide whether to accept or reject a given investment, calculate the ORR based on the benefits (or savings) and costs of making the investment versus not making it. If the ORR is greater than the investor's MARR, accept the investment.

Table 6-1 shows the ORR computed for two independent projects, X and Y. The investor's MARR (hence, the discount rate and reinvestment rate) is 12%. Project X is acceptable, provided that there is adequate funding. But project Y, which yields less than the required 12%, is unacceptable regardless of funding.

6.4.2 Design and Sizing Decisions

To use the ORR for design and sizing decisions, it is essential that you compare alternatives on the basis of their *incremental* rates of return. Perform incremental analysis in a series of pairwise comparisons of alternatives. The pairwise procedure is essentially the same as that for the BCR method described in Section 4.2.2.

The decision rule for designing or sizing with the ORR method is to increase the investment until there are no remaining alternatives with incremental ORRs greater than the MARR. The expansion, of course, must fit within the constraints of the budget.

Table 6-2 and Figure 6-1 illustrate the use of incremental ORRs to find the economic size of a project. There are five size alternatives for the project, increasing in order from size A to E. Read the table from left to right below the diagonal line. Suppose that the MARR is 8%. The incremental ORR on in-

[10]Either ORR or IRR indicates whether to accept or reject a given investment, based on the acceptance criterion that benefits or savings exceed costs. But ORR gives a more accurate measure of the rate of return over the study period.

Table 6-2. Incremental ORR for sizing decision: illustration

(1)	(2)	(3)	(4)	(5)
		Size Compared Against Baseline (to This Size)		
Baseline (From This Size)	B	C*	D	E
A	35	27	19	14
B		13	11	8
C			7	4
D				3

Note: Only incremental ORRs below the diagonal line are appropriate for sizing. We show the ORRs in the shaded region to emphasize that ORR may not reveal the economic size unless computed on the smallest possible increments.

*denotes economically efficient choice.
Source: Marshall (1983, p. 39).

creasing the project size from A to B is 35%, substantially higher than the 8% MARR. Increasing the project size from B to C results in an incremental ratio of 13%, still above the MARR. Increasing from size C to D results in an incremental ratio of 7%, a little below the MARR. Thus, size C is the economically efficient size.

Note that the ORRs in the shaded section of Table 6-2 are not appropriate for sizing because they are not computed on the smallest possible size increments. Look at the top row. The falling ORR (i.e., 35, 27, 19, 14) does not mean that size B, with the highest ORR, is best. Nor does an ORR greater than the MARR mean that every dollar spent on the project is worthwhile. For example, the ORR on expanding the project from size A to E is 14%, higher than the 8% MARR. We can conclude that E is more economic than A, but not necessarily that E is better than B, C, or D. We can see directly from the table, for example, that expansion beyond C is economically inefficient, because increments D and E earn less at the margin than the 8% MARR.

Figure 6-1 shows the potential size options in a network diagram. Each line in the network connects two lettered points and represents a potential sizing decision. Incremental analysis is accomplished by moving clockwise around the outside of the diagram from size A to B to C to D to E. The incremental ORR remains above the 8% MARR until the size is expanded beyond C.

Note that crossing the diagram, such as moving from A to D, or moving counterclockwise, such as from A to E, will not indicate the most economic size. Avoid the common mistake of selecting size B because it has the highest

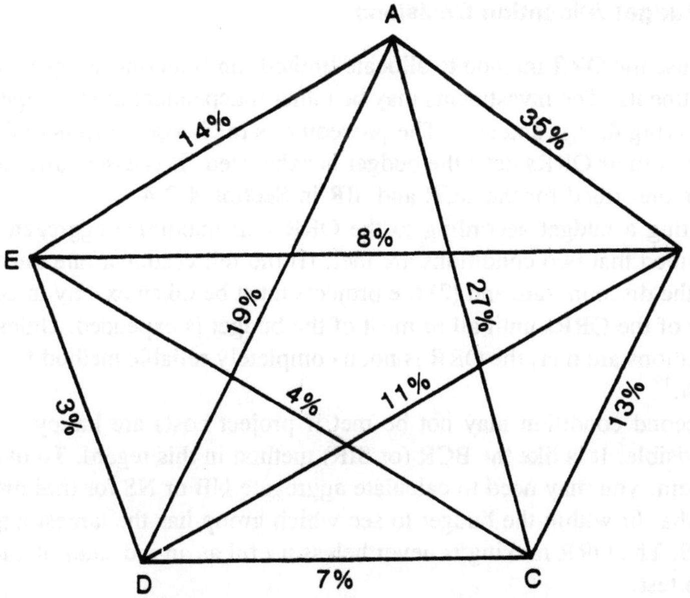

Figure 6-1. Incremental ORR for sizing decision: network diagram

ORR. Remember that additional expenditures are economically desirable as long as the yield on the increment exceeds the MARR.

6.4.3 Interdependent Decisions

You can also use incremental ORRs to guide selection of the economic combination of interdependent building systems of variable design and size. The approximation procedure is essentially the same as that for the BCR and SIR described in Section 4.2.3.

First compute ORRs for each interdependent building system in its least expensive size, without regard to interdependencies. Select the system with the highest ORR. Then recalculate the ORRs on all interdependent systems and increments to systems based on acceptance of the first option. In the second round, tentatively select the option with the next highest ORR. Recalculate the ORRs based on acceptance of the second option. Include the first-selected system in the recalculation to make sure that it soes not fall below the MARR if the second option is added. Continue this process until there are no more options available that have ORRs greater than the MARR.

6.4.4 Budget Allocation Decisions

You can use the ORR method to allocate limited funds among competing building investments. The investments may be either independent or interdependent[11] and of varying design and size. The procedure is to choose projects in descending order of their ORRs until the budget is exhausted. It is essentially the same procedure described for the BCR and SIR in Section 4.2.4.

Allocating a budget according to the ORR will maximize aggregate NB or NS, provided that two conditions are met: (1) the reinvestment rate must be set equal to the discount rate and (2) the projects must be taken exactly in descending order of the ORRs until all or most of the budget is expended. Unless these two conditions are met, the ORR is not a completely reliable method for budget allocation.[12]

The second condition may not be met if project costs are lumpy, i.e., not finely divisible. It is like the BCR (or SIR) method in this regard. To overcome the problem, you may need to calculate aggregate NB or NS for trial groups of projects that fit within the budget to see which group has the largest aggregate NB or NS. The ORR ranking is nevertheless useful as an indicator of candidate groups to test.

Table 6-3 illustrates the use of the ORR method to rank projects for budget priority. Projects A through H, listed in descending order of their ORR, are all cost effective. But their aggregate cost is \$67,000, greater than the available \$40,000 budget. With a budget of \$40,000, select projects A, B, C, D, E, and F. You can verify that this is the set of projects which maximizes aggregate NS by summing NS for each set of projects that fits within the budget. Compare Table 6-3 with Table 4-5 to see that the ORR method gives identical results to the BCR or SIR method.[13]

Also note that the ORR method, unlike the IRR, correctly indicates that project A is the economic choice of projects in the example illustrated by Table 5-1. The ORR for project A is 15.4% and for project B, 12.5%.[14]

---

[11]If projects are interdependent, take care to check project rankings each time an additional selection is made.

[12]The IRR, in contrast, is unreliable for maximizing total return because the reinvestment rate condition is not met. This is true even if the budget can be fully expended by taking projects according to their IRR ranking. See Table 5-1 for an illustration of the IRR failing to identify the investment that maximizes NB for a limited budget.

[13]The ORR and BCR (SIR) methods give consistent results provided that they are consistently formulated. Equation 6-1 is consistent with 4-1.

[14]$ORR_A = [(\$5{,}000(1 + .09)^4 + \$10{,}000(1 + .09)^3 + \$10{,}000(1 + .09)^2 + \$30{,}000(1 + .09)^1 + \$140{,}000(1))/\$100{,}000]^{1/5} - 1 = 0.154$. $ORR_B = [(\$60{,}000(1 + .09)^4 + \$30{,}000(1 + .09)^3 + \$30{,}000(1 + .09)^2 + \$10{,}000(1 + .09)^1 + \$10{,}000(1))/\$100{,}000]^{1/5} - 1 = 0.125$.

Table 6.3. Budget allocation by ORR: illustration

(1) Projects	(2) Investment Costs (PV $)	(3) Cumulative Investment Costs (PV $)	(4) Total Savings (PV $)	(5) Total Savings[a] (TV $)	(6) NS (PV $)	(7) ORR	(8) Ranking
A	4,000	4,000	50,000	107,946	46,000	39	1
B	2,500	6,500	16,000	34,543	13,500	30	2
C	8,000	14,500	32,000	69,086	24,000	24	3
D	2,500	17,000	6,000	12,954	3,500	18	4
E	15,000	32,000	33,000	71,245	18,000	17	5
F	8,000	40,000	12,000	25,907	4,000	12	6
G	2,000	42,000	2,500	5,397	500	10	7
H	25,000	67,000	27,500	59,370	2,500	9	8

Nomenclature: PV $ = present value dollars. NS is expressed as a single lump-sum amount occurring at the present. TV $ = terminal value dollars, a single lump-sum amount occurring at the end of the study period.

[a]Based on a reinvestment rate of 8%.

6.4.5 Summary of Points to Note in Applying Overall Rate-of-Return Method

How the ORR is formulated is important. Be sure to include in the denominator of equation 6-1 only those investment costs on which you wish to maximize the return. Otherwise you may bias your decisions in the same way as we discussed for the BCR in Section 4.1.

The ORR provides reliable guidance only if effects not measured in dollars are insignificant or uniform across alternatives. If any effects are not captured in the ORR, you need to take into account those excluded in some other way.

6.5 EXAMPLES OF CORRECT USE OF THE ORR METHOD

Example 1. A developer's MARR on speculative construction projects is 15%. The estimated ORR on a speculative office building is 11%. The developer correctly rejects the project.

Example 2. The facilities manager of a state university is planning to upgrade the energy efficiency of campus buildings. The budget is $300,000. The state legislature sets a MARR of 7% on capital budget expenditures. Economically efficient projects with a total investment cost of $500,000 have been identified. The manager allocates the budget fully by taking, in descending order of their ORR rank, candidates with ORRs greater than 7%. He correctly concludes that his allocation will provide maximum reductions in long-run building costs for the limited budget.

Example 3. A chemical plant engineer is designing a new process facility. A larger-capacity design has higher investment costs but lower operating costs than a smaller-capacity design. The corporation's MARR is 9.5%. The incremental ORR on the larger-capacity design relative to the smaller one is 7.8%. Hence, the engineer correctly chooses the smaller-capacity design.

6.6 EXAMPLES OF INCORRECT USE OF THE ORR

Example 1. A recently hired employee in a corporation estimates that a new product line will yield an ORR of 12%. Without checking company policy on its MARR, he recommends that the product line be approved. This is a faulty approach because it ignores the MARR for the company.

Example 2. An engineer is considering three alternative mechanical systems for a new retail store: (1) a single CV system, (2) a single VAV system, and (3) a high-efficiency VAV system with an economizer cycle. The owner's MARR is 10%. The ORR of system 2 compared with system 1 is 20%. The ORR of system 3 compared with system 1 is 16%. The engineer incorrectly concludes that system 2 is the cost-effective choice because it has the higher ORR. But if the incremental return on system 3 over system 2 is greater than the 10% MARR, and the budget allows, system 3 is the cost-effective choice.

REFERENCES

American Society for Testing and Materials (ASTM). 1985. *Standard Practice for Measuring Internal Rates of Return for Investments in Buildings and Building Systems*. E 1057-85. Philadelphia: ASTM.

Au, Tung, and Thomas P. Au. 1983. *Engineering Economics for Capital Investment Analysis*. Boston: Allyn and Bacon, Inc.

Beaves, Robert G. 1988. Net present value and rate of return: implicit and explicit reinvestment assumptions. *The Engineering Economist* 33(4):275–302.

Jaffe, Austin J. 1977. Is there a "new" internal rate of return literature? *AREUEA Journal* 5:482–502.

Lohmann, Jack R. 1988. The IRR, NPV and the fallacy of the reinvestment rate assumptions. *The Engineering Economist* 33(4):303–330.

Marshall, Harold E. 1983. *Recommended Practice for Measuring Net Benefits and Internal Rates of Return for Investments in Buildings and Building Systems*. NBSIR 83-2657. Gaithersburg, MD: National Bureau of Standards.

Ruegg, Rosalie T. 1989 (In Press). The internal rate of return and the reinvestment rate controversy: an attempt at clarification. *Persistent Problems in Investment Appraisal*. American Institute of Chemical Engineering Symposium Series.

Weston, J. Fred, and Eugene F. Brigham. 1981. *Managerial Finance*. Hinsdale, IL: Dryden Press.

White, John A., Marvin H. Agee, and Kenneth E. Case. 1984. *Principles of Engineering Economic Analysis*. New York: John Wiley & Sons, Inc.

7

Payback (PB)

The PB method measures how long it takes to recover investment costs. Benefits or savings, net of future costs, are accumulated year by year until the total is sufficient to offset investment costs. If, in computing the payback period, you ignore the time value of money (i.e., assume a zero discount rate), the method is called "simple payback (SPB)." If you take into account the time value of money (i.e., assume a positive discount rate), the method is called "discounted payback (DPB)." DPB is a more accurate measure of payback than SPB.

An estimated DPB shorter than the useful life of the building or system generally signals an economic investment. (A necessary condition is that no large costs occur after DPB, which makes the investment uneconomic.) SPB must meet an additional condition: It must be *substantially* shorter than the life of the building or system before you can safely conclude that an investment is economic.

Decision makers often specify a maximum acceptable payback period (MAPP) shorter than the useful life of the building or system. Then they consider an investment to be economically attractive only if its payback period is shorter than the MAPP.

An MAPP is sometimes set to reflect the investor's knowledge about specific future events that may jeopardize the ability to realize potential savings or benefits past some point. For example, if a building's lease will expire in 2 years, benefits from an investment in that building cannot be counted on after 2 years. Sometimes the MAPP is set arbitrarily to reflect the investor's strategy to preserve investment capital in the face of risks which are perceived to increase with time. Political instability and impending revolution, for example, would encourage investors to set low MAPPs to select only projects which pay back quickly.

The PB method—whether DPB or SPB—has serious shortcomings. Contrary to what is usually assumed by users of this method, the alternative with the shortest time to payback is not necessarily the project with the highest return. This deficiency is reflected in a nickname that has been used for the PB method—

the "fish bait" technique—so called because it focuses on recovering the bait (the initial investment) rather than on the size of the catch (the profitability of the investment).

PB indicates a project's liquidity and capital recovery rate.[1] It is the time required to break even. It is not a dependable measure of profitability. And if the analyst fails to take into account the time value of money by using SPB, the method fails even to give a correct measure of the time to break even.

Unfortunately, businesses and other decision makers often set very short MAPPs under the misconception that short payback periods will result in more profitable projects. They may instead realize a series of short-lived, low-profit investments. (See Section 7.2.5 for examples showing how use of the PB method can lead to uneconomic decisions.)

In its favor, the PB method is easy to understand. It has an innate appeal to a lay audience. And as a supplementary method, it is helpful in screening potential investment projects quickly and inexpensively.

Many managers like to know a project's PB, especially if their firm has liquidity problems and high financing costs. They may be skeptical about estimates of expected cash flows to be received later than several years in the future.

Chart 7-1 for quick reference summarizes key features of the PB method.

7.1 FORMULATION

7.1.1 The DPB Formula

DPB is usually formulated for the case where the initial outlay comprises total investment costs.[2] To determine the payback period, find the minimum solution value of Y in equation 7-1. The Y value found by equation 7-1 is called "discounted payback (DPB)" to indicate that cash flows are discounted.

$$\sum_{t=1}^{Y} \frac{B_t - \tilde{C}_t}{(1 + d)^t} = \bar{I}_0 \qquad (7\text{-}1)$$

where

Y = the minimum length of time (usually number of years) over which future net cash flows must be accumulated in order to offset initial investment costs, where the minimum solution value of $Y = DPB$,

[1]Clark et al. (1984).

[2]If, due to financing, investment costs consist of an initial outlay and a series of payments over time, payback does not occur until the accumulated amount (which is net of interim payments) is sufficient to cover the initial outlay plus any outstanding balance.

Chart 7-1. Quick Reference for payback method (simple and discounted)

Abbreviations Payback (PB) Discounted payback (DPB) Simple payback (SPB)	Unit of Measure	Years/months

Description

DPB measures how long it takes to recover investment costs with time-adjusted benefits or savings. SPB measures how long it takes to recover investment costs with benefits or savings that are *not* adjusted for time of occurrence.

Formulas

DPB is the minimum value of Y which satisfies the equation:

$$\sum_{t=1}^{Y} \frac{B_t - \bar{C}_t}{(1 + d)^t} = \bar{I}_0$$

SPB is found with the above equation by setting $d = 0$.

SPB is found with the following equation when cash flows are approximately uniform:

$$SPB = \frac{\bar{I}_0}{\overline{(B - \bar{C})}}$$

Equations for solving directly for DPB are given in the text, as well as a tabular approach and a graphical approach to estimating DPB.

Applications	Approach
1. Supplementary measure when project life is uncertain.	Solve for DPB (i.e., minimum breakeven life), and compare estimates of project life against it.
2. Identify feasible projects when the time horizon is constrained	Set a MAPP to reflect the constrained time horizon, and reject projects with DPBs greater than the MAPP.
3. Supplementary measure of how long capital is at risk	Solve for DPB and include it in descriptions of the project.
4. Rough guide to accept/ reject	If DPB is shorter—or SPB is significantly shorter—than the expected project life, accept the project.

Special Consideration

- PB is not a reliable guide for most decisions; projects with longer PB periods may be more profitable than projects with shorter PB periods.

Notation: PV = present value; Y = minimum number of years until cumulative benefits or savings equal initial investment cost; B_t = benefits in year t of A1 less those of A2; $\bar{C}_t$ = costs (excluding I_0) in year t of A1 less those of A2; d = discount rate; $\bar{I}_0$ = initial investment costs comprising total investment costs; SPB = simple payback; $\overline{(B - \bar{C})}$ = uniform net annual cash flow; and $(1 + d)^t$ = formula for single present value discount factor.

B_t = relevant benefits (i.e., positive cash flows such as revenues or other advantages which are assigned a dollar value) associated with a given alternative, A1, less relevant benefits of a mutually exclusive alternative, A2, at the end of period t,

$\tilde{C}_t$ = relevant costs, excluding initial investment costs, associated with a given alternative, A1, less those of a mutually exclusive alternative, A2, at the end of time t, and

$\bar{I}_0$ = initial investment costs of a given alternative, A1, less those of a mutually exclusive alternative, A2, which may be the alternative of doing nothing, where the initial outlay comprises total investment costs.

Direct solution of DPB using equation 7-1 is not possible. The common approach to calculating DPB when future cash flows are nonuniform is to accumulate the discounted value of future net cash flows year by year until the sum equals or exceeds the initial investment cost. The year in which this happens is DPB. You can use interpolation to estimate the month within the year that DPB occurs. The year-by-year accumulation approach is facilitated by worksheet 3 in series B of Appendix B.

7.1.2 The SPB Formula

If, in computing the payback period, you assume that there is no opportunity cost of capital and set the discount rate equal to zero, the result is SPB. It tends to understate the true time to payback.

With d set equal to zero, equation 7-1 becomes equation 7-2. The minimum solution value of y in equation 7-2 is SPB.

$$\sum_{t=1}^{y} (B_t - \tilde{C}_t) = \bar{I}_0. \qquad (7\text{-}2)$$

SPB is found by accumulating future net cash flows (undiscounted) year by year and comparing the cumulative amount at the end of each successive year to the initial outlay.

A shortcut calculation can be used if future cash flows are approximately uniform from year to year. In that case, SPB is found simply by dividing the initial investment cost by the yearly amount, according to equation 7-3. Worksheet 1 in series B of Appendix B accommodates this approach.

$$SPB = \frac{\bar{I}_0}{\overline{(B - \tilde{C})}} \qquad (7\text{-}3)$$

where

$\overline{(B - \tilde{C})}$ = a uniform annual net flow.

7.1.3 Calculating DPB from SPB

Because DPB is more accurate than SPB but SPB is simpler to compute, techniques have been devised for determining DPB from SPB. The techniques employ tables, graphs, and mathematical algorithms. We show how to use the techniques for converting from SPB to DPB for two cases: (1) future net cash flows are uniform, and (2) future net cash flows escalate at a constant rate. In either case, the first step is to calculate SPB.

7.1.3.1 Uniform Net Cash Flows

Mathematical Approach. For the case of uniform net cash flows, DPB can be determined mathematically from SPB with equations 7-4 and 7-5.[3]

$$DPB = \frac{\log\left[\dfrac{1}{1 - (SPB \cdot d)}\right]}{\log(1 + d)} \quad \text{when } d \neq 0 \qquad (7\text{-}4)$$

$$DPB = SPB \text{ when } d = 0 \qquad (7\text{-}5)$$

(DPB is undefined when $(SPB \cdot d) \geq 1$)

Tabular Approach. A second method of determining DPB from SPB for the case of uniform net cash flows is to use tables of uniform present value (UPV) discount factors, such as those in Appendix G. The procedure is to compute SPB using equation 7-3; find the UPV table for the applicable discount rate; look within the columns of the table to find the factor closest to the SPB solution value; and look across the row to find the corresponding year. To obtain a closer approximation, find the factors that bracket the SPB and solve for the DPB by interpolation.

---

[3]For a derivation of equations 7-4 and 7-5, see Marshall (1984, pp. A-1–A-2).

Graphic Approach. A third method of determing DPB from SPB for the case of uniform net cash flows is to use special graphs that are provided in Appendix C. A sample of these graphs is shown in Figure 7-1. It depicts DPBs with a time span of up to 10 years as a function of SPBs with a time span of up to 4 years. (Because SPB tends to understate the true time to payback, the DPB scale must cover a longer time span than the SPB scale.) SPBs are translated into DPBs through k curves. The k values each represent a combination of escalation rate and discount rate (including 0 values for either rate). You can either look up the applicable k value in appendix Table C-1 or compute it easily as $k = (1 + e)/(1 + d)$, where e = escalation rate and d = discount rate. We provide additional graphs for converting SPB to DPB in Appendix C.

Suppose that you want to find DPB for a \$12,000 investment earning \$4,500 a year for 6 years, assuming a discount rate of 10% and an escalation rate of zero percent. First, divide \$12,000 by \$4,500 to compute an SPB of 2.7 years. Find the k value in appendix Table C-1 for e = 0% and d = 10%. It is 0.91. Since the k value and the SPB value fall within the "range" of Figure 7-1, use

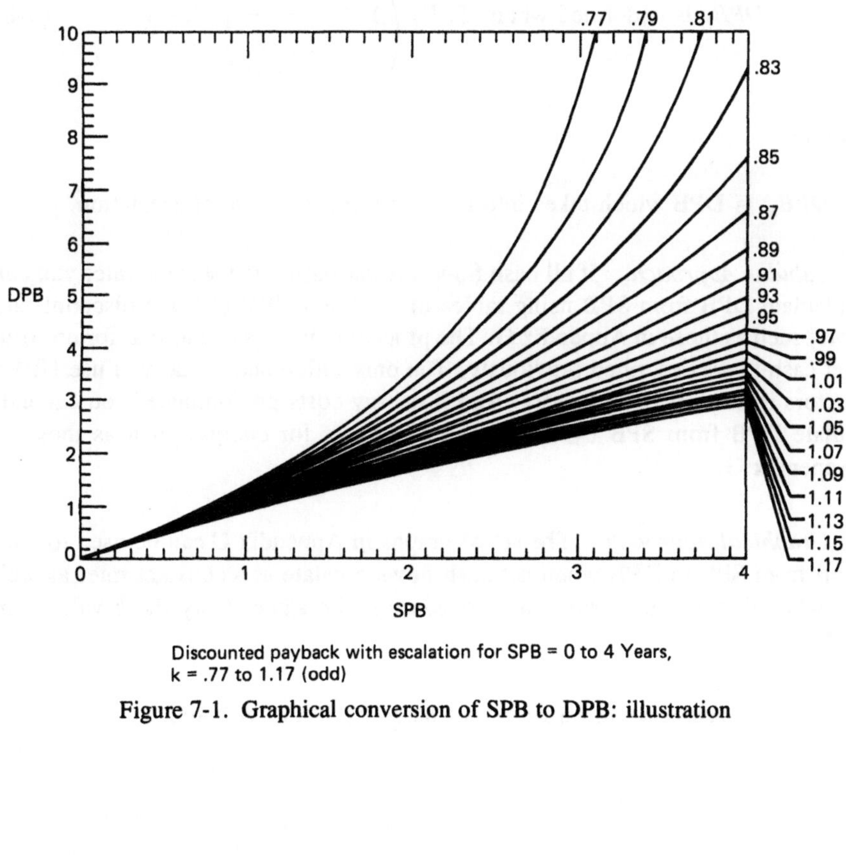

Figure 7-1. Graphical conversion of SPB to DPB: illustration

it to find DPB. Draw a vertical line from 2.7 on the SPB axis to the k curve labeled 0.91. Draw a horizontal line from the intersection to the vertical axis. The DPB is about 3.3.

7.1.3.2 Uniformly Escalating Net Cash Flow

Mathematical Approach. If net cash flows are expected to escalate at a constant rate, you can determine DPB from SPB using equations 7-6 and 7-7.[4]

$$DPB_E = \frac{\log\left[1 + (SPB)\left(1 - \dfrac{1}{\dfrac{1+e}{1+d}}\right)\right]}{\log\left(\dfrac{1+e}{1+d}\right)}, \quad \text{when } e \neq d \qquad (7\text{-}6)$$

$$DPB_E = SPB \text{ when } e = d \qquad (7\text{-}7)$$

$$DPB_E \text{ is undefined when } (SPB)\left(1 - \frac{1}{\dfrac{1+e}{1+d}}\right) \leq -1 \qquad (7\text{-}8)$$

where

DPB_E = DPB which takes into account a constant rate of escalation.

Tabular Approach. If all cash flows are escalating at the same rate, you can calculate DPB from SPB using tables of modified UPV (UPV*) discount factors, such as those in Appendix G. The procedure is the same as that for uniform net cash flows described previously. The only difference is that you use UPV* factors instead of UPV factors. And if energy costs predominate, you can calculate DPB from SPB by using UPV** factors for energy, such as those in Appendix G.

Graphical Approach. The set of graphs in Appendix C can be used to convert from SPB to DPB when net cash flows escalate at a constant rate, as well as when they are uniform. The procedure is the same. Only the k values are different.

---

[4]For a derivation of equations 7-6 and 7-7, see Marshall (1984, pp. A-2–A-4).

7.2 APPLICATIONS OF DPB

We recommend DPB only as a supplementary measure of economic performance, not as the sole criterion for making decisions about buildings. Used in conjunction with other economic measures, it can help determine the economic desirability of projects. But used alone, it can be misleading. We recommend against using SPB except as a "down-and-dirty" guide for simple accept/reject decisions and as a first cut at calculating DPB.

7.2.1 DPB as a Supplementary Measure When the Project Life Is Uncertain

As Chapter 15 explains, breakeven analysis is a technique for reflecting uncertainty. DPB is a form of breakeven analysis which is useful when the project life is uncertain. It finds the minimum project life necessary to recover investments costs. This provides a number against which to compare the estimates of project life. Rather than needing to know the exact project life, you need to know only whether or not it is likely to be longer or shorter than the computed time to payback.

Suppose, for example, that you wish to evaluate the cost effectiveness of a new roofing material. It costs more than an existing material of similiar appearance but has no established record of durability. Yet its cost effectiveness hinges on durability. DPB indicates how long the new material *must*, at a minimum, last to be economically viable. This provides a criterion against which to compare predicted life spans. Predictions might, for example, come from research studies of accelerated aging, from durability records of the material used in other applications, or from the results of field testing. A predicted life longer than the DPB suggests that the new material is economically viable.

7.2.2 DPB for Identifying a Feasible Set Of Candidate Projects When the Time Horizon Is Constrained

DPB is useful for identifying candidate projects when potential returns beyond a certain time are expected to be irrelevant. It provides information useful in the effort to preserve capital.

For example, an owner of a building expects to demolish the building in 5 years to use the land for another purpose. In the interim, the owner wishes to reduce the costs of operating the building. Projects which will not cover their costs and realize a return within the 5-year time frame are not viable candidates. In this case, setting an MAPP to less than 5 years and using the criterion DPB

≤ MAPP is a reasonable method for identifying candidate projects. Additional economic methods can then be used to choose the set of projects which will maximize the owner's NB.

DPB is also useful for identifying candidate projects when outside forces jeopardize their continued viability. For example, an investor in a foreign country where nationalization threatens to end outside control of assets might establish an MAPP as a criterion for project acceptance.

A similar application is to guard against losses threatened by changing market conditions. For example, suppose that a manufacturer of household furnishings has developed a new product line based on current fashions in home decorating. Shifting tastes and competitive products may threaten product demand with a short "shelf life," thereby requiring a quick payback. Likewise, an investor in billiard parlors in suburban locations may expect them to be a fad, and may screen prospective billiard projects using an MAPP which reflects the predicted length of the fad. In both examples, DPB is of keen interest to the investor, because it indicates the minimum time the product or service must remain marketable to avoid losses.

7.2.3 DPB as a Supplementary Measure to Indicate How Long the Investor's Capital Is at Risk

Risk is often viewed as an increasing function of time. Often an investor will wish to know how long it will be before the risk of losing capital is over and the "gravy" starts flowing. In this application an MAPP is usually not applied; the DPB is viewed as supplementary information. It helps to identify projects that have unusually large or small net cash flows early in their lives. Venture capitalists, bankers, and other financiers may request this information.

7.2.4 PB as a Rough Measure for Accept/Reject Decisions

It is safe to conclude that a project will be economic if it recovers its investment costs well before the end of its life and entails no negative cash flows after payback is achieved.

7.2.5 Unreliable Decision Rules

The following decision rules, although commonly used, may lead to uneconomic projects:

1. Accept a project if its DPB is less than the MAPP, and reject a project if its DPB is greater than the MAPP.

Table 7-1. Accepting and rejecting projects on the basis of PB can lead to uneconomic decisions

(1)	(2) $\bar{I}_0$	(3)	(4) $(B_t - \bar{C}_t)^a$	(5)	(6)	(7)	(8)	(9)
			Years			NB	SPB	DPB
		1	2	3	4	($)	(Years)	(Years)
Project	0							
A	−1,000	325	325	325	325	30	3.1	3.9
B	−1,000	800	500	−100	−300	−140	1.4	1.7

[a]Cash flows are discounted with a 10% discount rate.
Source: Marshall (1984, p. 43).

2. Select the alternative with the shortest DPB when designing or sizing projects.
3. When the budget is limited, select projects for funding in ascending order of their DPBs.

These rules are even more unreliable if SPB instead of DPB is used.

Table 7-1 illustrates how the first unreliable rule can lead to uneconomic decisions. Project A has an SPB of 3.1 and a DPB of 3.9, and project B has an SPB of 1.4 and a DPB of 1.7. Suppose that the MAPP were set at 3.0. According to the decision rule, you should reject project A and accept project B. But looking at column 7 of the table, you can see that project A yields positive net benefits, while project B yields net losses.

You may argue that a manager would not implement project B because it produces losses. But notice that it is necessary to look beyond the payback year to see this. If given no other measure of economic performance than PB, a manager might make an uneconomic decision such as this one.

Table 7-2 shows how the second unreliable rule can lead to uneconomic choices among alternative designs and sizes of projects or among other mutually exclusive alternatives. Design X requires a larger investment and has its benefits concentrated in the last 3 years of a 5-year period. Design Y has most of its benefits in the first 2 years. Design X has an SPB of 3.5 and a DPB of 4.1 years; design Y, an SPB of 1.5 and a DPB of 1.7 years. The faulty decision rule would support selection of design Y, with its shorter time to recover investment costs. But column 8 of the table shows that design X yields higher net benefits.[5] Choosing design Y on the basis of its DPB would result in an economic loss of $402 (i.e., $555 - $153).

[5]Applying DPB incrementally and comparing it with the project life, rather than with an arbitrarily constrained MAPP, would make DPB more useful.

Table 7-2. Designing and sizing projects on the basis of PB can lead to uneconomic decisions

(1)	(2) $\bar{I}_0$	(3)	(4)	(5) $(B_t - \bar{C}_t)^a$	(6)	(7)	(8)	(9)	(10)
		Years							
Design	0	1	2	3	4	5	NB	SPB	DPB
X	−1,500	0	0	1,000	1,000	1,000	555	3.5	4.1
Y	−1,000	700	600	10	10	10	153	1.5	1.7

[a]Cash flows are discounted with a 10% discount rate.

The PB is also unreliable for allocating a budget among competing projects. Table 7-3 illustrates how the third unreliable decision rule can lead to waste in selecting projects for a limited budget. There are five independent projects, H through L. Given a budget of $1,000, the PB decision rule supports funding projects J and L for a total NB of $625. But by following the appropriate BCR criterion instead and selecting project K, NB total $2,415. The economic loss from choosing according to the minimum PB is $1,790 (i.e., $2,415 − $625).

7.3 EXAMPLES OF CORRECT USE OF THE PB METHOD

Example 1. A homeowner is deciding whether to replace her existing oil-fired burner and boiler with a high-efficiency, oil-fired package boiler. The SPB on the new system is estimated to be 3 years. The new system is expected to last for at least 25 years, and the family expects to occupy the house for at least 15 more years. Future maintenance and repair costs for the new system are expected to be no more than those for the existing system over its life. The homeowner correctly concludes that the replacement is economically worthwhile.

Example 2. A new floor covering is being considered for use in locker rooms. It costs less to purchase and maintain than conventional coverings, but its durability is uncertain. Conventional coverings last an average of 12 years. The building manager wishes to know how long the new covering would have

Table 7-3. Allocating a budget on the basis of PB can lead to uneconomic decisions

(1)	(2) $\bar{I}_0$	(3)	(4) $(B_t - \bar{C}_t)^a$	(5)	(6)	(7)	(8)	(9)	(10)
			Years						
Project	0	1	2	3	4	NB ($)	BCR	SPB (Years)	DPB (Years)
H	−1,000	500	500	500	500	586	1.6	2	3
I	−2,000	1000	2000	0	0	562	1.3	2	2
J	−500	675	0	0	0	114	1.2	1	1
K	−1,000	0	0	0	5,000	2,415	3.4	4	4
L	−500	500	500	100	100	511	2.0	1	2

$1,000 budget allocation: DPB ranking: Choose J and L for *NB* = $625.
BCR ranking: Choose K for *NB* = $2,415.

[a]Cash flows are discounted with a 10% discount rate.

to last to be economically equivalent to the best conventional coverings. The DPB is 11 years. The manager chooses to stay with the conventional covering, because it is uncertain whether the new covering will last for at least 11 years.

7.4 EXAMPLES OF INCORRECT USE OF THE PB METHOD

Example 1. A corporate manager is reviewing PB analyses for two projects competing for limited funding. He sees that project A has a DPB of 2 years and project B, 6 years. The manager chooses project A without further consideration. This is a faulty decision because it turns out that project A has a life of only 4 years and NB of $2,000, while project B has a life of 20 years and NB of $20,000.

Example 2. An investor quickly computes the SPB by dividing investment costs by projected revenue for the first year. Determining the SPB to be 6 years and knowing that the project will probably last for 10 years, he concludes that it is an attractive investment. This is an invalid conclusion because it fails to take into account details of the investment. First, revenues might not be uniform. Second, a large replacement cost is likely in the seventh year. Third, income taxes will affect the cash flow. And fourth, the investor has an opportunity cost of 12% after taxes, which reduces considerably the present value of future amounts.

REFERENCES

Clark, John J., Thomas J. Hindelang, and Rober E. Pritchard. 1984. *Capital Budgeting: Planning and Control of Capital Expenditures*. Englewood Cliffs, New Jersey: Prentice-Hall, Inc.

Marshall, Harold E. 1984. *Recommended Practice for Measuring Simple and Discounted Payback for Investments in Buildings and Building Systems*. NBSIR 84-2850. Gaithersburg, MD: National Bureau of Standards.

Part 2

WHAT YOU NEED TO KNOW TO APPLY THE METHODS

8

Adjusting Dollar Amounts for Time of Occurrence

We introduced the concept of the time value of money in Part 1 and showed basic discounting operations in the method formulas. Here we explain why, for a valid economic evaluation, it is necessary to discount dollar amounts which occur at different times to time-equivalent amounts at a common time. We discuss the implications of discounting for building decisions. Then we explain how to do it. We begin the "how to" part with guidelines for selecting a common time and modeling cash flows. Then we show how to discount a variety of cash flows with eight time-equivalence formulas. Because discounting operations are often combined with cost estimation, we show how to combine the two.[1]

8.1 WHY ADJUST FOR TIME?

Given the choice of receiving \$100 today or \$100 a year from now (guaranteed), ask yourself which you would choose. Most people would prefer receiving \$100 today. A dollar in hand can be used for current purchase of goods and services, invested to earn a return, or used to pay off loans or avoid borrowing costs. Consider that \$100 invested today at, say, 10%, would amount to \$110 a year from now. There is an "opportunity cost" associated with deferring receipt of funds in that you give up the interim use of, or earnings on, the funds.

You would likely require a future amount greater than \$100 to be willing to forego receiving \$100 today. For a given level of risk, the amount required depends in part on expected inflation because inflation causes future dollars to buy less. It also depends on the rate of return you require over and above the inflation rate, i.e., on your required "real rate of return." The greater your earning opportunities from alternative investments, the higher will be your required rate of return.

[1]The explanation of basic concepts is expanded from a treatment by Ruegg (1989).

8.2 TIME-EQUIVALENT VALUES

By determining the future amount which makes you just willing to forego a present amount, it is possible to calculate your MARR. And knowing your MARR, it is possible, given a present amount, to calculate a future amount that you consider equally desirable as the present amount.[2]

Equation 8-1 shows how a person's or organization's MARR is used to convert a present amount, P, to a time-equivalent future amount, F. P is compounded forward at an annual interest rate equal to the MARR. This is a widely used time-equivalence formula called the "single compound amount formula."

$$F = P(1 + MARR)^n \tag{8-1}$$

where

P = present value,
$MARR$ = minimum acceptable rate of return expressed as a decimal,
F = future value for which the investor is willing to forego the present amount, and
n = number of periods until F occurs.

The higher your opportunity cost, the higher your MARR and the larger the future amount required to induce you to forego the present amount. Suppose that your two best uses of discretionary funds are to put them in a money market account, which yields 7% per year after taxes, or to make extra payments on a consumer loan outstanding at 11% after taxes. Your highest opportunity cost on receipts deferred is 11% in this case. Thus, we would expect it to take a relatively sure receipt[3] greater than $1,232, after taxes, in 2 years[4] to induce you to be willing to forego $1,000 at present (i.e., $\$1{,}000(1 + 0.11)^2$).

Equation 8-2, a rearrangement of equation 8-1, shows how the MARR can be derived if we know a present amount and a future amount which a person or organization likes equally well.

$$MARR = -1 + \left(\frac{F}{P}\right)^{1/n} \tag{8-2}$$

---

[2]A person's or organization's MARR may vary over time.

[3]Given a fixed loan rate, interest costs are highly certain; to be comparable, the future receipt must also be highly certain.

[4]We assume end-of-year cash flows in the examples of this chapter.

8.3 IMPLICATIONS FOR BUILDING DECISIONS

Knowing an individual's or organization's MARR enables one to make decisions for that entity which entail tradeoffs between present and future amounts. When the MARR is applied to a future amount to find its present value equivalent—the reverse of compounding—the rate is called the "discount rate"[5] and the process is called "discounting." The discounting operation to find the present value equivalent of a single future amount is shown in equation 8-3. It is called the "single present value (or worth)" formula.

$$P = \frac{F}{(1 + D)^n} \tag{8-3}$$

where

D = discount rate (and D is equal to the investor's MARR).

The ability to compute time-equivalent values enables you to compare present and future amounts on a consistent basis. Knowing the discount rate for yourself or for another individual or organization allows you to make investment decisions in keeping with their stated willingness to trade present dollars for future dollars.

The discounting process is integral to evaluating investments for buildings because costs and benefits usually extend far into the future. Ignoring timing in combining present and future amounts—merely adding and subtracting to find the net effect—can lead to uneconomic decisions. The higher the investor's MARR or discount rate, the smaller the present value of a future amount. And the more must be earned in future benefits or avoided in future costs to make a project worthwhile.

Suppose that you work for a firm which has a MARR of 14%. You estimate that by immediately repairing equipment at a cost of \$10,000, the company can avoid a major replacement in 6 years that will cost \$20,000. Should you recommend that the repair be made? Only if the present value equivalent of the future cost savings exceeds today's cost of \$10,000. The discounting operation is performed as follows:

$$P = \frac{\$20{,}000}{(1 + 0.14)^6} = \$9{,}112$$

[5]More information on determining the value of the discount rate is provided in Chapter 11. Here the emphasis is on what it means and how to use it.

We conclude that to this firm, paying $20,000 in 6 years is time equivalent to paying $9,112 today. Since the present value of the expected savings is less than the $10,000 cost, it is not cost effective for the firm to repair the equipment. You should not recommend it.

8.4 MODELING CASH FLOWS

8.4.1 Selecting a Common Time for Computing a Time-Equivalent Amount

You need to select a common time for expressing all costs and benefits for an economic evaluation. Common-time conventions are to express all cash flows as though they occur (1) as a lump sum at the present time (i.e., as present values), (2) as a series of annual amounts over the study period (i.e., as annual values), and (3) as a lump-sum amount at a designated future time (i.e., as future values). The present-value convention is prevalent, and the annual value convention is also often used. The future-value convention is seldom designated the common time in comparing investment alternatives.

8.4.2 Simplified Models of Cash Flows

Most analysts use simplified models of cash flows in evaluating building alternatives. The effort to describe with exactness the timing of costs and benefits over long study periods usually is not worthwhile. Inherent uncertainty in data make a high degree of precision in specifying the timing of cash flows of little significance. By representing complex flows of costs and benefits as conforming to simple patterns of occurrence, you can perform evaluations efficiently with little sacrifice in accuracy.

Here we describe the principal conventions for modeling cash flows. We also discuss briefly a closely related subject, compounding conventions.

8.4.2.1 Discrete Cash Flows

Common cash-flow conventions for compounding and discounting are to group cash flows occurring within a period in a discrete lump sum at the end of the period, at the beginning of the period, or at the middle of the period. The period specified for most capital investment decisions is a year. Hence, common conventions are beginning-of-year, middle-of-year, and end-of-year cash flows.

An additional simplifying cash-flow convention is to model costs or benefits which recur from one period to the next as a series. For example, rents are usually treated as a series recurring monthly or yearly, even though varying occupancy rates and delinquencies may cause both the amount and the timing of collections to vary from period to period. Routine maintenance cost is another example of a recurring cash flow.[6]

Usually multiple cash-flow models are combined in a project evaluation. For example, initial investment and replacement costs are often modeled as occurring in a lump sum at the beginning of the year. Repair and operating costs are often modeled as lump-sum amounts occurring at the end of each year.

The time-equivalence formulas presented in Section 8.5 and the factors presented in Appendix G are all based on discrete, end-of-year cash flows.

8.4.2.2 Continuous Cash Flows

An alternative to assuming that cash flows are discretely spaced over time is to model them as flowing continuously throughout the year at a uniform rate (continuous cash-flow model). Some argue that the continuous cash-flow model more closely fits the pattern of actual cash flows experienced by many organizations. But discrete cash flows and discrete compounding are prevalent in capital budget analysis.

From a practical standpoint, the difference between assuming discrete and continuous cash flows is generally insignificant, particularly given that the discount rate is usually expressed as an effective annual rate.

8.4.3 Compounding

Compounding refers to the frequency with which interest is credited. Daily compounding, for example, means that the interest earned on any given day is based on the investment principal plus interest earned and credited every preceding day since the investment was made. More frequent compounding of interest raises the effective interest rate or discount rate.

Note the distinction between frequency of cash flow and frequency of compounding interest. A person could make a single deposit a year into a savings account, but interest could be compounded continuously from the time of deposit.[7] Continuously compounding a discrete cash flow raises the effective an-

[6]Simplifying cash flows is more important when performing calculations manually. The use of high-speed computers reduces the computational advantage of describing cash flows as recurring series rather than as individual amounts.

[7]Tables of discount factors for continuous cash flows are given in White et al. (1984), as well as other engineering economics textbooks. Tables of continuous discount factors for application to discrete cash flows are also given in White et al. (1984), as well as Au and Au (1983). We give factors for discrete, end-of-year cash flows in Appendix G.

nual discount rate. Or a person could make frequent deposits during the year into a savings account, but interest might be compounded just once a year.

In economic evaluations, the frequency of compounding usually follows the frequency of the cash flow. Thus, a cash flow modeled as recurring annually is compounded annually using an effective annual discount rate, and a cash flow modeled as recurring quarterly is compounded quarterly using an effective quarterly discount rate.

There should be correspondence between the frequency of compounding and the way the discount rate is expressed. When there is a discrepancy between the quoted rate and the compounding frequency, you can compute the effective annual rate with equation 8-4.

$$i_e = \left(1 + \frac{i}{\bar{n}}\right)^{\bar{n}} - 1 \tag{8-4}$$

where

i_e = the effective annual interest rate, i.e., the actual yield received,
i = the quoted annual interest rate, i.e., the simple annual rate without regard to frequency of compounding, and
$\bar{n}$ = the number of compounding periods per year.

Suppose, for example, you wish to compute the effective annual discount rate based on a quoted annual rate of 8%, with quarterly compounding.

$$i_e = \left(1 + \frac{0.08}{4}\right)^4 - 1$$

$$= 0.08243$$

Now suppose that your *effective* annual discount rate is 10%, and you wish to model a cash flow as a recurring monthly payment. In this case, you need to convert the effective annual discount rate to an equivalent monthly rate. It is incorrect simply to use $i_e/\bar{n}$ (i.e., 0.10/12) as the monthly discount rate, because that would convert to an effective annual rate of 10.47 which is greater than the true effective rate of 10%.

You can use equation 8-4 to solve for the monthly rate. Set i_e equal to 0.10, the effective annual rate. Set $\bar{n}$ equal to 12, the number of compounding periods. Solve for the quoted annual rate, i. Divide the quoted rate by 12 to derive the monthly rate, which in this example is 0.0080. This is the monthly rate, which is equivalent to an effective annual rate of 0.10.

8.4.4 Cash-Flow Diagrams

We recommend the diagramming of cash flows prior to computing time-equivalent values. Diagramming serves as a visual checklist of relevant effects, as an organizer of data for discounting, and, *ex post*, as an informative part of project documentation. The cash-flow model of Figure 8-1 illustrates building-related cash flows. It has investment costs occurring at the beginning of the study period and all other amounts at the end of the year in which they occur.

To construct a cash-flow diagram, draw a horizontal time line as shown in Figure 8-1, divide it into periods, and indicate the timing of cash flows by drawing arrows perpendicular to the time line at the point where a cash amount is expected to occur. Cash inflows (benefits) and outflows (costs) can be distinguished by the direction of the arrows. Arrows pointing toward the line generally designate costs, and arrows pointing away from the line, benefits. Another way of distinguishing inflows and outflows is by placement of items above and below the line. Benefits are usually placed above the line and costs below the line. A vertical dollar scale may be added at the origin of the time line to indicate the size of dollar cash flows. But this level of detail is usually not necessary. Sometimes calendar dates are on the time line, but more typically, years are simply numbered.

8.5 CALCULATIONS

Looking back at the methods equations in Chapters 2 through 7, such as equation 2-1, you see the single present value formula used to discount period-by-period future cash flows. This form is used to provide a general expression of the discounting operations that the reader can easily follow from one method to the next.

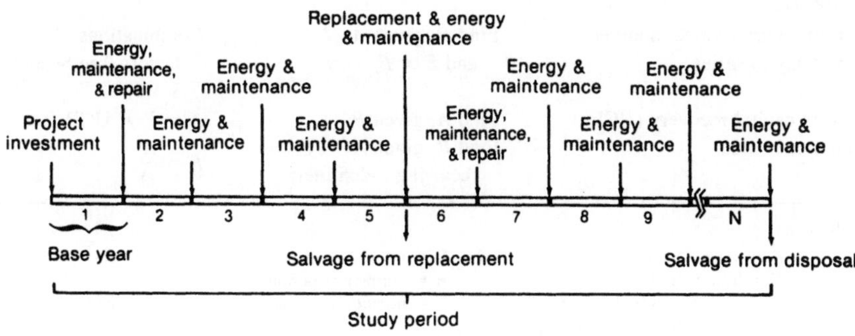

Figure 8-1. Cash-flow diagram.

But in practice, there are a variety of ways to structure a problem for discounting. Most investment decisions entail multiple cash flows having different patterns of occurrence. There are a number of specialized time-equivalence formulas for calculating present values, annual values, or future values of cash flows. The choice of formula depends on the characteristics of the cash flow to be discounted and the common time used to compare alternatives. You can also choose between using formulas written in full or in an abbreviated form using precalculated discount factors.

A practical way to compute the measures of economic performance given in Chapters 2 through 7 is to group costs and benefits with similar patterns of occurrence; apply the appropriate discounting operations to each group; and combine the time-equivalent amounts according to the relationships specified by the method formulas.

Our purpose in this section is to present the main formulas you will need for adjusting cash flows to make them time equivalent. Table 8-1 summarizes them.

Table 8-1. Quick reference for time-equivalence formulas

Formula Name/Abbreviation	Application	Formula Using Discount Factors
	For a given D and n:	
Single compound amount (SCA)	Find F, given P	$F = P \times$ SCA factor
Single present value (SPV)	Find P, given F	$P = F \times$ SPV factor
Uniform present value (UPV)	Find P, given A	$P = A \times$ UPV factor
Uniform present value modified for constant rates of change (UPV*)	Find P, given A_0 and E	$P = A_0 \times$ UPV* factor
Uniform present value modified for variable rates of change (UPV**)	Find P, given A_0 and E_i's	$P = A_0 \times$ UPV** factor
Uniform present value modified for delay in series	Find P, given A_0, L, and E or E_i's	Combinations of factors (see Section 8.4.6)
Uniform capital recovery (UCR)	Find A, given P	$A = P \times$ UCR factor
Capitalized value	Find P, given A recurring indefinitely	$P = \frac{A}{D}$

Notation:

P = present value
F = future value
A = recurring periodic value
A_0 = initial value of a recurring amount changing over time
E = constant rate of change
E_i's = varying rates of change
i or D = interest or discount rate
n = number of periods
L = number of elapsed periods before A begins

For quick reference, it gives the formula name in the first column, a shorthand description of the application in the second column, and the abbreviated form of the equation using factors in the third column.

We present each formula in more detail and illustrate its use. When a formula is commonly used in other ways than to find time-equivalent values, we identify the other applications.

8.5.1 Single Component Amount (SCA)

The SCA formula is used to calculate the time-equivalent future value of a single present amount. This is illustrated in the following schematic:

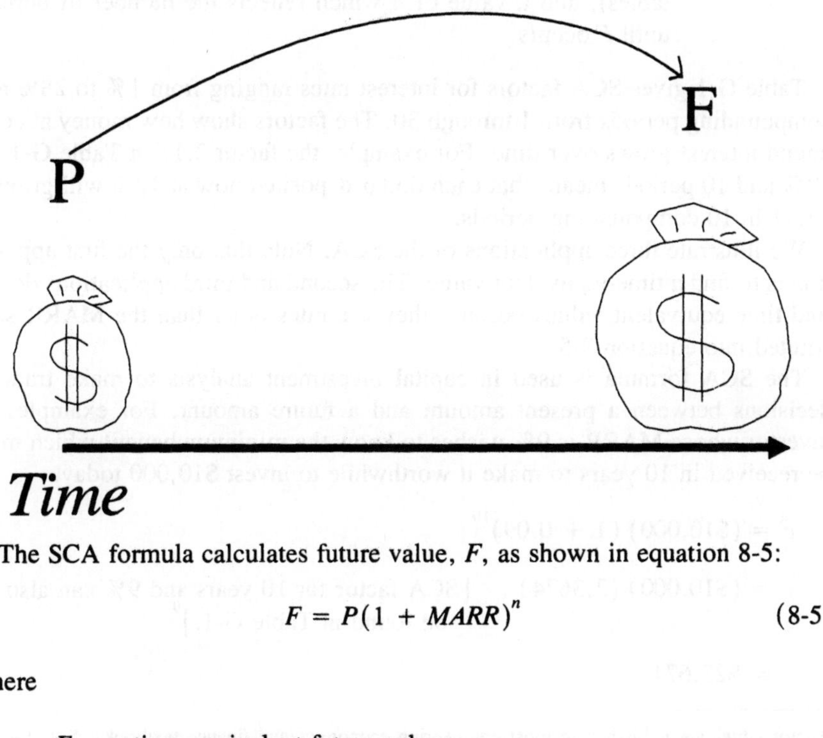

The SCA formula calculates future value, F, as shown in equation 8-5:

$$F = P(1 + MARR)^n \tag{8-5}$$

where

F = a time-equivalent future value,
P = a present value,
$MARR$ = the investor's minimum acceptable rate of return expressed in decimal form, and
n = the number of periods until F occurs.

Discount factors are calculated from a portion of the time-equivalence formula and published to reduce computational requirements.[8] The factors are based on specified discount rates and specified lengths of time. A cash amount is simply multiplied by the appropriate factor to find its time-equivalent value.

An SCA factor can be substituted for $(1 + \text{MARR})^n$ in equation 8-5, as shown in equation 8-6:

$$F = (P)\,(\text{SCA factor}) \tag{8-6}$$

where

SCA factor is based on the appropriate value of MARR (usually denoted "i" for interest rate or "d" (or "D") for discount rate in factor tables), and a value of n which reflects the number of periods until F occurs.

Table G-1 gives SCA factors for interest rates ranging from 1% to 28% and compounding periods from 1 through 30. The factors show how money at compound interest grows over time. For example, the factor 3.11 in Table G-1 for 12% and 10 periods means that each dollar deposited now at 12% will grow to $3.11 in 10 compounding periods.

We illustrate three applications of the SCA. Note that only the first application is to find a time-equivalent value. The second and third applications do not find time-equivalent values because they use rates other than the MARR substituted into equation 8-5.

The SCA formula is used in capital investment analysis to make tradeoff decisions between a present amount and a future amount. For example, an investor whose MARR is 9% wishes to know the minimum benefit which must be received in 10 years to make it worthwhile to invest $10,000 today.

$$F = (\$10{,}000)\,(1 + 0.09)^{10}$$

$$= (\$10{,}000)\,(2.3674)$$ [SCA factor for 10 years and 9% can also be found in Table G-1.][9]

$$= \$23{,}674$$

---

[8]Factor tables are published in most engineering economics and finance textbooks. See, for example, Grant and Ireson (1982), White et al. (1984), and Weston and Brigham (1981). The National Institute of Standards and Technology and the American Society of Testing and Materials (ASTM) publish extensive sets of factor tables. See Lippiatt and Ruegg (1988), Petersen (1989), and ASTM (1984). Examples of specialized factors not widely published, together with selected, commonly published factor tables, are presented in Appendix G. A computer program (DISCOUNT) referenced in Appendix D calculates discount factors.

[9]The factor is here calculated from the second term $(1 + 0.09)^{10}$ in the equation. Alternatively it could be found in Table G-1 rounded to two decimal places.

To this investor, \$23,674 in 10 years is time equivalent to \$10,000 today. Unless the projected future benefit exceeds \$23,674, the investor would prefer to keep the \$10,000 and to forego the future benefit.[10]

The SCA formula also is used in banking and finance to calculate the ending amount of a deposit invested at a given interest rate for a specified number of periods. (Note that the interest rate is *not* necessarily equal to the investor's MARR. If it is not, the ending amount is not a time-equivalent value to the investor.) For example, a depositor places \$5,000 in a bank account at a quoted annual interest rate of 7%, which is compounded monthly. She wishes to know how much she will have at the end of 5 years.

$$F = (\$5{,}000)\left(1 + \frac{0.07}{12}\right)^{5 \times 12} \qquad \text{[Effective annual rate = 7.23\%]}$$

$$= (\$5{,}000)\,(1.4176) \qquad \text{[SCA factor for 60 compounding periods and a quoted annual rate of 7\%.]}$$

$$= \$7{,}088.13$$

Given the precise nature of the data, the results are normally stated to the nearest penny. The depositor can plan on receiving exactly \$7,088.13 in 5 years. (Data for making building decisions are seldom this precise.)

The SCA also is used by analysts to estimate future costs and benefits based on amounts today and a projected rate of change. (Note that the rate of change is *not* the investor's MARR; hence, the ending amount is *not* a time-equivalent value to the investor.) What, for example, is the projected price in 15 years of a building component which is priced at \$1,000 today and is projected to increase at an average annual rate of about 5%?

$$F = (\$1{,}000)\,(1 + 0.05)^{15}$$

$$= (\$1{,}000)\,(2.0789) \qquad \text{[SCA factor for 15 years and can also be found in Table G-1]}$$

$$= \$2{,}079$$

The estimated cost of the building component is expected to be about \$2,000 in 15 years. This information might be used in an economic evaluation.

---

[10] We recognize that imprecision in the input data should be reflected by expressing output values in significant digits. But our focus here is on the calculations, and we give insufficient information in most of the examples to judge significance. Therefore, we have not applied significant digits rules.

8.5.2 Single Present Value (SPV)

The SPV formula is used to calculate the present value of a single future amount. This is illustrated in the following schematic:

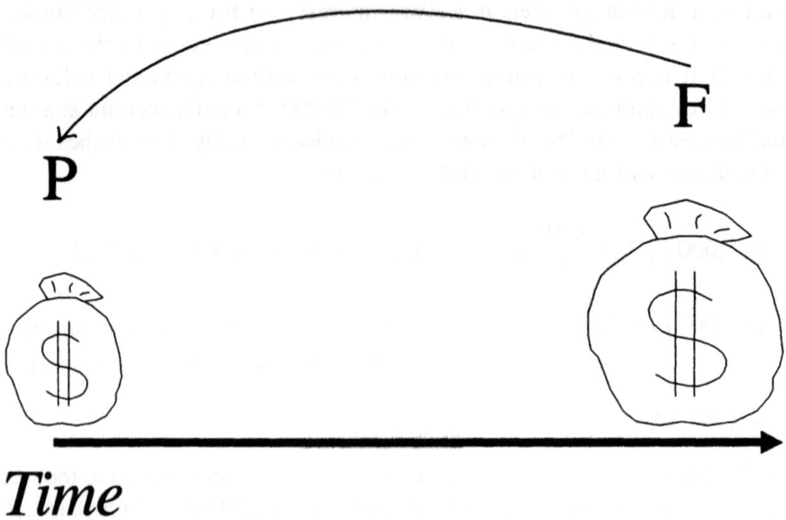

The SPV formula calculates present value, P, as shown in equation 8-7:

$$P = \frac{F}{(1 + D)^n} = F\left[\frac{1}{(1 + D)^n}\right] \tag{8-7}$$

where

D = the investor's discount rate (which equals the investor's MARR)

An SPV factor can be substituted for $[1/(1 + D)^n]$ in equation 8-7, as shown in equation 8-8:

$$P = (F)\,(\text{SPV factor}) \tag{8-8}$$

where

SPV factor is based on D and n.

Table G-2 gives SPV factors for discount rates of 1% through 28% and compounding periods from 1 through 30. The factors can be used to determine what \$1.00 received or paid in the future is worth today. For example, the factor

0.32, found in Table G-2 for 10 periods and a 12% discount rate, means that an investor whose MARR is 12% per year will view a dollar of cost in 10 years as equivalent to $0.32 now. Note that the SPV factor is the inverse of the SCA factor.

We illustrate two applications of the SPV. The first application is to find a time-equivalent value. The second application uses the SPV formula to estimate the selling price of an asset based on its future value. The estimated selling price is not a time-equivalent value, because a rate other than the investor's discount rate is used.

A building manager must decide whether it is cost effective to pay an additional $3,000 for a higher-quality floor covering to avoid a replacement cost in 4 years of $6,000. The owner has an MARR (hence, a discount rate) of 8%.

$$P = \frac{\$6{,}000}{(1 + 0.08)^4}$$

$$= (\$6{,}000)\,(0.7350) \qquad \text{[SPV factor for 6 years and 8\% can also be found in Table G-2.]}$$

$$= \$4{,}410$$

To the building manager, paying $6,000 in 4 years is time equivalent to paying $4,410 today. Therefore, it is cost effective to pay the additional $3,000 up front for the higher-quality floor covering in order to avoid the future cost.

The SPV formula also can be used to estimate the present selling price of an asset based on its future value. (Note that the discount rate used in the calculation does *not* equal the seller's MARR; hence, for the seller, the estimated selling price is not time equivalent to its face value.) Suppose, for example, you hold a promissory note (low risk) with a face value of $20,000, payable in 7 years. You are considering whether to sell it in the market today. You require a price of at least $15,000 to be willing to sell. Notes of similar risk offer an average return in the market of 6.5%. Are you likely to obtain at least $15,000 for the note?

$$P = \frac{\$20{,}000}{(1 + 0.065)^7}$$

$$= (\$20{,}000)\,(0.6435) \qquad \text{[SPV factor for 7 years and 6.5\%.]}$$

$$= \$12{,}870$$

No, the current estimated selling price is only $12,870.

8.5.3 Uniform Present Value (UPV)

The UPV formula is used to calculate the present value of a uniform series of future amounts. This is illustrated in the following schematic:

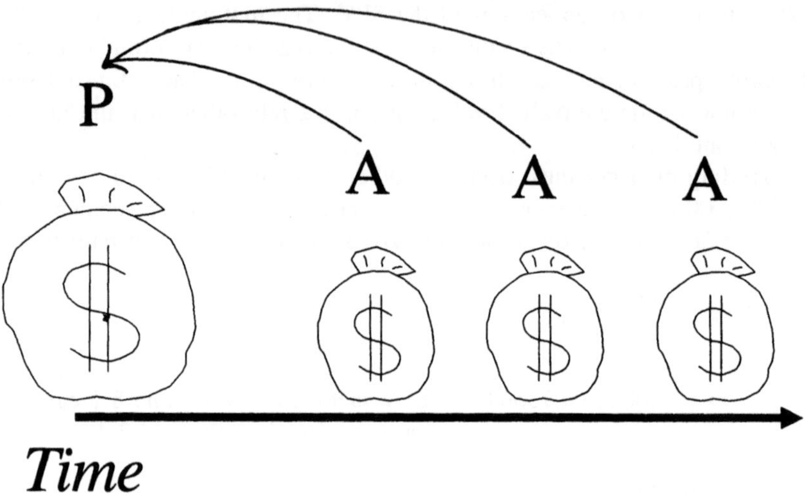

The UPV formula calculates present value, P, as shown in equation 8-9:

$$P = A\left[\frac{(1 + D)^n - 1}{D(1 + D)^n}\right] \qquad (8\text{-}9)$$

where

A = an end-of-period amount in a uniform series over n periods.[11]

A UPV factor can be substituted for $((1 + D)^n - 1)/(D(1 + D)^n)$ in equation 8-9, as shown in equation 8-10:

$$P = (A)\,(\text{UPV factor}) \qquad (8\text{-}10)$$

where

UPV factor is based on D and n.

---

[11] "A" typically means "annual amount." But in equation 8-9 "A" means a uniform amount that recurs periodically over any designated time period, such as yearly, monthly, or semiannually.

Table G-3 gives UPV factors for discount rates of 1% through 28% and compounding periods from 1 through 30. UPV factors can be used to determine what \$1.00 payable (or receivable) periodically is worth today. For example, the factor 5.65, found in Table G-3 for 10 periods and a 12% discount rate, means that to an investor whose MARR is 12%, paying (or receiving) \$1.00 at the end of each of the next 10 periods is equivalent to paying (or receiving) \$5.65 now.

We give two examples of using the UPV to find time-equivalent values. The first shows it used for a capital investment decision. The second example shows it used to compare alternative financing plans.

A designer is considering a new exterior building material which costs more than an alternative material but reduces painting costs by \$2,000 a year over the 20-year anticipated life of the building. The owner's discount rate is 11%.

$$P = \$2{,}000\left[\frac{(1 + 0.11)^{20} - 1}{(0.11)(1 + 0.11)^{20}}\right]$$

$= (\$2{,}000)(7.9633)$ [UPV factor for 20 years and 11%. To find an 11% UPV factor in Table G-3, interpolation between 10% and 12% is necessary.]

$= \$15{,}927$

Paying \$2,000 a year for 20 years for painting is time equivalent to paying about \$16,000 today. If the cost of the new material is no more than about \$16,000 greater than the alternative, it will be cost effective.

A nonprofit organization needs to secure temporary quarters for the next 5 years. It has the option of paying \$10,000 at the end of each year or paying a one-time lump sum of \$35,000 up front. Excess funds are put into Treasury bills at a rate of 7%.

$$P = \$10{,}000\left[\frac{(1 + 0.07)^{5} - 1}{(0.07)(1 + 0.07)^{5}}\right]$$

$= (\$10{,}000)(4.100)$ [UPV factor for 5 years and 7% can also be found in Table G-3.]

$= \$41{,}000$

Paying \$10,000 a year for 5 years is time equivalent to paying \$41,000 up front. Therefore, it is more cost effective to secure the space by paying \$35,000 up front.

8.5.4 Modified Uniform Present Value (UPV*)

The UPV* formula is used to calculate the present value of a series of recurring amounts projected to change at a constant annual rate, i.e., a geometric series.[12] This is illustrated schematically as follows:

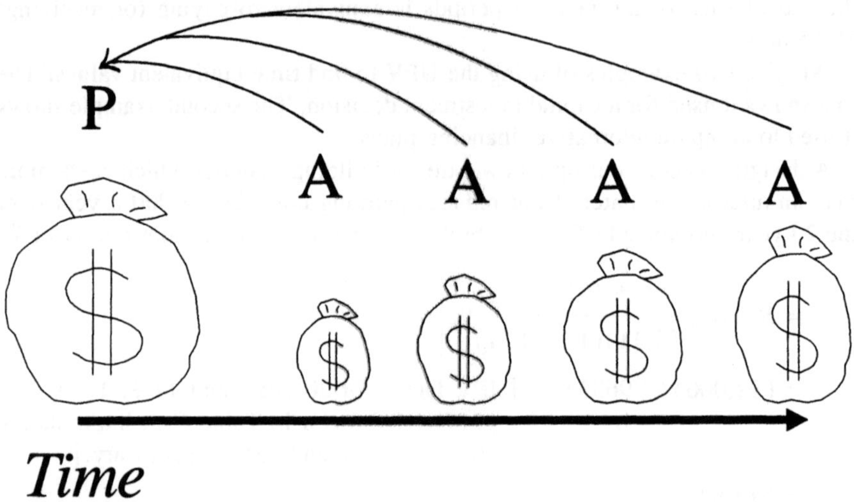

The UPV* formula calculates present values, P, as shown in equations 8-11 through 8-13:

$$P = A_0 \sum_{j=1}^{n} \left(\frac{1 + E}{1 + D} \right)^j \tag{8-11}$$

$$P = A_0 \left[\left(\frac{1 + E}{D - E} \right) \left(1 - \left(\frac{1 + E}{1 + D} \right)^n \right) \right] \tag{8-12}$$

when $E \neq D$

$$P = (A_0)(n) \qquad \text{when } E = D \tag{8-13}$$

where

A_0 = initial value of a periodic amount which occurs over n periods, and
E = constant periodic rate of change (escalation).

[12] Another type of series is a gradient series whereby the periodic amount changes each period by a constant amount, rather than by a constant rate. It is not presented here, but the formula and factors can be found in most engineering economics textbooks, such as White et al. (1984).

A UPV* factor can be substituted for $[((1 + E)/(D - E))(1 - ((1 + E)/(1 + D))^n)]$ in equation 8-12, as shown in equation 8-14:

$$P = (A_0)(\text{UPV* factor}) \tag{8-14}$$

where

UPV* factor is based on D, E, and n.

Tables G-5 through G-21 give UPV* factors for discount rates of 1% through 24% and compounding periods from 1 through 40. Each table has factors for a single discount rate paired with compound rates of change in cash flow (E values) from 1% to 10%. The factors show what money growing at a rate of E% per period and payable (or receivable) periodically over n periods is worth today. For example, in Table G-12, the UPV* factor 6.80 is for 10 periods, a 12% discount rate, and a periodic escalation rate of 4%. The factor indicates that to an investor whose MARR is 12%, paying (or receiving) an amount which begins at \$1.00, escalates 4% annually, and recurs at the end of each of the next 10 years is equivalent to paying (or receiving) \$6.80 now.

The UPV* formula is used in capital investment evaluations to make tradeoff decisions between a series of recurring future amounts changing at a constant rate over time and a present amount. For example, a designer must decide whether to specify water-conserving fixtures and appliances at an added cost of \$800. At today's costs, savings in the first year are estimated at \$100. Water costs are estimated to increase at an annual rate of 4%. Savings are projected over the next 15 years. The owner's MARR is 12%.

$$P = \$100\left[\left(\frac{1 + 0.04}{0.12 - 0.04}\right)\left(1 - \left(\frac{1 + 0.04}{1 + 0.12}\right)^{15}\right)\right]$$

$= (\$100)(8.7227)$ [UPV* for 15 years, an escalation rate of 4%, and a discount rate of 12% can also be found in Table G-16.]

$= \$872$

Water savings over the next 15 years are time equivalent to \$872 today. Present value savings slightly exceed the \$800 cost; hence the water-conserving fixtures and appliances are estimated to be cost effective. (Net savings of \$72 are estimated over the required 12% rate of return.) But the designer may be reluctant to raise first costs by \$800 for a small, and uncertain, net savings.

8.5.5 Uniform Present Value Modified for Variable Rates of Change (UPV**)

The UPV** formula is used to calculate the present value of a series of recurring amounts projected to change at varying annual rates. This is illustrated by the following schematic:

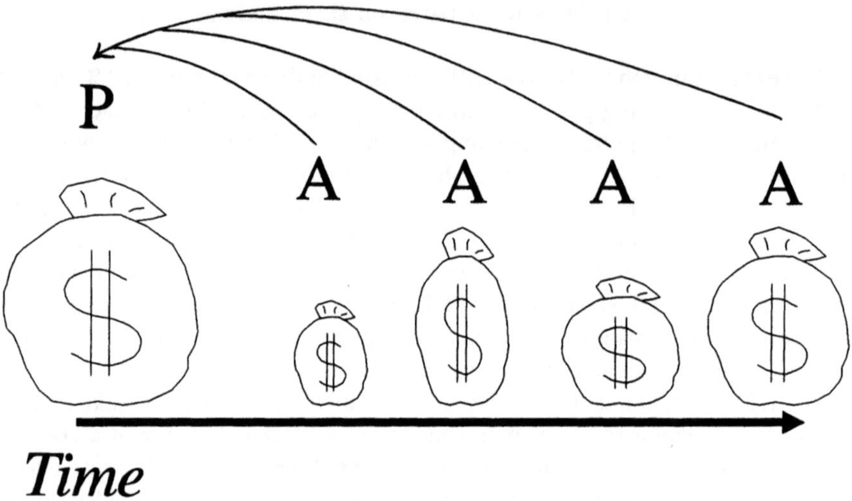

The UPV** formula calculates present value P, as shown in equation 8-15:

$$P = A_0\left[\sum_{j=1}^{v_1}\left(\frac{1+E_1}{1+D}\right)^j + \left(\frac{1+E_1}{1+D}\right)^{v_1}\sum_{j=1}^{v_2}\left(\frac{1+E_2}{1+D}\right)^j + \cdots + \left(\frac{1+E_1}{1+D}\right)^{v_1}\left(\frac{1+E_2}{1+D}\right)^{v_2}\cdots \left(\frac{1+E_{k-1}}{1+D}\right)^{v_{k-1}}\sum_{j=1}^{v_k}\left(\frac{1+E_k}{1+D}\right)^j\right] \tag{8-15}$$

where

A_0 = initial value of a periodic payment or receipt,
E_i = average rate of change within a given time interval i,

v_i = number of discounting periods within interval i over which a given rate of change, E_i, is assumed to hold, and $\sum_{i=1}^{k} v_i = n$ (total number of compounding periods),

k = number of intervals over which there are different rates of change, and

$$\sum_{j=1}^{v_i} \left(\frac{1 + E_i}{1 + D}\right)^j = \left[\left(\frac{1 + E_i}{D - E_i}\right)\left(1 - \left(\frac{1 + E_i}{1 + D}\right)^{v_i}\right)\right] \tag{8-16}$$

A UPV** factor can be substituted for terms within the brackets in equation 8-16, as shown in equation 8-17:

$$P = (A_0)(\text{UPV** factor}) \tag{8-17}$$

where

UPV** factor is based on D, multiple E values, and n.

Tables of UPV** factors based on varying rates of change are specialized tables designed to support analyses of particular types of investment. For example, special UPV** tables support analyses of federal energy-related projects. Tables G-22 through G-26, give samples of federal UPV** factors for 1988. (They are updated annually.) To use the factors, locate the table for the appropriate region of the country; locate the section of the table for the appropriate type of building; locate the column for the applicable type of fuel; and then find the factor in the row for the number of periods over which costs or savings are recurring. The illustrative UPV** factors are helpful for estimating the present value of energy costs escalating at projected rates and payable (or receivable) at the end of each of N years.[13]

The UPV** formula used with federal UPV** factors is for evaluating investment decisions which affect energy consumption. To illustrate, a federal manager of energy conservation on a military base in Montana wants to know if it is cost effective to pay $125 to install electronic thermostats in base housing. The price of electricity at the time of the analysis in late 1988 was $19.12 per million Btu. Over the next 20 years (the estimated life of the thermostats), the price of electricity is projected to change at rates slightly lower than those of prices in general until 1990 and slightly higher thereafter.[14] The federal dis-

[13]For further description of the UPV** factors in Tables G-22 through G-26, see Section 12.5.

[14]The U.S. Department of Energy makes energy price projections using an econometric model of demand and supply. The forecasts predict how energy prices will move relative to prices in general.

count rate for energy conservation decisions in 1988 was 7%.[15] Annual electricity savings were expected to run about 2 million Btu.

$$P = (2{,}000{,}000 \text{ Btu})\left(\frac{\$19.12}{1{,}000{,}000 \text{ Btu}}\right)(10.94)$$
$$= \$418$$

[UPV** for region 4, residential pricing, electricity, 20 years, and discount rate of 7% can be found in Table G-25.]

To the federal government, the stream of future energy savings is time equivalent to a lump-sum amount today of $418. Hence, present value savings exceed the $125 in initial cost, and the thermostats are estimated to be cost effective.

8.5.6 Uniform Present Value Modified for Delay in Onset of the Series

The UPV, UPV*, and UPV** formulas can be modified to allow for a delay in the onset of the series. Finding the present value of a delayed series is illustrated in the following schematic:

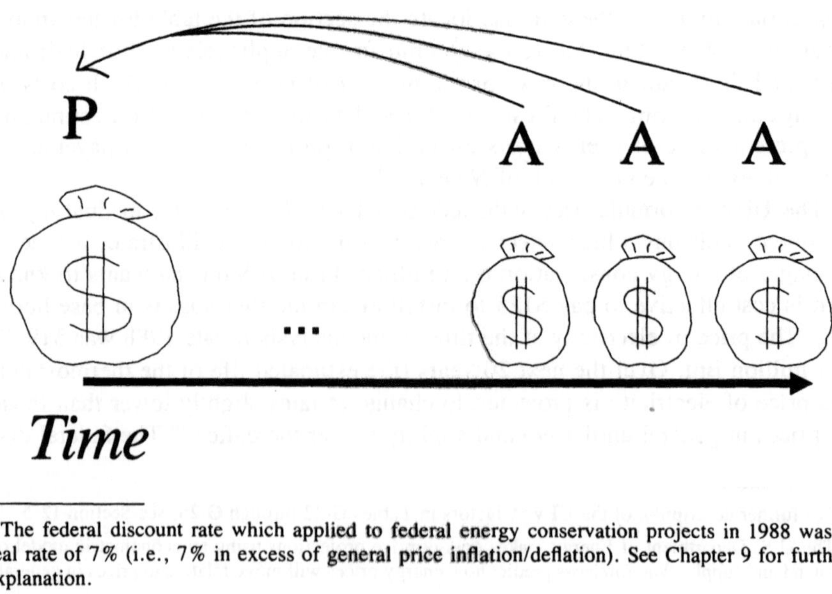

[15]The federal discount rate which applied to federal energy conservation projects in 1988 was a real rate of 7% (i.e., 7% in excess of general price inflation/deflation). See Chapter 9 for further explanation.

Modified versions of equations 8-9, 8-11, and 8-15 can be used to find P when there is a time delay before the series begins. Each of the following pairs of equations accommodates a delay and fits a different assumption about the rate of change in amount. Each pair of equations first gives the formula in full and then using discount factors. Note that the delay is accomplished by combining the SPV, SCA, or both, with the series formula.

1. Present value of a delayed series when there is no change in the recurring amount A. Equation 8-18 is equation 8-9 adjusted for the delay interval, L, and equation 8-19 is equation 8-10 adjusted for the delay.

$$P = A\left[\frac{1}{(1 + D)^L}\right]\left[\frac{(1 + D)^n - 1}{D(1 + D)^n}\right] \tag{8-18}$$

or

$$P = (A)(SPV_L)(UPV) \tag{8-19}$$

2. Present value of a delayed series when the initial amount A_0 changes at rate $\overline{E}$ during the delay interval L, with no change in the recurring amount thereafter. This is equation 8-18 adjusted for change during the delay interval.

$$P = A_0\left[\left(\frac{1 + \overline{E}}{1 + D}\right)^L\right]\left[\frac{(1 + D)^n - 1}{D(1 + D)^n}\right] \tag{8-20}$$

or

$$P = (A_0)(SCA)(SPV)(UPV) \tag{8-21}$$

3. Present value of a delayed series when the initial amount A_0 changes at rate $\overline{E}$ during the delay interval L, and thereafter the series changes at rate E (the two rates can be equal). This is equation 8-11 adjusted for delay.

$$P = A_0\left(\frac{1 + \overline{E}}{1 + D}\right)^L\left[\left(\frac{1 + E}{D - E}\right)\left(1 - \left(\frac{1 + E}{1 + D}\right)^n\right)\right] \tag{8-22}$$

or

$$P = (A_0)(SCA)(SPV)(UPV^*) \tag{8-23}$$

4. Present value of a delayed series when A_0 changes at rate $\overline{E}$ during the delay interval L, and thereafter the series changes at rates E_i through E_k. This is equation 8-15 adjusted for delay.

$$P = A_0\left[\left(\frac{1+\overline{E}}{1+D}\right)^L \sum_{j=1}^{\nu_1}\left(\frac{1+E_1}{1+D}\right)^j + \cdots + \left(\frac{1+\overline{E}}{1+D}\right)^L\left(\frac{1+E_1}{1+D}\right)^{\nu_1} \cdots \left(\frac{1+E_{k-1}}{1+D}\right)^{\nu_{k-1}} \sum_{j=1}^{\nu_k}\left(\frac{1+E_k}{1+D}\right)^j\right] \quad (8\text{-}24)$$

or

$$P = (A_0)((UPV^{**}{}_{n+L}) - (UPV^{**}{}_L)) \quad (8\text{-}25)$$

The delay formulas are used in capital investment decisions to make tradeoffs between a series of future amounts whose onset is delayed and a present amount. For example, a building manager must decide whether it is cost effective to add labor-saving features in a building. The decision must be made now, but because of the budget cycle and construction period, the project will not be completed for 2 years. Labor costs are incurred as lump-sum, end-of-year payments. Based on today's salaries, savings are estimated at \$20,000 per year. But labor costs are projected to surge over the next 2 years at an annual rate of 8% per year, and thereafter to slow to a rate of increase of about 3% per year. The owner's MARR is 10% per year, based on market rates of return available elsewhere. The building will be in use for 10 more years, and the equipment will last at least that long. Taking into account the delay, present value construction costs are \$45,000.

Using either equation 8-22 or 8-23, calculate the present value labor savings and compare them with the present value construction costs:

With equation 8-22:

$$P = \$20{,}000\left(\frac{1+0.08}{1+0.10}\right)^2\left[\left(\frac{1+0.03}{0.10-0.03}\right)\left(1-\left(\frac{1+0.03}{1+0.10}\right)^8\right)\right]$$

$$= (\$20{,}000)(5.802)$$

$$= \$116{,}040$$

With equation 8-23:

$$P = (\$20{,}000)(SCA)(SPV)(UPV^*)$$

$= (\$20{,}000)(1.17$ from Table G-1$)(0.83$ from Table G-2$)(6.02$ from Table G-14$)$

$= \$116{,}920$ (Note that use of factors introduces rounding errors.)

Present value labor savings in the range \$116,000 to \$117,000 exceed present value construction costs of \$45,000. The project is estimated to be cost effective.

8.5.7 Uniform Capital Recovery (UCR)

The UCR formula is used to convert a present amount into a series of uniform amounts over N periods. This is illustrated in the following schematic:

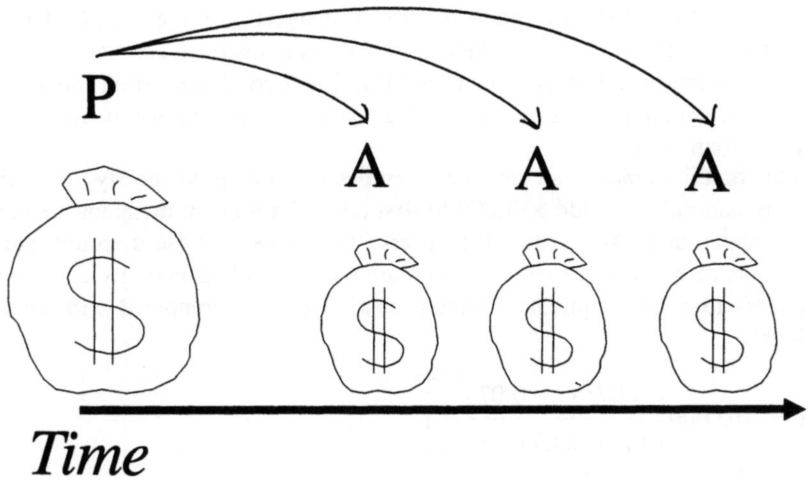

The UCR formula is used to find A, given P, as shown in equation 8-26:

$$A = P\left[\frac{D(1 + D)^n}{(1 + D)^n - 1}\right] \tag{8-26}$$

where

A = end-of-period amount in a uniform series over n periods, and
D = discount rate.

A UCR factor can be substituted for $D(1 + D)^n/((1 + D)^n - 1)$ in equation 8-26, as shown in equation 8-27:

$$A = (P)(\text{UCR factor}) \tag{8-27}$$

where

UCR factor is based on D and n.

Table G-4, Appendix G, gives UCR factors for discount rates of 1% through 28% and compounding periods from 1 through 30. You can use the factors to find the size of an annual payment over *n* years which is time equivalent to $1.00 today. For example, the UCR factor 0.18, found in Table G-4 for 10 years and a discount rate of 12%, means that paying (or receiving) $1.00 today is time equivalent to paying (or receiving) $0.18 at the end of each of the next 10 years to an investor whose MARR is 12%.

We illustrate three applications of the UCR. The first demonstrates the primary use of the UCR in capital investment analysis to express LCC, NB, or NS as time-equivalent annual values or to express investment costs as an annual value for comparison against annual savings. The second illustration shows how to calculate an annuity based on an initial payment. The third illustration calculates a loan repayment.

In the first illustration, an architect must decide whether to specify low-maintenance materials that add $10,000 to first costs. Savings in maintenance costs are estimated at $500 a year for 25 years. The owner, whose discount rate is 7%, wants to know how first costs compare with annual savings. By converting first costs to a time-equivalent annual cost, it can be compared with annual savings:

$$A = \$10{,}000\left[\frac{0.07(1 + 0.07)^{25}}{(1 + 0.07)^{25} - 1}\right]$$

$$= (\$10{,}000)(0.0858)$$ [UCR factor for 25 years and 7% can also be found in Table G-4.[16]]

$$= \$858$$

Estimated first costs are equivalent to $858 each year, $358 more than the estimated savings each year of $500. Therefore, the low-maintenance materials are not estimated to be cost effective.

In the second illustration, the owner of a large estate has donated it to the state for an art museum, together with $500,000 in cash to be used to operate the building over the next 10 years. The state administrator can invest the funds at an annual rate of 12%. She wants to know how much she can budget for operating expenses in uniform annual amounts over the next 10 years, depleting the fund by the end of the 10-year period.

The annual budget can be estimated by amortizing the donation over 10 years at 12% interest:

---

[16]Factors in Appendix G are rounded and are therefore not identical to those computed using the formula.

$$A = \$500{,}000\left[\frac{0.12(1 + 0.12)^{10}}{(1 + 0.12)^{10} - 1}\right]$$

$= (\$500{,}000)(0.1770)$ [UCR factor for 10 years and 12%. Also available from Table G-4.]

$= \$88{,}500$

In the third illustration, a prospective buyer of a building wishes to estimate the yearly payments required to pay off a \$350,000 loan at 8% over 15 years in uniform end-of-year payments.

$$A = \$350{,}000\left[\frac{0.08(1 + 0.08)^{15}}{(1 + 0.08)^{15} - 1}\right]$$

$= (\$350{,}000)(0.1168)$ [UCR factor for 15 years and 8%. See Table G-4.]

$= \$40{,}880$

The annual loan payment is \$40,880.

8.5.8 Capitalized Value of a Perpetual Series

The capitalization formula is used to find the present value of a uniform series recurring indefinitely. This is illustrated in the following schematic:

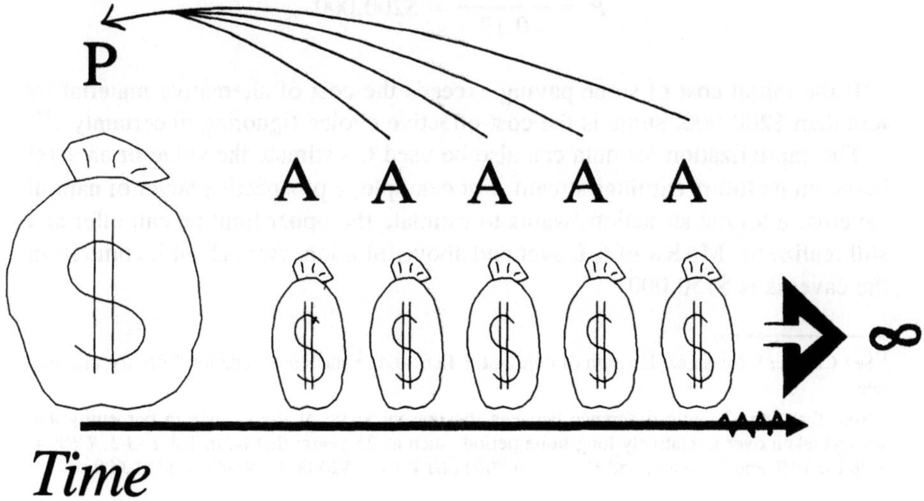

The capitalization formula is used to find P, given $\bar{A}$ and D, as shown in equation 8-27:

$$P = \frac{\bar{A}}{D} \tag{8-27}$$

where

$\bar{A}$ = end-of-period amount in a uniform series recurring indefinitely
D = discount rate.

We illustrate three applications. The first is a typical capital investment application. The second is a real estate valuation application. The third shows how to compute the amount required to establish an annuity.

The capitalization formula is used for capital investment decisions to make tradeoffs between an indefinitely recurring series of future amounts and a present amount. For example, a state government has commissioned the design of a monument which is to have an indefinite life. The designer estimates that using stone for floors and stairways will reduce annual maintenance costs by $20,000 (constant dollars) over a proposed alternative paving material. Is stone the cost-effective choice? The applicable discount rate is 10% (excluding inflation).[17]

Capitalizing annual savings will indicate the maximum additional amount that stone can cost and remain cost effective.

$$P = \frac{\$20{,}000}{0.10} = \$200{,}000.$$

If the initial cost of stone paving exceeds the cost of alternative material by less than $200,000, stone is the cost-effective choice (ignoring uncertainty).[18]

The capitalization formula can also be used to estimate the value of an asset based on its future earnings stream. For example, a prospective buyer of natural caverns, a tourist attraction, wants to estimate the upper limit he can offer and still realize his MARR of 6% over and above inflation. Annual net income from the caverns is $150,000.

---

[17]See Chapter 9 for an explanation of constant dollars and exclusion of inflation from the discount rate.

[18]Note that there is little difference between the present value of the savings in perpetuity and savings taken over a relatively long finite period, such as 25 years; that is, in Table G-3, UPV = 9.08 for 10% and 25 years, and P = ($20,000) ($UPV$) = ($20,000) (9.08) = $181,600.

$$P = \frac{\$150,000}{0.06} = \$2,500,000.$$

Capitalizing net income provides an estimate of the maximum price that this buyer can pay for the caverns and still meet his 6% MARR. It is $2.5 million. Future resale value is captured in the measure.

The capitalization formula can be used to establish a perpetual annuity, i.e., to find the present amount which must be deposited into a fund at interest rate i per period in order to generate a periodic amount of size A which can be withdrawn each period in perpetuity. (Note that i is not necessarily equal to the discount rate; hence, the result is *not* a present value.) For example, a wealthy person wishes to endow a graduate school of economics at a local university. Estimated annual expenses are $1.8 million. What must be the amount of the endowment, given that it will earn 9% per year?

$$P = \frac{\$1,800,000}{0.09} = \$20,000,000$$

An endowment of $20 million is required.

REFERENCES

American Society for Testing and Materials (ASTM). 1988. *Discount Factor Table; Adjunct to Standard Practice for Measuring Life-Cycle Costs of Buildings and Building Systems*. E 917. Philadelphia: ASTM.

Au, Tung, and Thomas P. Au. 1983. *Engineering Economics for Capital Investment Analysis*. Newton, Massachusetts: Allyn and Bacon, Inc.

Clark, J. J., T. J. Hindelang, and R. E. Pritchard. 1984. *Capital Budgeting: Planning and Control of Capital Expenditures*. Englewood Cliffs, New Jersey: Prentice-Hall, Inc.

Grant, Eugene L., and W. Grant Ireson. 1982. *Principles of Engineering Economy*. New York: The Ronald Press Company.

Lippiatt, Barbara C., and Rosalie T. Ruegg. 1988. *Energy Prices and Discount Factors for Life-Cycle Cost Analysis*. NISTIR 85-3273-3. Gaithersburg, Maryland: National Institute of Standards and Technology.

Petersen, Stephen R. 1989. *Discount Factor Tables for Life-Cycle Cost Analyses*. NISTIR 89-4203. Gaithersburg, Maryland: National Institute of Standards and Technology.

Ruegg, Rosalie T. 1989. *Life-Cycle Costing for Energy Conservation in Buildings: Instructor's Guide*. NISTIR 89-4129. Gaithersburg, MD: National Institute of Standards and Technology.

Weston, J. Fred, and Eugene F. Brigham. 1981. *Managerial Finance*. Hinsdale, Illinois: The Dryden Press.

White, John A., Marvin H. Agee, and Kenneth E. Case. 1984. *Principles of Engineering Economic Analysis*. New York: John Wiley & Sons, Inc.

9

Treatment of Price Changes

We give attention to the question of how to deal with changes in price level because it is a much misunderstood and confusing issue. First, we demonstrate the need to adjust for changes in the purchasing power of the dollar.[1] Next, we show how to measure changes in the price of a good or service in current dollars and in constant dollars. Finally, we describe two valid approaches for dealing with inflation in an economic evaluation: (1) working in current dollars and removing inflation as part of the discounting operation and (2) working in constant dollars and excluding inflation at the outset.

9.1 WHY ADJUST FOR CHANGES IN PRICE LEVEL?

A valid economic analysis requires that the monetary unit in which all costs and benefits are measured (e.g., dollars, pounds sterling, or yen) have a constant purchasing power. This requirement is analogous to the requirement for a constant unit for measuring the physical dimensions of a building (e.g., a standard foot or meter). Inflation reduces the purchasing power of the monetary unit, and deflation increases it, making measurement of economic effects in inflated or deflated dollars meaningless.

The need to exclude purely inflationary or deflationary changes from economic analyses is illustrated by the gross national product (GNP) data in Figure 9-1. GNP is a measure of the nation's annual output of final goods and services. The dark area of the figure shows U.S. GNP from 1967 through 1988, expressed in actual dollars of those years (using economics terminology, in "current dollars"). In current dollars, GNP increased from about \$800 billion to about \$5 trillion, or about by six times over the 21-year period from 1967 through 1988.

But now look at the light-shaded area of the figure, which shows GNP in

[1]The dollar is used as the monetary unit for illustration purposes.

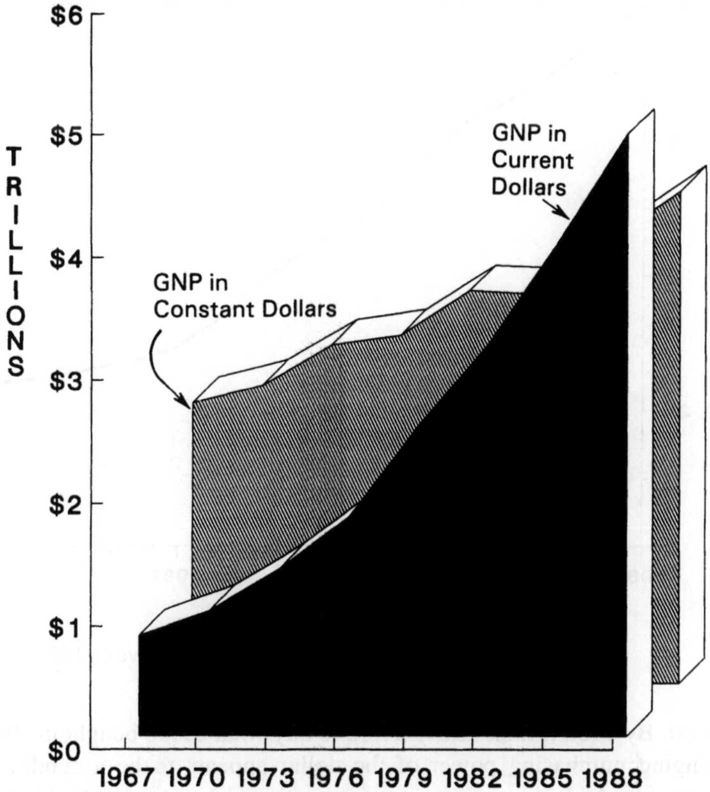

Figure 9-1. GNP in current and constant dollars. (Source: Data taken from *Economic Report of the President.*)

each year expressed in 1982 dollars. Measured in dollars with an unchanging purchasing power (using economics terminology, in "constant dollars"),[2] the value of the nation's output less than doubled. The extra rise in the current dollar measure over the constant dollar measure reflects price inflation. Only by looking at the constant dollar measure can we deduce real changes in the nation's output.

The diminished purchasing power of the dollar over time is demonstrated in Figure 9-2. It uses 1967 as the reference year for setting the dollar's value equal

---

[2]In this example, 1982 is the base year for fixing the value of the dollar equal to $1.00. Amounts for all other years are expressed in 1982 dollars.

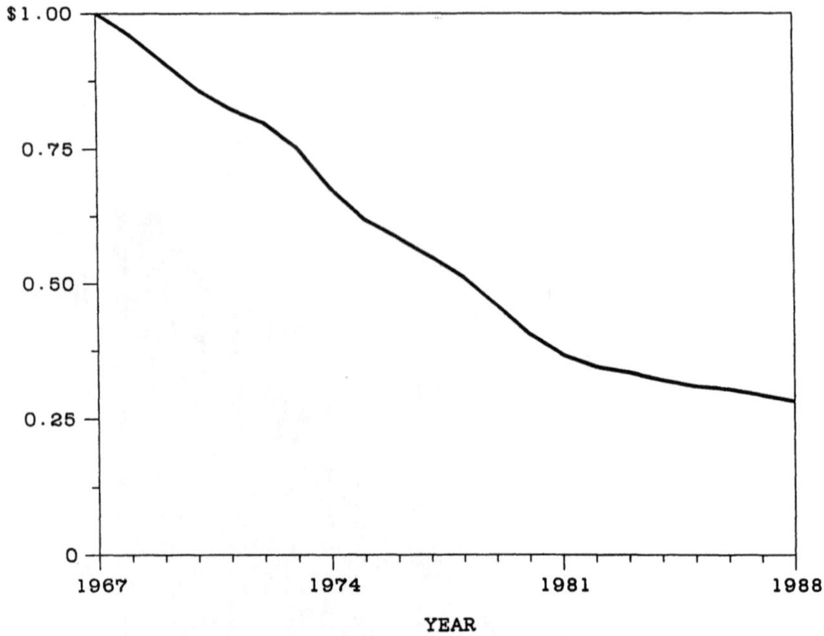

Figure 9-2. Purchasing power of the dollar relative to 1967.

to $1.00. By 1988 it took nearly $4.00 to buy what $1.00 bought in 1967. Since a changing purchasing power of the dollar appears to be a condition of the economy in which we operate, it is imperative to know what to do about these changes when estimating benefits and costs.[3]

9.2 MEASURING CHANGES IN PRICE[4]

Figures 9-3 through 9-6 show the rates of change in actual prices (E), year by year, of four building-related items between 1967 and 1988. The items are (1) maintenance and repair (M&R), (2) fuel, (3) construction materials, and (4) house furnishings. Each figure also shows the year-by-year rate of change in

[3]Emphasis is on inflation, rather than deflation, because inflation has been prevalent in the latter half of the twentieth century, and many economists believe that the modern mixed economy tends to be inflationary prone. See, for example, Heilbroner and Thurow (1982).

[4]Figures 9-2–9-6 and Sections 9.2 through 9.4 provide expanded treatments of the subject of price changes presented by Ruegg (1989) and Ruegg and Fuller (1990).

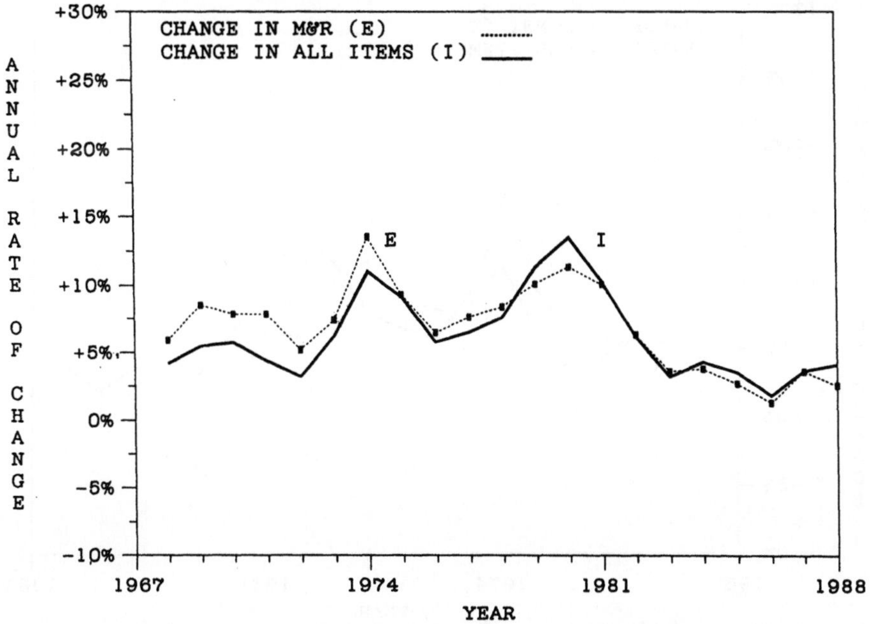

Figure 9-3. Rate of change in the prices of maintenance and repair and of all items.

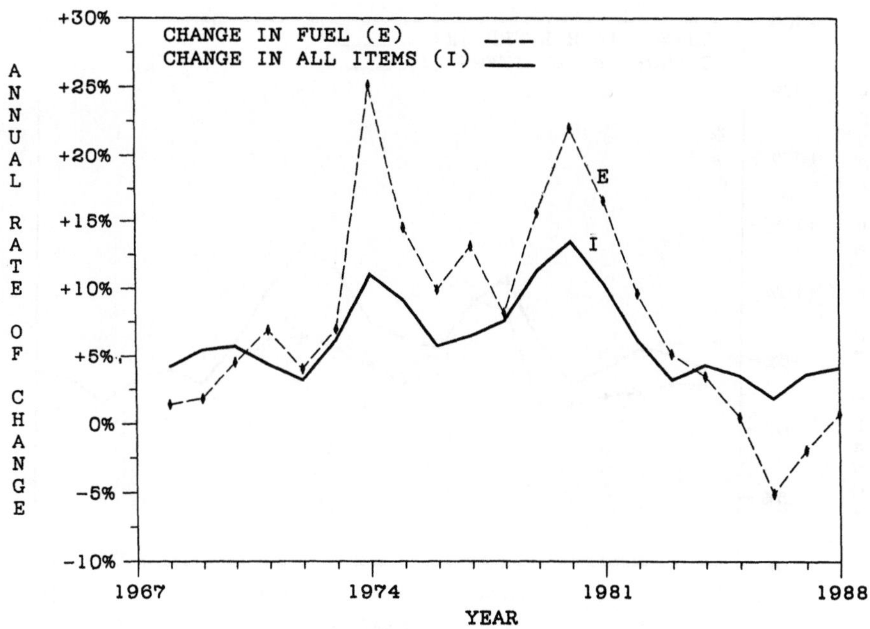

Figure 9-4. Rate of change in the prices of household fuel and of all items.

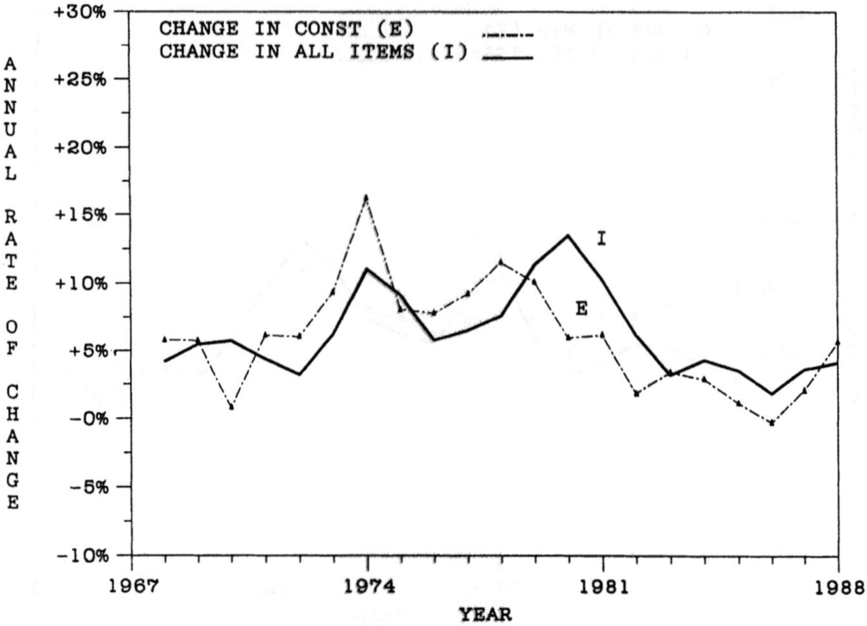

Figure 9-5. rate of change in the prices of construction materials and of all items.

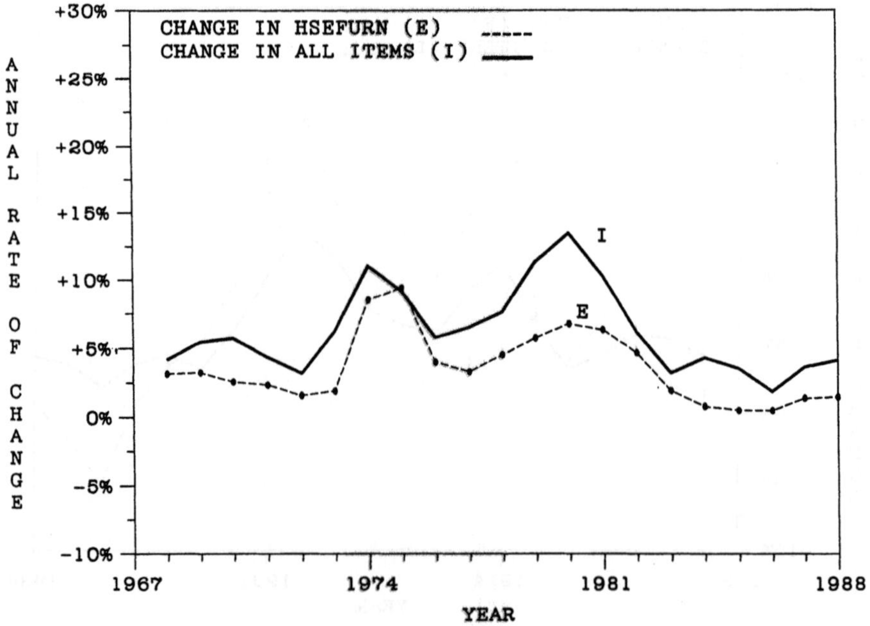

Figure 9-6. Rate of change in the prices of house furnishings and of all items.

average prices of a fixed market basket of consumer goods and services[5] labeled "all items." This rate of change (I) is a measure of general price inflation/deflation. We focus first on the building-related items. Our purpose in giving these historical data is to show how rates of change differ for specific items and how they vary over time.

Yearly rates of change in maintenance and repair prices, shown in Figure 9-3, averaged 6.8% over the 21-year period. This reflected annual rates of change averaging about 7% from 1967 through 1972, 9% from the end of 1972 through 1982, and 3% from the end of 1982 through 1988.[6]

Yearly rates of change in household fuel prices (including fuel oil, natural gas, and electricity), shown in Figure 9-4, averaged 7.5%. Price rises accelerated in 1974 and 1980 and prices declined in the mid-1980s.

Yearly rates of change in construction material prices, shown in Figure 9-5, averaged 5.9%. Annual rates were as low as 0.08% and as high as 16%.

Yearly rates of change in house furnishing prices, shown in Figure 9-6, averaged 3.5% over the 21-year period. The rate of increase was low from 1967 through 1973, about 9% in 1974–75, and relatively low again during the 1980s.

Projected escalation rates in actual prices can be used to estimate future prices of individual items. Because E values are projected changes in actual prices, they include general price inflation. The resulting price estimates are therefore in current dollars. For example, future maintenance costs per hour in current dollars can be estimated by applying the single compound amount formula to today's cost per hour, using the projected rate of change, E, in place of an interest or discount rate, as shown in equation 9-1.[7]

$$FP = TP(1 + E)^n \tag{9-1}$$

where

FP = projected future price in current dollars,
TP = today's price,

[5]The consumer price index (CPI) measures the average change in prices over time in a fixed market basket of goods and services purchased by consumers. The Bureau of Labor Statistics, which prepares the CPI, also provides the producer price index (PPI). The PPI measures the average change in prices of all commodities at all stages of processing produced for sale in primary markets in the United States.

[6]Note that even though the price function dips down in many years, this means only that the rate of increase is less than that of the year before. In fact, in Figure 9-3, the rate of change is always positive, indicating that prices went up every year.

[7]See Section 8.5.1 for more about the formula.

E = projected escalation in the actual price of the item in question, where E is an annual compound rate, and
n = number of periods over which the price is escalated.

For illustration, suppose that you expect maintenance and repair prices to change over the next 4 years at the same average rate of 6.8% experienced from 1967 to 1988. (Recognize, of course, that the past is not always a good predictor of the future.) Further, suppose that today's wage rate for maintenance and repair labor is \$15/hour. Therefore the estimated hourly charge 4 years hence is \$19.52 (i.e., $FP = \$15 \times (1 + 0.068)^4 = \$19.52$). This is the expected wage rate stated in dollars of the future year, i.e., in current dollars.

9.3 MEASURING RELATIVE CHANGES IN PRICES

As an alternative to projecting the total change in the prices of individual goods and services, it is possible to project just the differential rate of change in prices *relative* to general price inflation (I). Differential change relative to general price inflation may be seen in Figures 9-3 through 9-6 by comparing the graphs for individual items with the graph for "all items." The annual rate of inflation in the prices of the fixed market basket of goods and services, labeled "all items," averaged 6.2%.

Maintenance and repair prices (Figure 9-3) changed at close to the same rate as prices in general and remained essentially unchanged in relative affordability over the 21-year period from 1967 to 1988. Fuel oil prices (Figure 9-4) increased on the average more than prices in general, causing fuel to become less affordable. Construction prices (Figure 9-5) changed on the average slightly less than prices in general, meaning that the relative affordability of construction materials improved marginally. But, as you can see in the figure, the two rate of change lines are divergent over most of the period, with material prices first rising more, then less, than most other prices. Prices of house furnishings (Figure 9-6) rose at a lower rate than prices in general and, in relative terms, house furnishings became more affordable.

If there were no shifts in underlying demand and supply conditions, we would expect all prices to change at the rate of general price inflation. In relative terms prices would remain unchanged. But demand and supply conditions do shift. Thus, prices of individual items may change faster or slower than prices in general.

We call the rate of change in relative prices the "differential" rate of change and use the notation e to distinguish it from the total rate of change, E, for a given item. The differential annual rate of price change (e) and the total annual rate of price change (E) are related through the general price inflation rate (I). We can compute e from E or E from e, as shown in equations 9-2 and 9-3.

$$e = \left[\frac{1 + E}{1 + I}\right] - 1 \tag{9-2}$$

$$E = [(1 + e)(1 + I)] - 1 \tag{9-3}$$

where

e = differential rate of change between the price of a particular item and prices in general,
E = absolute rate of change in the price of the same item, and
I = rate of general price inflation.

Thus, if you know the total rate of change in the price of an item, together with the rate of general price inflation, you can calculate the differential rate of change. For example, when fuel oil prices rose at an average annual compound rate of 7.5%, prices in general rose at a rate of only 6.2%. Part of the rise in fuel oil prices was due to general price inflation, but part was also due to scarcity of oil. Apart from general price inflation—i.e., in real terms—fuel oil prices rose at an annual rate of about 1.2%. That is $e = (1 + E)/(1 + I) - 1 = (1 + 0.075)/(1 + 0.062) - 1 = 0.012$, where e = the real or differential rate of change, E = the total rate of change, and I = general inflation rate. We should point out, however, that fuel price changes are a significant factor in general price inflation during this period. The rate of general price inflation and the rate of change in fuel prices are highly correlated. Therefore the real rate of change in fuel prices tends to be underestimated by the technique we use here.

The projected values of differential escalation rates, e, can be used to estimate future costs and benefits in constant dollars, i.e., excluding effects of general price inflation. Simply substitute e for E in equation 9-1, as shown in equation 9-4.

$$\overline{FP} = TP(1 + e)^n \tag{9-4}$$

where

$\overline{FP}$ = projected future price in constant dollars,
TP = today's price,

e = differential annual compound rate of change in the item in question, as computed by equation 9-2, and
n = number of periods over which the price is to be projected.

Suppose, for example, that you expect wage rates for maintenance and repair tasks, which are assumed to be \$15/hour today, to change over the next 6 years at about the same rate as prices in general. The projected e value is zero. The estimated wage rate 6 years hence, expressed in today's dollars, is \$15/hour, the same as the charge rate today [i.e., $\overline{FP} = (\$15/\text{hour})(1 + 0)^6 = \$15/\text{hour}$].

Suppose that you expect the price of fuel oil, which is \$0.82/gallon today, to change over the next 6 years at a rate 1.2% faster than prices in general. The projected e value is 1.2%. The estimated price of fuel oil in 6 years expressed in today's dollars (i.e., constant dollars) is \$0.88/gallon [i.e., $\overline{FP} = (\$0.82/\text{gallon})(1 + 0.012)^6 = \$0.88/\text{gallon}$]. The estimated price does not reflect general price inflation, so you don't have to remove it.

9.4 TWO APPROACHES FOR TREATING INFLATION OR DEFLATION

Whether costs are expressed in current or constant dollars dictates whether the discount rate should include or exclude general price inflation. If cash flows are in current dollars, use a market discount rate,[8] i.e., a rate that, like the current dollar cash flows, includes inflation. If cash flows are in constant dollars, use a "real" discount rate, i.e., a rate that, like the constant dollar cash flows, excludes inflation. You will obtain identical results using the "current dollar/market discount rate approach" or the "constant dollar/real discount rate approach" provided you are consistent in how you treat inflation in cash flows and the discount rate.

9.4.1 Current Dollar/Market Discount Rate Approach

If you express cash flows in current dollars, use a market discount rate to perform discounting. Equation 9-5 illustrates the discounting operation for a single future amount, with inflation removed as part of discounting.

---

[8]Market discount rates are also called "nominal" discount rates.

$$P = \frac{F_{cur}}{(1 + D)^n} \tag{9-5}$$

where

F_{cur} = future amount in current dollars, i.e., $F_{cur} = (FP)(Q)$, where FP = future price in current dollars and Q = future quantity, and
D = market discount rate (including inflation).

9.4.2 Constant Dollar/Real Discount Rate Approach

If you express cash flows in constant dollars, including only the differential price change, use a real discount rate (excluding inflation/deflation) to perform discounting. To state cash flows in constant dollars, choose a reference date for fixing the value of the dollar and express all future amounts in dollars with that same value. The reference date is usually chosen to coincide with the date the economic analysis is performed, but it could be any date.

Equation 9-6 illustrates the discounting operation for a single future amount expressed in constant dollars.

$$P = \frac{F_{con}}{(1 + d)^n} \tag{9-6}$$

where

P = present value,
F_{con} = future amount in constant dollars,[9] i.e., $F_{con} = (\overline{FP})(Q)$, where $\overline{FP}$ = future price in constant dollars and Q = future quantity, and
d = real discount rate (excluding inflation).

Avoid the common error of mixing the approaches. That is, *do not* discount current dollar estimates of costs and benefits with real discount rates, or constant dollar estimates with market discount rates. Also, use the same inflation rate when estimating costs and benefits and when setting the discount rate.

[9]Note that costs and benefits may be expressed as constant dollars at the outset. Or they may be stated initially in current dollars and converted to constant dollars using a price deflator index prior to discounting with a real discount rate. Thus an alternative to equation 9-6 is the following:

$$P = (F_{cur} \times PDI)/(1 + d)^n$$

where PDI = price deflator index and d = real discount rate (excluding inflation).

9.5 ILLUSTRATING CONSISTENCY OF THE TWO APPROACHES

9.5.1 Data and Assumptions

Suppose that you wish to know the present value of replacing a roofing system in 10 years. Further, suppose that the general inflation rate, I, is 5% per year, and the cost of the roofing system is expected to increase at an annual rate 2% faster than the general inflation rate due to anticipated shortages of roofing materials, i.e., $e = 2\%$. Assume that the roofing system would cost \$10,000 if replaced today. Also assume that the investor's real discount rate is 4%. The projected market discount rate equals the real discount rate plus compensation for inflation. The following data are given:

$$I = 0.05 \qquad e = 0.02 \qquad d = 0.04$$

Substitute into equation 9-3 to find the total rate of annual price change, E:

$$E = [(1 + 0.02)(1 + 0.05)] - 1 = 0.071.$$

The market discount rate, D, is related to the real discount rate, d, through the inflation rate, just as E is related to e. D includes general price inflation; d excludes it. Equation 9-7 is used to compute D, given d.

$$D = [(1 + d)(1 + I)] - 1 \tag{9-7}$$

Substitute into equation 9-7 to find the market discount rate:[10]

$$D = [(1 + 0.04)(1 + 0.05)] - 1 = 0.092.$$

9.5.2 Illustrating the Current Dollar/Market Discount Rate Approach

With estimates of the total rate of price change, E, the single compound amount formula can be used to estimate future replacement costs in current dollars. With estimates of the nominal discount rate, D, the single present value formula can be used to compute the present value of the replacement. The two operations can be combined, as shown in equation 9-8.

---

[10] Note that D and E are not simply the sum of d and I or E and I. Summing, although prevalent in practice, will understate D and E slightly.

$$P = C_0\left[\frac{1 + E}{1 + D}\right]^n \tag{9-8}$$

where

C_0 = cost today, i.e., $C_0 = (TP)(Q)$, where TP = today's price and Q = quantity,
E = projected total rate of change in the item, i.e., change in actual prices,
D = market discount rate, and
n = number of periods elapsed until the future amount occurs.

Substitute the values of E and D into equation 9-8 to find:

$$P = \$10{,}000\left[\frac{1 + 0.071}{1 + 0.092}\right]^{10} = \$8{,}235$$

9.5.3 Illustrating the Constant Dollar/Real Discount Rate Approach

Alternatively, with estimates of e and d, find the present value of replacement costs using equation 9-9. It is equation 9-8 with inflation removed.

$$P = C_0\left[\frac{1 + e}{1 + d}\right]^n \tag{9-9}$$

Substitute the values of e and d into equation 9-9 to find:

$$P = \$10{,}000\left[\frac{1 + 0.02}{1 + 0.04}\right]^{10} = \$8{,}235$$

Note that identical present value results are obtained using the current dollar/market discount rate approach or the constant dollar/real discount rate approach. If inflation is included in cash flows, it is removed through the discounting operation.

9.5.4 Pros and Cons of the Two Approaches

An advantage of the constant dollar/real discount rate approach is that it is streamlined. It does not require first putting in and then taking out purely inflationary or deflationary changes.

An additional advantage of the constant dollar/real discount rate approach for the federal government is that it does not require a forecast of general inflation. There is always the possibility that, in predicting inflation, the federal government will actually contribute to its occurrence; that is, there may be a "self-fulfilling prophecy" effect. This is sometimes given as a reason for performing government analyses using the constant dollar/real discount rate approach.

The current dollar/market discount rate approach also offers advantages, particularly to businesses. First, it facilitates treatment of income taxes, because taxes are computed in current dollars. Second, it allows one to express the discount rate as a market rate of interest, and costs and benefits in projected actual dollars of each year. This is an advantage because market rates of interest and current dollar costs and benefits are generally more familiar to clients and managers, and correspond with budget estimates which must be stated in current dollars.

REFERENCES

Heilbroner, Robert, and Lester Thurow. 1982. *Economics Explained*. Englewood Cliffs, New Jersey: Prentice-Hall, Inc.

Ruegg, Rosalie T. 1989. *Life-Cycle Costing for Energy Conservation in Buildings: Instructor's Guide*. NISTIR 89-4129. Gaithersburg, MD: National Institute of Standards and Technology.

Ruegg, Rosalie T. and Sieglinde K. Fuller. 1990. *Economic Analysis for MILCON Design: Concepts, Techniques, and Applications for the Analyst*. NISTIR 90-4255. Gaithersburg, MD: National Institute of Standards and Technology.

U.S. Office of the President. (1988). *Economic Report of the President*. Washington DC, U.S. Government Printing Office.

10

Setting the Study Period

The study period defines the time over which costs and benefits are considered in a decision. Setting it correctly is critical to obtaining a valid solution. In this section we state the criterion for setting it, identify factors which determine its length, give examples of setting study periods under a variety of circumstances, discuss the need for replacements and residuals, and, in summary, provide a set of guidelines.

10.1 CRITERION FOR SETTING THE STUDY PERIOD

The study period should be a length of time which will give a correct decision when evaluating projects. We call this out in a separate section for emphasis. Although the study period is often assumed to be synonymous with the investor's time horizon or with the project's life, the terms are not always equivalent, as we demonstrate in Section 10.3.

10.2 FACTORS AFFECTING THE LENGTH OF THE STUDY PERIOD

The study period can range from short to long, depending on the circumstances. Key to how long the period should be are these factors: (1) the investor's time horizon, i.e., the length of time of concern to the investor; (2) the life of the building, project, system, or equipment; (3) whether the decision is to accept or reject a project or to make a choice among project alternatives; and (4) whether the perspective is that of an individual investor or of society. The following examples illustrate how each of these factors influences the length of the study period.

10.3 EXAMPLES OF STUDY PERIODS

10.3.1 Study Period Equal to Investor's Holding Period

In the case where you intend to hold a building for a specific time before selling or demolishing it, the holding period may be your study period. Suppose, for example, that you plan to hold a building for 12 years and are concerned about owning and operating costs over that period, i.e., your time horizon is 12 years. An energy conservation device can be added to the building. The device normally lasts for 20 years. Over what period should you evaluate the device?

You should set the study period equal to your holding period of 12 years, despite the fact that the device lasts for 20 years. You can take into account the longer life of the device by estimating its residual value as of the 12th year, when the building is sold.

10.3.2 Study Period Equal to Project Life

To decide whether to accept or reject a project, you often can set the study period equal to the life of the project. Suppose, for example, that you manage post-office automation and are considering whether to invest in new automated mail-sorting equipment. Sorting is now done manually. The building life is very long, and so is your time horizon.

The proposed equipment has a useful life of 10 years. Future innovation is expected in the technology of mail-sorting equipment, but there is no specific forecast as to when the next generation of equipment will be available. What study period should you use to evaluate the equipment?

A study period of 10 years is recommended. There is nothing to be gained by assuming renewals of the investment over additional years, particularly when the equipment may be obsolete after 10 years.

10.3.3 Study Period Equal to a Common Multiple of the Lives of Alternatives

A hard and fast rule is to use the same study period for alternatives when comparing them with LCC, NB, or NS in present value dollars. The study period must be identical because using different study periods means taking benefits and costs over a different number of years. Even alternatives having exactly the same cash flows will have different measures if their cash flows are taken over different numbers of years.

One way to have an equal study period for alternatives is to set it equal to the minimum common multiple of their lives. But you can do this *only* if the common multiple is either equal to the investor's time horizon or divides evenly into it.

Suppose, for example, that you intend to hold a building for 20 years, and this is your time horizon for making decisions about the building. Suppose you need to choose between system A, having a life of 5 years, and a mutually exclusive alternative, system B, having a life of only 2 years. The systems are state of the art, and you see no reason to believe that their NB streams cannot be replicated by replacements over your holding period.

Using a present value model, you can set the study period equal to 10 years. Ten works because it is the minimum common multiple of their lives and divides evenly into 20, the investor's holding period. (A study period of 20 can also be used, but it entails additional replacements.)

10.3.4 Study Periods Equal to Uneven Lives of Alternatives

Analysts often assume that they can evaluate and compare alternatives over their individual uneven lives, provided that they use an annual value discounting model. But note that this is correct *only* if the investor's time horizon is very long or is a common multiple of lives of the alternatives, or if the residual value fully compensates for shortened life. Put another way, a condition is that the economic efficiency of each alternative would be the same if the alternatives were evaluated for an identical period equal to the investor's time horizon.

The example in Section 10.3.3 meets the condition for using an annual value model and using the individual, uneven lives of alternatives as their study periods. Set the study period equal to 5 years to evaluate system A and 2 years to evaluate system B, and calculate their annual values. This approach avoids the need to calculate system renewals. Note, however, that it is correct only because the holding period is a common multiple of the lives of the alternatives and their cash flows can be replicated identically over that period.

10.3.5 Study Period Equal to the Investor's Time Horizon in Evaluating Alternatives

If cash flows of alternatives cannot be replicated identically over the investor's time horizon, it is necessary to set the study period equal to the time horizon. Suppose, for example, that the problem is to choose between floor covering A which last for 5 years, and floor covering B, which lasts for 7 years. The building is to be demolished in 11 years, and no salvage value is expected. Assume that the initial choice of A or B will determine the choice over the remainder of the building's life. For example, initially choosing wood flooring instead of vinyl tile over concrete will limit the future choice to wood flooring. What study period should be used?

The answer is 11 years for both A and B, because that is the time horizon. This is true whether a present value or an annual value discounting model is used. Two replacements of alternative A and one replacement of alternative B must be modeled.

10.3.6 Study Period Equal to the Longest Life of Alternatives

What if you have a limited budget and want to choose between two projects with unequal lives, both within the investor's time horizon? Suppose that these are two limited partnerships in real estate. They are one-time, mutually exclusive investments that cannot be replicated. They are mutually exclusive because you only have enough funds to invest in one of them. They cannot be replicated because the partnerships are unique associations which are dissolved at the end of their lives.

Partnership A pays out over 5 years; partnership B, over 7 years. Suppose that you intend to retire in 15 years and are interested in investments with lives of up to 15 years, i.e., 15 years is your time horizon. You estimate the rate at which recovered capital can be reinvested as equal to your discount rate, since that is a measure of your opportunity cost. What study period should you use to compare the partnerships?

We suggest using a 7-year study period. For the investment with the 5-year life, assume that net receipts are reinvested for years 6 and 7. (Since the earnings rate on the reinvested net receipts normally equals the discount rate, the net present value of the cash flows for years 6 and 7 will normally equal zero.) Note that the yield obtained by the recommended approach will differ from that obtained by taking each partnership over its respective life.

10.3.7 Study Period Equal to Building Life

For certain public policy decisions, a strong case based on society's best interest can be made for setting a study period longer than the one an individual investor might choose. By "society" we mean the collective interest of all who own and live in a house over its life-cycle from inception to demolition.[1] A case

[1]In this discussion, we do not address underinvestments which may arise from market externalities such as pollution effects.

may be made, for example, for a code-setting body to set a long study period in the economic analysis used to develop requirements for energy conservation in housing codes.

The argument for taking a long view is as follows: The developer's time horizon is usually extemely short, often less than 1 year. The homeowner's time horizon in the United States averages only about 7 years because people on the average hold their houses for only about 7 years before selling them. Thus it will not be in the developer's or homeowner's interests to take a long view in their decisions regarding energy conservation unless they can capture the long-run benefits from the conservation investment when they sell the property. But both initial demand for conservation features and resale demand may be weak because homebuyers lack information, and lenders are unwillingness to take into account lower operating costs when computing loan eligibility.

Once a lower than socially optimal level of energy conservation has been provided, it is not easy to rectify the problem. Steps that can be taken easily during construction become more intrusive and expensive after construction. As a result of the initial underinvestment, a series of individuals own a house, each paying more for energy costs than they should.

One way to overcome the problem is through building codes which set requirements for energy conservation. If the government chooses to take this route, what study period should be used for the economic evaluation underlying the establishment of building codes? We think that the study period should be set to maximize social NB, and we think that this period is long, given that most houses remain in the housing stock for many years.[2]

10.4 REPLACEMENTS AND RESIDUALS

When the study period is longer or shorter than the lives of alternatives, take into account replacements and possible residual values to reconcile the lives with the study period. The shorter the study period, the more important it is to account for residual values at the time of project termination. (See Section 12.4 for a discussion of estimating procedures.)

10.5 GUIDELINES FOR SETTING THE STUDY PERIOD

In summary, we recommend these guidelines for setting the study period. Each guideline is illustrated in Section 10.3.

---

[2]This argument is based in part on Ruegg (1989).

- Set the study period equal to the investor's holding period or another identifiable time horizon when it is shorter than the lives of assets being evaluated or when evaluating alternatives which cannot be replicated identically over the period. (See Sections 10.3.1 and 10.3.5)
- Set the study period equal to the project life for accept/reject decisions when the time horizon is very long. (See Section 10.3.2)
- Set the study period for alternatives equal to the minimum common multiple of their lives, provided that the common multiple is either equal in length to or evenly divides into the investor's holding period or other identifiable time horizon. (See Section 10.3.3.)
- Set the study period for evaluating each alternative equal to the life of that alternative, using an annual cost or annual value model, but only under the condition that the economic efficiency of each alternative will be the same as it would be if all the alternatives were evaluated for a common period equal to the investor's time horizon. (See Section 10.3.4.)
- Set the study period equal to the life of the alternative with the longest life (not to exceed the investor's time horizon) when comparing one-time investments that cannot be replicated. (See Section 10.3.6.)
- Set the study period long (such as equal to the life of the building) for public policy decisions affecting society's best interest. (See Section 10.3.7.)

REFERENCE

Ruegg, Rosalie T. 1989. Benefits, costs, and time perspectives. Paper presented at the policy forum, Making Housing More Affordable through Energy Efficiency, 4-6 October 1989, in Washington, DC.

11

Selecting a Discount Rate or MARR

The discount rate, or MARR,[1] imposes a condition of minimum profitability which a project or project increment must meet to qualify for acceptance. Because it affects whether a project will be accepted or rejected and how much will be spent on it, the value of the discount rate is a key ingredient in an economic evaluation. If it is set too high, some projects which are economic will be rejected; if too low, some projects which are uneconomic will be accepted.

Chapter 8 explains, in the context of the time value of money, what the discount rate is and why it is needed to convert costs and benefits to time-equivalent values. This section provides guidance in determining the appropriate value of the discount rate for various applications.

11.1 THEORETICAL VALUE OF THE DISCOUNT RATE

The discount rate should reflect the rate of return available on the next best investment opportunity of similar risk to the project in question. The numerical value of the discount rate should reflect the "opportunity cost" that investors experience when they forego the return on the next best investment to invest in a given activity or project.

If your best alternative investment opportunity yields 10%, for example, you should require a project of similar risk to yield more than 10% to prefer it to the best alternative. Setting the discount rate at 10% requires, in effect, that at

[1]Most of the economic evaluation methods treated incorporate the minimum acceptable rate of return in the form of a discount rate. The internal rate of return and the overall rate of return methods instead compare the solved-for rate of return against the investor's MARR. The discount rate and the MARR are equivalent in theory and value; the terminology differs depending on how the rate is used. Rather than repeat both terms each time, we use "discount rate" in this chapter.

least a 10% return be realized before a project appears desirable on economic grounds.

11.2 AFTER-TAX DISCOUNT RATES

In setting the discount rate, you should be aware of the difference between after-tax and before-tax rates. If a project evaluation is performed on a before-tax basis, with no adjustment of cash flows for taxes, the discount rate should be on a before-tax basis. If the valuation is on an after-tax basis—which is the recommended approach when taxes apply—the discount rate or MARR should also be on an after-tax basis. Analysis of business decisions requires an after-tax discount rate.

There are few, if any, tax effects in evaluating cost-saving investments for the homeowner. Yet the use of an after-tax discount rate is also usually appropriate for the homeowner because savings in housing costs for the homeowner are like tax-free income in that no taxes are owed on them. In effect, these savings are after-tax income.

The after-tax discount rate is calculated from a before-tax rate using equation 11-1.

$$D_{at} = (D_{bt})(1 - T) \qquad (11\text{-}1)$$

where

D_{at} = after-tax discount rate,
D_{bt} = before-tax discount rate, and
T = combined federal and state marginal income tax rate (see Section 13.2.1).

11.3 REAL AND MARKET DISCOUNT RATES

In setting the discount rate, you should also keep in mind the distinction between "real" discount rates (those which exclude purely inflationary or deflationary changes in the general price level) and "market" discount rates (those which include changes in the general price level).

You can remove the inflation component from a market discount rate and convert it to a real discount rate by using equation 11-2.

$$d = \left[\frac{1 + D}{1 + I}\right] - 1 \qquad (11\text{-}2)$$

where

d = real discount rate which does not include the rate of general price inflation,
D = market discount rate which does include the rate of general price inflation, and
I = projected rate of general price inflation.

Conversely, you can use equation 11-3 (which is identical to equation 9-7) to derive a market discount rate from a real discount rate by including projected inflation.[2]

$$D = [(1 + d)(1 + I)] - 1 \tag{11-3}$$

The guideline for choosing between real and market discount rates is to choose a real rate if project cash flows are stated in constant dollars and a market discount rate if project cash flows are stated in current dollars. (See Chapter 9 for an explanation of constant and current dollars.)

11.4 DISCOUNT RATES AND RISK

The higher an investment's level of risk, the greater its expected rate of return must be in order to appeal to investors. This concept is known as the "risk-return relationship." This is the rationale for using the risk-adjusted discount rate technique that raises discount rates to evaluate higher-risk projects. (See the treatment of risk-adjusted discount rates in Section 16.1.)

[2]The relationship between real and market discount rates implied by equation 11-2 is consistent with the dominant view that market interest rates incorporate the anticipated rate of general price inflation; that is, market rates contain an "inflation premium." Thus, a 3% increase (or decrease) in the anticipated rate of price inflation is assumed to result in a general increase (or decrease) in market interest rates of approximately equal amount. The inflation-premium theory, credited primarily to Fisher (1954), emphasizes that the real interest rate is directly related to the productivity of physical capital and tends to be relatively stable over time. Hence, an investor who requires a real annual rate of return of 3% will require a market rate of about 5% if inflation is 2% annually and about 12% if inflation is 9% per year.

An opposing theory to Fisher's inflation-premium theory, implied by Keynes (1936), is that market interest rates are determined by the supply of and demand for money without reference to the inflation rate. A rise in the inflation rate results in a drop in the real rate of return on bonds rather than a rise in the market interest rate—at least in the short run—and a rise in the price of inflation-hedged assets. Empirical research on inflation and interest rates in the post-World War II period, as summarized by LeRoy (1973), shows mixed results and lends support to each theory in different periods.

An alternative approach is to omit risk from the discount rate altogether and use a risk-free discount rate. In this approach, a separate risk adjustment is made to cash flows rather than to the discount rate. (See the treatment of the certainty equivalent technique in Section 16.2.)

When a firm uses its average weighted cost of capital as the discount rate, the discount rate reflects the average riskiness of all of the firm's investments. In this case, the discount rate does not impose a high enough required rate of return on projects with higher-than-average risk. And it imposes too high a required rate on projects with lower-than-average risk. (See Section 11.7.) The Capital Asset Pricing Model has the advantage of deriving the rate of return appropriate for the given level of risk, and, therefore, has promise as a method of establishing discount rates with the correct risk adjustment. (See Sections 11.7 and 17.7.)

11.5 HISTORICAL RATES OF RETURN

We illustrate historical rates of return because past rates of return are of interest in projecting future discount rates. Actual rates of return have varied over time.

Holland and Myers (1977) estimated average market and real rates of return over all grades of nonfinancial common stocks and long-term corporate bonds from profit rate data published by Ibbotson and Sinquefield (1976). Over the period from 1946 to 1975, the average market rate of return was estimated at 9.85%. Over the same period, the average real rate (computed by subtracting percentage changes in the CPI from the market rates) was estimated at 6.00%.[3] The averages mask large year-to-year swings.

Figure 11-1 shows yields on U.S. Treasury notes and bonds with maturities of 5 years, 10 years, and greater than 10 years (composite) from 1970 to 1988. For these maturities, rates ranged from a low of 5.6% on long-term bonds in 1972 to a high of 14.2% on 5-year notes in 1981. These are market rates; hence, they reflect inflationary expectations.

Insight into anticipated real rates is provided by Drexel Burnham Lambert (1988). The firm determined 10-year inflation expectations by survey. It estimated expected real pretax yields on 10-year Treasury bonds by subtracting the expected inflation rate from the 10-year Treasury bond yield coincident with the time of the survey. From the late 1970s to the late 1980s, expected real pretax yields ranged from a low of 1.6% in 1980 to a high of 8.3% in 1981. This sharp swing came when market rates rose as the Federal Reserve Board

---

[3]Ruegg, *et al.* (1982, p. 88) summarize the Holland-Myers results and compute averages over selected periods.

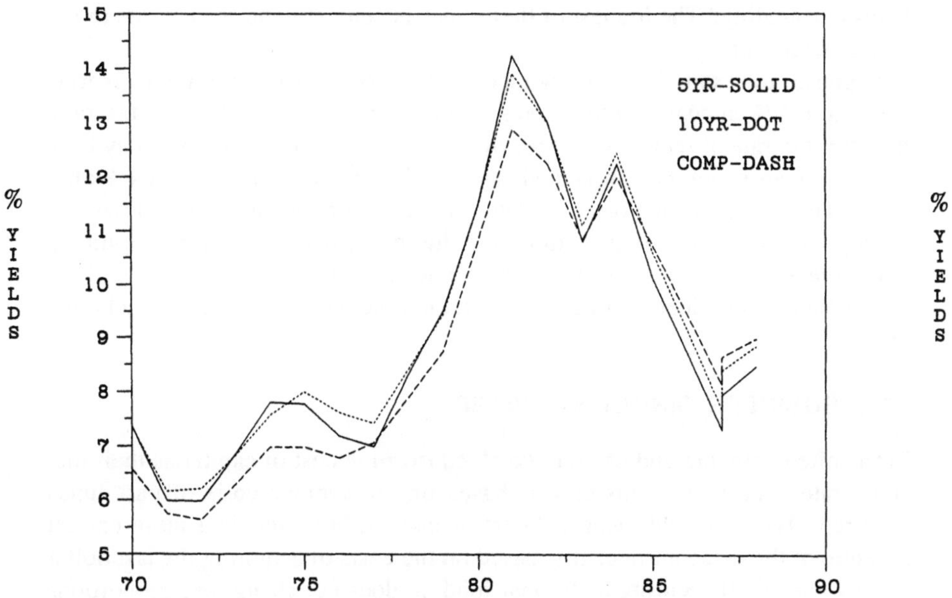

Figure 11-1. Percentage yields on Treasury securities maturing in 5 years, 10 years, and over 10 years, 1970–1988. (Source: Ruegg, 1990)

tightened credit and a recession at the same time moderated inflationary expectations. From 1986 through 1988, expected real pretax yields on 10-year Treasury bonds ranged between about 2% and 4%.

Anticipated real rates are more stable than actual real rates. Zero or negative real rates may occur when actual inflation exceeds anticipated inflation. High real rates may occur when actual inflation is lower than anticipated.[4]

11.6 HOMEOWNER DISCOUNT RATES[5]

To specify discount rates, homeowners should ask: (1) What is the highest rate of return, after taxes, they can earn on investments of comparable risk to the building-related investment? (2) If the project is to be financed, what is the after-tax loan interest rate? and (3) What is the after-tax rate they are paying on

[4]LeRoy (1973) discusses the relationship between real rates, inflation, and market rates both theoretically and empirically.

[5]The approach to selecting a homeowner discount rate that is described here is developed in detail in Ruegg and Petersen (1987).

loans outstanding? The highest of these rates is generally the correct rate to use as a discount rate.[6]

Suppose, for example, that a homeowner can put money in a money market account at 9% annually before taxes. With a 28% marginal income tax rate, the after-tax rate of return is 6.48%. Suppose that another option is to pay back part of a loan which has an annual rate of 12% before taxes and 8.64% after taxes. The recommended discount rate for evaluating the purchase of low-risk, energy-conserving building features for this homeowner is 8.64%, assuming that there are no alternatives with higher rates of return.

Appendix B provides a worksheet for estimating the discount rate for a homeowner.

11.7 BUSINESS DISCOUNT RATES

Firms often compute and use the weighted *average* cost of capital as their discount rate. The rate in this case is based on the average costs of past funds acquired. The weighted *marginal* cost of capital, however, is a more correct measure of the discount rate. It is based on the costs of acquiring the last dollar of new capital. If securing additional funding does not change the proportions of debt and equity (i.e., their weights) or the costs of each, the weighted marginal cost of capital will be the same as the weighted average cost of capital. If the proportion of debt increases, this may cause the market price of the company's stock to drop, thereby raising the cost of equity funds.

The weighted cost of capital is computed by finding the after-tax cost of each source of funding to the firm—equity and debt issue—and applying weights based on the proportion of funds obtained from each source. The calculation method[7] is illustrated in abbreviated form below and is accommodated in detail by Appendix B, series A, worksheet 2, part II.

Source of Funds	Weight[8]	x	After-Tax Rate[9]	=	Cost of Capital
Equity	30%		10%		3.0%
Debt	70%		7%		4.9%
			Weighted Cost of Capital		7.9%

[6]Faced with various interest rates, individuals often make apparently inconsistent decisions. For example, a person may put money in a savings account at 5%, finance a car at 10%, carry credit card balances at 21%, and make extra payments to reduce a mortgage loan at 8%.

[7]Ruegg and Petersen (1987 pp. 37–38).

[8]Weight is found by computing funds from each source as a percent of total funds.

[9]The cost of acquiring equity funds by issuing common stock is computed by dividing the dividend by the current market price of the stock adjusted for flotation costs and adding the annual growth

The weighted average cost of capital reflects the average riskiness of the firm's past activities; the weighted marginal cost of capital reflects the average riskiness of incremental activities. Neither reflects the risk associated with specific investment projects. Neither is likely to equal exactly the theoretically correct discount rate. But the weighted cost of capital can serve as a proxy for the correct rate, with the weighted *marginal* cost of capital being closer to the correct rate.

The rationale for using the weighted cost of capital as the discount rate is as follows: The firm must obtain an overall yield on its investments at least sufficient to cover its cost of capital in order to prevent the market price of its outstanding securities from falling.

The following example illustrates how a firm can compute the weighted marginal cost of capital to use as a discount rate. It shows (in Table 11-1) how to use worksheet 2, part II, from appendix B to calculate the rate.

Suppose that a manufacturing firm needs to raise $10 million for a new factory to expand its product line. It plans to finance the expansion by obtaining 50% of the funding from an issue of new common stock and 50% from a bond issue. The firm's marginal tax rate is 34%.

The new stock can be sold at the current market price of the company's outstanding stock of $50 per share. Flotation costs are expected to be 2% of the market price per share. The dividend is approximately $6.00 annually. The annual growth rate of the dividend is about 3%. The cost of new common stock equity, therefore, is 15.2% annually (calculated by using the worksheet, line B(4)).

The bonds will be sold in $1,000 increments. Flotation costs are expected to be 1% of the face amount. Yearly interest of $100 per face value of $1,000 will be paid. The after-tax cost of new long-term debt, therefore, is 6.7% annually (calculated using the worksheet, line B(2)).

The weighted marginal cost of capital is 10.9% (Worksheet line A (g)). This is the discount rate the firm will use to evaluate the profitability of the new plant.

A criticism sometimes made of using the weighted cost of capital as a discount rate is that it does not provide for accepting investments having a rate of return lower than the weighted cost of capital, even though such investments may be desirable because their level of risk is lower than the average risk of the firm. Many companies, for example, hold U.S. Treasury bonds having yields below the average cost of capital.

---

rate of the dividend to the results. The cost of acquiring funds by borrowing is computed by dividing the after-tax yearly interest owed by the issue price of the debt instrument adjusted for flotation costs. See series A, worksheet 2, part II, of appendix B for step-by-step calculation procedures.

Table 11-1. Computing weighted marginal cost of capital with worksheet

Worksheet 2.

Part II. ***Firm's Discount Rate***

A traditional approach to determining a proxy for the firm's appropriate discount rate is to compute it as a weighted average of the marginal costs associated with each source of funds to the firm.

A Complete the table below for the Applicable Sources of Funds:

Components of Capital for Investment (1)	Market Values ($ Amounts to be Raised) (2)	Market Weights (Market Values ÷ Total New Capital) (3)	After-Tax Cost of Capital (As Computed in Sec. B ~~next page~~) (4)	Weighted After-Tax Cost of Capital (5)=(3)×(4)
(a) New Current Liabilities	____	____ (a)(2)÷(f)(2)	____ B(1)	____
(b) New Long-Term Debt	$5.0 mill	0.5 (b)(2)÷(f)(2)	0.067 B(2)	0.033
(c) New Preferred Stock	____	____ (c)(2)÷(f)(2)	____ B(3)	____
(d) New Common Stock	$5.0 mill	0.5 (d)(2)÷(f)(2)	0.152 B(4)	0.076
(e) Other Sources	____	____ (e)(2)÷(f)(2)	____ B(5)	____
(f) Total New Capital	$10.0 mill			
(g) Marginal Cost of Capital, Sum col. (5) (This is a proxy for your after-tax, nominal discount rate.				0.109

B Compute After-Tax Cost of Capital and enter results in Section A, Column (4), of preceding page[a]:

(1) Cost of New Current Liabilities[b] (Enter in A(a)(4)) = 0.____ (interest rate to be paid) × (1−0.____) (composite marginal income tax rate[c]) = 0.____

(2) Cost of New Long-Term Debt (Enter in A(b)(4)) = [$100.00 (yearly interest to be paid) × (1−0.34) (composite marginal income tax rate[c])] ÷ [$1000.00 (sale price of debt) × (1−0.01) (flotation cost as a fraction of sales price)] = 0.067

(3) Cost of New Preferred Stock Equity (Enter in A(c)(4)) = [$____ (stated dividend on new issue)] ÷ [$____ (sale price of new preferred issue) × (1−0.____) (flotation cost as a fraction of sales price)] = 0.____

(4) Cost of New Common Stock Equity[d] (Enter in A(d)(4)) = [$6.00 (expected dividend in the next period) ÷ [$50.00 (current market price of stock) × (1−0.02) (flotation cost as a fraction of market price)]] + 0.03 (annual growth rate of dividend) = 0.152

[a] Depreciation is often omitted from the cost of capital calculation on grounds that the calculated cost of capital *is* the cost of depreciation; hence separate treatment adds nothing.

[b] Assumes no flotation or service charge.

[c] Computed by multiplying the State tax rate times the remainder of one minus the Federal tax rate, and adding the product to the Federal tax rate, as shown on Worksheet 2, Part I "Homeowner's Discount Rate," B(3).

[d] Note that the value of common stock is assumed to reflect retained earnings and a separate calculation for retained earnings is not made. As an alternative to the estimation approach shown here, the cost of common stock equity capital is sometimes estimated according to the Capital Asset Pricing Model. (See John J. Clark, et al., ***Capital Budgeting***, 2nd ed., pp. 267–273.)

Adjusting the weighted cost of capital to reflect the risk of various projects is one approach used to address the problem. Another approach is to set the discount rate using the capital asset pricing model (CAPM) approach.[10] Higher return is associated only with higher nondiversifiable (systematic) risk in an efficient market. The CAPM indicates what the expected return on a given investment should be, based on its risk, in order to be acceptable. But, as Section 17.7 explains, there is little experience with the CAPM applied to building-related investments.

11.8 DISCOUNT RATES FOR PUBLIC DECISIONS

Discount rates influence the allocation of public funds and are therefore important tools of public policy. In the United States, specification of discount rates for evaluating public projects dates from the 1930s, when federal agencies were directed to discount the benefits and costs of water resources projects.

It is widely agreed by economists that the appropriate discount rate for evaluating public investments is a rate set to maximize social NB. Thus the rate is often called the "social discount rate." This view, and term, gained favor in the 1960s as an aspect of welfare economics. What should be the appropriate numerical value of the discount rate, or rates, for evaluating public investments is, however, a topic of controversy and ongoing debate. It is a complex question with many unresolved issues.

Although a full treatment of the topic is beyond the scope of this book, we discuss alternative conceptual approaches to setting public discount rates and identify some of the key issues in Sections 11.8.1 through 11.8.4. Then we describe actual discount rates currently used to evaluate capital investments by the federal government.

11.8.1 Setting the Social Discount Rate Equal to the Opportunity Cost of Capital in the Private Sector

Mikesell (1977) presents the position held by many economists that the appropriate social rate of discount is the opportunity cost of capital in the private sector. He argues for determining social benefits and cost streams of public

---

[10]Ruegg and Petersen (1987, p. 38). See Fuller (1981) for a comprehensive explanation of the CAPM, and section 17.7 for a brief treatment.

projects directly and he argues against "manipulating" their present values with discount rates chosen to achieve particular social welfare goals.

We agree in theory with this position. The logic is that the social discount rate should reflect the fact that public funds are obtained from the private sector by taxation and borrowing. Resources should not be transferred from private investment to public investment with a lower rate of return.

Economists who share the position that the social discount rate should be based on the private opportunity cost of capital nevertheless differ as to the appropriate measure of private opportunity cost. One view is that public projects should be discounted at a rate corresponding to the average cost of capital in the private sector for projects of similar risk. A second view is that the social discount rate should be based on the empirically measured average rate of return in the private sector for projects of similar risk. A third view is that the social discount rate should be based on the average cost of government borrowing.

The third view is based on the argument that market yields on long-term government bonds are a measure of society's opportunity cost.[11] The private sector buys government bonds. Price and yield are responsive to market demand.[12] Because government bonds are considered risk free, their market yields are understandably lower than yields on other investments. The argument is that in a well-functioning market, the rate of interest on government bonds will equal the marginal rate of return on private capital after allowance for risk; therefore, the yield on government bonds represents the private opportunity cost of capital if there were no risk.

If the government bond rate is used as the measure of opportunity cost, there is the question of whether the rate should be adjusted to incorporate risk. One argument is that pooling effects result when the government undertakes a very large number of projects, and that these effects fully diversify risks and cause government projects to be appropriately considered riskless. Another argument is that public projects should be discounted at a rate corresponding to the rate employed in the private sector for projects of comparable risk.[13] Yet another

---

[11]Mikesell (1977) discusses some of the questions which have been raised about government bond rates as a measure of private opportunity cost, such as: Do collective time preferences determine market rates? Do market rates of interest reflect the collective willingness of society to trade present consumption for future consumption?

[12]It may also be argued that bond rates are affected by monetary policy more than private sector demand.

[13]The pooling argument is that when a large number of investments are made, each with probabilities of various rates of return, the actual rate of return from the group as a whole should approximate the weighted average of expected returns from the individual projects. See Arrow (1966) for support of the idea that government projects can be considered relatively riskless. See Hirshleifer

position accepts government projects as risky. But rather than use a risk-adjusted discount rate, the approach is to adjust expected costs and benefits for risk and use a risk-free discount rate for discounting, i.e., to use the certainty equivalent technique described in Section 16.2.

11.8.2 Setting the Social Discount Rate Lower than the Private Opportunity Rate in the Interests of Future Generations

Some economists, such as Pigou (1932), Eckstein (1957), and Marglin (1963), have given reasons for setting the social discount rate lower than the opportunity cost of capital in the private sector. One reason is to correct what may be described as the "myopic vision of the present generation," which results in inadequate consideration for the welfare of future generations. Taking this view, the social discount rate is used actively as an instrument of social welfare. It is deliberately set below the opportunity cost rate in the private sector, and is applied to social benefit and cost streams that typically take into account effects far into the future.

The concern is that setting a high government discount rate will (1) discourage government projects which generate benefits far in the future, and (2) favor government projects which have delayed costs over those which have short-run costs. Hence, arguments for low social discount rates often center on public projects which may affect the natural environment, the consumption of exhaustible resources, the preservation of the nation's heritage, and the rate of economic growth.[14]

The social welfare approach to setting a public discount rate has been criticized on a number of grounds, which include the following: (1) There is no agreement as to what the rate should be. (2) A low discount rate may transfer money from people who pay for a project today to people who enjoy the benefits of the project in the future. (3) A low social discount rate leads to selection of low-yielding public projects over those with higher returns. Fisher and Krutilla (1975) are among those who have pointed out that obtaining *high* rates of return from both public and private projects will contribute most to future welfare.

---

and Shapiro (1969) for support of the idea that public projects should be discounted at rates which incorporate risk.

[14]For further treatment, consult literature in the field of economics of natural resources. For a bibliography, see Fisher and Peterson (1976).

They questioned the notion that it is good policy to set an artificially low rate of social discount to conserve natural resources, given that the primary effect would be inefficient use of resources.

11.8.3 Setting the Social Discount Rate Equal to the Rate on Government Securities

Another position on setting the social discount rate is that a government body, like a firm in the private sector, should set its discount rate to reflect its average cost of capital.[15] An advantage of this approach is that the rate can be observed directly in the market and is a pragmatic solution to a difficult question. But the cost-of-government-borrowing philosophy can be criticized on the grounds that implicit in it is the view of government as a profit-maximizing organization. The concern is that using the cost of borrowing as a discount rate is more compatible with the government's seeking to maximize its own net benefits rather than society's as a whole.[16]

The position that a government body should maximize its own NB rather than society's lacks credibility among most economists. But, as Section 11.8.1 explains, there are other conceptual grounds for setting the social discount rate equal to the cost of government borrowing. The argument, which has its critics, is that in a well-functioning market, the rate of interest on government securities equals the marginal rate of return on private capital after allowance for risk and, therefore, represents the private opportunity cost of capital. (See Section 11.8.1.)

11.8.4 Setting the Social Discount Rate through the Political Process

According to Somers (1971), public policy makers, guided by voters, should make decisions about public projects. The appropriate discount rate will then be implicitly revealed through each decision. There may be a uniquely determined discount rate for each decision.

This position is not widely supported by economists. A weakness in the concept is that voters are unlikely to express their views successfully about nu-

---

[15]Brown and Yanuck (1980) are among those who argue that the discount rate for evaluating public programs should be based on the cost of government borrowing, i.e., on the average market rate on government securities. They do not identify the conceptual basis for their position.

[16]U.S. Congress, Joint Economic Committee (1968).

merous public projects by voting for a given candidate. Furthermore, public policy makers are not in a position to make the detailed investment decisions that are required on a project-by-project basis.

11.8.5 Social Discount Rates in Practice

In practice, there is not just one federal discount rate in the United States; there are at least four. They reflect a mix of the approaches described.

1. A 10% (real) discount rate is specified by the U.S. Office of Management and Budget (1972) for evaluating most federal capital investments not specially exempted. This rate is determined administratively and reflects in concept the opportunity cost in the private sector.
2. The discount rate governing water resources projects (such as dams) is calculated annually, on the basis of U.S. Treasury bond rates, by the U.S. Treasury according to a formula specified in CFR 704.39.
3. For lease-buy evaluations (which are to be performed after it is decided to obtain an asset), the discount rate is a market rate, calculated project by project according to U.S. Treasury bond rates of similar maturity. A formula is specified in OMB Circular A-104 (rev. June 1986).
4. For energy conservation and renewable energy projects in federal buildings, the discount rate is based on U.S. Treasury bond rates and is determined annually by the U.S. Department of Energy as specified in CFR 436.14(a), and described by Ruegg (1990).[17]

We use the 10% federal discount rate in Chapter 23 to evaluate a roadway and parking lot paving project for a VA hospital. In Chapter 20, we use the 7% federal discount rate that was effective in 1989 for federal energy conservation and renewable energy projects to evaluate a solar energy project on a military base.

REFERENCES

Arrow, K. J. 1966. Discounting and public investment criteria. *Water Resources Research*. (Edited by A. V. Kneese and S. C. Smith.) Baltimore: Johns Hopkins University Press.

[17]The Federal Energy Management Improvement Act of 1988 (P.L. 100-615) called for the Secretary of the Department of Energy to establish the discount rate for evaluating federal energy conservation and renewable energy projects. Prior to that, the discount rate was set at 7% (real) by the National Energy Security Act of 1980. And prior to that, the discount rate was set at 10% (real) in accordance with Office of Management and Budget Circular A-94.

Board of Governors of the Federal Reserve System. 1988. *Annual Statistical Digest 1987.* Washington, DC: Publication Services, Board of Governors.

Board of Governors of the Federal Reserve System. 1988. *Federal Reserve Statistical Release H.15.* Weekly Series for 1988. Washington, DC: Publications Services, Board of Governors.

Brown, Robert J. and Rudolph R. Yanuck. 1980. *Life Cycle Costing: A Practical Guide for Energy Managers.* Atlanta, Georgia: The Fairmont Press, Inc.

Drexel Burnham Lambert. 1988. Decision-makers real yield. Research report prepared by Hoey, Richard.

Eckstein, Otto. 1957. Investment criteria for economic development and the theory of intertemporal welfare economics. *Quarterly Journal of Economics* LXXI(1): 56–85.

Fisher, Irving. 1954. *The Theory of Interest.* New York: Kelly and Millman.

Fisher, A. C., and J. V. Krutilla. 1975. Resource conservation, environmental preservation, and the rate of discount. *Quarterly Journal of Economics* 89(3): 358–370.

Fisher, A. C., and F. M. Peterson. 1976. The environment in economics: a survey. *Journal of Economic Literature* 14(1): 1–33.

Fuller, Russell J. 1981. *Capital Asset Pricing Theories—Evolution and New Frontiers.* Monograph No. 12. Charlottesville, Virginia: The Financial Analysts Research Foundation.

Hirshleifer, J., and D. Shapiro. 1969. The treatment of risk and uncertainty. *Analysis and Evaluation of Public Expenditures: The PPB System.* Subcommittee on Economy in Government, Joint Economic Committee, 91st Congress, 1st session.

Holland, Daniel M. and Stewart C. Myers. 1977. Trends in Corporate Profitability and Capital Costs. WP 937-77. Cambridge, MA: Massachusetts Institute of Technology, Sloan School of Management.

Ibbotson, Roger G. and Rex A. Sinquefield. 1976. Stocks, bonds, bills, and inflation: Year-by-year historical returns (1926–74). *Journal of Business* 49: 11–47.

Keynes, J. M. 1936. *General Theory of Employment, Interest and Money.* New York: Harcourt, Brace and World.

Krutilla, J. V., and A. C. Fisher. 1975. *The Economics of Natural Environments.* Baltimore: Johns Hopkins University Press.

LeRoy, Stephen F. 1973. Interest rates and the inflation premium. *Monthly Review of the Federal Reserve Bank of Kansas City.* May: 11–18.

Marglin, Stephen A. 196[illegible]. The social rate of discount and the optimal rate of investment. *Quarterly Journal of Economics.* LXXVII(1): 95–111.

Mikesell, Raymond F. 1977. *The Rate of Discount for Evaluating Public Projects.* Washington, DC: American Enterprise Institute.

Pigou, A. C. 1932. *The Economics of Welfare.* London: Macmillan Publishing Co.

Ruegg, Rosalie T. 1990 (in press). *Life-Cycle Costing Manual for the Federal Energy Management Program* (Revised). NIST Handbook 135. Washington, DC: US Government Printing Office.

Ruegg, Rosalie T., G. Thomas Sav, Jeanne W. Powell, and E. Thomas. 1982. *Economics Evaluation of Solar Energy Systems in Commercial* Buildings. NBSIR 82-2540. Gaithersburg, MD: National Institute of Standards and Technology.

Ruegg, Rosalie T. and Stephen R. Petersen. 1987. *Comprehensive Guide to Least-Cost Energy Decisions.* SP 709. Gaithersburg, MD: National Institute of Standards and Technology.

Somers, H. M. 1971. On the demise of the social discount rate. *Journal of Finance.*

U.S. Code of Federal Regulations. Part 704, Section 39.

U.S. Code of Federal Regulations. 10 Part 436, Section 14.

U.S. Congress. *Federal Energy Management Improvement Act of 1986* (P.L. 100-615).

U.S. Congress, Joint Economic Committee, Subcommittee on Economy in Government. 1968. *Hearings on Economic Analysis of Public Investment Decisions: Interest Rate Policy and Discounting Analysis*. 84th Congress, 2nd session.

U.S. Congress. *Energy Security Act of 1980* (P.L. 96-294, 94 STAT 611, Sec. 405).

U.S. Office of Management and Budget. 1972. Circular A-94 (Revised). Discount rates to be used in evaluating time-distributed costs and benefits. Washington, DC: Office of Management and Budget, March 27.

U.S. Office of Management and Budget. 1986. Circular No. A-104 (Revised). Evaluating leases of capital assets. Washington, DC: Office of Management and Budget, June 1.

12

Estimating Costs and Benefits

In this chapter we provide guidelines for estimating typical costs and benefits resulting from building decisions. We focus on acquisition costs, capital replacements, residual value, energy costs, maintenance and repair costs, functional use costs, and downtime costs. We also discuss rental income and other benefits. Less emphasis is placed on benefits estimation, but many benefits are in the form of savings computed from cost reductions, and therefore are treated. We also show how to use several published databases. Table 12-1 lists representative types of some of the published data sources we discuss.

12.1 ESTIMATING FUTURE AMOUNTS BASED ON TODAY'S AMOUNTS

Often you may think that you have little or no basis for estimating future costs and benefits. But observable values in today's marketplace provide a useful point of departure for estimating future amounts. Start with today's values, and ask if there are any reasons to expect fundamental changes in demand and supply conditions over time in the goods and services in question.

If you have no sound basis for assuming otherwise, it is accepted practice to assume that a particular good or service will change in price over time at approximately the same rate as prices in general; that is, its total rate of change, E, is equal to the rate of general price inflation, I, and its differential rate of change, e, equals zero.[1] This assumption greatly facilitates cash-flow estimation: The constant dollar price of the item in future years will be identical to the initial price, and the current dollar price will be equal to the initial price escalated at the rate of general price inflation. (As Chapter 9 explains and demonstrates, estimates of future values may be in either constant or current dollars.)

Be careful to distinguish between costs and benefits which are expected to be

[1]See Chapter 9 for a discussion of E, I, and e.

Table 12-1. Sample published data sources for estimating costs and benefits

Type of Cost or Benefit	Sample Published Data Sources
Construction	*Design Cost and Data* *Means System Assembly Costs Manual* *Means Square Foot Costs Manual* *Means Building Construction Cost Data Book*
Maintenance and repair	CERL Maintenance and Repair Reports *BOMA Experience Exchange Report* *ULI Report on Shopping Centers*
Energy prices	Energy Prices and Discount Factors for Life-Cycle Cost Analysis
Rental income	*BOMA Experience Exchange Report* *ULI Report on Shopping Centers*

unchanged in amount over time in constant dollars and those unchanged in current dollars. Suppose, for example, that the future cost of a replacement is fixed by contract at today's price of $10,000. Also suppose that general price inflation is 5%. In constant dollars, the cost of the future replacement declines 5% per year to $5,987 in 10 years [i.e., ($10,000) $(1 + (-0.05))^{10}$ = $5,987]. But in current dollars, the cost remains unchanged at $10,000 because it is set by contract at that amount.

If you think that the price of a good or service will change faster or slower than the rate of general price inflation, you need forecasts of E or e in order to estimate future prices based on today's prices. These are not published for most industrial products and construction materials. One approach to developing projections is to examine the CPI or the PPI to see the past pattern of price changes in the item or in similar items. A simple trend projection may be feasible. Or large cyclical swings of changing duration and amplitude may suggest the need for a more sophisticated forecasting approach. You may also look to suppliers and utilities for guidance on future price changes. Contractual agreements may contain written provisions regarding escalation. (See Section 12.5 for published projected rates of change in energy prices.)

12.2 ESTIMATING ACQUISITION COSTS

Acquistion costs may be easier to estimate than other kinds of costs because they occur closer to the present time. For an "off-the-shelf" system, a reliable cost quote often may be obtained directly from a supplier at little expense or effort. When actual quotes are not available early in the decision process, you may want to use the following estimating techniques.

12.2.1 Statistical Cost Estimating Relationships

For preliminary project screening, you might base estimates of construction costs on statistical relationships derived from historical data. Kelly (1984) has demonstrated how to use databases derived from such relationships to develop the sketch design that is most economic. Examples of statistically derived cost-estimating relationships are costs as a function of capacity, costs as a function of physical dimension, and, more rarely, costs as a function of weight.

The cost-capacity estimating technique applied to construction specifies how construction costs change as production capacity changes. This type of cost-estimating relationship is widely used for estimating preliminary costs of industrial plants and facilities. For example, the costs of a water treatment plant may be estimated on the basis of gallons of water to be treated per day; costs of an aluminum plant, on tons per year; costs of a steam boiler plant, on pounds of steam per hour; and costs of an electric generating plant, on megawatts of capacity. Cost–capacity relationships are also sometimes used to estimate the costs of nonindustrial facilities. For example, the cost of a hospital might be a function of the number of hospital beds, and the costs of a school a function of the number of pupils.

Statistical relationships between cost and physical dimensions, such as cost per square foot or per square meter of floor area, often provide the basis for preliminary estimates of construction costs for commercial and residential buildings. Actual cost data for thousands of buildings are condensed to represent unit costs of buildings by type of design, size, and quality of the construction. Location multipliers adjust the resulting estimates for cost variation by location.

Costs based on physical dimensions are published in construction cost-estimating guides. Examples are *Means Square Foot Costs* published by the R.S. Means Company, Inc.; and the *Boeckh Building Valuation Manual*, published by the American Appraisal Company.

Statistical relationships between cost and weight are sometimes used for preliminary estimates of equipment costs. For example, mechanical/electrical equipment may have a similar cost per pound. Precision equipment sometimes has similar costs per unit of weight. And costs of functional equipment may be approximated according to weight.

12.2.2 Semi-detailed and Detailed Cost Estimates

For economic analysis studies beyond the screening state, a higher level of accuracy may be required. Semidetailed or conceptual estimates of construction costs can achieve this. Generally you can develop estimates by securing quotations on major items or components of construction, and for the remaining items, adding unit costs obtained from construction cost-estimating guides, such as the *Means Building Construction Cost Data Book*. Particularly helpful in this early stage of design are system cost manuals such as the R.S. Means Company's *Means System Assembly Costs*, and Allan Thompson Publishers' *Design Cost and Data*.[2]

Detailed estimates of construction costs are generally not necessary for economic analyses. First, they are costly to prepare. More importantly, they should be made *after* an economic analysis has been performed to decide if a project is economically worthwhile, not before! If you are using detailed construction cost data in design decisions, it is likely that you have missed the best opportunities to create cost-effective designs.

12.2.3 Standard Format for Cost Estimating

For large projects we encourage you to use a standard format for organizing construction cost data. It assists in the retrieval of information and promotes understanding by others who may wish to use or review the data.

In the United States, the Construction Specifications Institute (CSI) publishes a Manual of Practice which provides a masterlist of titles and numbers for organizing construction information. It is called "MASTERFORMAT." It is a functional-oriented cost system and provides for organizing construction cost estimates into 16 divisions including, for example, (Division 3) concrete, (Division 4) masonry, and (Division 9) finishes. MASTERFORMAT is used principally for design manuals and detailed construction estimating.

Better for the purpose of building design and cost comparison and analysis is a systems approach to formatting cost data. The U.S. General Services Administration's cost estimating system called "UNIFORMAT" is such a system. It organizes costs into 4 levels for each of 12 categories, which include, for example, (Category 03) superstructure, (Category 06) interior construction, and (Category 12) site work.

Systematic approaches to data collection similar to UNIFORMAT are also

[2]These manuals, as well as additional sources of data, are listed in Appendix F.

used in other contries such as Canada (Royal Architectural Institute of Canada), Britain (Royal Institute of Chartered Surveyors), Japan (The Society of Japanese Quantity Surveyors), Germany, and Belgium.

12.3 ESTIMATING CAPITAL REPLACEMENT COSTS

The number of capital replacements is a function of the estimated life of the system and the length of the study period. To estimate the cost of a future replacement, a reasonable starting point is what it costs to make the replacement today. Then use the estimating procedure described in Section 12.1.

12.4 ESTIMATING RESIDUAL VALUES

Here we use the term "residual value" as an umbrella term to include resale, reuse, salvage, scrap, and unused value remaining when the study period is shorter than the asset's life. You may need to estimate residual values because you plan to sell an asset, trade it in, convert it to another use, scrap it, or continue using it past the end of the study period. The figure you want will be net of selling costs, removal costs, conversion costs, or other costs of disposal or change in use.

Resale value is an estimate of the dollar price for which an asset will sell at a designated time. Methods of estimating resale value include the following:

1. Base the estimate on the price at which similar, comparably aged property is selling in commercial markets at the time of the study. If the analysis is in current dollars, adjust the estimate to reflect projected inflationary changes until the planned time of sale, using equation 9-1. If the analysis is in constant dollars, no adjustment should be necessary unless differential price change is expected. To incorporate differential change, use equation 9-4.
2. If the asset generates annual income, estimate market value, as of the time of resale, as the capitalized value of the future stream of income remaining at the time of resale. If the income stream is anticipated over a specific number of years, capitalized value can be calculated using equation 12-1.

$$CV = \sum_{t=n}^{EL} \frac{Y_t}{(1 + i)^t} \qquad (12\text{-}1)$$

where

CV = capitalized value of the asset at the time of resale,
Y_t = income in year t,
n = the time of resale,
EL = the end of the asset's life, and
i = the prevailing rate of return on comparable assets at the time of resale.[3]

If the income stream is relatively stable in amount and is anticipated to continue indefinitely, its capitalized value is calculated using equation 12-2.

$$CV = \frac{\overline{Y}}{i} \tag{12-2}$$

where

$\overline{Y}$ = average yearly income.

3. If you think that the depreciation schedule accurately reflects the decline in market value, calculated book value at the time of resale may be appropriate as an estimate of remaining market value. Since book value does not include inflation, adjust it for inflation if the analysis is in current dollars. If the analysis is in constant dollars, state book value dollars in the same base-time dollars used elsewhere in the analysis.

If an asset is to continue in use past the study period, it may be desirable to estimate and assign a "pseudo" residual value to the asset at the end of the study period. This pseudo residual value is sometimes estimated as a percent of the original cost of the asset. One way to estimate the percent is to prorate the initial value of the system or component over its estimated life, and to take the amount remaining at the end of the study period as the constant dollar residual value. Adjust for inflation if the analysis is in current dollars.

Residual values at the end of long study periods often have little weight in economic analyses for three reasons: (1) value declines due to deterioration or

[3]Note that i in this case is not the investor's discount rate. Because resale depends on how the market values the future stream of income, make i the prevailing market rate of return on comparable assets.

obsolescence, (2) disposal costs offset residual value, and (3) discounting diminishes residual value. Therefore, when study periods are long, it is often less worthwhile to improve estimates of residual values than to improve estimates of other costs and benefits.

After 10 years, for example, a building system with an original service life of 20 years would have a residual value with present value equal to only about one-fourth of the initial value based on the following assumptions: a straight-line decline in the constant dollar value of the system over the service life of 20 years, a discount rate of 7%, and no disposal costs. [(0.5) (0.51, the SPV discount factor for a 7% discount rate) = 25.5%.]

When calculating the residual value of entire buildings, it is a good idea to separate the building from the land. Land is generally not subject to deterioration and obsolescence and, due to demand and supply conditions, its value may increase while the building's value is decreasing.

12.5 ESTIMATING ENERGY AND UTILITY COSTS

Choices among building designs and systems affect energy consumption. To evaluate building projects, you need forecasts of prices and quantities of energy consumed. Forecasts of energy prices are available from several sources. The U.S. Department of Energy (DoE) provides detailed energy-price forecasts by region of the country, fuel type, and categories of buyers.[4] Projected energy prices are also published on a subscription basis by private organizations such as Data Resources, Inc.,[5] and sometimes by utilities.

DoE energy price forecasts support three approaches to estimating energy costs for use in economic evaluations: (1) yearly price indexes, (2) average annual escalation rates for 5-year intervals, and (3) uniform present value discount factors modified to include yearly differential escalation and based on federal discount rates.

12.5.1 Estimating Energy Costs with Yearly Price Indexes

Table 12-2 shows sample energy price indexes reported by Lippiatt and Ruegg (1988). The sample is for 1988 residential energy prices in the western part of

---

[4]The DoE forecasts are updated periodically. Tables of energy price data and discount factors for economic evaluations are prepared from DoE forecasts and reported by the National Institute of Standards and Technology in annual editions of NISTIR 85-3273 (Lippiatt and Ruegg).

[5]Data Resources publishes a quarterly review called *The Energy Review*.

Table 12-2. Projected average fuel price indexes tied to the value of the dollar in 1988, by end-use sector and major fuel (Census region 4: Montana, Idaho, Wyoming, Colorado, New Mexico, Arizona, Utah, Nevada, Washington, Oregon, California, Alaska, Hawaii)

	Projected Average Fuel Price Indexes (1988 = 1.00)												
Sector/Fuel	1988	1989	1990	1991	1992	1993	1994	1995	1996	1997	1998	1999	2000
Residential													
Electricity	1.00	1.00	1.00	1.00	0.99	0.99	1.00	1.01	1.02	1.03	1.05	1.07	1.08
Distillate fuel	1.00	1.06	1.07	1.10	1.12	1.15	1.19	1.24	1.30	1.35	1.41	1.46	1.50
Liquified petroleum gas	1.00	1.06	1.07	1.08	1.09	1.11	1.15	1.20	1.26	1.31	1.36	1.39	1.42
Natural gas	1.00	1.01	1.03	1.04	1.06	1.09	1.12	1.16	1.20	1.25	1.30	1.34	1.37
Commercial													
Electricity	1.00	1.00	1.00	1.00	0.99	0.99	1.00	1.01	1.02	1.03	1.05	1.07	1.07
Distillate fuel	1.00	1.08	1.10	1.13	1.16	1.21	1.27	1.33	1.42	1.49	1.57	1.64	1.70
Residual fuel	1.00	1.04	1.04	1.05	1.05	1.09	1.17	1.27	1.37	1.48	1.58	1.64	1.68
Natural gas	1.00	1.01	1.02	1.02	1.03	1.05	1.08	1.11	1.14	1.18	1.22	1.25	1.27
Steam coal	1.00	1.03	1.06	1.09	1.13	1.16	1.20	1.23	1.26	1.30	1.34	1.38	1.42
Industrial													
Electricity	1.00	1.00	0.99	0.99	0.99	0.99	1.01	1.01	1.02	1.04	1.07	1.10	1.11
Distillate fuel	1.00	1.08	1.10	1.14	1.17	1.22	1.28	1.35	1.43	1.51	1.60	1.67	1.72
Residual fuel	1.00	1.05	1.05	1.05	1.06	1.11	1.19	1.30	1.42	1.55	1.65	1.72	1.76
Natural gas	1.00	1.02	1.05	1.07	1.11	1.16	1.22	1.30	1.36	1.46	1.54	1.61	1.67
Steam coal	1.00	1.03	1.06	1.08	1.11	1.15	1.18	1.21	1.23	1.27	1.30	1.34	1.38
Transportation													
Motor gasoline	1.00	1.02	1.04	1.04	1.05	1.08	1.14	1.20	1.27	1.34	1.40	1.44	1.47

Source: Lippiatt and Ruegg (1988, p. 28). Updated annually to reflect revised projections by the Energy Information Administration of the U.S. Department of Energy.

Note: These indexes are used to compute projected energy prices or costs in base-year dollars. For example, to compute the projected price of distillate fuel oil at commercial rates in 1995, in 1988 dollars, multiply the 1988 price by 1.33. To compute 1995 costs in 1988 dollars, multiply the annual quantity of fuel oil consumed by the 1988 price, and the product by 1.33

Year-by-year *e* values in decimal form may be derived from the indexes by dividing the index for year t by the index for year t-l and subtracting 1.0 from the quotient.

the United States (i.e., in census region 4). It takes four steps to calculate present value energy costs:

1. Locate the appropriate set of indexes by finding in the table title the (a) region in which the building is located, (b) type of building, and (c) type of fuel.
2. Multiply the initial price of energy by the index for each year in the study period to estimate future energy prices in constant dollars.
3. Multiply the quantity of energy in each year by the future energy prices found in step 2 to estimate future energy costs.
4. Multiply the SPV factor (see Section 8.5.2) for each year by the future energy costs found in step 3 to estimate present value energy costs for each year, and sum them for total present value.

12.5.2 Estimating Energy Costs with Escalation Rates

Table 12-3 shows sample escalation rates for 5-year intervals. It takes three steps to calculate present value energy costs:

1. Locate the appropriate table of escalation rates (revised annually) for the region, building, and type of energy.
2. Calculate the estimated cost of energy for 1 year (A_0) by multiplying the initial energy price by the annual quantity of energy. (This approach requires that the annual quantity of energy be constant over the study period.)
3. Substitute the escalation rates in decimal form and A_0 into equation 8-12 and solve for present value energy costs.

12.5.3 Estimating Energy Costs with UPV** Factors

Table 12-4 shows sample UPV** factors which include differential escalation in energy prices and are based on a 7% real discount rate. These factors simplify the calculation of present value energy costs.[6] There are three calculation steps:

1. Locate the appropriate table of factors (revised annually) for the region, building, and type of energy.

---

[6]UPV** factors that incorporate the DoE energy price projections but allow users to specify their own discount rate and general price inflation can be calculated using the computer program, "Discount," which is described in Appendix D.

Table 12-3. Projected average fuel price escalation rates exclusive of general price inflation by end-use sector and major fuel (Census region 4: Montana, Idaho, Wyoming, Colorado, New Mexico, Arizona, Utah, Nevada, Washington, Oregon, California, Alaska, Hawaii)

Sector/Fuel	1988 to 1990	1990 to 1995	1995 to 2000	2000 to 2005	2005 to 2010	2010 to 2013
	Percentage change compounded annually					
Residential						
Electricity	−0.20	0.20	1.30	0.30	0.20	0.24
Distillate fuel	3.60	2.90	3.90	1.40	0.70	0.68
Liquefied petroleum gas	3.60	2.20	3.40	1.40	0.70	0.69
Natural gas	1.20	2.60	3.30	1.40	0.70	0.69
Commercial						
Electricity	−0.20	0.20	1.30	0.30	0.20	0.25
Distillate fuel	5.00	3.90	4.90	1.40	0.70	0.68
Residual fuel	2.00	4.00	5.80	1.40	0.70	0.66
Natural gas	0.80	1.80	2.80	1.40	0.70	0.71
Steam coal	3.20	3.00	2.90	1.40	1.40	1.40
Industrial						
Electricity	−0.30	0.30	1.80	0.30	0.20	0.24
Distillate fuel	5.20	4.10	5.00	1.40	0.70	0.70
Residual fuel	2.20	4.50	6.30	1.40	0.70	0.71
Natural gas	2.20	4.40	5.20	1.40	0.70	0.70
Steam coal	2.90	2.70	2.60	1.40	1.40	1.38

Source: Lippiatt and Ruegg (1988, p. 35). Updated annually to reflect revised projections by the Energy Information Administration of the U.S. Department of Energy.

Note: These rates are *e* values described in Chapter 9, expressed in percentage terms. Therefore, to get the *e* value needed in the single compound amount formula for escalation of energy prices, divide the percentage rate by 100; e.g., the *e* value for natural gas (residential sector) for the period 1995 to 2000 is 0.0330.

2. Calculate the estimated cost of energy for 1 year (A_0) by multiplying the initial price by the annual quantity of energy.
3. Multiply A_0 by the UPV** factor for the applicable study period. (Using the UPV** factors as published requires that the annual quantity of energy be constant over the study period.)

12.6 ESTIMATING MAINTENANCE AND REPAIR COSTS

Expenditures for maintenance and repair are more often subject to the personal judgment of owners and managers than are other building costs. Maintenance and repair costs vary considerably even for similar types of buildings of similar

Table 12-4. UPV discount factors adjusted for average fuel price escalation exclusive of general price inflation by end-use sector and major fuel[a] (Census region 4: Montana, Idaho, Wyoming, Colorado, New Mexico, Arizona, Utah, Nevada, Washington, Oregon, California, Alaska, Hawaii)**

	Discount Rate = 7%															
	Residential				Commercial					Industrial					Transportation	
N	Elec.	Dist.	LPG	Nat. Gas	Elec.	Dist.	Resid.	Nat. Gas	Coal	Elec.	Dist.	Resid.	Nat. Gas	Coal	Gaslne	N
1	0.94	0.99	0.99	0.95	0.94	1.01	0.97	0.95	0.96	0.94	1.01	0.98	0.95	0.96	0.96	1
2	1.81	1.93	1.93	1.84	1.81	1.97	1.88	1.84	1.89	1.81	1.98	1.89	1.87	1.89	1.86	2
3	2.62	2.82	2.81	2.69	2.62	2.89	2.74	2.67	2.79	2.62	2.90	2.75	2.74	2.77	2.71	3
4	3.38	3.67	3.65	3.50	3.38	3.78	3.54	3.46	3.65	3.37	3.79	3.56	3.59	3.62	3.52	4
5	4.09	4.49	4.44	4.27	4.09	4.64	4.32	4.20	4.48	4.08	4.66	4.35	4.41	4.44	4.29	5
6	4.76	5.29	5.21	5.02	4.76	5.48	5.10	4.92	5.28	4.75	5.51	5.14	5.22	5.23	5.04	6
7	5.39	6.06	5.95	5.74	5.38	6.31	5.89	5.61	6.05	5.38	6.35	5.95	6.03	5.99	5.79	7
8	5.98	6.82	6.69	6.44	5.97	7.14	6.69	6.27	6.78	5.98	7.19	6.78	6.83	6.70	6.53	8
9	6.54	7.55	7.40	7.12	6.53	7.95	7.49	6.92	7.49	6.54	8.01	7.62	7.62	7.39	7.26	9
10	7.07	8.27	8.09	7.78	7.07	8.75	8.29	7.54	8.17	7.09	8.82	8.46	8.40	8.05	7.98	10
11	7.58	8.97	8.75	8.41	7.57	9.53	9.07	8.13	8.83	7.61	9.61	9.28	9.17	8.69	8.66	11
12	8.06	9.63	9.38	9.02	8.05	10.29	9.82	8.70	9.46	8.10	10.38	10.06	9.91	9.30	9.31	12
13	8.50	10.26	9.98	9.60	8.50	11.00	10.52	9.23	10.05	8.56	11.11	10.80	10.61	9.88	9.93	13
14	8.92	10.86	10.54	10.14	8.92	11.67	11.19	9.74	10.62	8.99	11.79	11.50	11.28	10.43	10.51	14
15	9.32	11.42	11.07	10.66	9.31	12.31	11.82	10.22	11.16	9.40	12.44	12.17	11.91	10.95	11.06	15
16	9.69	11.96	11.58	11.15	9.68	12.92	12.42	10.68	11.67	9.78	13.06	12.80	12.51	11.44	11.59	16
17	10.03	12.47	12.06	11.61	10.02	13.49	12.99	11.11	12.15	10.13	13.64	13.40	13.08	11.91	12.08	17
18	10.36	12.95	12.51	12.05	10.34	14.03	13.52	11.51	12.61	10.47	14.19	13.96	13.61	12.35	12.55	18
19	10.66	13.39	12.94	12.46	10.65	14.54	14.03	11.89	13.04	10.78	14.71	14.49	14.11	12.77	12.99	19
20	10.94	13.82	13.34	12.84	10.93	15.02	14.50	12.25	13.45	11.07	15.19	14.98	14.58	13.17	13.40	20
21	11.21	14.21	13.71	13.21	11.20	15.47	14.94	12.59	13.84	11.34	15.65	15.45	15.02	13.55	13.79	21
22	11.46	14.59	14.06	13.55	11.44	15.89	15.36	12.91	14.21	11.60	16.08	15.89	15.44	13.90	14.16	22
23	11.69	14.94	14.40	13.87	11.68	16.29	15.75	13.21	14.56	11.84	16.48	16.30	15.83	14.24	14.50	23
24	11.91	15.27	14.71	14.17	11.90	16.66	16.12	13.49	14.89	12.06	16.86	16.69	16.20	14.56	14.82	24
25	12.12	15.58	15.00	14.46	12.10	17.01	16.47	13.75	15.20	12.28	17.22	17.06	16.55	14.86	15.13	25

[a]Elec. = electricity; Dist. = distillate; LPG = liquefied petroleum gas; Nat. Gas = natural gas; Resid. = residual; Coal = steam coal; and Gaslne = gasoline.

Source: Lippiatt and Ruegg (1988, p. 11). Updated annually to reflect revised projections by the Energy Information Administration of the U.S. Department of Energy.

age. Standards of maintenance differ among owners and users; occupants vary in their treatment of buildings; and the availability of maintenance funds varies from time to time. Despite these difficulties, estimating methods and databases can help.

12.6.1 Estimating Methods

12.6.1.1 Direct Quotes

Obtaining quotes directly from contractors and vendors is one method of estimating maintenance and repair costs. For example, if you have established the desired frequency of routine cleaning, request cost quotes from contractors supplying cleaning services. Base the frequency of cleaning on past practice, organizational standards, or prevalent practices for similar buildings. Some analysts base repair estimates on the manufacturer's recommended service and parts replacement schedules. They cost out these requirements for the initial year by obtaining direct quotes from suppliers and base future costs on initial costs. If costs are expected to rise as the system ages, this can be taken into account through the use of an escalation factor.

Another direct method for estimating the costs of maintaining and repairing equipment is to use the costs of service and maintenance contracts provided by manufacturers or vendors. These contracts are typically offered on an annual basis, and their price usually increases for each succeeding year of equipment life.

12.6.1.2 Statistical Cost-Estimating Relationships

Maintenance and repair costs may also be based on historical averages for similar buildings. Functional relationships may be derived by performing regression analyses with historical data. For example, the cost of routine maintenance for a school may be a function, among other things, of the age of the school and the age and socio-economic background of the pupils.

12.6.1.3 Time-Motion Study Data

An approach to estimating labor costs for maintenance and repair is to estimate the time required for given tasks and to apply labor rates to the estimated times. First, list the maintenance and repair tasks to be performed. Second, use time-motion study[7] data to estimate the time required to perform the tasks. Third,

[7]Time-motion studies divide complex tasks into elements and measure the time required on average for laborers to perform each element of the task.

identify the type of labor required, and apply labor rates and quantities to the task data to estimate costs.

12.6.2 Selected Maintenance and Repair Databases

12.6.2.1 Maintenance and Repair Database Developed by the Construction Engineering Research Laboratory[8]

The Construction Engineering Research Laboratory[9] has developed a database for estimating the long-run costs of maintaining building components. The database is based on time-motion studies. It supports LCC analysis, resource planning and budgeting, and is also used by maintenance and repair shops to estimate job times.

The database covers the four major building systems: (1) architectural, (2) electrical, (3) plumbing, and (4) HVAC. At the aggregate level, the database provides indexes for estimating the LCCs of keeping selected building components in good service condition, such that they continue to fulfill their original purpose. At the disaggregate level, the database provides raw data for about 3,000 typical tasks needed to maintain and repair building components.

Each of the four major building systems is divided into components. For example, architectural systems are divided into exterior and interior wall finishes, roofing, flooring, and windows and doors. The database has maintenance and repair cost indexes for alternative components. For roofing alternatives there are indexes for built-up roofing, slate, cement asbestos, tile, shingles, metal, fiberglass, and others. For flooring alternatives there are indexes for masonry and tile products such as ceramic tile, quarry tile, brick, and marble; wood products such as wood parquet and plywood; synthetic surfaces such as linoleum, vinyl tile, rubber tile, and cork tile; concrete surfaces, both finished and unfinished; terrazzo; and carpet. For exterior wall alternatives there are indexes for a variety of finishes such as adobe, clay brick, concrete brick, and structural clay tile.

---

[8]This section is based in part on material presented by Ruegg and Fuller (1990).

[9]The Construction Engineering Research Laboratory (CERL) is a research laboratory of the U.S. Army Corps of Engineers located in Champaign, Illinois. For additional information about the maintenance and repair database, contact the Facility Systems Division of CERL, P.O. Box 4005, Champaign, Illinois 61820-1305.

Table 12-5. Indexes for computing present value maintenance and repair costs

Component Description	Unit	Present Value Indexes for d = 10% and 25 Years		
		By Resources		
		Labor[a]	Material[b]	Equipment[c]
Architecture				
Roofing				
Roof covering				
Builtup roofing	SF	0.03990	0.37220	0.02000
Mod. bit./thermoplastic	SF	0.02440	0.33090	0.01180
Thermosetting	SF	0.01680	0.23950	0.00850
Slate	SF	0.01850	0.10440	0.00890
Cement asbestos	SF	0.01820	0.24340	0.00870
Tile	SF	0.01550	0.20990	0.00740
Roll roofing	SF	0.07140	0.42700	0.03640
Shingles	SF	0.02210	0.22150	0.01170
Metal	SF	0.01460	0.11060	0.00740
Fiberglass rigid stp. roof	SF	0.02190	1.15340	0.01080
Concrete, sealed panel roof	SF	0.04300	0.11750	0.02120
Concrete, sealed panel RF4	SF	0.03900	0.08410	0.02020
Concrete sealed poured	SF	0.09830	0.63020	0.04950
Fiberglass, rigid roof	SF	0.03800	1.15340	0.01930

[a]Indexes in this column give the labor hours per SF of roofing to maintain, repair, and replace it as needed over 25 years. Multiply the labor index by your labor wage rate for roofers and by the number of SF of roofing to estimate present value labor costs.
[b]Indexes in this column give material costs per SF of roofing in Washington, DC as of July 1985. Multiply the material index times an Area Cost Factor to adjust for your location. Then multiply by an inflation index to update costs to the current year. And, finally, multiply by the number of SF of roofing to estimate present value material costs.
[c]Indexes in this column give the equipment hours per SF of roofing. Multiply the equipment index by your hourly equipment rate and by the number of SF of roofing to estimate present value equipment costs.
Source: Ruegg and Fuller (1990). Only a portion of the table is shown here. It is a draft table provided originally by the U.S. Army Corps of Engineers, Construction Engineering Research Laboratory.

Table 12-5 gives a sample of indexes for estimating present value maintenance and repair costs of alternative roofing covers over 25 years using a discount rate of 10%. Look, for example, at the roofing alternative shingles. The labor index 0.02210 is used to compute your present value labor cost for maintaining 1 square foot (SF) of shingled roof over 25 years. If your labor rate is

$14.00/hour and your roofing area is 10,000 SF, present value labor costs for maintenance and repair over 25 years are estimated at $3,094 [i.e., (0.02210 hours/SF)($14.00/hour)(10,000 SF) = $3,094]. Present value material and equipment costs for maintenance and repair of a shingle roof are also computed from their respective indexes. Material costs are computed by multiplying the material index by a location factor, a time update multiplier, and the number of square feet. Equipment costs are computed by multiplying the equipment index by an equipment hourly rate and the number of square feet. The summation of present value labor, equipment, and materials costs is the estimated LCC of maintaining a shingled roof over 25 years, given that your discount rate is 10%.

The database has been tested by a number of military installations and contractors. It is accessible by cost reports and by computer software. It is the most comprehensive, consistent, and well-documented database we know for estimating maintenance and repair costs. Because the military has a diverse population of buildings, including offices, houses, storage facilities, heating plants, hangers, schools, hospitals, and gymnasiums, the list of building components is comprehensive. The heart of the labor portion of the database is time-motion data to which you apply your own wage rates. And, you apply adjustment factors to update materials costs and make them location specific. These features make the database suitable for evaluating a variety of buildings in both the public and private sectors.

12.6.2.2 BOMA Experience Exchange Report

The Building Owners and Managers Association International (BOMA)[10] publishes results of annual samples of building operating costs in *The Downtown and Suburban Office Building Experience Exchange Report* (*EER*). The report includes average owning and operating costs per square foot. Wages, supplies, contract services, and miscellaneous expenses are broken out. Costs are published by age of building, by geographic location, by downtown versus suburban location, by number of stories, by number of square feet in the building, and by city size.

Maintenance and repair costs reported by the BOMA include (1) routine cleaning costs; (2) costs of maintaining the electrical system throughout the building, including repairs and ordinary replacements; (3) costs of maintaining the heating system; and (4) costs of maintaining the air conditioning and ventilating system. Average costs are stated as cents per square foot.

[10]The BOMA is located at 1221 Massachusetts Avenue, N.W., Washington, DC 20005.

12.6.2.3 ULI Report on Shopping Centers

The Urban Land Institute (ULI)[11] conducts a study of income and expenses in shopping center operations every 3 years. The study provides data on several types of shopping centers, in 6 geographic areas, and for 4 age groupings. Costs are presented per square foot of gross leasable area. Maintenance and housekeeping costs are broken down by (1) cost of building maintenance; (2) cost of maintaining parking lot, mall, and other common areas; (3) cost of office area services; and (4) cost of central utility systems (including energy costs).

12.6.2.4 Other Maintenance and Repair Cost Data

Manuals can also be helpful for estimating repair and remodeling costs for commercial and residential buildings. The R.S. Means Company, for example, publishes such a manual. A "unit price section" lists data for estimating repair and remodeling costs for specific components, and a "systems section" lists data for estimating repair and remodeling costs for whole systems. Labor rates are averages from 30 major cities. You adjust the rates by cost indexes for your area.

Most maintenance and repair data are ad hoc in nature. Flanagan and Norman (1983), for example, analyzed the LCCs, including maintenance and repair costs, of a primary school in the United Kingdom. Even though such a case study is not identical to your own, it may provide the best data available given limited resources.

12.7 ESTIMATING FUNCTIONAL USE COSTS

Choices of building designs and building systems may affect functional use costs related to the performance of intended functions. Labor costs and productivity of workers within the building may be affected, as well as the quality of services.

The layout of a hospital floor, for example, may affect the response time of the nursing staff. If remodeling of hospital nursing stations reduces the response time, this might allow a reduction in staff while keeping the same performance. A corresponding reduction in labor costs might result. The reduction in functional use costs should be included in an economic evaluation of remodeling.

Ward (1987) provides a comprehensive survey of functional use costs measured by leading researchers in the field. He includes functional use costs among

[11] The ULI is located at 1010 Vermont Avenue, N.W., Washington, DC 20010.

other major building costs. He demonstrates the significance of functional use costs by computing a functional use ratio (FUR). This is the ratio of the sum of initial investment cost, operating and maintenance costs, and functional use costs to initial investment costs. Ward's (1987, pp. 116–118) survey shows that a conservative average FUR would be 20.

Investments which decrease operating and maintenance costs but impair the environmental quality of a building can raise its functional use costs by resulting in "sick building syndrome" (SBS) and "building related illness" (BRI). SBS and BRI hamper productivity, contribute to absenteeism, and increase claims for workers' compensation and medical insurance. Woods (1989) questions the wisdom of current design criteria that have the objective of satisfying at least 80% of occupants with regard to environmental conditions. In face of evidence that even small negative effects on occupant productivity can outweigh cost savings from lower operating and maintenance costs, he asserts that "healthy buildings" are likely to result in minimum life-cycle costs.

12.8 ESTIMATING DOWNTIME COSTS

Downtime costs occur when a building investment causes a slowdown or shutdown in operations. The loss in revenue or output, net of operating costs avoided, should be included in the economic analysis of the investment.

12.9 ESTIMATING RENTAL INCOME AND OTHER BENEFITS

Estimates of rental income are often based on market rents for buildings with comparable features. They are sometimes based on historical averages. The BOMA publishes average rental income per square foot of space in downtown and suburban office buildings in its annual *Experience Exchange Report*. Income is published separately for office areas, ground floor store areas, storage areas, and special areas such as dining or assembly rooms, theaters, clubs, and garages. The ULI publishes average rental income per square foot of space in shopping centers in *Dollars and Cents of Shopping Centers*.

Benefits from increased productivity, improved comfort, security, safety, and aesthetics tend to be difficult to measure directly. When increased productivity generates higher revenue, the dollar gain can be used as a measure of productivity changes. When improved comfort, security, safety, and aesthetics reduce labor turnover or occupancy vacancy rates, these benefits may also be reflected in increased revenue and reduced costs. Improved aesthetics may be valued by looking at comparative rents of buildings with and without similar qualities.

Vandell and Lane (1988) are among those attempting to evaluate empirically the contribution of architectural quality to the value of buildings. We need both improved measures of these hard-to-quantify benefits and costs and greater use of the measures in economic evaluations of buildings. This is a major area in building economics which requires more research.

REFERENCES

American Appraisal Company. *Boeckh Building Valuation Manual.* Milwaukee, Wisconsin: Boeckh Division, The American Appraisal Company.

Building Owners and Managers Association International (BOMA). *The Downtown and Suburban Office Building Experience Exchange Report (EER).* Washington, D.C.

Data Resources, Inc. *The Energy Review* (quarterly). Lexington, Massachusetts.

Design Cost and Data. Glendora, California: Allan Thompson.

Flanagan, Roger, and George Norman. 1983. *Life Cycle Costing for Construction.* Prepared for the Quantity Surveyors Division of the Royal Institution of Chartered Surveyors. London: Harris Print Ltd.

Kelly, John. 1984. A method for the formulation of a cost database for use during the evolution of the sketch design of a building. *Proceedings of the International Symposium on Building Economics.* Vol. 6. Ottawa, Canada: National Research Council of Canada.

Lippiatt, Barbara C., and Rosalie T. Ruegg. 1988. *Energy Prices and Discount Factors for Life-Cycle Cost Analysis.* NISTIR 85-3273-3. Gaithersburg, Maryland: National Institute of Standards and Technology.

R.S. Means Company, Inc. *Means Building Construction Cost Data Book.* Kingston, Massachusetts.

Ruegg, Rosalie T. and Sieglinde K. Fuller. 1990. *Economic Analysis for MILCON Design: Concepts, Techniques, and Applications for the Analyst.* NISTIR90-4256 Gaithersburg, MD: National Institute of Standards and Technology.

Urban Land Institute. *Dollars and Cents of Shopping Centers.* Washington, D.C.

U.S. Army Corps of Engineers, Construction Engineering Research Laboratory. 1989. *CERL Maintenance and Repair Reports* (Draft). Champaign, Illinois.

Vandell, Kerry D., and Jonathan S. Lane. 1988. The economics of architecture and urban design: some preliminary findings. Paper presented at the Allied Social Sciences Associations annual meetings, New York.

Ward, Robertson. 1987. Office building systems performance and functional use costs. *CIB Proceedings of the Fourth International Symposium on Building Economics,* Session A. Copenhagen: Danish Building Research Institute.

Woods, James E. (1989) Cost avoidance and productivity in owning and operating buildings. *Occupational Medicine: State of the Art Review* 4(4): 753–769.

13

Accounting for Taxes and Financing

Taxes can dramatically alter the profitability of a business investment in buildings.[1] In this chapter we describe taxes which affect building investments and show how to include them in economic analyses. We also explain when and when not to include financing in economic analyses.

Because tax laws are complex, changeable over time, and variable by state and country, we do not provide comprehensive treatment of specific tax laws. Rather, we provide general concepts and procedures to help you deal with any given set of tax laws, regulations, and rates.[2]

13.1 TAXES AFFECTING BUILDINGS

Buildings used in conjunction with profit-making businesses, such as commercial office buildings, factories, warehouses, and residential rental properties, are affected by a number of taxes. Table 13-1 provides a quick reference to typical taxes and tax allowances, their impacts, and the sections in the text where they are treated. Some taxes decrease the attractiveness of a building investment by reducing after-tax revenue or by increasing the owning and operating costs borne by the investor. We describe these taxes in Section 13.1.1. Some taxes improve an investment's attractiveness by decreasing the owning and operating costs borne by the investor. We describe these tax allowances in Section 13.1.2. Then we show how to calculate both taxes and tax allowances in Section 13.2.

[1]Taxes redistribute benefits and costs. When the government taxes revenue, it shares a portion of the benefits; when it allows businesses to deduct operating expenses from taxable income, it shares a portion of the costs.

[2]The illustrations are based, for the most part, on provisions of the 1986 United States Tax Reform Act as set forth by Commerce Clearing House Tax Law Editors (1986). The treatment is not intended to provide the in-depth knowledge of tax law needed to comply with Internal Revenue Service requirements. We advise you to obtain up-to-date tax information from your tax advisor before you evaluate investments with tax implications.

Table 13-1. Quick reference guide to taxes and tax allowances affecting building decisions

(1) Tax or Tax Allowance	(2) Impact	(3) Section
Taxes		
Income tax	Reduces revenue	13.1.1
Sales tax	Increases costs	13.1.1
Excise tax	Increases costs	13.1.1
Property (ad valorem) tax	Increases costs	13.2.7
Capital gains tax	Reduces revenue	13.2.5
Depreciation recapture tax	Reduces revenue	13.2.5
Tax allowances		
Deduction of operating expenses	Reduces taxable income	13.2.3
Deduction of depreciation	Reduces taxable income	13.2.4
Deduction of loan interest	Reduces taxable income	13.2.6
Tax credits	Reduces taxes directly	13.2.2

13.1.1 Taxes Which Decrease Profitability

We include in this category income taxes, sales taxes, excise taxes and duties, property taxes, capital gains taxes, and depreciation recapture taxes.

Rental income generated by buildings, as well as revenues from sales of products or services produced within buildings, are reduced by the amount of federal income tax owed. In addition, revenue may be subject to state income tax. Income taxes make building investments which generate revenue less attractive on an after-tax basis than they are on a before-tax basis.

Purchases of construction materials and, in some cases, construction labor may be subject to state sales taxes. Excise taxes or duties may be imposed on the manufacture, sale, or consumption of commodities. Sales taxes, excise taxes, and duties raise the acquisition costs of buildings.

Property, or ad valorem, taxes are levied locally and annually on the appraised value of property. They raise the costs of ownership.

When property is sold, any gain on the sale (i.e., excess of the selling price over the remaining undepreciated cost) is subject to taxation.[3] Part of the gain may be taxed at a capital gain rate; part or all may be taxed at ordinary income

[3]Exemptions from or deferral of capital gains taxes are provided to homeowners under certain conditions.

tax rates. Depreciation recapture provisions may require payment of taxes previously avoided by depreciating capital expense. If investment tax credits were taken at the time of the initial investment, there also may be provisions to recapture, at the time of sale, all or part of these tax credits.

13.1.2 Tax Allowances Which Increase Profitability

We include in this category deductions from taxable income and tax credits. Deductions from taxable income include current business operating expenses, depreciation, and interest payments.

Current business operating costs are deductible from taxable income in the year they are incurred. They may include maintenance costs; utility costs for energy, water, and sewage; insurance fees; real estate taxes; state and local income taxes; and current lease payments.

To be depreciable, property must (1) be used in business or held for production of income, (2) have a useful life longer than 1 year, and (3) be something that wears out, decays, is used up, becomes obsolete, or loses value from natural causes. Investment in a depreciable asset is treated as a prepaid expense which is deducted over time for income tax purposes. Depreciation is not itself an actual cash flow; it is an artificial allocation. But a cash flow does result from the dollar savings in income taxes generated by depreciation deductions.

Shortening the depreciation recovery period increases the present value of writeoffs because any given amount of writeoffs is discounted over a shorter period. Taking larger deductions earlier in the depreciation period by using an accelerated depreciation method also increases the present value of writeoffs. Straight-line depreciation, in contrast, takes deductions in equal amounts over the depreciation period, which means that the present value of writeoffs is less than for an accelerated method.

Businesses take depreciation allowances according to prescribed schedules which vary by class of depreciable property. Special provisions may allow the deduction of specified amounts of capital costs immediately in the year they are incurred instead of through the depreciation schedule.

Sales tax paid on the purchase of a depreciable asset is generally included in the property's depreciable cost basis for tax purposes rather than expensed in the year incurred. Interest on business loans is fully deductible from taxable income in the year it is paid. Interest on residential mortgage loans is also deductible in the year it is paid, subject to prescribed limits.

Tax credits are sometimes allowed for investments in depreciable property. Remember that a tax credit is *not* the same as a tax deduction. Tax credits are particularly potent because they provide a direct dollar-for-dollar offset to taxes owed. Thus, a tax credit of $1.00 reduces taxes owed by $1.00, whereas a deductible expense of $1.00 reduces taxes owed by only $0.28 (assuming a marginal tax rate of 28%). Tax credits have been used in the United States to encourage investment in low-income housing, historic buildings, energy conservation and renewable energy systems, and equipment and machinery used in research and production.

13.2 HOW TO CALCULATE TAX EFFECTS

Tax effects are usually incorporated into an economic evaluation in either of two ways: (1) by adjusting each benefit and cost item for applicable taxes, such that it is modeled on an after-tax basis, or (2) by creating additional categories of benefits and costs for taxes. These are equivalent treatments.

The specific impact of a tax depends on the amount of the cost or benefit subject to the tax, tax deduction, or tax credit; on the marginal tax rate; and on the applicable tax rules. The larger the subject amount, the higher the tax rate, and the sooner the tax is applied, the larger will be the tax impact, other factors being the same.

13.2.1 Calculating a Combined Marginal Tax Rate

To compute most tax effects, you must know the marginal income tax rate. The marginal rate is the rate which applies to incremental income. For example, a corporation with a marginal income tax rate of 34% has its tax liability reduced by $0.34 for each additional $1.00 of deductible expenses.

To simplify after-tax analysis, combine the federal and state rates at the margin, as shown in equation 13-1.

$$T = T_s + T_f(1 - T_s) \tag{13-1}$$

where

T = combined (state and federal) marginal income tax rate,
T_s = state marginal income tax rate, and
T_f = federal marginal income tax rate.

Equation 13-1 is based on the typical case where state income taxes are an allowable deduction from taxable income at the federal level, but federal taxes are not deductible at the state level.[4]

[4]It is advisable to verify the relationship between federal and state income taxes for the state in which the investment occurs before using the equation.

13.2.2 Calculating Investment Costs Adjusted for Tax Credits

Tax credits usually are linked to capital investment costs and described in terms of a percentage rate applicable to "eligible" capital investment costs. To perform an after-tax analysis of an investment which is eligible for an investment tax credit, compute investment costs on an after-tax basis as follows:

$$IAT = NEI + \left[(EI)(1 - (Tc)(SPV_{D,n}))\right] \tag{13-2}$$

where

IAT = present value investment costs adjusted for a tax credit,
NEI = capital investment costs not eligible for the tax credit,
EI = capital investment costs eligible for the tax credit,
Tc = the tax credit as a percentage of eligible investment costs (expressed as a decimal), and
$SPV_{D,n}$ = single present value discount factor for a specified market discount rate, D, and the number of elapsed years, n, until the tax credit is taken.

Suppose, for example, that you want to compute IAT for a certified historic structure under the following conditions: costs for the unrestored building which are not eligible for the tax credit are \$50,000; eligible rehabilitation costs are \$100,000; the market, after-tax discount rate is 9%; a tax credit is available equal to 20% of eligible costs; and there is a 1-year lag until the tax credit is realized ($SPV_{9\%,\,1\,\text{year}} = 0.92$).[5] Further, assume that all investment costs occur at the beginning of the study period. Modeled on an after-tax basis, present value investment costs adjusted for the tax credit are computed according to equation 13-2 as follows:

$$IAT = \$50{,}000 + \left[(\$100{,}000)(1 - (0.20)(0.92))\right] = \$131{,}600$$

13.2.3 Calculating Business Operating Expenses After Taxes

Calculate present value business operating expenses, after taxes, as follows:

$$OEAT = \sum_{t=1}^{n} \left[OE_t(1 - T)(SPV_{\bar{d},t})\right] \tag{13-3}$$

[5]The 1986 Tax Reform Act allows a 20% tax credit for rehabilitation costs of certified historic structures. Note that end-of-year cash flows are assumed for all items except initial investment costs.

where

$OEAT$ = present value business operating expenses after-taxes,
OE_t = business operating expenses in year t,
T = marginal combined income tax rate (which is shown as a constant but could be specified as a variable rate), and
$SPV_{\bar{d},t}$ = single present value discount factor for a specified discount rate (which is a market rate, D, if business operating expenses are stated in current dollars or a real rate, d, if business expenses are stated in constant dollars) and the year, t, in which the expenses are incurred.

Alternatively, if OE_t is a uniform amount, OE, and the tax rate is constant, the UPV discount factor can be used to find $OEAT$, as shown in equation 13-4. If OE_t is expected to change at a constant rate, the UPV* discount factor can be used.

$$OEAT = [OE(1 - T)(UPV_{\bar{d},n})] \tag{13-4}$$

where

$UPV_{\bar{d},n}$ = uniform present value discount factor for a specified discount rate (which is a market rate, D, if business expenses are stated in current dollars, or a real rate, d, if business expenses are stated in constant dollars) and number of periods, n.

Suppose, for example, that you wish to compute the present value, after taxes, of maintenance costs for a building over 10 years under the following conditions: annual maintenance costs are estimated at $30,000 (constant dollars), the combined marginal income tax rate is 34%, and end-of-year cash flow is assumed. The real, after-tax discount rate is 11%. The present value of maintenance costs, after taxes, is computed according to equation 13-4 as follows:

$$OEAT = (\$30{,}000)(1 - 0.34)(5.89) = \$116{,}622$$

13.2.4 Calculating Depreciable Capital Costs

Straight-line depreciation provides for a uniform writeoff, calculated with equation 13-5:

$$SLD = \frac{CB - S}{V} \tag{13-5}$$

where

SLD = the annual amount of depreciation deduction (in current dollars) allowed by the straight-line method,
CB = original depreciable costs of the asset, i.e., the "cost basis,"[6]
S = salvage value (which tax rules may allow to be set equal to zero), and
V = the depreciation period.

Declining balance depreciation provides an accelerated writeoff of assets as calculated by equation 13-6:

$$AD_t = (F)(B_{t-1}) \tag{13-6}$$

where

AD_t = the amount of depreciation deduction (in current dollars) in year t, beginning with $t = 1$,
F = the depreciation fraction, where F is a constant fraction of the previous year's book value and is calculated as follows: for 200% declining balance depreciation, $F = 2/V$, and for 150% declining balance depreciation, $F = 1.5/V$, and
B_{t-1} = book value in $t - 1$; that is, the original cost basis of the depreciable asset less depreciation taken through year $t - 1$.[7]

Equation 13-7 shows how to compute present value investment costs adjusted for straight-line depreciation over a specified period:

$$IAD = I - [(SLD)(T)(UPV_{D,w}] \tag{13-7}$$

where

IAD = present value investment costs after tax adjustment for depreciation writeoffs,
I = investment costs,
SLD = the annual amount of depreciation deduction allowed by the straight-line method and calculated with equation 13-5,

---

[6]The cost basis is the investment which can be depreciated over the allowable depreciation period. It is generally the cost of the asset plus the costs of any additions to it, including installation costs.
[7]These are the costs remaining to be depreciated at the beginning of year t.

T = marginal combined discount rate, and

$UPV_{D,w}$ = uniform present value discount factor for market discount rate, D, and years, w, where for a study period N, less than the depreciation period, $w = N$, and for a study period equal to or longer than the depreciation period, $w = V$, the depreciation period.

Suppose, for example, that a warehouse is purchased on January 1 for \$600,000, including \$100,000 for land costs. You want to compute the present value of investment costs adjusted for depreciation deductions over the entire depreciation period under the following conditions: the depreciation period is 31.5 years, depreciation is taken according to the straight-line method, salvage value is assumed to be zero, the building is to be kept indefinitely, the after-tax market discount rate is 12%, and the combined marginal income tax rate is 34%. (A market rate is used because the cost basis is fixed and the current dollar depreciation deduction declines in the face of inflation.)

$$IAD = \$600{,}000 - \left[\left(\frac{\$500{,}000}{31.5}\right)(0.34)(8.10)\right] = \$556{,}286$$

13.2.5 Calculating Resale Value After Tax Adjustment for Capital Gains and Depreciation Recapture

For the case where all gain on the sale of depreciated property is taxed at ordinary income tax rates, the present value of a future resale, after taxes, can be calculated by equation 13-8:

$$RAT = \left[R_t - \left(\left(R_t - (L + B_t)\right)(T)\right)\right](SPV_{D,t}) \qquad (13\text{-}8)$$

where

RAT = present value of resale after tax adjustment for capital gains and depreciation recapture at ordinary income tax rates,

R_t = resale price of the property (in current dollars), including land and structures, at the time of resale in year t,

L = original land costs,[8]

B_t = book value of depreciable costs (i.e., the original cost of the depreciable asset (exclusive of land costs) less the

[8]Remember that land is not depreciable.

amount taken in depreciation) at the time of resale in year t,

$R_t - (L + B_t)$ = gain on the sale in year t,

T = combined marginal income tax rate, and

$SPV_{D,t}$ = single present worth discount factor for discount rate, D (market), and year in which property is sold, t.

Suppose that you are evaluating the profitability of purchasing a commercial building, holding it for 10 years, and then selling it. As part of the evaluation, you need to estimate the present value of anticipated proceeds from resale, adjusted for taxes on the gain.

The following conditions must be taken into account. The purchase price of the building is \$600,000, of which \$400,000 is for the building (the adjusted cost basis) and \$200,000 is for the land. The building can be depreciated over 31.5 years using the straight-line method. In 10 years, when the building is sold, it is expected that 100% of land costs (measured in constant dollars) and 50% of the original purchase price of the building will be recouped in the sales price. That is, the estimated resale in constant dollars is \$400,000 [i.e., \$200,000 + (0.5)(\$400,000) = \$400,000].

The prices of the land and the building are expected to increase at the rate of general price inflation, projected at 5% per year. This means that in current dollars the resale value 10 years hence will be \$651,558 [i.e., \$400,000 $(1 + 0.05)^{10}$]. The book value of depreciable costs, plus land costs, at that time will be \$473,016 [i.e., \$200,000 + \$400,000 − ((\$400,000/31.5)(10)]. Therefore the estimated gain on the sale is \$178,542 (i.e., \$651,558 − \$473,016).

The after-tax market discount rate is 12%. The marginal tax rate is 34%. The present value of resale, adjusted for taxes on the gain, can be estimated according to equation 13-8 as follows:

$$RAT = [\$651{,}558 - ((\$651{,}558 - \$473{,}016)(0.34))](0.32) = \$189{,}073$$

13.2.6 Calculating Tax Saving from Mortgage Interest Deductions[9]

In economic evaluations, mortgage loans are typically amortized over the life of the loan in a series of uniform *yearly* payments.[10] Each succeeding payment

[9] See section 13.3 for guidelines on when to take financing into account in economic evaluations.

[10] Note that the level of accuracy required for an economic analysis tends to be less than that required by financial institutions for analyzing returns on financial instruments such as bonds. For the latter purpose, it is important to specify the terms precisely, such as monthly or daily com-

consists of an increasing share of principal payment and a decreasing share of interest over time. Present value tax savings from mortgage interest deductions can be computed with equation 13-9:

$$TSMI = \sum_{j=1}^{z} \left[\frac{(RP_j)(i)(T)}{(1 + D)^j} \right] \tag{13-9}$$

where

$TSMI$ = present value of tax savings from deductions of mortgage interest from taxable income,

z = years over which interest payments are made within the study period, where for a study period (N) which is less than the mortgage loan period (m), $z = N$, and for a study period equal to or longer than the mortgage loan period, $z = m$,

RP_j = remaining principal in year j, where, for $j = 1$,

$$RP = I \times (1 - dp), \tag{13-10}$$

and for $j = 2$ through z,

$$RP_j = RP_{j-1} - \left[((UCR_{i,m})(RP_{j=1})) - ((RP_{j-1})(i)) \right], \tag{13-11}$$

I = investment costs,

dp = fraction of the investment costs comprising the down payment,

$UCR_{i,m}$ = uniform capital recovery factor based on an annual loan interest rate (i) and loan term (m),

i = annual interest rate on the mortgage loan,

T = combined marginal income tax rate, and

D = market discount rate (market rate because the mortgage payments and the interest share of those payments are in current dollars).

---

pounding and payments on the first of every month. For the purpose of evaluating a building investment, modeling cash flows on a yearly basis, as shown in equation 13-9, usually provides sufficient accuracy.

Table 13-2. Present value of tax savings from loan interest deductions: illustration

(1) Year	(2) Remaining Principal RP_j (\$)	(3) Interest i (\$)	(4) Present Value Tax Savings[a] $TSMI$ (\$)
1	100,000	10,000	3,091
2	83,600	8,360	2,349
3	65,560	6,556	1,675
4	45,716	4,572	1,062
5	23,888	2,389	504
		Total present value tax savings	\$8,681

[a]Calculated with equation 13-9, assuming a loan of \$100,000 over 5 years at an annual interest rate of 10%, a marginal income tax rate of 34%, and a discount rate of 10%. (The fact that the loan rate and the discount rate are equal is coincidental.) $UCR_{10\%}$, 5 years = 0.264

Suppose, for example, that you want to compute the tax savings from deducting interest on a \$100,000 real estate loan to be repaid in even payments over 5 years at an annual rate of 10%. The payment schedule calls for end-of-year payments of principal and interest. The marginal income tax rate is 34%, and the discount rate is 10%. Table 13-2 lists present value tax savings year by year and in total for the \$100,000 loan. The present values are calculated using equation 13-9.

13.2.7 Calculating Property Taxes Adjusted for Income Tax Deductions

The present value of property taxes after adjusting for the fact that they can be included in expense deductions from taxable income is calculated with equation 13-12:

$$PTAT = \sum_{t=1}^{n} [(MI_t)(a_t)(p_t)] [(SPV_{D,t-1}) - ((T)(SPV_{D,t}))] \quad (13\text{-}12)$$

where

$PTAT$ = present value of property taxes, adjusted for income tax deductions,

MI_t = recorded market value of property on tax records in year t.

a_t = the rate of assessment, computed as the ratio of assessment value in year t to recorded market value, and expressed as a decimal,

Table 13-3. Present value of property taxes adjusted for income tax deductions

(1) Year	(2) Recorded Market Value MI_t ($)	(3) Assessed Value ($)	(4) Property Taxes Owed ($)	(5) Income Tax Savings ($)	(6) Present Value[a] of Property Taxes After Income Taxes $PTAT$ ($)
1	100,000	50,000	2,000	680	1,370
2	103,000	51,500	2,060	700	1,307
3	106,090	53,045	2,122	721	1,247
4	109,273	54,637	2,185	743	1,188
5	112,551	56,276	2,251	765	1,134
	Present value property taxes after income tax deductions = $6,246				

[a]Calculated with equation 13-12, and assuming a 3% annual rate of increase in recorded market value; an assessment rate of 50% of the recorded market value; a property tax rate of 4%; a marginal income tax rate of 34%; a discount rate of 8%; and recording of market value and payment of property taxes at the beginning of each year, and payment of income taxes at the end of each year.

p_t = property tax rate per assessed dollar value in year t,
T = combined marginal income tax rate, and
$SPV_{D,t}$ = single present value discount factor for market discount rate, D, and year t.

This formulation assumes that property taxes are paid at the beginning of each year and deductions from taxable income are taken at the end of each year.

Suppose, for example, that the market value of a rental house on tax records is $100,000 now and is scheduled to increase at a rate of 3% each year. Suppose that the property is assessed at a rate of 50% of the recorded market value; the property tax rate per assessed dollar value is 4%; the marginal income tax rate is 34%; the owner's discount rate is 8%; property taxes are paid at the beginning of each year; and they are deductible from taxable income at the end of each year. Table 13-3 shows the computation of present value property taxes, adjusted for income tax effects, over the next 5 years. Computations are made using equation 13-12.

13.3 WHEN AND HOW TO INCLUDE FINANCING

Here we explain when to include and when to leave out financing effects. We also describe calculations needed to include financing in an economic analysis.

13.3.1 When to Include Costs of Financing

Include financing costs in an economic evaluation if the terms of financing differ among alternative investments. You should realize, however, that if financing costs are included, the evaluation results will reflect *both* the economic performance of the project and the attractiveness of the financing. For example, one building investment may yield a higher return than another in part because the seller provides financing at a rate several points below the market rate. The seller is, in effect, offering a package price consisting of two parts: the sale price of the building and the interest price to borrow the money.

13.3.2 When Not to Include Financing Effects

When financing terms are not unique for the investments to be evaluated, financing costs generally should be omitted from economic analyses.[11] Separating the financing decision from the project choice avoids confusing the economic attractiveness of a project with the desirability of borrowing money.

Capital investment analysis normally separates financing considerations (i.e., sources of funds) from project evaluation (i.e., uses of funds). Taking this traditional approach, the entire acquisition cost is usually treated as an initial cash outlay, regardless of whether the acquisition is financed or paid out of equity funds.

The reason for separating project evaluation from financing is this: Typically, the appropriate discount rate will be greater than the borrowing rate because there will be a greater opportunity cost to the proposed use of funds than avoiding borrowing costs.[12] Including financing in an economic analysis when the discount rate exceeds the loan interest rate means that the present value of investment cost will be lower than the investment cost prior to taking financing into account. Tax savings from interest deductions further reduce after-tax investment costs. As a result, the project is more attractive with financing than without it. In fact, a project which is not cost effective on its own merits may appear cost effective when financing is included. The reason is simply that the profit from borrowing money under the assumed differential rates outweighs the economic loss on the investment per se. The inclusion of financing clouds the economic merits of the project. For this reason, it is a general principle of financial analysis to separate the financing decision from the project decision.

---

[11] See, for example, Weston and Brigham (1981).

[12] Another reason the discount rate might exceed the borrowing rate is that the borrower has expectations of a higher inflation rate than the lender.

13.3.3 How to Include Financing Effects in an Economic Analysis

We provide equation 13-13 for those situations when it is appropriate to include financing effects in economic analysis. Investment costs are treated in three parts: (1) the part paid up front out of equity funds (usually the down payment), (2) the part borrowed and paid back over the loan period, and (3) any unpaid principal remaining at the end of the study period.

For a loan amount amortized in a series of even yearly payments over the life of the loan, you can estimate the present value of investment costs, before tax effects[13] and after financing, with equation 13-13.

$$FI = (dp)(I) + \left[\left((I)(1 - dp)\right)(UCR_{i,m})(UPV_{D,z})\right] + RPB \quad (13\text{-}13)$$

where

FI = present value of investment costs (before taxes), taking financing into account,
dp = the down payment as a fraction of investment costs,
I = total investment costs,
$UCR_{i,m}$ = uniform capital recovery discount factor for yearly payments at an annual loan rate (i) over the loan period (m years).
$UPV_{D,z}$ = uniform present value factor for a market discount rate (D) and period (z) where for a study period (N) less than the mortgage loan period (m), $z = N$, and for a study period equal to or longer than the mortgage loan period, $z = m$, and
RPB = the present value of the remaining principal when the study period is shorter than the life of the loan (i.e., when $z = N$), calculated with equation 13-14 as follows:[14]

$$RPB = (I)(1 - dp)\left[1 - \left(\frac{UCR_{i,m} - i}{UCR_{i,N} - i}\right)\right](SPV_{D,N}) \quad (13\text{-}14)$$

[13]Treatment of tax effects is presented separately from financing effects. See Section 13.2.6 for treatment of tax savings from mortgage interest deductions. By subtracting equation 13-9 from equation 13-13, it is possible to calculate present value investment costs after deduction of mortgage interest expenses from taxable income.

[14]Equation 13-14 is adapted from equation (7) in Powell (1980, p. 25).

where

$UCR_{i,N}$ = uniform capital recovery discount factor for an annual loan rate (i), over a loan period equal to the study period (N), and other terms are as previously defined.

Suppose, for example, that you want to calculate the present value of capital costs (before taxes and after financing) under the following conditions: You buy a building for $400,000, making a down payment out of equity funds of $40,000 and borrowing $360,000 from the seller at an annual rate of 8% for 10 years. Principal and interest are to be repaid in 10 even annual payments at the end of each year. You plan to sell the building in 5 years. Therefore, you will, in fact, make the loan payments only over 5 years and will repay the balance at the end of 5 years. The discount rate is 12%. The study period is 5 years.

Apply equation 13-13 to compute a present value investment cost of $354,643. This is made up of the down payment of $40,000, already in present value dollars; the present value of loan payment of $193,372;[15] and the present value of the principal balance remaining at the end of the study period of $121,271.[16] The implication is that financing the building purchase contributes $45,357 to present value benefits, before taxes (i.e., $400,000 − $354,643 = $45,357).

Thus, if this project were a loser without financing incentives, it could still be a winner with preferential financing.

REFERENCES

Commerce Clearing House Tax Law Editors. 1986. *Explanation of Tax Reform Act of 1986.* Chicago, Illinois: Commerce Clearing House, Inc.

Powell, J. W. 1980. *An Economic Model for Passive Solar Designs in Commercial Environments.* National Bureau of Standards Building Science Series 125. Gaithersburg, Maryland: National Bureau of Standards.

Weston, J. Fred, and Eugene F. Brigham. 1981. *Managerial Finance.* Hinsdale, Illinois: The Dryden Press.

---

[15]The present value of loan payments = ($360,000)(0.149)(3.605) = $193,372, where 0.149 is the UCR factor for 8% and 10 years, and 3.605 is the UPV factor for 12% and 5 years.

[16]Present value of the principal balance remaining at the end of 5 years = $360,000 (1 − ((0.149 − 0.08)/(0.250 − 0.08)) × 0.567 = $121,271, where 0.149 is the UCR factor for 8% and 10 years, 0.08 is the loan interest rate, 0.250 is the UCR factor for 8% and 5 years, and 0.567 is the SPV factor for 12% and 5 years.

Part 3

TECHNIQUES FOR TREATING UNCERTAINTY AND RISK

14

Risk Exposure and Risk Attitude

Investments in long-lived projects such as buildings are characterized by uncertainties regarding project life, operation and maintenance costs, revenues, and other factors that affect project economics. Since the exact values of these variable factors are usually unknown, it is difficult to make economic evaluations with a high degree of reliability.

A common approach to project investment analysis is to apply the economic methods discussed in Part 1 to "best-guess" estimates of project input variables as if they were certain estimates and to present results in single-value, deterministic terms. Yet failure to account for uncertain input variables means that decision makers have insufficient information to know their "risk exposure," i.e., the probability of a project's having an economic outcome less favorable than what is economically acceptable.[1]

Decision makers faced with an investment choice under uncertain conditions also confront a second aspect of risk which is their attitude towards it (called "risk attitude" or "risk perference"). Risk attitude can be measured by the willingness of a decision maker to take a chance or gamble on an investment of uncertain outcome. The implication of different risk attitudes is that a given investment of known risk might be economically acceptable to an investor who is a risk taker but unacceptable to another investor who is risk averse.

Risk analysis is the body of theory and practice that has been developed to help decision makers assess their risk exposures and risk attitudes so that they can select investments that are best for them.

The remainder of the chapter describes how to measure and interpret risk exposure and risk attitude to help you understand the techniques for treating

[1]Measuring the probability of the project's economic worth being less than an economically acceptable value reveals the risk of accepting an uneconomic project. Another type of risk exposure that some decision makers are concerned about is the probability of passing up a good investment. For example, measuring the probability of the project's economic worth being greater than the minimum acceptable value reveals the risk of rejecting an economic project. The risk of accepting an uneconomic project is focused on here.

uncertainty and risk that we present in the following chapters.[2] Section 14.1 describes the typical investment situation of decision makers who do not know with certainty the outcome of their investment and illustrates with probability distributions a way of measuring risk exposure. Section 14.2 shows conceptually the conditions under which measures of risk exposure by themselves will fail to indicate the best of competing projects and introduces the concept of risk attitude or preference as a way to find the best project. The use of utility functions to determine a decision maker's risk attitude is also explained.

14.1 RISK EXPOSURE

One way to illustrate risk exposure is in a probability distribution of the measure of economic worth. The probability profile quantifies risk exposure by showing probabilities of achieving different values of economic worth.

Figure 14-1 is a discrete probability distribution that shows graphically for a building investment the profile of probabilities for a benefit-to-cost ratio (BCR).[3] Each bar of the histogram shows on the vertical axis the probability of the

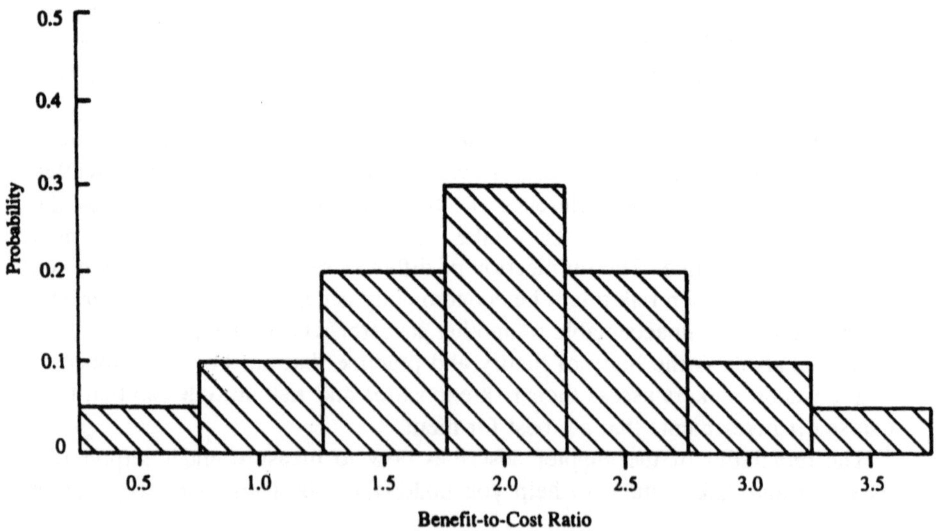

Figure 14-1. Discrete probability distribution of the BCR.

[2]The treatment of risk and uncertainty in this chapter and in those that follow in Part 3 draws in part from Marshall (1988).

[3]See Chapter 4 for a complete description of the BCR.

investment's achieving the corresponding BCR on the horizontal axis. The mean (expected value) of the BCR is 2.0. This suggests that the most likely measure of worth for the investment will likely exceed the 1.0 BCR that is normally regarded as the minimum hurdle necessary for project acceptance. Other values for the BCR are possible, however, including a value less than 1.0.

Having the standard deviation and mean for the distribution helps the decision maker determine the likelihood that the actual BCR is within acceptable bounds around the mean. The smaller the spread of the distribution, as measured by the standard deviation, the tighter the distribution is around the mean value and the smaller is the risk exposure associated with the project.

It is known that in a normal distribution the probability is 68.26, 95.46, and 99.73%, respectively, that the actual value will be within one, two, and three standard deviations of the mean. Assuming that the discrete probability distribution in Figure 14-1 approximates a normal distribution, we can estimate the probability of the BCR's being within any one of the standard deviation ranges. The standard deviation for Figure 14-1 is found to be 0.72 by the following equation:

$$\sigma = \left(\sum_{s=1}^{K} (BCR_s - \mu)^2 \cdot P_s \right)^{1/2} \tag{14-1}$$

where

σ = standard deviation,
s = possible state,
K = number of possible states,
BCR_s = BCR in the sth state,
μ = mean or expected value of the distribution, and
P_s = probability of the sth state.

Thus we know, for example, that there is a 68.26% probability that the BCR will lie in the range of 1.28 to 2.72 (i.e., 2.0 ± 0.72).

Although the probability distribution in Figure 14-1 does not reveal directly the probability of choosing a project having a BCR less than some target value (say, less than the expected value of 2.0 in this case), it is easily transformed to the cumulative distribution function in Figure 14-2 which does.[4] Any percent on the vertical axis in Figure 14-2 is read "less than." The function relating

[4]See Section 20.5 for a detailed description of how to build a cumulative distribution function from a probability density function.

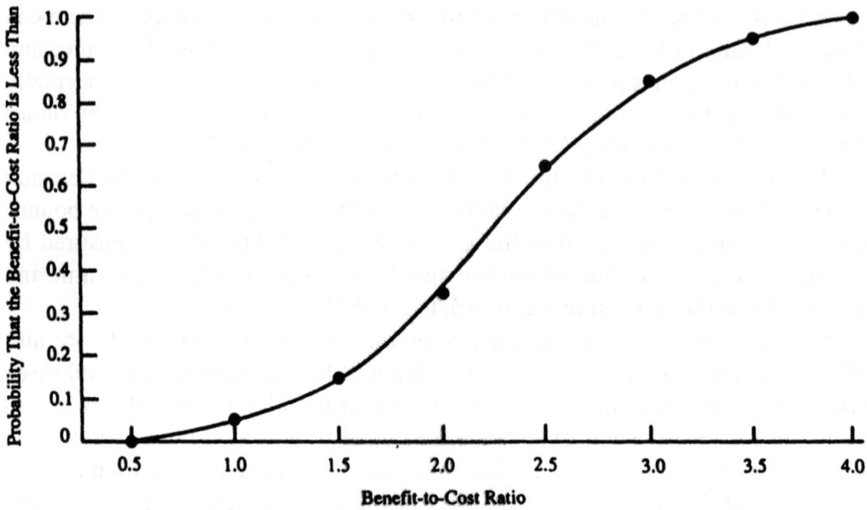

Figure 14-2. Cumulative distribution function of the BCR.

BCRs to cumulative probabilities is upward sloping, indicating that the probability of the BCR's being less than any given BCR value on the horizontal axis increases as the given BCR value increases.

The probability (or risk of exposure) of the BCR's being less than 1.0 is 5% in Figure 14-2. Or, said another way, the probability of the project's earning positive net benefits or at least breaking even is 95%. The probability that the BCR is less than the expected value of 2.0 is 35%.

Probability and cumulative distribution functions provide information about risk exposure lacking in deterministic approaches that assume certainty and provide single-value measures of project worth. But the functions do not reveal risk attitude. They show only risk exposure. Different decision makers, whether individuals or institutions,[5] may respond differently to any given profile of risk exposure. Thus, to make efficient choices when investment outcomes are uncertain, decision makers also need to consider their unique risk attitudes.

[5]An entity such as a corporation has a risk attitude just as an individual does. To determine it, however, may be difficult. One approach is to use the risk attitude of the chief executive officer as a proxy for the corporation. Another is to use the risk attitude of the board of directors. For a description of how to quantify risk attitudes for an individual and for an organization, see Spetzler (1968).

14.2 RISK ATTITUDE

There are two general approaches that decision makers might follow to incorporate risk attitudes in their project evaluations. First, they can make a decision on the basis of their subjective or intuitive perception of whether they are prepared to accept the degree of risk exposure indicated by the distribution profile, mean, and standard deviation of the measure of project worth. This approach (see Section 20.5 for an example) allows for the informal consideration of risk attitude but lacks any standard procedure for measuring personal or institutional risk attitude when making a choice. If the investment decision is to accept or reject a single project, this approach will often be adequate. For example, the project described by Figure 14-2 is likely to be deemed cost effective by all but the most risk-averse decision makers, since there is little probability of a BCR less than 1.0.

Even where several projects are competing for a limited budget, the informal approach for considering risk attitude may be adequate. While the preferred choice may not be obvious from an examination of probability density functions[6] for individual projects, it may become obvious when functions for alternative projects are superimposed, as shown in Figures 14-3 and 14-4. Here the probability profiles are a good index to project choice because project A clearly has "stochastic dominance" over project B. That is, for every BCR value in Figures 14-3 and 14-4, there is as high or higher probability that project A will exceed that BCR than will project B. In other words, for every BCR, there is as high or higher probability that project B will provide a lower or equal BCR compared to project A. Thus the project alternative whose function is farthest to the right is the preferred alternative. (Note that if LCCs of alternatives were measured on the horizontal axis instead of the BCRs, the alternative farthest to the left would be preferred because the objective function would be to minimize LCC rather than to maximize the BCR.)

The second approach for considering risk attitude is formal in that it employs a standard procedure for measuring the decision maker's attitude toward risk and then uses that measure in evaluating the economic worth of a risky project. The need for a formal technique is illustrated by the intermingled distributions shown in Figures 14-5 and 14-6. Although project D has the larger mean, it also has the larger variance. That is, the project with greater expected return also has greater variance or risk exposure. There is no clear indication of stochastic dominance or project preference.

---

[6]The probability density function is the derivative of a continuous, cumulative distribution function. The area under the probability density function must equal 1.0.

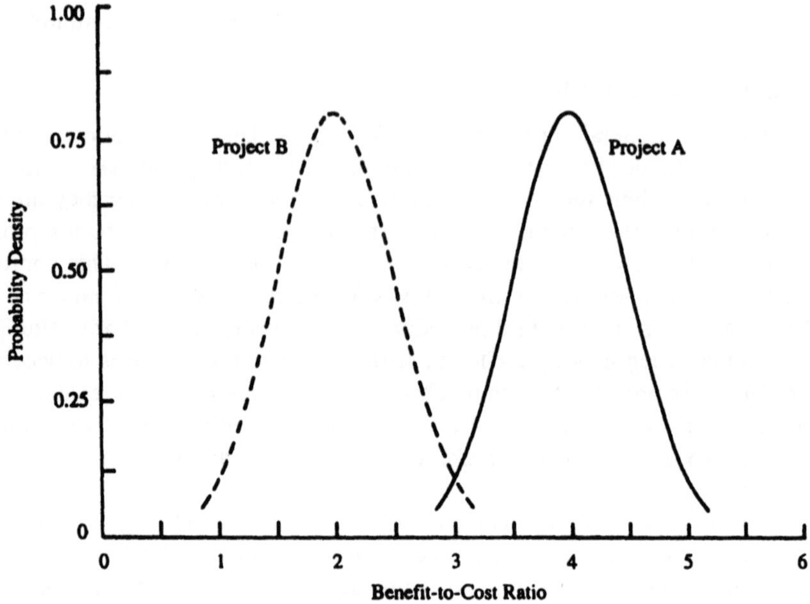

Figure 14-3. Probability density functions of the BCR for projects A and B.

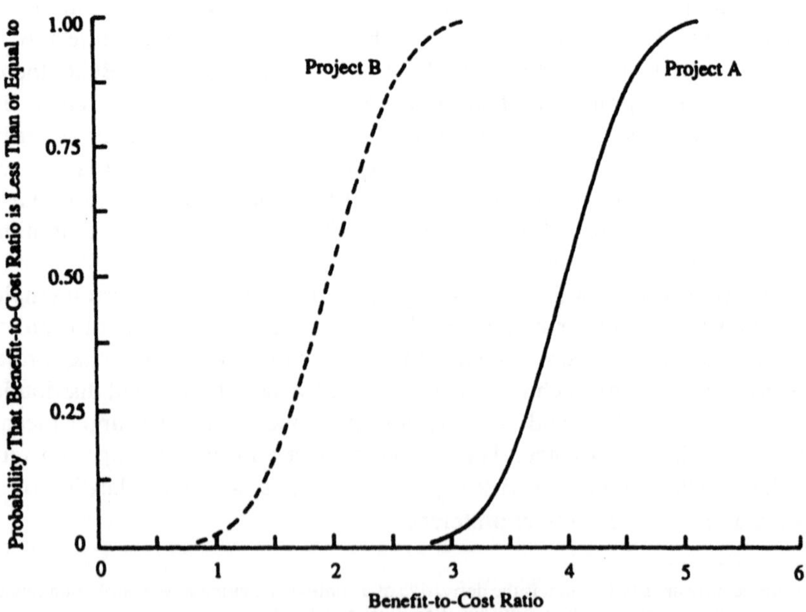

Figure 14-4. Cumulative distribution functions of the BCR for projects A and B.

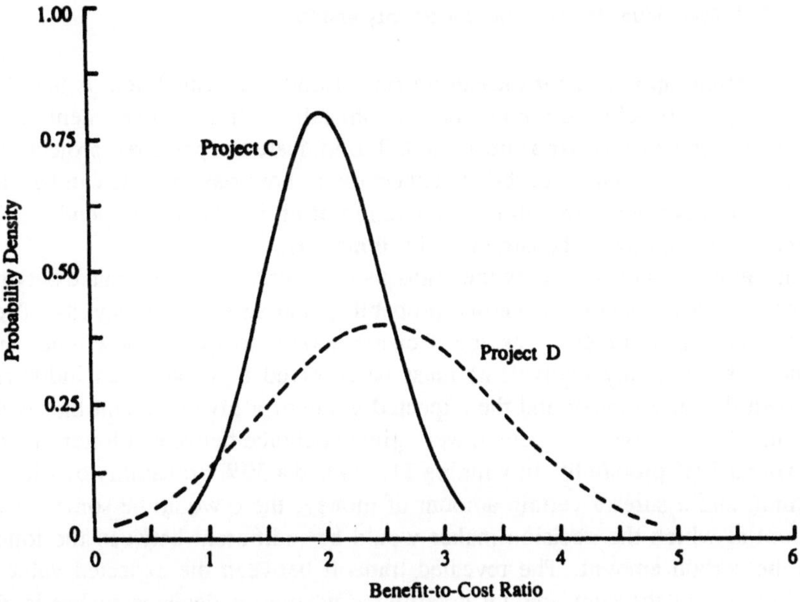

Figure 14-5. Intermingled probability density functions of the BCR for projects C and D.

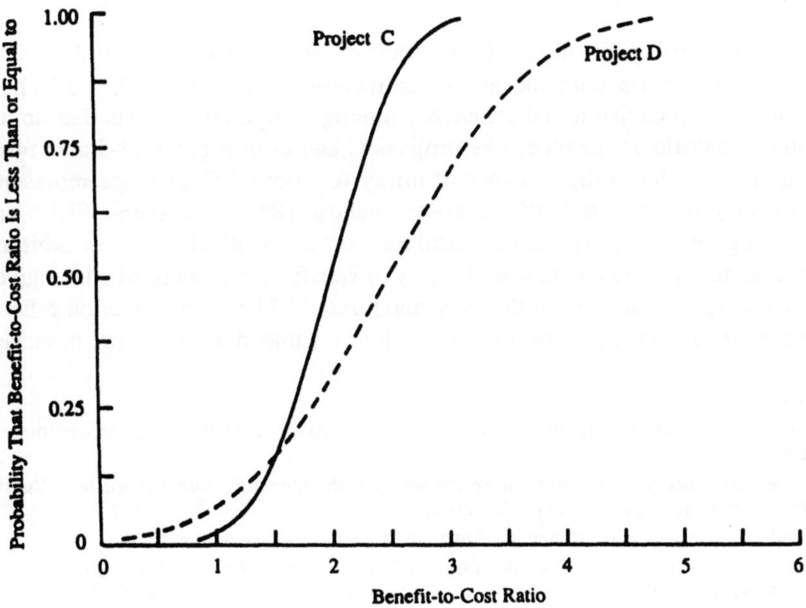

Figure 14-6. Cumulative distribution functions of the BCR for projects C and D.

The formal approach for measuring risk attitude described here is based on utility theory. It helps determine the economically preferred investment choice when measures of risk exposure alone fail to indicate the preferred project. The remainder of this chapter describes conceptually how risk attitude can be measured and interpreted through the application of utility theory. Appendix E describes in detail how to construct utility functions.

Risk attitude is measured by the tradeoffs decision makers will make between uncertain money payoffs of known probability and sure money payoffs. These tradeoffs are determined by asking decision makers to specify how much sure money (the certainty equivalent) must be received to make them indifferent between the sure money and the expected value of a given amount that is uncertain. For example, if a person were given a choice between a lottery ticket, say with a 50% probability of winning $1,000 and a 50% probability of winning nothing, and a sure or certain amount of money, there would be some certain amount at which the decision maker would be indifferent between the lottery and the certain amount. The revealed tradeoff between the expected value of the lottery and the sure amount determines whether a decision maker is risk neutral, risk averse, or a risk taker. With several tradeoffs specified, a relationship between money and utility can be determined.[7] A graph of the money-utility relationship, called a "utility function," shows the decision maker's risk attitude.[8]

There are formal techniques (e.g., decision analysis) for using utility functions in conjunction with the economic methods described in Part 1. These techniques help decision makers choose among competing projects that do not exhibit stochastic dominance, like projects C and D in Figures 14-5 and 14-6.

Figure 14-7 shows three shapes of utility functions.[9] Each shape represents one of three different risk attitudes—risk neutral (RN), risk averse (RA), and risk taking (RT). Utility values, displayed on the vertical axis, are arbitrary units used to measure the degree of utility or satisfaction associated with a given amount of money shown on the horizontal axis.[10] The utility function reflects a particular relationship between satisfaction, a subjective value, and monetary

---

[7]"Utility" here means simply the satisfaction a decision maker receives from given quantities of money.

[8]Because utility theory is used to measure risk attitude, the literature often uses the term "utility theory" when treating the subject of risk attitude.

[9]Appendix E describes in detail how the functions in Figure 14-7 were derived.

[10]Note that relative utility—not absolute utility—is being measured. An individual could assign any two utility values on the vertical axis to any two dollar amounts on the horizontal axis. However, once those values have been assigned, there exists a unique utility value for every other dollar amount.

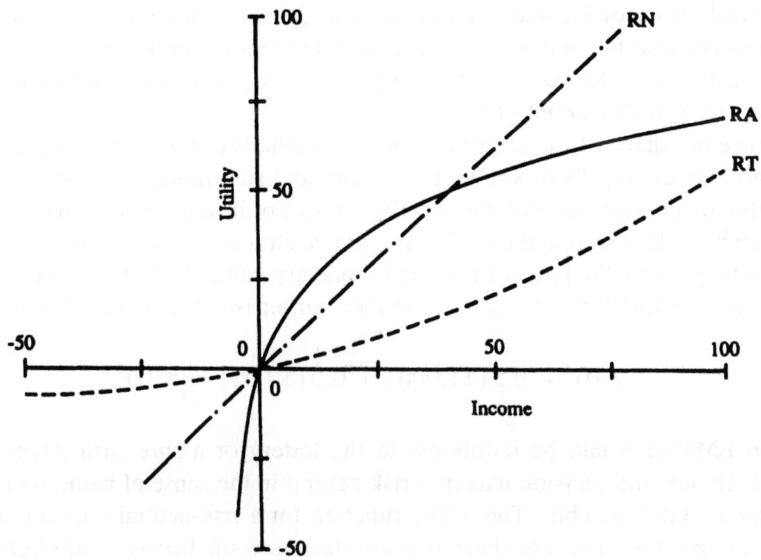

Figure 14-7. Three types of risk attitude.

amounts. Thus the utility function is unique to one individual, firm, or institution. Each decision maker is likely to have a different utility function.

In using the utility function in an economic analysis, it is assumed that a decision maker will be indifferent among investments with the same expected utility and will prefer investment x to investment y if the expected utility is greater for x than for y.[11] The procedure for using the utility function to choose among alternative investments can be described as follows:[12] (1) find from the utility function, for each investment alternative, utility values that correspond to each dollar-valued outcome in the probability distribution of potential outcomes; (2) find the expected utility value (the sum of outcome utilities weighted by outcome probabilities) for each investment; and (3) select the investment with the maximum expected utility.

Look again at the three utility functions in Figure 14-7 and consider how each can be interpreted. For the straight-line utility function (RN), each additional, fixed increment of income yields a constant increase in utility, i.e., the

[11]The utility theory described in this chapter evolved from that described by von Neuman and Morgenstern (1944).

[12]This procedure is most descriptive of decision analysis, one of the techniques described in Chapter 17. Note that if the measure of worth is not dollars (e.g., a BCR measure), the projected benefits and costs are first transformed to utility units before the measure of worth is computed.

marginal utility of income is constant. The decision maker is considered risk neutral because the gain or loss of a large amount of money would yield the same increase or decrease, respectively, in utility per dollar as would the gain or loss of a small amount of money.

Since the shape of the utility function is dependent on tradeoffs between uncertain money payoffs of known probability and sure money payoffs, it is also helpful to consider risk attitude directly in terms of how a decision maker reacts to gambles. Decision makers who are risk neutral are called "EMV'ers" because they act on the basis of expected monetary value (EMV).[13] For example, the worth or EMV of the lottery described earlier is calculated as follows:

$$EMV = 0.5(\$1{,}000) + 0.5(\$0.00) = \$500$$

An EMV'er would be indifferent to the lottery or a sure cash payment of \$500. Hence, the decision maker is risk neutral in the sense of being willing to accept a "fair" gamble. The utility function for a risk-neutral decision maker is a straight line, because there is a constant tradeoff between satisfaction in utility and dollar amounts. An implicit assumption in many economic analyses is that decision makers are EMV'ers. Thus there is no explicit consideration of risk attitude because maximizing the expected value of NB is assumed to be equivalent to maximizing expected utility.

If the utility function bends over to the right (RA in Figure 14-7), the decision maker is risk averse. Increasingly large dollar amounts are required to achieve constant increments of utility; that is, the marginal utility of income is diminishing. This means that a decision maker would prefer a sure payment that is *less* than the expected value of a lottery to the chance of participating in the lottery. In the lottery described earlier, the decision maker would prefer a sure cash payment less than \$500 to the lottery, because of aversion to the risk of the lottery's outcome. This implies that decision makers regard marginal payoffs to be worth less (i.e., to be of less utility) per dollar as total payoffs increase. Thus, to determine the true value of investments for risk-averse decision makers, economic analyses must account for decreasing satisfaction of higher payoffs with corresponding decreases in marginal utility. Since empirical studies have shown that most corporate decision makers are in the risk-averse category,[14] accounting for risk attitude is a significant consideration in project evaluation.

---

[13]See Raiffa (1970, p. 9) for a discussion of EMV'ers.

[14]See, for examples, Spetzler (1968) and Swalm (1966).

If the utility function bends upward to the left (RT in Figure 14-7), the decision maker is a risk taker. Successively smaller dollar amounts are required to achieve constant increments in utility; that is, the marginal utility of income is increasing. This implies that the decision maker would actually pay a premium for a lottery ticket, i.e., a value greater than the expected value of the lottery. In the lottery example, the decision maker would prefer the lottery ticket to a sure amount greater than $500. The reason is that the decision maker regards project payoffs to be worth more (i.e., to have more utility) per dollar as the total payoffs increase. Thus, to determine the true value of investments for risk takers, economic analyses must account for increasing marginal satisfaction of higher payoffs with corresponding increases in marginal utility.

For completeness, we describe all three risk attitudes. Most investors, however, are likely to be risk averse. The utility principles apply both to the typical case where investors have different degrees of risk aversion and to the atypical case where some investors are risk takers.

The behavior of decision makers implies that more than one risk attitude is possible for a given person, depending on the monetary stakes. For example, many low-income persons are willing to buy insurance at a premium greater than the expected value of a loss without insurance (the sign of a risk averter) and at the same time to play the lottery at worse than fair odds (the sign of a risk taker). This suggests a utility function with risk-averting and risk-taking segments, as shown in Figure 14-8.[15]

Expected utility analysis based on subjective utility functions will not always predict the way decision makers will actually choose among alternative investments. Individual decision makers cannot be expected to act rationally and consistently in every investment situation with respect to their revealed utility–money functions. It is even more unlikely that a group of executives representing a firm will always agree upon and act consistently according to a corporate utility function. Friedman and Savage (1952, p. 86) point out that decision makers do not, in fact, calculate utilities before making every choice. Utility analysis is useful, nevertheless, provided decision makers generally act as if they compare expected utilities and know the odds for the economic choices being evaluated. Under these conditions, a firm or institution can use utility theory in a normative or prescriptive role to establish risk policy that will direct management to investments that support the firm's or institution's risk attitude.

Several other factors limit the application of utility theory in practice. First, many decision makers consider the technique impractical for their institution. This may be due in part to a lack of understanding and in part to an unwilling-

[15]Friedman and Savage (1952, pp. 57–96) discuss at length various shapes of utility functions.

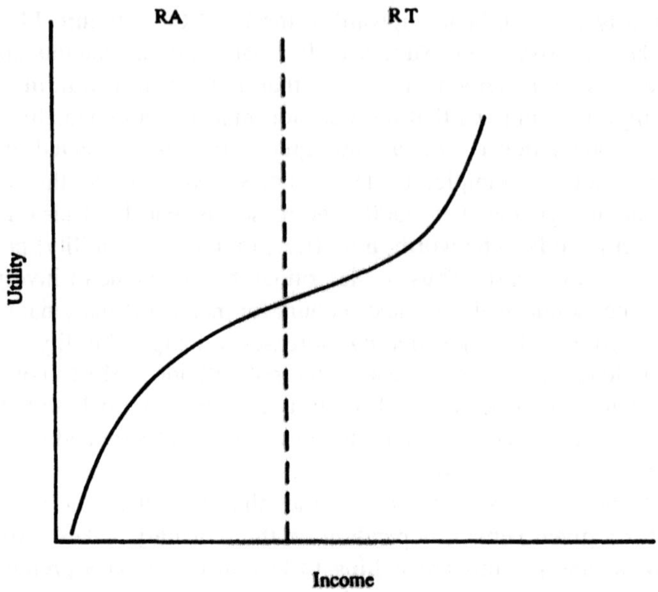

Figure 14-8. Utility function showing both risk-averse (RA) and risk-taking (RT) attitudes.

ness to give up opportunities to use personal judgment in project evaluation. Second, there is often considerable difficulty in determining an organization's risk attitude. One reason is that individual decision makers may be unwilling to cooperate in defining the organization's risk policy because they do not want to be bound by that risk policy. Another reason is that they may have difficulty in identifying an institutional risk attitude that is risk-taking because they are risk averters in their personal frame of reference.

The use of utility theory in project evaluation, however, does have merit. It has a sound theoretical basis. And a firm or institution that can establish a risk policy that directs management to investments that are consistent with the firm's or institution's risk attitude will select better projects over the long run. Competent professional assistance in developing and implementing a risk policy could overcome many of the limiting factors.

In the remaining chapters of Part 3, we describe and illustrate various techniques of treating risk exposure and risk attitude. Some techniques, such as the mathematical/analytical technique, focus on risk exposure, whereas others, such as the risk-adjusted discount rate, account for risk attitude as well as risk exposure. Confusion is often created by the literature because the term "risk analysis" is used sometimes to describe exclusively either risk exposure or risk attitude and at other times to describe both. Thus, in these illustrations, it will be made explicit in each case whether risk exposure, risk attitude, or both are treated.

REFERENCES

Baumol, William. 1977. *Economic Theory and Operation Analysis.* 4th edition. Englewood Cliffs, New Jersey: Prentice-Hall, Inc.

Friedman, Milton, and L. J. Savage. 1952. The utility analysis of choices involving risk. In *Readings in Price Theory*, ed. George J. Stigler and Kenneth E. Boulding, pp. 57–96. Homewood, Illinois: Richard D. Irwin, Inc.

Marshall, Harold E. 1988. *Techniques for Treating Uncertainty and Risk in the Economic Evaluation of Building Investments.* SP757. Gaithersburg, MD: National Institute of Standards and Technology.

Neuman, John von, and Oskar Morgenstern. 1944. *The Theory of Games and Economic Behavior.* Princeton, New Jersey: Princeton University Press.

Raiffa, Howard. 1970. *Decision Analysis: Introductory Lectures on Choices Under Uncertainty.* Reading, Massachusetts: Addison-Wesley Publishing Company.

Spetzler, Carl S. 1968. Establishing a corporate risk policy. *1968 Proceedings of the American Institute of Industrial Engineers.* Norcross, Georgia: AIIE.

Swalm, Ralph O. 1966. Utility theory–insights into risk taking. *Harvard Business Review* 44(6):123–136.

15

Conservative Estimating, Breakeven Analysis, and Sensitivity Analysis

This chapter and the rest of Part 3 describe techniques that can be used with capital budgeting methods to treat uncertainty and risk in project evaluation. We examine techniques that use single-value, multiple-value, and probability function estimates of input data and measures of worth. The techniques are generally in order of increasing complexity in Chapters 15 through 17. The set of techniques in any given chapter, however, tends to require a similar level of technical sophistication except for Chapter 17, where we include some less often used techniques that vary in complexity. We describe advantages and disadvantages of each technique and the extent to which each accounts for risk exposure and risk attitude. Chapter 18 treats in detail how to select the best technique for handling uncertainty and risk in evaluating an investment.

15.1 CONSERVATIVE BENEFIT AND COST ESTIMATING

A simplistic approach to accounting for uncertainty of input values is to be conservative in their estimation so that any error in prediction will likely be an underestimation of project economic worth. To accomplish this, intentionally estimate benefits on the low side, costs on the high side, or both. Or, when estimating the values of parameters on which benefits and costs depend, such as the discount rate or project life, estimate parameters in the direction that lowers expected benefits and raises expected costs.

The more conservative are benefit and cost estimates, the more assurance one has that a project estimated to be cost effective will in fact be so. This technique is appropriate only when a decision maker suffers a penalty for selecting an uneconomic project but incurs no penalty for rejecting an economic project. However, if a firm used this technique to determine its bids on construction projects, it would be unlikely to win many contracts in a competitive market. The technique provides no measure of the risk of making an uneconomic decision, and it encourages decision makers to reject some projects that would be economic.

The greater the risk exposure perceived by the decision maker, and the more risk averse the decision maker, the more conservative will be the benefit and cost estimates. With this technique there is no clear distinction between adjustments for handling risk exposure and risk attitude.

15.2 BREAKEVEN ANALYSIS

When an uncertain variable is critical to the economic success of a project, decision makers frequently want to know the minimum or maximum value that variable can reach and still have a breakeven project, i.e., a project where benefits (savings) equal costs. For example, the breakeven value of an input *cost* variable is the maximum amount one can afford to pay for the input and still break even relative to the benefits earned. A breakeven value of an input *benefit* variable is the minimum amount the project can produce in that benefit category and still cover the estimated costs of the project.

To perform a breakeven analysis, take the following steps:

1. Construct an equation that sets benefits equal to costs for a given investment project.
2. Specify the values of all inputs except the breakeven variable.
3. Solve for the breakeven variable algebraically.

Suppose that a decision maker is deciding whether or not to invest in a piece of energy-conserving equipment. The derivation of the formula for computing breakeven investment costs for the equipment is as follows:[1]

$$S = C$$

$$C = I + O\&M + R$$

$$S = I + O\&M + R$$

$$I = S - O\&M - R$$

where

S = savings (benefits) in reduced energy costs from using the equipment,

[1]All costs and benefits are in present value dollars.

C = all costs associated with the equipment except energy costs,
I = initial investment costs of the equipment,
$O\&M$ = operation and maintenance costs of the equipment, and
R = replacement costs required to keep the equipment functional over the study period.

By rearranging terms, the breakeven investment "unknown" is isolated on the left side of the equation. Substitution of known values for the terms on the right side allows the analyst to solve for the breakeven value. For example, if S = \$20,000, $O\&M$ = \$2,500, and R = \$1,000, then

$$I = \$20{,}000 - \$2{,}500 - \$1{,}000$$

or

$$= \$16{,}500$$

This means that \$16,500, the breakeven value, is the maximum amount that can be paid for the energy-conserving equipment and still recover all costs through energy savings.

An advantage of breakeven analysis is that it can be computed quickly and easily with limited information. It also simplifies project evaluation in that it gives decision makers just one value to use as a benchmark for comparison against the predicted performance of that uncertain variable. Breakeven analysis helps decision makers assess the likelihood of achieving the breakeven value and thereby contributes implicitly to the analysis of project risk.

A disadvantage is that it provides no probabilistic picture of input variable uncertainty or of project risk exposure. Furthermore, it includes no explicit treatment of risk attitude.

See Section 20.1 for a case study that applies breakeven analysis in the evaluation of a building project.

15.3 SENSITIVITY ANALYSIS

Sensitivity analysis measures the impact on project outcomes of changing a key input value about which there is uncertainty. For example, a pessimistic, an expected, and an optimistic value can be chosen for an uncertain variable. Then an economic analysis can be performed for each of the three values to see how the outcome changes as they change, with other things held the same.

Sensitivity analysis also applies to different combinations of input values. That is, several variables can be altered at once. Then a measure of worth can be computed for each combination. For example, one scenario might include a combination of all pessimistic values, another all expected values, and a third some combination of pessimistic and expected values. Examining different combinations is required if uncertain variables are interrelated.

The following illustration of sensitivity analysis treats an accept/reject decision. Consider a decision on whether or not to install a programmable time clock to control HVAC equipment in a commercial building. The time clock reduces electricity consumption by turning off that part of the HVAC equipment that is not needed when the building is unoccupied. Using the BCR as the economic method, the time clock is acceptable on economic grounds if its BCR is greater than 1.0. The energy reduction benefits from the time clock, however, are uncertain. They are a function of three factors: the initial price of energy, the rate of change in energy prices over the life cycle of the time clock, and the number of kilowatt-hours (kWh) saved. Assume that the initial price of energy and the number of kWh saved are relatively certain, and that the sensitivity of the BCR is being tested with respect to the following three values of energy price change: a low rate of energy price escalation (slowly increasing benefits from energy savings), a moderate rate of escalation (moderately increasing benefits), and a high rate of escalation (rapidly increasing benefits). These three assumed values of energy price change might correspond to our projections of pessimistic, expected, and optimistic values. Three BCR estimates result from repeating the BCR computation for each of the three energy price escalation rates. For example, BCRs of 0.8, 2.0, and 4.0 might result. Whereas a deterministic approach might have generated a BCR estimate of 2.0, now it is apparent that the BCR *could be* significantly less than 2.0 and even less than 1.0. Thus accepting the time clock could lead to an inefficient outcome.

There are several advantages of sensitivity analysis. First, it shows how significant a single input variable is in determining project outcomes. Second, it recognizes the uncertainty associated with the input. Third, it gives information about the range of output variability. And fourth, it does all of these when there is little information, resources, or time to use more sophisticated techniques.

Major disadvantages of sensitivity analysis are that it gives no explicit probabilistic measure of risk exposure and it includes no explicit treatment of risk attitude. The findings of sensitivity analysis are ambiguous. How likely is a pessimistic or expected or optimistic value, for example, and how likely is the corresponding outcome value? Hillier (1969, p. 444) maintains that sensitivity analysis can in fact be misleading if all pessimistic assumptions or all optimistic assumptions are combined in calculating economic measures. Such combinations of inputs would be most unlikely in the real world. Furthermore, he em-

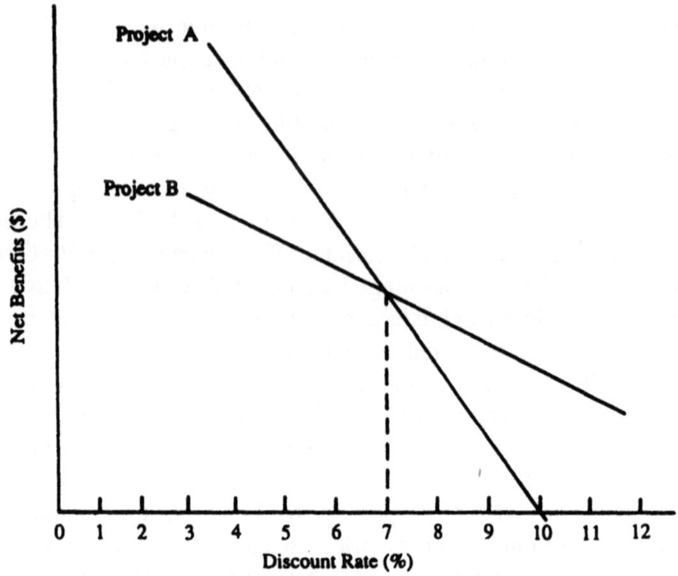

Figure 15-1. Sensitivity of the NB of projects A and B to discount rate.

phasizes that no conclusions can be reached about the impact of combinations of errors in the input estimates.[2]

Sensitivity results can be presented in text, tables, or graphs. One type of graph that is useful in showing the sensitivity of project worth to a critical variable is illustrated in Figure 15-1. NB for projects A and B decrease as the discount rate increases. The slopes of the functions show that NB are more sensitive to discount rate changes for project A than for project B, assuming that other variables remain unchanged. These functions also help in determining which project is more cost effective. At a discount rate below 7%, for example, project A has greater NB. At a rate above 7%, project B yields greater NB. And at 7%, the two projects provide identical NB.[3]

Note that the functions indicate potential values of NB *if* different values of the discount rate occur. If decision makers have some idea about the likelihood of specific discount rates, the graph will help them evaluate the NB implications for these two projects. The sensitivity graph in this sense contributes to an implicit description of risk exposure. Yet the graph fails to provide a quantitative measure of the probability of any given NB occurring.

---

[2]For additional information on sensitivity analysis, see Hendrickson (1984).

[3]Note that the graph also shows the breakeven discount rate (i.e., 10% for project A).

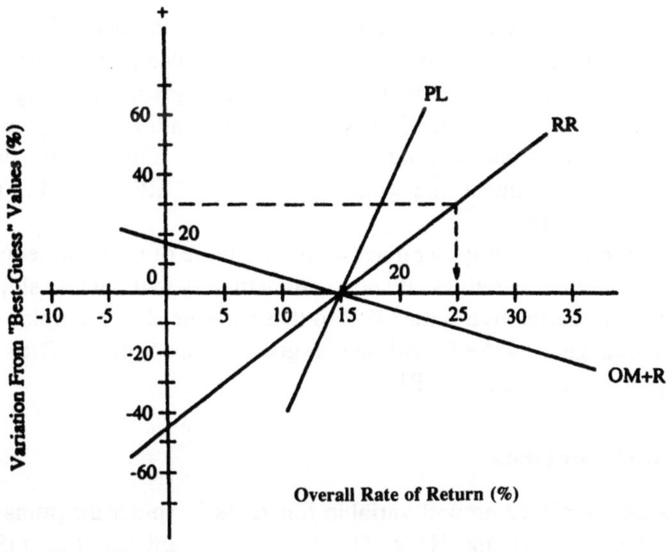

Figure 15-2. Spider diagram showing the sensitivity of the ORR to variations in uncertain variables.

15.3.1 Spider Diagram

Another special graph for sensitivity analysis that presents a snapshot of potential impacts of uncertain input variables on project outcomes is the spider diagram.[4] The one illustrated in Figure 15.2 shows for a prospective commercial building investment the sensitivity of the ORR to three variables: operation, maintenance, and replacement costs (OM+R); project life (PL); and the reinvestment rate (RR). A labeled function for each variable shows what values of the ORR result from different values of that uncertain variable.[5] For example, the downward-sloping OM+R function indicates that the ORR is inversely proportional to OM+R costs. By design, the OM+R function (as well as the other two functions) passes through the horizontal axis at the best-guess estimate of the ORR (15% in this case), based on the best-guess estimates of the three uncertain variables.[6] Since each of the variables is measured by different units

[4]For descriptions of spider diagrams used in the analysis of construction and building projects, see Perry and Hayes (1985) and Flanagan et al. (1987).

[5]The functions are linear here to simplify exposition. In practice, they may not be linear.

[6]Other variables, for example occupancy rate, will affect the ORR, but these are assumed to be known for the purpose of this illustration.

(money, years, and percentage), the vertical axis is denominated in positive and negative percentage changes from the best-guess values fixed at the horizontal axis. The value for the ORR that corresponds to any given percentage variation indicated by a point on the function is found by extending a line perpendicular to the horizontal axis and reading directly the value. Thus a 30% increase in the best-guess reinvestment rate would yield a 25% ORR, assuming that other values remain unchanged.[7]

The contribution of the spider diagram is its picture of the relative importance of different uncertain variables. It shows immediately that the less slope a function has, the more sensitive is the ORR to that variable. For example, any given percentage change in OM+R will have a greater impact on the ORR than will an equal percentage change in RR or PL.

15.3.2 Contour Lines

Wrapping contour lines around variable functions in spider diagrams has been recommended by Perry and Hayes (1985) and by Flanagan et al. (1987). Because contour lines are based on probabilities, their use might appear to overcome the perceived limitation ın sensitivity analysis of failing to describe uncertain input variables with probabilities. However, it is shown here that wrapping contours around spider diagrams adds *no* additional information and may actually mislead analysts.

To evaluate this approach, we describe how to wrap a contour around the spider diagram shown in Figure 15-3. (This is the same spider diagram shown in Figure 15-2, except that now it has a contour.) Assume that uncertain variables are normally distributed and that we know their means and standard deviations. A probability must be selected on which to base the contour. In this case, we choose 68% because it represents about one standard deviation from the mean in a normal distribution. Thus the range of values within which each uncertain variable is expected to be found 68% of the time can be calculated (i.e., a high of $\mu + \sigma$ to a low of $\mu - \sigma$). This range is indicated in Figure 15-3 by the segment of the vertical axis that corresponds to the vertical distance between the two x's on each function. Joining these x's with a continuous line forms the 68% contour.

According to Perry and Hayes (1985, p. 504), such a contour implies that 68% of the possible outcomes fall inside the contour and this "is then a subjective opinion by the estimator of the likelihood of occurrence." This might

[7]Note that if the measure of ORR were also given in percentage differences, the best-guess ORR would be at the origin.

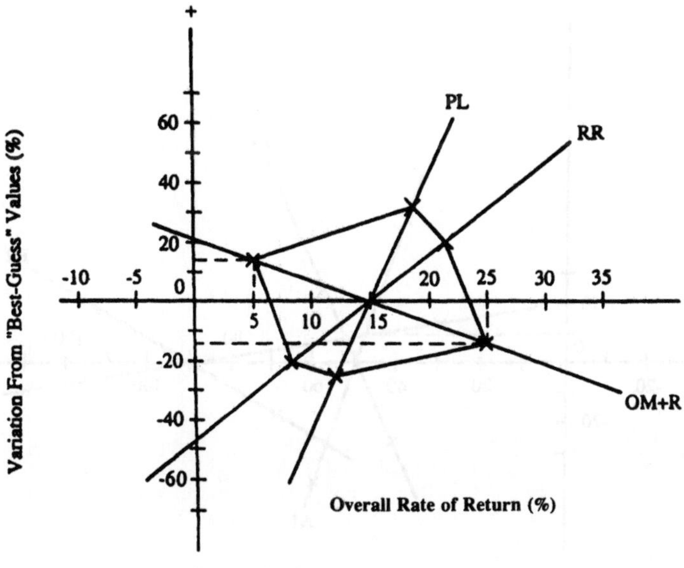

Figure 15-3. Spider diagram with a contour.

be true for values of each of the input variables, but it is *not* true of the measure of worth, i.e., the ORR. The contour reveals nothing about the likelihood of obtaining any single value of the ORR or the range of possible values of the ORR, because it applies only to the maximum variation in an individual variable where all other variables are held constant at their best-guess values.

The most that can be inferred from Figure 15-3 is the range in ORR values that might be expected 68% of the time for variation in any one of three variables while the other two are held constant. This can best be seen, however, by ignoring the contour and simply finding the corresponding ORR values for the *x* marks on each variable function. For example, in Figure 15-3 there is a 68% probability that OM+R will be in the range bounded by the two *x*'s of 14% above to 14% below the best-guess estimate. Assuming that all other variables are at their best-guess values, this translates into a range in the ORR of 5% to 25%, as shown by the dotted lines to the horizontal axis.

15.3.3 Spider Diagrams for Competing Projects

Spider diagrams are helpful when comparing competing projects as long as the decision maker keeps in mind that extreme values of the measure of worth reflect variations in one variable only. For example, look at the spider diagram

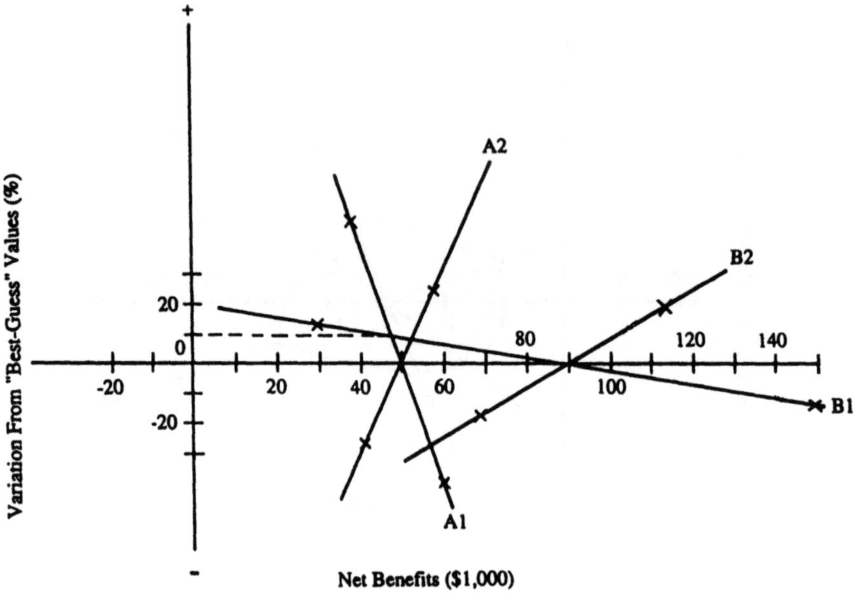

Figure 15-4. Spider diagrams for competing projects.

for projects A and B in Figure 15-4. The NB of project A is a function of variables A1 and A2, and the NB of project B is a function of variables B1 and B2. The horizontal axis suggests that project B has a higher present value NB ($90,000) than project A ($50,000). That is, if only best-guess values were used in a single-value, deterministic approach, B would be the preferred project. However, if we assign, say, a 50% confidence interval about the uncertain variables A1, A2, B1, and B2, as shown by x's on the functions, there appears the possibility that project A could yield a higher NB than project B. That is, within that confidence interval, if the extreme B1 value to the left were to occur, project B would yield a smaller NB than would project A for A1 or A2 extreme values to the left. Furthermore, if A1 and B1 were the same input variable, we would know that project A would be preferred at values of A1 and B1 above 10% over the best-guess value, and project B would be preferred at values of A1 and B1 below 10%.

Once again, however, sensitivity analysis gives no indication of the probability of any given value of NB. Furthermore, because only one variable is allowed to change at a time, and because NB is a function of more than one variable, sensitivity analysis gives an incomplete description of possible outcomes.

See Sections 19.1 and 23.1 and Chapters 21 and 22 for case studies that apply sensitivity analysis in the evaluation of building decisions.

REFERENCES

Flanagan, Roger, Adrian Kendell, George Norman, and Graham Robinson. 1987. Life-cycle costing and risk management. *CIB Proceedings of the Fourth International Symposium on Building Economics*. Copenhagen: Danish Building Research Institute.

Hendrickson, Robert G. 1984. *A Survey of Sensitivity Analysis Methodology*. National Bureau of Standards Interagency Report 84-2814. Germantown, Maryland: National Bureau of Standards.

Hillier, F. 1969. The evaluation of risky interrelated investments. *Studies in Mathematical and Managerial Economics*, ed. Henri Theil. Amsterdam: North Holland Publishing Company.

Marshall, Harold E. 1988. Techniques for Treating Uncertainty and Risk in the Economic Evaluation of Building Investments. SP757. Gaithersburg, Maryland: National Institute of Standards and Technology.

Perry, J. G., and R. W. Hayes. 1985. Risk and its management in construction projects. *Proceedings Institution of Civil Engineers* 78 (June): 499–521.

16

Risk-Adjusted Discount Rate and Certainty Equivalent Techniques

16.1 RISK-ADJUSTED DISCOUNT RATE

One technique used by the business community to account for risk is the risk-adjusted discount rate (RADR). The objective of using the RADR technique is to increase the likelihood that the investor will earn a return over time sufficient to compensate for the extra risk associated with specific projects.

Projects with anticipated high variability in distributions of worth have their returns discounted at higher rates than projects with low variability. Thus, in computing the NB or BCR, the discount rate is higher for benefit streams of risky projects than for those with certain outcomes. Or, when applying rate-of-return methods, the MARR is raised above its normal level to compensate for the higher variability of returns in risky projects.

The RADR can be found by equation 16.1:

$$RADR = RF + AR1 + AR2 \tag{16-1}$$

where

RF = risk-free rate,
$AR1$ = adjustment for normal risk encountered in the firm's operations,
$AR2$ = adjustment for extra risk above or below normal risk, with all terms expressed as percentages.[1]

The risk-free rate (RF) component is required to account for the time value of money. It is what might be earned, for example, on U.S. Treasury bills, the closest thing to a riskless investment available to most investors. The adjust-

[1]For additional information on the RADR, see Clark et al. (1984, pp. 180–189) and Baumol (1977, pp. 620–621).

ment for normal risk (AR1) is the risk premium that a firm might impose to cover the average riskiness of its normal operations. The sum of RF and AR1 should equal the MARR that the firm requires on typical investments. The AR2 component adjusts for projects with more or less risk than what is normally associated with the firm. You can adjust for risk exposure and/or risk attitude. The adjustment can be positive or negative.

For discounting *benefit* streams, AR2 is an increasing function of the (1) perceived variability in project outcomes (risk exposure) and the (2) degree to which the decision maker is risk averse (risk attitude). For *cost* streams, AR2 is a decreasing function of those same risk factors.

When using the RADR, discount each benefit and cost stream with a unique RADR that includes AR1 and AR2 values that describe that stream's uncertainty.[2] For benefit or savings streams, adjust AR1 and AR2 upward as perceived risk increases. That is, as future benefits become more uncertain, the RADR technique requires the discount rate to be raised to make the project look less desirable. For cost streams, adjust AR1 and AR2 downward as perceived risk increases. That is, as future costs become more uncertain, correct application of the RADR technique requires the discount rate to be lowered to make the project look less desirable. This point is rarely made in the literature.[3] It follows then that the appropriate adjustment for uncertain cost streams when using LCC analysis is a decrease in the discount rate to make project costs appear higher.[4] Raising the discount rate for uncertain cost streams will bias decision makers toward projects with a greater risk of higher than anticipated costs.

Let us look once again at the BCR of the time clock (described in Section 15-3) for an illustration of the RADR technique when making an accept/reject decision. If no unusual risk is associated with the time clock, the discount rate

---

[2]An alternative RADR approach is to use a single discount rate with a constant risk adjustment to evaluate positive or negative net cash flows. The constant risk adjustment value for AR1 and AR2 is, in effect, a composite of the risk adjustments that might be made individually on benefit and cost elements about which there is uncertainty in a given project. Instead of discounting each benefit and cost stream with a unique discount rate, an implicit weighted average of the uncertainty is incorporated in the RADR value assigned a given project. This approach, albeit common, is *not recommended* because it distorts the risk adjustment mechanism.

[3]The only statement in the literature the authors have found that recommends lowering discount rates when evaluating uncertain cost streams is in Arrow and Lind (1970, p. 378). They say, "if net benefits to an individual are negative, this requires discounting expected returns at a rate lower than the certainty rate. Raising the rate of discount only reduces the present value of net benefits when they are positive."

[4]If values in a project's cost stream are selected from the higher end of the distribution of possible costs, lowering the discount rate via the RADR technique will further raise cost estimates for the uncertain stream and bias project selection toward possibly less economic alternatives.

is equal to the sum of RF and AR1, as shown in equation 16-1. Suppose the BCR for the time clock is 2.0 computed with a discount rate that reflects RF and AR1. This would be an acceptable project.

Now assume instead that the economic performance of the time clock is more risky than average. This situation might arise, for example, from the impact of uncertain kWh reductions or uncertain future energy prices. Furthermore, assume that the decision maker is risk averse. Using the RADR technique, we raise the discount rate for evaluating energy cost savings by some positive value of AR2. If the resulting BCR falls below 1.0, the project no longer appears economically acceptable.[5]

Advantages of the RADR technique are that it is relatively simple to understand; it is easy to compute; and it accounts, to some extent, for the uncertainty of inputs, risk exposure, and risk attitude.

A major limitation of the RADR is the lack of any accepted procedure for establishing the RADR value. Decision makers typically estimate it on the basis of their best judgment. One common approach is to simply lump projects into "risk categories," each of which has an assigned RADR. There is little fine tuning. Furthermore, there is no distinction between adjustments for handling risk exposure and risk attitude.

Another shortcoming is that analysts typically apply a constant AR2 over the entire study period, even when there are periods for which no special adjustment is necessary for risk above or below what is considered normal risk. Furthermore, discounting with an RADR that includes a constant AR2 over the whole study period implies, in effect, that returns become exponentially more uncertain over time.[6] Thus a discount rate that includes a constant AR2 severely reduces the weight of NB accrued in later years, regardless of the certainty of their occurrence. This biases selection toward projects with early payoffs.[7]

See Chapter 22 for a case study that applies the RADR in the evaluation of a lease or buy decision.

---

[5]The RADR technique as described here generates a single measure of project worth. The decision maker incorporates implicitly the probabilities of uncertain inputs and risk preference attitudes in the determination of a value for the RADR. However, the RADR technique can be coupled with the mathematical/analytical technique (Section 17.3.3) to generate probability density functions and/or cumulative distribution functions of measures of project worth. In this coupling of techniques, risk exposure is indicated explicitly by the distributions of project worth, and risk attitude is treated explicitly in the adjustment component to the RADR.

[6]In many cases, it may be reasonable to assume that estimates for distant years are of lower quality than those for near years.

[7]For an explanation and illustration of how perceived risk is implicitly assumed to increase over time with a constant RADR, see Weston and Brigham (1981, pp. 479–480 and 482–484).

16.2 CERTAINTY EQUIVALENT TECHNIQUE

With the certainty equivalent technique, the decision maker determines a certainty equivalent factor by which project returns are adjusted to reflect risk exposure, risk attitude, or both. Adjusting estimated returns of an investment by a certainty equivalent factor yields a "certainty equivalent" amount that a decision maker finds equally acceptable to making the investment with its uncertain outcome.

Consider how the certainty equivalent technique is used with the NB method. We compute NB in the conventional manner,[8] except for multiplying net cash flows in each period by a certainty equivalent factor (CEF) to account for risk. The certainty equivalent version of NB is found with equation 16-2:[9]

$$NB_{ce} = \sum_{t=0}^{N} \frac{CEF_t\,(B_t - \overline{C}_t)}{(1 + RF)^t} \qquad (16\text{-}2)$$

where

NB_{ce} = certainty equivalent NB,
CEF_t = certainty equivalent adjustment factor in time period t, and
RF = risk-free discount rate.

The discount rate is risk free since risk is now handled through the CEF. As perceived risk exposure increases, CEF_t becomes smaller, making NB_{ce} smaller. And as aversion to risk increases, CEF_t becomes smaller, likewise making NB_{ce} smaller. Any CEF_t less than 1.0 implies that an investment will be worth less than the estimated NB because of the decision maker's assessment of risk.[10]

Two approaches are used to derive CEFs. The first derivation of CEFs is based on risk attitude. Suppose that a decision maker specifies a certain payment (the certainty equivalent) that must be received to make him or her indifferent between that certain payment and the expected value of a return that is uncertain. For example, assume that a person has a choice between a certain payment

[8] See equation 3-1 for the conventional formulation of NB.

[9] If B_t-$\overline{C}_t$ is negative, a CEF_t that is reduced in value to reflect risk will make NBs larger instead of smaller. Thus the certainty equivalent technique does not handle negative NB cash flows appropriately.

[10] For an LCC analysis, the *CEF* adjustment for uncertain cost streams would be in the opposite direction. That is, as risk aversion and exposure increase, *CEF* becomes larger, making LCC_{ce} larger.

of money and a lottery with a 50% probability of winning $1,000 and a 50% probability of winning nothing. For some amount of certain payment, the decision maker will be indifferent between the lottery and that certain payment. Once this tradeoff is specified, the CEF can be derived using equation 16-3.

$$CEF = \frac{\text{certain payment}}{\text{expected value of uncertain return}} \tag{16-3}$$

If a decision maker accepts a certain payment of $500 in lieu of the lottery described above, then the decision maker is risk neutral and, according to equation 16-3, $CEF = 1.0$ (i.e., $500/$500). If the decision maker is indifferent between the lottery or a sure payment of $550, the decision maker is a risk taker and $CEF = 1.1$ (i.e., $550/$500). On the other hand, if the decision maker is indifferent between the lottery or a sure payment of $400, the decision maker is risk averse and $CEF = 0.8$ (i.e., $400/$500).

Although the decision maker might have some vague notion of the level of investment risk using this approach, risk exposure is not included explicitly in the derivation of the CEF.

The second approach to deriving CEFs is based on risk exposure. This approach seems particularly appropriate for private firms that have historical records of business performance for different types of investments. The records can be used to establish a table of CEF values for evaluating each type of investment. The first step in deriving a CEF table is to divide the firm's investments into categories such as "new plant" and "equipment replacement," as shown in Table 16-1. The next step is to divide these general categories into subcategories based on their historical risk-return records, as indicated by their coefficient of variation.[11] The final step is to assign CEFs by year according to the coefficients of variation.

The greater the coefficient of variation of cash flows for a given investment subcategory, the greater is the risk exposure associated with that subcategory, and therefore the smaller will be the subjectively determined CEF assigned to each year for that subcategory. Looking at Table 16-1, for example, for each year CEF_t for "new plant B" would be smaller than for "new plant A." Note also that, since the performance of replacement equipment is likely to be more certain than that of new equipment, the CEFs would typically be larger for the replacement equipment.

[11]The coefficient of variation is the standard deviation divided by the mean value. See Section 17.3.2 for a discussion of the coefficient of variation.

Table 16-1. A firm's CEF table for use in adjusting net cash flows from uncertain investments

Investment Categories	Coefficient of Variation	Certainty Equivalent Factor		
		Year 1	Year 2	Year 3
New plant A	$v \leq 0.15$	0.94	0.90	0.84
New plant B	$v > 0.15$	0.90	0.86	0.79
Equipment replacement A	$v \leq 0.15$	0.98	0.94	0.89
Equipment replacement B	$v > 0.15$	0.94	0.88	0.81

The procedure for using CEFs established in Table 16-1 in an NB_{ce} evaluation is as follows:

1. Determine for the proposed investment the appropriate investment category and its coefficient of variation.
2. Find the CEF_t for each year for that investment category and its coefficient of variation.
3. Use equation 16-2 to determine NB_{ce}.

To illustrate the impact on project choice of handling risk with the certainty equivalent technique, consider once again the time clock. Two decision makers are individually examining the same time clock investment. Assume that one views the weighted average of CEF_t as equal to 1.0 for the time clock. Assume further that, for this case, NB, and therefore NB_{ce}, are positive. Thus the decision maker finds the time clock economically acceptable. Suppose that the second person assigns a weighted average of CEF_t that is less than 1.0. Using the same NB value, NB_{ce} for the second person will be lower than it was for the other person. The adjustments for risk make the project appear less attractive economically to the second person than to the first person.

One advantage of the certainty equivalent technique is that it accounts for risk by a factor that can include both the decision maker's risk attitude and assessment of risk exposure. If the CEF is based on risk attitude (equation 16-3), an additional adjustment can be made to account for risk exposure. If, on the other hand, the CEF is based on risk exposure (historical risk-return records), an additional adjustment can be made to account for risk attitude.

Another advantage of the certainty equivalent technique is that it separates discounting that accounts for time value of money from adjustments for risk, both of which are combined in the RADR technique. Thus the certainty equivalent technique allows for differential risk "weighting" over time. This is probably more appropriate than the increasingly heavy discounting for risk over time implicit in the RADR technique. Finally, the certainty equivalent technique can be used with probabilistic data and formal utility models.

A limitation of the certainty equivalent technique is that it is "data hungry," and the procedure for "feeding" it is to a large extent subjective. Another limitation of this technique is the lack of a rigorous, theoretically defensible, mathematical expression for establishing a CEF that combines risk attitude and risk exposure. While it is an advantage of the CEF that risk attitude and risk exposure can be treated in one factor, combining the two types of risk adjustment subjectively is not acceptable to many decision makers. The first approach, using equation 16-3, fails to measure explicitly risk exposure. The second approach, based on historical experience, treats risk exposure explicitly but attitude only implicitly. A certainty equivalent technique that *explicitly* treats exposure and attitude would assure a more comprehensive, consistent, and objective treatment of risk in project evaluation.

REFERENCES

Arrow, Kenneth J., and Robert C. Lind. 1970. Uncertainty and the evaluation of public investment decisions. *The American Economic Review* LX(3):364–378.

Baumol, William. 1977. *Economic Theory and Operations Analysis*. 4th edition. Englewood Cliffs, New Jersey: Prentice-Hall, Inc.

Clark, J. J., T. J. Hindelang, and R. E. Pritchard. 1984. *Capital Budgeting: Planning and Control of Capital Expenditures*. Englewood Cliffs, New Jersey: Prentice-Hall, Inc.

Marshall, Harold E. 1988. Techniques for Treating Uncertainty and Risk in the Economic Evaluation of Building Investments. SP757. Gaithersburg, Maryland: National Institute of Standards and Technology.

Weston, F. J., and E. F. Brigham. 1981. *Managerial Finance*. Hinsdale, Illinois: The Dryden Press.

17

Decision Analysis, Simulation, and Other Techniques

17.1 DECISION ANALYSIS

Decision analysis is one of the few techniques for making economic decisions in an uncertain environment that formally treats both risk exposure and risk attitude. It provides a methodology that allows a decision maker to include alternative outcomes, risk attitudes, and subjective impressions about uncertain events in an evaluation of investments.[1]

Decision analysis typically uses decision trees to represent all possible outcomes, costs, and probabilities associated with a given decision problem. A decision tree is a decision-flow diagram that serves as a road map to clarify possible alternatives and outcomes of sequential decisions. We use a decision tree in this section to illustrate how it helps bring order to complex decisions about risky investments.

Procedures for using decision analysis are documented extensively in the literature.[2] In general, the decision analysis approach has the following three steps:

1. Structure the problem. This includes defining variables, describing their relationships with models, assigning values to possible outcomes, and measuring the importance of the variables through sensitivity analysis.
2. Assign probabilities to important variables and possible outcomes, and find the "best-bet" alternative. This includes describing uncertainty with probability distributions,[3] describing risk attitude with a utility function

[1]Decision analysis as defined here is based on Bayesian analysis; that is, uncertainty regarding input variables and outcomes is quantified probabilistically, and statistical conclusions are interpreted in the direct probability sense. For a comprehensive argument that supports Bayesian decision theory, see Berger (1985).

[2]See, for examples, Berger (1985), Matheson and Howard (1977), Holstein (1977), and Raiffa (1970).

[3]In many cases, the decision analyst will have to develop subjective probability distributions. To do this, the analyst asks the decision maker or an expert(s) designated by the decision maker a series of probing questions designed to reveal the best judgments available on the likelihood of uncertain events.

(Appendix E), and finding the alternative that is expected to yield the greatest economic return (or utility if the decision maker is not risk neutral).

3. Determine whether obtaining additional information is worth the cost. If it is, collect the information and repeat steps 1 and 2.

This three-step procedure is not required for every decision analysis. In fact, several of the operations just described are not included in the following illustration. However, the procedure is useful as a reference when deciding which operations are necessary for applying decision analysis to a specific problem.

17.1.1 Decision Analysis of New Product

Figure 17-1 is a decision tree. It describes for a building equipment manufacturer the possible events in developing and marketing a new product that preliminary research shows to be promising. The purpose of the decision tree is to help the decision maker decide if it is more economical to pursue a policy of product development and marketing or to abandon further investment in the product.

The decision tree is read from left to right, starting with the left-hand box on Figure 17-1. The box, like all boxes in a decision tree, represents a "decision juncture" or "node," and the line segments represent "alternative branches" on the decision tree. Circles represent chance nodes. The objective of the equipment manufacturer is to choose the branch sequence expected to maximize the present value of NB. The first decision node shows two alternatives: (1) abandon product development (top branch) and (2) develop the product (lower branch). The $2 million cost projected for product development appears beneath

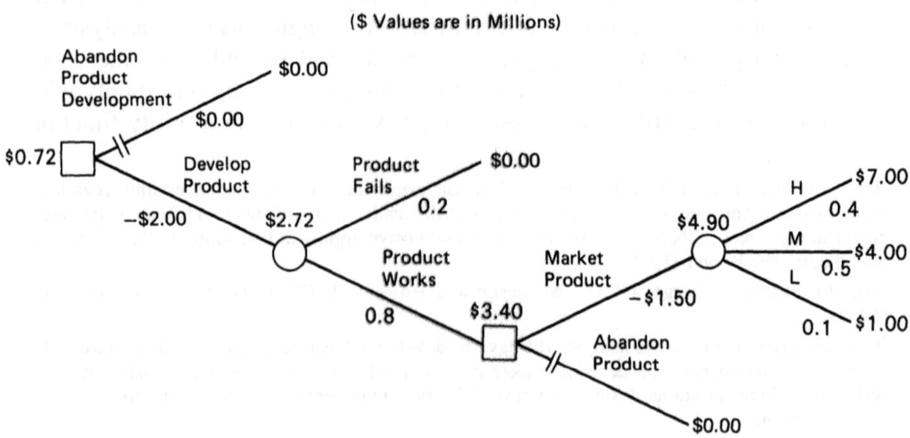

Figure 17-1. Decision tree for a new building product.

the line. Abandoning product development would cost nothing (amount below the line) and yield nothing (amount at the tip of the line).

The outcome or payoff depends on a sequence of certain chance events and on choices by the equipment manufacturer. Hence, at the end of the development branch is a "chance node" or "juncture" followed by two "alternative outcomes." The top line represents product failure, which yields a zero return, as shown at the tip of the branch. The manufacturer assigns a probability of 20% (shown beneath the line) that the product will fail to meet performance goals. The lower line segment represents success (with an 80% probability) that the development phase will yield a product that meets performance goals.

The payoff from developing the technically feasible product depends on another decision (the box to the right) and the likelihood of commercial success. Thus, if the product is abandoned, there will be a zero return, as shown at the tip of the abandon branch. If, on the other hand, the choice is to market the product, at a cost of $1.5 million, the payoff depends on which of three possible levels of commercial success result: high (H) level of success, with earnings of $7 million (40% probability); moderate (M) level, with earnings of $4 million (50% probability); and low (L) level, with earnings of $1 million (10% probability).

To establish the probabilities, earnings, and costs shown in Figure 17-1, the analyst must work with the firm's managers, potential retailers and users, financial advisors, and others who know the industry. The values shown at each decision and chance node, on the other hand, must be calculated. The following steps describe the calculation of the node values and the selection of the sequence of branches that maximizes NB.[4] Note that the decision tree is solved from right to left.

First, starting from the right-hand side of the tree, average out the expected value for each chance node; that is, calculate a weighted average for each probability fan by summing the products of the possible outcomes weighted by their respective probabilities. Taking the probability fan in Figure 17-1 for level of marketing success (the far right), we compute the expected value (*EV*) as follows:

$$EV = 0.4(\$7{,}000{,}000) + 0.5(\$4{,}000{,}000) + 0.1(\$1{,}000{,}000)$$
$$= \$4{,}900{,}000$$

Write the expected value atop the chance node (circle), as shown in Figure 17-1.

---

[4]The objective here is to maximize the expected value of NB. If the decision maker is risk neutral, maximizing NB will be consistent with maximizing the expected value of utility. The procedure for using decision analysis when the decision maker is not risk neutral is elaborated later in this section.

The next step is to fold or roll back to the next preceding stage. That is, evaluate alternative branches at each decision node for their expected benefits and costs, and select the one with the highest NB (all benefits and costs are assumed to be in present value terms). Write the maximum value of NB atop the decision node box. Looking again at the right-hand side of Figure 17-1, we subtract the $1.5 million of marketing costs from the $4.9 million expected value of earnings to arrive at an expected value of net income of $3.4 million. Since the other path, abandonment of the product, returns zero dollars of net income, it is a less desirable path. Parallel slash marks are used to truncate this branch and indicate that it is not the most economic alternative.

Moving to the left to the next chance node, once again there is an averaging out of expected value. Given a 20% chance of the product failing to meet performance goals and an 80% chance that it will meet goals, compute expected value as follows and write it atop the chance node.

$$EV = (0.2 \cdot \$0.00) + (0.8 \cdot \$3{,}400{,}000) = \$2{,}720{,}000$$

Rolling back one more time to the starting decision node, we subtract the $2 million development cost from the $2.72 million expected value to arrive at an NB of $0.72 million. This is the value written next to the decision node because it is higher than the zero dollar return on the alternative branch of abandoning project development. Slash marks on the "abandon" branch indicate that it is a less desirable path. Thus the decision tree in Figure 17-1 displays the economically efficient branch to be development of the product, and the expected value of NB from taking that branch to be $0.72 million.

The decision tree in Figure 17-1 does not account for risk attitude. No treatment of risk attitude is necessary if the decision maker is risk neutral, because the present value of NB is an accurate index of utility. If the decision maker is risk averse or a risk taker, however, maximizing expected value NB is insufficient. In this case, convert projected earnings and costs associated with each decision branch to utility values. A utility function (see Chapter 14 and Appendix E) shows utility values corresponding to specific benefits and costs. The averaging out to find expected values (now expected utility values) and the rolling-back process is the same as described earlier. Once a maximum utility value is identified, its certain equivalent dollar value is found on the utility function. The certain equivalent value shows what the investment is worth, taking into consideration the decision maker's risk attitude.

17.1.2 Advantages and Disadvantages of Decision Analysis

An advantage of decision analysis is that it describes multiple investment alternatives in one series of calculations. For example, to evaluate each decision path sequence in Figure 17-1 as a separate project requires six separate NB calculations. This can be done. The decision tree, however, provides a structure for solving directly and efficiently for the specific alternative that is likely to yield the greatest NB. Since we truncate inefficient alternative branches in the decision tree, the analysis requires fewer calculations to arrive at the single NB value for the most efficient path sequence.

Another advantage of the decision tree structure is that it provides perspective in defining investment alternatives. It forces analysts to examine how the different alternatives and chance events relate to one another. And it helps them iteratively to embellish, prune, and refine alternative branches before beginning the process of calculating the most efficient investment.

A disadvantage of decision analysis as shown here is that NB values of truncated investment paths customarily do not appear on the tree. Thus the relative economic worth of alternative designs may not be apparent to the decision maker who sees only evaluation results. Lack of this information could be important if other considerations make the preferred investment an alternative that is *not* most economically efficient. The tree structure, however, contains all the information necessary to generate a measure of project worth for each path sequence if a comprehensive picture is desired.

Another possible problem with the decision tree analysis as shown in Figure 17-1 and discussed here is that there is no explicit measure of the risk of choosing an uneconomic project. The probability fans are for benefits only, instead of NB measures of worth. Thus it is not obvious what the NB distributions look like. This should not make any difference to decision makers who are confident in their statements of probability, because risk exposure and attitude are both handled in the computation to find the preferred alternative. However, if desired, the data presented in the decision tree can be used to generate probability functions and cumulative distribution functions of NB for each design.[5] With these functions, decision tree analysis provides a profile of risk exposure.

See Section 19.2 for a case study that applies decision analysis in the evaluation of a building design decision.

17.2 SIMULATION

Simulation is a technique commonly used to determine risk exposure from an investment decision.[6] To perform a simulation, probability functions of the input variables are required. The simulation process for building a probability

[5]See Section 19.2.4 for an illustration of how data in the tree can be converted to probability functions and cumulative distribution functions that describe risk exposure.

[6]Simulation is a useful technique for doing things other than measuring risk exposure, but we focus exclusively on that use.

density function and a cumulative distribution function of the measure of project worth is as follows:

1. Draw randomly a value for each input variable from its probability function.
2. Substitute the set of input values for that round of draws into the measure-of-worth formula and compute the measure.
3. Repeat the process over and over until a probability density function and a cumulative distribution function can be constructed for the measure of worth.

For example, in analyzing the time clock, the initial energy price, rate of energy price escalation, and kWh savings are uncertain input variables. By describing each of these inputs with a probability distribution and performing a simulation, we can arrive at a probability distribution of the time clock's BCR (or some other measure of worth). Select randomly each of the three variables and combine them with constant inputs to compute a BCR. Repeat this random sampling over and over, typically 500 to 1,000 times. Compute the BCR for each combination. From the resulting set of BCRs, generate a probability density function and a cumulative distribution function for evaluating the economic efficiency of the time clock.

Newnan (1980), Pouliquen (1970), and Hertz (1964) discuss in detail the application of simulation in measuring the risk of capital investments. Here we illustrate how simulation helps determine construction contingencies. The example has a special twist in that the cumulative distribution generated by the simulation describes the probability of a cost overrun rather than of a measure of economic worth. We present the contingency example because it illustrates the risk analysis capability of simulation with a problem familiar to many people in the building and construction industries. An elaborate illustration of simulation that is applied to a measure of worth is presented in Section 20.4.

17.2.1 Construction Contingency Simulation Example

Cost engineers use contingency analysis routinely in estimating the costs of construction projects. A contingency is a cost element included in project cost estimation to cover costs that have some likelihood of occurrence but whose amounts cannot be predicted with certainty. By adding the contingency to line-item estimates of project costs, the cost engineer hopes to project the most likely final cost. Typical uses of contingencies are to cover possible increases in material or labor costs beyond normal escalation, unanticipated developments in applying a first-time technology, changes in project scope due to omission or error, or unforeseen work disruptions when operating in a politically volatile foreign country.

Analysts often estimate a contingency simply as a percentage of the base estimate of project cost. Historical data on the differences between actual and estimated costs for similar projects are sometimes useful in determining an average percentage of underestimation (or overestimation). The percentage can apply to the overall project or to specific project elements which are estimated separately. This simple percentage approach is typical in estimating the cost of small projects. There is no distinction, however, between accounting for risk exposure and risk attitude in the contingency estimate.

For large construction projects having many uncertain variables, a risk-analysis technique based on simulation is sometimes used to estimate contingencies. It provides decision makers with probabilities of cost overruns (i.e., risk exposure) associated with every possible contingency markup in the relevant range. When the decision maker chooses a contingency markup and thereby establishes an acceptable probability of overrun, risk attitude is also accounted for. We have adapted the following example from an article (Zaheer, 1983) that appeared in the *Cost Engineers' Notebook*. It illustrates how to use simulation to measure risk exposure when making a cost estimate for a specific construction project.

The objective is to estimate construction costs for project X. Expected costs, exclusive of contingencies, are $140 million. Of the $140 million, $60 million are spent dollars or firm commitments. Being relatively certain, they require no consideration for contingency. The other $80 million are uncertain and make up the base on which the contingency is calculated.

The process for carrying out a contingency/risk analysis is as follows. Generate probability distributions for every activity that is deemed particularly uncertain. The distributions describe the percent of estimated costs of these activities, where the midpoint is 100% of the estimated value. Enter these data, along with estimated dollar costs of both certain and uncertain activities, into a computer simulation package. The package generates a probability distribution of the contingency percent of the total project cost and a graph that plots probability of cost overrun against contingency percent and amount.

Figure 17-2 shows how the probability of a cost overrun (i.e., risk exposure) varies with the contingency adjustment for construction project X. To use this contingency/risk analysis to help select a single cost estimate, the decision maker considers risk exposure and risk attitude. Risk exposure, as indicated by the rising probability of a cost overrun, increases as the percent contingency markup goes down. Risk exposure decreases as the percent contingency

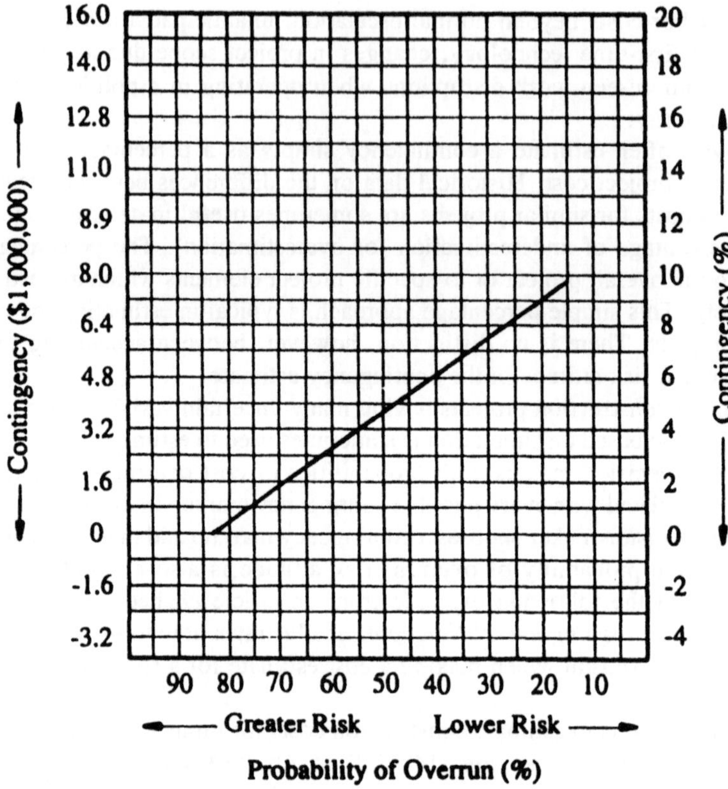

Figure 17-2. Contingency/risk graph for construction project X. (Adapted from Zaheer, 1983, p. 13.)

markup increases. Risk attitude enters when the decision maker chooses a contingency amount, thereby establishing a probability of overrun that will be acceptable.

The risk-neutral decision maker will choose the most likely cost estimate of $144 million, which includes the $140 million without contingency plus a contingency of $4 million (0.05 · $80 million of uncertain costs). That is, the vertical axis on the right side of the graph shows the most likely contingency percent (where the probability of overrun is 50%) to be about 5%. This assumes an underlying probability distribution which is symmetric.

A more risk-averse decision maker might opt for a lower risk of overrun by choosing a larger contingency. For example, if an overrun probability of only 20% is acceptable, the contingency will be $7.2 million (0.09 · $80 million) and the total cost estimate $147.2 million ($140 million plus $7.2 million contingency).

A risk taker, on the other hand, might choose a relatively low contingency. For example, if an overrun probability of 70% is acceptable, the contingency will be $1.6 million (0.02 · $80 million) and the total cost estimate $141.6 million ($140 million plus $1.6 million contingency). Note that Figure 17-2 is really a cumulative distribution function similar to that described in Figure 14-2, with two exceptions. First, Figure 17-2 reverses the axes. The item whose distribution is being measured (contingency percent and amount) is on the vertical axis instead of the horizontal axis, and the cumulative distribution measure (probability of overrun) is on the horizontal axis. Second, the cumulative distribution function in Figure 17-2 measures cumulative probabilities of a value being "greater than" instead of "less than," as in Figure 14-2. Thus, for example, there is about a 35% probability of a cost overrun with a contingency of 7%. That is, there is a 35% probability that the necessary contingency to avoid an overrun is greater than 7%.

17.2.2 Advantages and Disadvantages of Simulation

Simulation has several advantages when applied to risk analysis. First, simulation works with any kind of distribution of input variables, so it is not limited to certain classes of well-behaved distributions.

Second, it can handle interdependencies between inputs. That is, where one input relates to another, the two can be tied together. For example, if input A depends on input B, first draw a value for input B. Then search among the several probability density functions of input A for the one that, in its range, includes values for input A corresponding to the value drawn for input B. There might be several probability density functions of input A that together would map to every possible value of input B.

Third, simulation provides the necessary data for generating probability density functions and cumulative distribution functions for any of the economic methods described in Chart 1-3.

Fourth, while it is true that cumulative distribution functions generated by simulation describe only risk exposure, they also facilitate incorporation of risk attitude by the decision maker in the decision process. That is, when the decision maker selects projects on the basis of cumulative distribution functions, inherent in that selection is an implicit assessment of risk attitude.

The disadvantage of simulation is that it requires many calculations, thereby making it practical only when used with a computer.

See Section 20.4 for a case study that applies simulation in the evaluation of an accept/reject decision.

17.3 OTHER TECHNIQUES

17.3.1 Input Estimation Using Expected Values[7]

Expected value analysis (also known as "probability analysis") is one technique for estimating uncertain input data. The "expected value" of an input is the sum of products of possible input values and their respective probabilities of occurrence. The values of any number of inputs can be estimated by this technique. Expected input values ultimately become data for computing a single-value measure of project worth.

Again we use the time clock for illustration. To calculate the BCR for the time clock, we examine uncertainty regarding the critical kWh input. The kWh savings resulting from the time clock determine in part its BCR. Taking, for example, an annual after-hours energy consumption of 230,000 kWh, we might expect one of the following three levels of kWh reductions in consumption: 184,000, 138,000, or 92,000 kWh. Table 17-1 shows how the single expected value of kWh reduction can be found from potential kWh reductions and their estimated probability weights.

The expected reduction of 152,000 kWh is unlikely to be the exact reduction achieved. Yet, if expected values were used over repeated applications, the differences between actual values and predicted values would tend to be less

Table 17-1. Expected value of reduction in kWh

Reduction in Energy (1,000 kWh) (1)	Probability (2)	Expected Value of Reduction (3) = (2) · (1)
184	0.4	74
138	0.5	69
92	0.1	9
	Expected kWh reduction	152

[7]Note that expected values can be used with other techniques for handling uncertainty and risk. Expected value analysis applies, moreover, to measures of project worth as well as to inputs. The expected value technique, as used here, however, refers specifically to the weighted average of an input value and is described in this section solely as a way of handling uncertainty associated with input variables.

than if simple ''best-guess'' estimates were used. This is the primary advantage of using expected value analysis to estimate input values.

To compute the monetary benefits of the time clock, combine the expected value of energy reductions with estimates of values (point estimates or expected values) for the energy escalation rate and initial energy price. Uncertainty regarding kWh reductions is treated to the extent that several possible values, with their associated probabilities of occurrence, are considered.

A major shortcoming of this technique is that a single BCR provides no measure of risk exposure. Also, there is no treatment of risk attitude. Finally, applications are limited to problems where analysts can develop input distributions.

17.3.2 Mean-Variance Criterion and Coefficient of Variation

Comparing the mean values and standard deviations of the measure of project worth helps decision makers evaluate the returns and risk exposure of one project versus another. It also helps determine stochastic dominance.[8] If two projects competing for limited funds are compared on the basis of BCRs, for example, the mean-variance criterion dictates that the one with the higher mean (i.e., expected value) and lower standard deviation be chosen. This presumes that decision makers prefer higher rather than lower BCRs and less rather than more risk.[9]

If one project has a higher mean and higher standard deviation of the measure of project worth, then the choice is not clear with the mean-variance criterion. In this case, compute the coefficient of variation to determine the relative risk of the two projects. The coefficient of variation is found by dividing the standard deviation by the mean, as shown in equation 17-1:

$$CV = \frac{\sigma}{\mu} \tag{17-1}$$

where

CV = coefficient of variation,
σ = standard deviation, and
μ = mean or expected value.

[8]See Chapter 14 for a discussion of the mean, standard deviation, and stochastic dominance.

[9]Both the mean-variance criterion and the coefficient of variation are typically applied under the assumption that the decision maker is risk averse.

The project with the lower coefficient of variation has the lesser risk per unit of return or of project worth. It will be preferred by risk-averse decision makers.[10]

An advantage of the coefficient of variation is that it provides an explicit measure of relative risk exposure. The decision maker implicitly incorporates risk attitude when choosing among alternative projects on the basis of their coefficients of variation. The major limitation is in acquiring σ and μ values for the measure of project worth.

17.3.3 Mathematical/Analytical Technique

Hillier (1963) developed the mathematical/analytical (M/A) technique to obtain probability functions for economic measures of worth without the repeated trials of simulation. The M/A technique allows one to generate probability density functions and cumulative distribution functions of LCC, NB, and IRR measures.[11]

Equations 17-2 and 17-3 show how to compute the aggregated, discounted mean and variance for the NB measure of project worth. The input data are the mean and variance of NB cash flows for each year of the investment's study period.

$$\mu_{\text{NB}} = \sum_{t=0}^{N} \frac{E(B_t - \overline{C}_t)}{(1 + d)^t} \tag{17-2}$$

and

$$\sigma^2_{\text{NB}} = \sum_{t=0}^{N} \frac{\text{Var}(B_t - \overline{C}_t)}{(1 + d)^{2t}} \tag{17-3}$$

where

μ_{NB} = the mean of NB,
σ^2_{NB} = the variance of NB,
$E(B_t - \overline{C}_t)$ = the expected value (mean) of NB cash flows in time period t, and
$\text{Var}(B_t - \overline{C}_t)$ = the variance of NB cash flows in time period t.

---

[10]Another technique sometimes used for evaluating alternatives with just σ and μ is the "gambler's indifference map." Indifference curves, each of which connect equally attractive combinations of σ and μ, help the decision maker select the preferred alternative. For a description and case example of this technique, see Au et al. (1972, pp. 185–187).

[11]Park (1983) extended Hillier's model to develop a method for computing probability distributions of BCRs.

By using the mean and variance of NB earned on the investment and the probability tables for the normal distribution, we can derive probability density functions and cumulative distribution functions for the NB of the investment. We can also compute directly the probability of the investment yielding NB less than, equal to, or greater than zero (or some other specified target value).

The M/A technique requires that input values be normally distributed to generate probability density functions and cumulative distribution functions that describe measures of worth. Even without normal input distributions, however, the technique produces means and variances of economic measures of worth. And these statistics by themselves can be useful.

17.3.3.1 Advantages and Disadvantages of M/A Technique

A major advantage of the M/A technique is that you can calculate directly both means and variances (and therefore probability density functions and cumulative distribution functions) of project worth without repeated simulations.

Another advantage is that the decision maker implicitly incorporates risk attitude into the selection process when selecting a project based on cumulative distribution functions.

A disadvantage of the M/A technique is that it presumes that the underlying probability distributions associated with the inputs are normally distributed. This condition is necessary to make precise probability statements about measures of worth. Hillier (1963, p. 446) maintains, however, that subjective probability distributions that analysts generate in describing uncertain events are often symmetrical and resemble normal distributions. Furthermore, even where distributions are not normal, the mean and variance of the measure of worth, as computed by equations 17-2 and 17-3, will still be accurate and provide helpful information in evaluating alternative investments.

A possible disadvantage of the M/A technique is that interdependence is ignored. First, consider interdependence among a single project's measures of worth over time. For example, the level of interdependence among net cash flows in, say, years 1, 2, and 3 might range from total independence (zero correlation among net cash flows in different years) to total interdependence (perfect correlation among net cash flows in different years). Sensitivity analysis is one approach to handling this interdependence. Calculate the variance first with the assumption of zero correlation and then with the assumption of perfect correlation. A more accurate approach is to assume that net cash flows in some years are related and in other years they are not. Hillier (1969) presents a mathematical model for calculating the mean and standard deviation under conditions of varying interdependence.

A second interdependence issue arises when evaluating interrelated projects. The economic worth of one project may vary as a function of the worth of competing or complementary projects. For example, the returns from investment A correlate with the returns on investments B, C, and F in a group of projects because common factors affect the correlated projects. Hillier (1969) has developed techniques for evaluating large, risky, interrelated investments.

Another disadvantage of the M/A technique is that input distributions in each period must be of the measure of economic worth. That is, uncertainty expressed in input distributions pertains only to the final measure of project worth for that period. Thus, with the NB measure, for example, the distribution in each period is of NB cash flows. This means that analysts have to determine how all uncertain factors that affect benefits and costs interact in a given period before they can generate that period's NB cash flow distribution. The simulation technique, on the other hand, accommodates separate distributions for each of the uncertain inputs and provides for their systematic combining to form a final distribution of the measure of project worth. Note, however, that with simulation, input variables are all discounted to present or annual values, whereas with the M/A approach, there is a separate distribution for each period in the project's study period. Thus the M/A technique provides more opportunity for fine tuning estimated project performance from period to period.

See Section 20.5 for a case study that applies the M/A technique to an accept/reject decision.

17.3.3.2 M/A Technique Used with the RADR and the Certainty Equivalent Technique

The M/A technique can be used with the RADR technique and the certainty equivalent technique in computing NB. Uncertainty associated with variables used to compute NB cash flows period by period is captured in period-by-period probability distributions of NB cash flows. Thus the mean and variance of NB, plus probability density functions and cumulative distribution functions generated from those statistics, indicate ***risk exposure***. The certainty equivalent or RADR technique in this case incorporates the sole adjustment to net benefits for ***risk attitude***.

When using an RADR, its value, as found from equation 16-1, substitutes for discount rate, d in equation 17-2. Equation 17.3 is used as shown.[12] The more risk averse the decision maker, the higher will be the RADR.[13] When used with the M/A technique, the RADR adjusts *only* for risk attitude. Otherwise there is a double adjustment for risk exposure.

[12]The RADR and CEF adjustments apply to the mean, but not to the variance, when used with the M/A technique.

[13]This presumes that benefit streams are being discounted. If cost streams are being discounted, the RADR is adjusted downward to account for a risk-averse attitude.

The advantage of using a joint M/A and RADR analysis over using the M/A or simulation technique alone is that it treats risk attitude in a formal manner. Typically the mean and standard deviation (and probability density functions and cumulative distribution functions) reflect only risk exposure, and the decision maker has to make a subjective determination of risk attitude when considering each project alternative. When using the techniques jointly, on the other hand, the distributions already have risk attitude incorporated in them. That is, the NB values have been adjusted by the RADR to reflect risk attitude. Thus the decision maker's selection rule is simply to choose the project with the highest expected value of adjusted NB.

If decision makers are not comfortable choosing on the basis of expected values adjusted for risk attitude, additional selection rules might be used in determining which project is preferred after applying equations 17-2 and 17-3. Where the probability density function and cumulative distribution function show clear stochastic dominance, they can be the guide to project choice. Where stochastic dominance is not obvious, the preferred project might be the one with the lower coefficient of variation.

When using the certainty equivalent technique with the M/A technique, insert the CEF into equation 17-2 as a multiplicative factor times $E(B_t - \overline{C}_t)$. Use equation 17-3 as shown. Discount with a risk-free rate of interest.[14] In this case, the CEF should adjust for risk attitude only. Thus, the more risk averse the decision maker, the lower the CEF and the lower will be the mean of NB.[15] Be aware that if the certainty equivalent technique also adjusts for risk exposure while using the M/A technique, there will be an overadjustment for risk exposure.

The same selection rules for determining the preferred project discussed for the RADR in conjunction with the M/A technique apply to the certainty equivalent technique in conjunction with the M/A technique. And, as with the RADR technique, when used with the M/A technique, the certainty equivalent technique provides a formal treatment of risk attitude. Thus the decision maker whose objective function is to maximize expected value of utility simply chooses the project with the highest expected value of adjusted NB.

The two following techniques for evaluating uncertain investments are gaining increasing attention in the building field. The first, portfolio analysis, evaluates the risk and return of combinations or portfolios of projects in which interactions among the projects affect the overall riskiness of the portfolio through diversification. The second, the capital asset pricing model, determines

---

[14]It is appropriate to use risk-free rates of interest for discounting whenever risk is being accounted for by some means other than the RADR.

[15]This presumes the NB cash flows in each period are positive. The certainty equivalent technique does not handle negative values appropriately.

what the expected return of an individual asset should be, given its systematic risk.

17.3.4 Portfolio Analysis

Portfolio analysis helps the analyst find the combination of assets with the maximum return for any given degree of risk (i.e., variance of the return) or the minimum risk for any given rate of return.[16] A firm can reduce its overall investment risk by diversifying its assets so that returns are not perfectly positively correlated. Risk reduction from this diversification of assets is called the "portfolio effect."

Portfolio theory originated in stock analysis. The model is useful, however, in the analysis of other assets. Evaluating building investments seems a logical application of portfolio theory. Yet few applied efforts are on record.[17] One reason is that it is difficult to get data to measure the correlation of building investments. Another reason portfolio analysis is applied infrequently to building investments is that these investments involve large chunks of capital, large and lumpy chunks of real estate, cumbersome markets, lengthy transaction times, and large transaction costs. Kangari and Riggs (1988) describe in some detail the pitfalls of applying portfolio theory in construction.

17.3.5 Capital Asset Pricing Model

The capital asset pricing model (CAPM), the best-known model of risk and return among asset pricing models, is an extension of portfolio theory.[18] It explains how an asset should be priced, based on the additional risk it will add to the risk of a market portfolio of assets. Said another way, it indicates what the return on an asset should be, based on that asset's risk.

Investors in an asset experience two types of risk. One is "unsystematic risk," i.e., risk associated with random events, such as fires and strikes, that are unique to the investment. The other is "systematic risk" (also called "market risk") resulting from market variations to which all investments are subjected.[19]

Investors can eliminate unsystematic risk with a diversified portfolio, but not

---

[16]For the landmark treatment of portfolio analysis, see Markowitz (1959).

[17]An academic study of portfolio analysis in construction is available in Vegara's (1977) thesis. It combines probabilistic estimating, portfolio theory, and utility theory in the development of models for helping construction firms select an optimal mix of new projects.

[18]For the landmark treatment of the CAPM, see Sharpe (1964).

[19]Examples of events that affect market risk are interest rate changes, war, and economic expansion and contraction.

systematic risk. The systematic risk of a given asset is a function of how sensitive its return is to changes in market conditions. This sensitivity to market conditions is measured by "beta." The beta coefficient is the covariance of the asset's returns with the market portfolio's returns. According to the CAPM, an asset's beta measures the risk that is compensated for in the marketplace. An asset with a beta of 1.0 has average market risk and, therefore, one would expect an average return for it. A beta below 1.0 indicates proportionately less than average market risk and therefore a lower return. And a beta above 1.0 indicates proportionately more than average market risk and therefore a higher return.

We can define risk and return relationships using the CAPM. Then we can estimate the expected rate of return on an asset based on its systematic risk as defined by beta. The return on an asset should equal the risk-free rate of return plus a risk premium in proportion to the asset's beta. An asset with an estimated return higher than the expected return based on its beta is a good buy. Over time, however, increased demand for it will raise its price and lower its relative return in a perfect market until its estimated return is in line with the asset's beta.[20]

Portfolio analysis and the CAPM show promise theoretically for the evaluation of building investments. The data needed to apply these techniques, however, are not readily available. Estimating the beta for a real estate investment, for example, is much more complicated than for a stock. Thus widespread application of these techniques to building investments awaits additional research.

REFERENCES

Au, T., Richard M. Shane, and Lester A. Hoel. 1972. *Fundamentals of Systems Engineering: Probability Models*. Reading, Massachusetts: Addison-Wesley Publishing Company.

Berger, James O. 1985. *Statistical Decision Theory and Bayesian Analysis*. 2nd edition. New York: Springer-Verlag.

Bierman, H., Jr., and S. Smidt. 1984. *The Capital Budgeting Decision: Economic Analysis of Investment Projects*. 6th edition. New York: Macmillan Publishing Company.

Fuller, Russell J. 1981. *Capital Asset Pricing Theories—Evolution and New Frontiers*. Monograph No. 12. Charlottesville, Virginia: The Financial Analysts Research Foundation.

Hertz, D. 1964. Risk analysis in capital investment. *Harvard Business Review* 42(1):95–106.

Hillier, F. 1963. The derivation of probabilistic information for the evaluation of risky investments. *Management Science* 9(3):443–457.

Hillier, F. 1969. The evaluation of risky interrelated investments. *Studies in Mathematical and Managerial Economics*, ed. Henri Theil, pp. 1–111. Amsterdam: North-Holland Publishing Company.

Holstein, C. S. von. 1977. A tutorial in decision analysis. *Readings in Decision Analysis*, ed.

[20]For a comprehensive treatment of the CAPM and other asset pricing theories, see Fuller (1981).

Ronald Howard, James E. Matheson, and Katherine L. Miller, pp. 119–145. Menlo Park, California: SRI International.

Kangari, R., and L. S. Riggs. 1988. Portfolio management in construction. *Construction Management and Economics* 6(2):161–169.

Marshall, Harold E. 1988. Techniques for Treating Uncertainty and Risk in the Economic Evaluation of Building Investments. SP757. Gaithersburg, Maryland: National Institute of Standards and Technology.

Markowitz, Harry. 1959. *Portfolio Selection: Efficient Diversification of Investment*. New York: John Wiley and Sons, Inc.

Matheson, J. E., and R. A. Howard. 1977. An introduction to decision analysis. *Readings in Decision Analysis*. ed. Ronald Howard, James E. Matheson, and Katherine L. Miller, pp. 5–43. Menlo Park, California: SRI International.

Newnan, D. 1980. *Engineering Economic Analysis*. San Jose, California: Engineering Press, Inc.

Park, C. 1983. Probabilistic benefit-cost analysis. *The Engineering Economist* 29(2):83–100.

Pouliquen, L. Y. 1970. *Risk Analysis in Project Appraisal*. Baltimore: Johns Hopkins University Press.

Raiffa, Howard. 1970. *Decision Analysis: Introductory Lectures on Choices Under Uncertainty*. Reading, Massachusetts: Addison-Wesley Publishing Company.

Sharpe, W. F. 1964. Capital asset prices: A theory of market equilibrium under conditions of risk. *Journal of Finance* 19(4):425–442.

Vegara, J. A. 1977. Probabilistic estimating and applications of portfolio theory in construction. Ph.D. Thesis, Graduate College of the University of Illinois at Urbana-Champaign, Urbana, Illinois.

Zaheer, S. H. 1983. Contingency and capital cost estimates. *Cost Engineers' Notebook*. A-1.200(March):1–14. Morgantown, West Virginia: American Association of Cost Engineers.

18

Selecting the Best Technique

There is no single "best" technique for treating uncertainty and risk in all investment decisions. What is best will depend on the following items of information that you, as an analyst or a decision maker, should consider before selecting a technique.

First, assess your resources. How much time is available to evaluate the project? How much money is available for staff and computer support? Does the staff have the technical capability to apply each technique? For example, can they encode utility functions from inferred risk attitudes and develop probability distributions for uncertain variables? And finally, are computer software and hardware available for applying each technique? Asking these questions will help you define techniques that are feasible, given resource availability.

Second, identify the audience that will use the analysis and their likely reactions to each feasible technique. Will the management team understand the information a given technique generates? Will they accept the type of information it produces? If they want a deterministic, single-value estimate of project worth, for example, it might be inappropriate to present a cumulative density function using the simulation technique.

Third, consider the size of the investment relative to the institution's budget or portfolio of other investments. If an investment is small relative to the total portfolio, for example, and it affects total profitability only slightly if it yields a poor return, then a sophisticated technique, such as the mathematical/analytical technique, is unnecessary. If the investment is relatively large, on the other hand, and a poor return could bankrupt the institution, then a sophisticated technique might be appropriate.

Fourth, consider risk attitude. If the decision maker is risk neutral, for example, decisions should be made on the basis of maximizing expected value of net benefits instead of utility. Procedures to account for risk attitude, such as encoding utility functions, are unnecessary. If the decision maker is risk averse, on the other hand, techniques that adjust for risk attitude, such as decision analysis and the RADR, are appropriate.

Table 18-1 is a quick-reference guide to data and measures-of-worth char-

Table 18-1. Characteristics of techniques for treating uncertainty and risk

	(1)			(2)			(3)		(4)	
	Form of Input Data			Form of Measure of Worth			Consider Risk Exposure		Consider Risk Attitude	
Technique	($\underline{S}$)	($\underline{M}$)	($\underline{D}$)	(S)	(M)	(D)	(Ex)	(Im)	(Ex)	(Im)
Breakeven Analysis[a]	*			*				*		*
Sensitivity Analysis		*			*			*		*
Risk-Adjusted Discount Rate[b]	*		*	*		*	*	*	*	*
Certainty Equivalent[b]	*		*	*		*	*	*	*	*
Input Estimation Using Expected Values[c]			*	*						
Mean-Variance/ Coefficient of Variation			*			*	*			*
Decision Analysis			*	*		*	*		*	
Simulation			*			*	*			*
Mathematical/ Analytical			*			*	*			*

($\underline{S}$) = single value for each data input
($\underline{M}$) = multiple values for each data input
($\underline{D}$) = distribution of values for each data input
(Ex) = explicit, numerical measure of risk
(Im) = implicit consideration of risk
(S) = single-value measure of worth
(M) = multiple-value measure of worth
(D) = distribution of measure of worth

[a]Breakeven analysis solves for the value of the uncertain input value that will just make the investment break even (e.g., NB = 0, BCR = 1.0, or ORR = MARR). Thus the economic measure of worth would be a single value.

[b]The RADR and the certainty equivalent technique have both single values and distributions of values indicated for the form of data input and for the form of measure of worth. Distributions are cited because both techniques can be used with the mathematical/analytical technique.

[c]Input estimation using expected values does not treat risk exposure or risk attitude. Expected values are computed only for inputs, yielding no information in the single-value measure of worth to help the decision maker consider even implicitly risk attitude.

acteristics of most of the techniques described in Chapters 15–17 for treating uncertainty and risk. Column 1 indicates the form in which input data are accepted in the analysis. The possibilities include single values ($\underline{S}$), multiple values ($\underline{M}$), and probability distributions ($\underline{D}$). Breakeven analysis, for example, requires only single-value data inputs, whereas the simulation technique requires probability distributions. On the other hand, the certainty equivalent technique can be used with either single values or probability distributions.

Column 2 indicates the form that the measure of worth takes. It can be expressed as a single value (S), multiple values (M), or distribution of values (D). For example, input estimation using expected values displays the measure of worth as a single value, whereas the mathematical/analytical approach yields a cumulative distribution of worth values.

Column 3 treats risk exposure. Explicit (Ex) consideration of risk exposure means that the technique provides a numerical adjustment for or graphical measure of risk exposure. The cumulative distribution of project worth given by the simulation and mathematical/analytical techniques are explicit measures of risk exposure. An implicit (Im) treatment considers risk exposure, but fails to treat it quantitatively or graphically. Sensitivity analysis, for example, suggest a possible range of output values and thereby implies something about risk exposure but does not quantify it.

Column 4 treats risk attitude. Explicit consideration requires a numerical adjustment in the measure of worth to account for risk attitude. Using utility functions in decision analysis is one example. Implicit treatment considers risk attitude but provides no measure of it in project evaluation. For example, if a cumulative distribution function of project worth is constructed through simulation, implicit in the decision based on that function is the decision maker's attitude towards risk. That is, a risk averse decision maker might make a different decision than a risk taker, based on the same profile of risk exposure.

Note that the RADR and the certainty equivalent technique can be used to treat risk exposure and risk attitude either explicitly or implicitly.

Once you answer the questions asked earlier about resources, management acceptance, and risk attitude, Table 18-1 can help you select the appropriate technique. Suppose, for example, you want to know the range of possible ORRs an investment might take and at least enough information for the implicit consideration of risk exposure and risk attitude. Furthermore, assume that you are uncomfortable with the process of generating probability functions of uncertain events and that you prefer deterministic over non-deterministic answers. Looking at Table 18-1, the single technique that satisfies you is sensitivity analysis. It provides multiple (but deterministic) ORR values. The array of possible ORR values helps you consider risk exposure and risk attitude implicitly in the making of a decision. Any other technique that meets your criteria violates the constraint that probability functions not be used.

Techniques for risk analysis are becoming increasingly common in the economic evaluation of buildings and building components. First, current, widespread education in these techniques is yielding practitioners that understand and want to use the techniques in building evaluations. Second, the increased availability, decreased cost, and better understanding of computers and applied software makes computations much easier. Third, managers are increasingly better educated in advanced analysis techniques and are requesting risk analysis as a routine part of economic evaluations. And fourth, professional and technical societies, such as the American Association of Cost Engineers and the American Society for Testing and Materials, respectively, are encouraging the development of standard methods for measuring the risk associated with building decisions.

REFERENCES

Marshall, Harold E. 1988. Techniques for Treating Uncertainty and Risk in the Economic Evaluation of Building Investments. SP757. Gaithersburg, Maryland: National Institute of Standards and Technology.

Part 4

APPLICATIONS OF ECONOMIC EVALUATION METHODS AND RISK ANALYSIS TECHNIQUES

19

Selection of Building Designs and Building Components

Designers must make many choices in planning a building, ranging from global building design to detailed selections of mechanical systems and furnishings. The first case study illustrates the use of LCC, NS, SIR (applied incrementally), and sensitivity analysis to evaluate the choice of alternative building designs. The second case study illustrates the use of NB and decision analysis to decide which package of retrofit designs for energy conservation is economically preferred.

19.1 CHOOSING THE COST-EFFECTIVE BUILDING DESIGN: EVALUATED WITH LCC, NS, SIR, AND SENSITIVITY ANALYSIS

In this section we show how to determine which of two designs is more cost effective for a proposed office building.[1] Three methods, LCC, NS, and SIR provide complementary measures of cost effectiveness. Sensitivity analysis takes into account uncertainty in costs.

19.1.1 Design Problem

Assume that you manage the building stock for a large nonprofit organization that has decided to construct a building. The architect submits two preliminary designs of the building. One is conventional. The other is energy conserving, using innovative as well as proven approaches to save energy. In support of your recommendation to management, you want to present the LCCs of the two designs and the NS of one relative to the other. There is uncertainty about the quantity of energy savings and, therefore, about the monetary savings that will result from the conserving design. So you must also account in some way for uncertainty when you compare the two designs.

[1] This case study is adapted from a problem in Ruegg (1987, pp. 73-81).

19.1.2 Approach

Cost estimates have been developed for the two designs, and you choose the LCC method as a straight-forward, simple way to present the alternatives.

You think management may better follow your presentation of LCCs if you include *all* major construction and operating costs rather than only those relevant to the choice. (We indicated in Chapter 2 that only *relevant* costs need be included.) You are concerned, however, that the difference between the two LCCs, i.e., the NS, may appear puny and unimportant in comparison to total LCC.

In this case, in addition to the LCC and NS measures, we recommend that you give an incremental SIR to emphasize the return on the extra costs required for the conserving design. Performing sensitivity analysis (Section 15.3) will indicate how vulnerable the selection is to variations in energy savings.

19.1.3 Evaluation

19.1.3.1 Data and Assumptions

The conventional and energy-conserving designs are for an office building to be erected in Madison, Wisconsin. The two designs are approximately equivalent both aesthetically and functionally. Each has two underground levels for parking, seven office floors, and a mechanical house. Each has a gross floor area of approximately 176,000 ft^2.

The two designs have different orientations, building configurations, envelopes, lighting systems, and HVAC systems. The energy-conserving design is elongated on the east-west axis for greater exposure of the south side to solar radiation. The window area of the energy-conserving design is 25% of the wall area, and most windows are on the south side. The conventional design has 40% of the wall area in windows, uniformly distributed on all sides. The energy-conserving design has extra mass in exterior surfaces, extra insulation, and window fins to reduce summer cooling loads. The north wall of the energy-conserving design is earth bermed up to the second floor. The energy-conserving design has advanced technology lighting and HVAC systems.

The organization plans to occupy the building for at least 30 years. Thus a 30-year study period is reasonable. According to the financial officer, the organization's opportunity cost of capital is 10% without inclusion of inflation. Assume that the analysis is performed in 1987 dollars, using a real discount rate of 10%. Since the analysis is of a non-profit institution, you need not make any adjustments in the analysis for tax effects.

Table 19-1 shows, for both designs, the costs of construction, nonfuel operation and maintenance, and major repair, as projected by the architectural firm's cost estimator. Construction costs are higher for the energy-conserving

Table 19-1. Cost and energy data for conventional and energy-conserving building designs

	Conventional Design	Energy-Conserving Design
Cost Items		
Construction	\$9,130,000	\$9,880,000
Annual, nonfuel operation and maintenance	90,000	70,000
Major repairs (every 10 years)	100,000	60,000
	Annual Energy Consumption (Btu)	
Natural gas	$4{,}980 \times 10^6$ Btu	$2{,}290 \times 10^6$ Btu
Electricity	$7{,}277 \times 10^6$ Btu	$3{,}886 \times 10^6$ Btu
	Energy Prices (\$)	
Natural gas	\$4.72/10^6 Btu	\$4.72/10^6 Btu
Electricity	\$22.63/10^6 Btu	\$22.63/10^6 Btu

design, as expected, because of the design differences described earlier. The land has already been purchased, and it will accommodate either design.

The table shows separately annual energy consumption projected by the architectural firm and prices for energy currently paid by the organization. Anticipated Btu consumption is significantly lower for the energy-conserving design. Gas and electricity prices, on the other hand, are the same regardless of the design selected, i.e., the demand charge component is not affected. Future energy prices are based on U.S. Department of Energy projections.

19.1.3.2 Calculate LCCs and NS

To determine which building design is less costly, compute the LCCs. As explained earlier, LCCs are here based on the sum of total construction, operation and maintenance, repair and replacement, and energy costs, all discounted to present value. Construction costs, as shown in Table 19-1, are assumed to occur at the present, so discounting is unnecessary. Operation and maintenance costs, on the other hand, are recurring annual expenses and can be discounted by the appropriate UPV factor. For a study period of 30 years and a real discount rate of 10%, the UPV factor is 9.43. Calculate present value operation and maintenance costs (PVO&M) as follows:

PVO&M conventional design = (\$90,000) (9.43) = \$848,700

Table 19-2. Major repair costs for the conventional and energy-conserving building design

Building Design (1)	Year (2)	Repair Costs (3)	SPV (4)	Present Value Repair Costs (5)=(3)·(4)
Conventional	10	$100,000	.39	$39,000
	20	100,000	.15	15,000
				$54,000
Energy-conserving	10	$60,000	.39	$23,400
	20	60,000	.15	9,000
				$32,000

and

$$\text{PVO\&M energy conserving design} = (\$70{,}000)\,(9.43) = \$660{,}100$$

Since major repairs are anticipated about every 10 years, you must plan for two cash outlays over the 30-year study period—one at the end of year 10 and one at the end of year 20. Table 19-2 shows how to calculate present value repair costs. They total $54,000 for the conventional design and $32,000 for the energy-conserving design.

Table 19-3 shows how to compute present value energy costs. Multiply the estimated quantity to be consumed of each energy type by its respective price and UPV** factor. The UPV** factors in column 5 combine into one multiplicative number the effects of discounting and projected price escalation for electricity and gas in Wisconsin.[2] For the conventional design, they total $1,663,231; for the energy-conserving design, $865,588.

The final step in calculating LCCs is to sum for each design its present value costs, as shown in Table 19-4. LCC for the energy-conserving design, $11,437,688 is less than for the conventional design, $11,695,931. The difference in LCCs between the two designs, $258,243, is the NS from choosing the energy-conserving design over the conventional design.

[2]The UPV** factors in this problem are for a 30-year study period and a 10% real discount rate. They are from specialized tables in a report prepared for the DoE (Lippiatt and Ruegg, 1987, p.20). In the source report the factors are designated "UPW*" factors, but here we refer to them as "UPV**" factors to distinguish them from factors with a constant rate of change which we designate "UPV*." The factors are based on differential rates of energy escalation, which are required when performing analyses in constant dollars. See Section 12.5.3 for a discussion on how to use the factors, and Appendix G for sample UPV** factors for 1988.

Table 19-3. Present value energy costs for the conventional and energy-conserving designs

Building Design (1)	Energy Type (2)	Annual Quantity (10^6 Btu) (3)	1987 Price per 10^6 Btu (4)	UPV** (5)	Present Value Costs (6)=(3)(4)(5)
Conventional	Electricity	7,277	\$22.63	8.25	\$1,358,598
	Gas	4,980	4.72	12.96	304,633
					\$1,663,231
Energy-conserving	Electricity	3,886	\$22.63	8.25	\$725,506
	Gas	2,290	4.72	12.96	140,082
					\$865,588

19.1.3.3 Perform Sensitivity Analysis

Before making a recommendation, you want to see how sensitive NS is to variations in projected energy consumption. The energy-conserving design incorporates innovative technologies which increase uncertainty. The consulting firm's thermal engineer believes that energy consumption with the energy-conserving design will likely vary no more than plus or minus 25% from the best-guess estimate. This suggests sensitivity testing using a 25% reduction and a 25% increase in the estimated consumption of electricity and gas for the energy-conserving design. Table 19-5 shows LCCs based on low, best-guess (the level of consumption used in Table 19-4), and high levels of energy consumption for the energy-conserving design and, for comparison, LCC of the conventional design, as computed in Table 19-4. NS is also shown for the same three levels of energy consumption.

No matter which of the three levels of energy consumption occurs, the LCC of the conventional design is estimated to exceed the LCCs of the energy con-

Table 19-4. LCCs of conventional and energy-conserving designs in present value dollars

Cost Items (1)	Conventional Design (2)	Energy-Conserving Design (3)
Construction	\$9,130,000	\$9,880,000
Nonfuel operation and maintenance	848,700	660,100
Major repairs	54,000	32,000
Energy	1,663,231	865,588
LCCs	\$11,695,931	\$11,437,688

Table 19-5. Sensitivity of LCCs and NS to projected energy consumption of the energy-conserving design

	Conventional Design	Energy-Conserving Design		
		Low	Best Guess	High
Nonenergy costs[a]	$10,032,700	$10,572,100	$10,572,100	$10,572,100
Energy costs[b]	1,663,231	649,191	865,588	1,081,985
LCCs	11,695,931	11,221,291	11,437,688	11,654,085
NS	...	474,640	258,243	41,846

[a]Nonenergy costs are the sum of the first three costs listed in Table 19-4.

[b]The energy cost at the low level was computed by multiplying 0.75 by $865,588. The energy cost at the high level was computed by multiplying 1.25 by $865,588.

serving design. This means that the energy-conserving design has positive NS for each energy consumption scenario.

19.1.4 Recommend Which Design to Choose

You will probably recommend selection of the energy-conserving design because it appears cost effective even under conservative estimates of energy savings. Consider, however, how management might respond to your recommendation. They may not find it convincing because projected NS is estimated as low as $42,000 (high consumption level), in comparison with total construction costs ranging from $9,000,000 to $10,000,000. In fact, decision makers may feel that NS is too insignificant to influence a major decision on building design.

One approach for making management more aware of the economic contribution of the project is to compute the SIR on incremental costs and savings. It measures the impact of making the $750,000 construction investment for the energy-conserving design. This increment is the extra, up-front money required for the energy-conserving design. Total savings generated by the energy-conserving design include savings from reductions in energy ($797,643 for the best-guess consumption level), reductions in operation and maintenance ($188,600), and reductions in repairs ($22,000). Dividing the sum of these savings by the extra investment yields an SIR of 1.34 (i.e., $1,008,243/$750,000). This means that $1.34 is expected back from every dollar invested, over and above the 10% required rate of return.

An SIR of 1.34 might look more persuasive to management than the NS figures presented earlier. While the energy-conserving design has little economic impact relative to total building costs, its savings appear robust relative to its additional costs.

19.2 DECIDING WHICH RETROFIT DESIGN FOR ENERGY CONSERVATION IS ECONOMICALLY PREFERRED: EVALUATED WITH NB AND DECISION ANALYSIS

This case study shows how to evaluate several proposed energy conservation retrofits to public buildings.[3] We demonstrate the NB method in conjunction with the decision analysis technique. A decision tree represents the possible outcomes, costs, and probabilities associated with the decision.

19.2.1 Design Problem

Suppose that you are a consultant who has been hired by a state government to evaluate alternative designs for retrofitting two state buildings for energy conservation. An energy consultant for the state has already identified six possible designs. The main purpose of the conservation project is to demonstrate to private companies that energy conservation is profitable. Your job is to identify which design package among the six possible packages has greatest profit potential.

19.2.2 Approach

Using decision analysis helps account for uncertainty associated with performance of the retrofits. A decision tree is useful to display the possible building retrofits, their costs, and the probability of achieving specific levels of energy savings. Computing the expected NB for various design packages will give an indication of their profitability. The package with the highest value of expected NB will likely be the most profitable. If none of the packages appears to yield a profit, the recommendation will be to choose none of them.

Finding the design package with the highest expected NB does not provide a measure of the risk of choosing an uneconomic project (i.e., of risk exposure). However, you can use the data presented in the decision tree to generate probability distributions and cumulative distribution functions of NB for each design package. These will give you a picture of risk exposure.

19.2.3 Evaluation

19.2.3.1 Data and Assumptions

You must consider two types of retrofit costs. The first is a fixed investment cost to be incurred in each building, regardless of which design package is

[3]This case study is adapted in part from Marshall (1988, pp. 37–43).

Table 19-6. Costs of alternative design packages for retrofitting public buildings I and II for energy conservation (costs in $1,000,000)

Building I				Building II			
Fixed Costs	Design Package Costs			Fixed Cost	Design Package Costs		
F1	D1	D2	D3	F2	D4	D5	D6
0.5	2.0	3.0	4.0	0.8	3.0	4.0	4.5

chosen. The second is the cost of implementing individual design packages. Present value fixed investment costs (F1 and F2) and design retrofit package costs (D1–D6) are shown in Table 19-6. Assume costs to be certain.

The predicted benefit outcomes (energy savings in present value dollars) are uncertain for the different design packages. Table 19-7 shows estimates of possible benefit outcomes with their respective probabilities of occurrence. Assuming that the state government is risk neutral and acts so as to maximize the expected monetary value of its investments, you do not need to account for risk attitude with utility measures of outcomes. Furthermore, since the state does not pay taxes, they are excluded from the analysis.

19.2.3.2 Construct a Decision Tree and Find the Design Package That Maximizes Expected Value of NB

The decision tree in Figure 19-1 shows possible building choices and design packages, as well as their respective costs. You must assign dollar values to potential outcomes (shown at the ray tips of the six probability fans) and probability values to each outcome (shown over each ray). You accomplish this by discussing with engineers, architects, building managers, equipment manufacturers, and other knowledgeable people the implications of alternative design packages in buildings I and II. Probability assignments to the fans are based on anticipated potential impacts of changes in uncertain input variables, including energy prices, length of system life, performance of energy conservation designs, and quantity of energy saved.

Starting from the right-hand side of the tree, calculate the weighted average for each probability fan by summing the products of possible outcomes weighted by their respective probabilities. For example, compute the expected value of the probability fan D1 as follows:

$$D1_{EV} = 0.9(\$3{,}000{,}000) + 0.1(\$2{,}000{,}000) = \$2{,}900{,}000$$

Write the expected value atop each chance node, as shown in Figure 19-1.

Table 19-7. Possible benefit outcomes and their estimated probabilities of occurrence for the six design packages

(1) Design Retrofit Packages	(2) Possible Benefit Outcomes ($1,000,000)	(3) Estimated Probabilities
D1	3.0	0.9
	2.0	0.1
D2	4.5	0.6
	3.0	0.3
	−1.0	0.1
D3	6.0	0.7
	4.0	0.2
	2.0	0.1
D4	4.0	0.8
	3.0	0.1
	2.5	0.1
D5	7.0	0.5
	4.5	0.4
	4.0	0.1
D6	7.0	0.5
	4.5	0.3
	1.0	0.2

Next, fold back to the next preceding stage. That is, at each square decision node, compare the alternative branches with respect to their costs and expected benefits. Choose the one with the highest expected NB and write it atop the decision node box. For example, if you fold back the decision node values on the building II path sequence, expected NB values (before subtracting the $800,000 building II fixed cost) for design packages D4–D6 are as follows:

$$D4_{NB} = \$3{,}800{,}000 - \$3{,}000{,}000 = \$800{,}000$$

$$D5_{NB} = \$5{,}700{,}000 - \$4{,}000{,}000 = \$1{,}700{,}000$$

$$D6_{NB} = \$5{,}100{,}000 - \$4{,}500{,}000 = \$600{,}000$$

The preferred alternative branch is D5, because it has the highest expected NB. Write its value, $1,700,000, atop the decision node box. Truncate the other two paths by parallel slash marks to indicate that they are less economic choices.

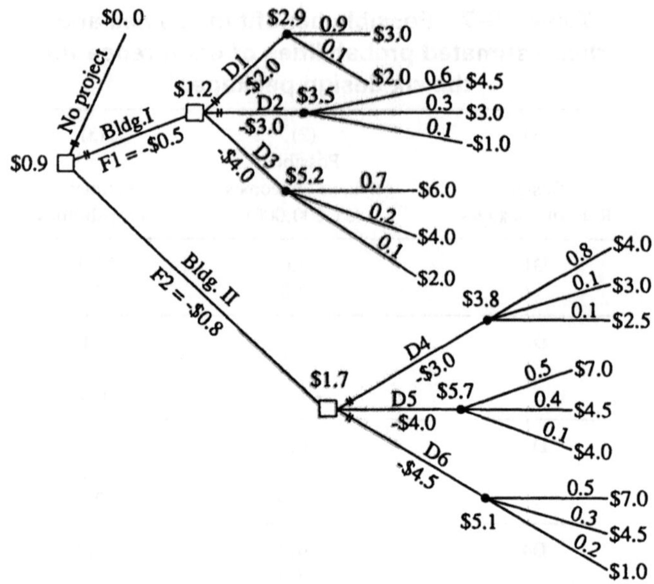

Figure 19-1. Decision tree for conservation investment (values are in millions of dollars). □ = decision node; ● = chance node; D = design retrofit package.

The final step is to fold back one more time. Design package D5 in building II is the most efficient choice. It has expected value NB of $900,000 (i.e., $1,700,000 − $800,000) compared to $700,000 (i.e., $1,200,000 − $500,000) for design package D3 in building I and zero dollars for having no project. Enter the maximum expected value at the initial decision node box at the far left of the decision tree. Use parallel slash marks to truncate the "no project" and building I alternatives. The decision tree, once all values are written in, shows explicitly the economically efficient path sequence (building II/D5 in this case) and the expected value of NB ($900,000) for that path sequence.

19.2.4 Recommendation

You will probably want to recommend design package D5 in building II. Its expected value of NB is $200,000 greater than that of the next best retrofit. Since the state government is presumed to be risk neutral, you have handled risk attitude by choosing according to expected value.[4]

[4]Decision analysis handles risk attitude by substituting utility values for dollar values, as described in Chapter 17 and Appendix E. However, since the state is presumed to be risk neutral, maximizing expected monetary value is the same as maximizing utility.

You might also provide a cumulative probability distribution if your client wants to see the explicit likelihood of different NB values. The probability fans in Figure 19-1 are for benefits only instead of NB measures of worth. However, you can use data in the decision tree to generate probability functions and cumulative distribution functions of NB for each design.[5] By generating these functions, decision tree analysis provides a profile of risk exposure.

To illustrate the generation of the functions, we focus on just one package, D2, in Figure 19-1. First, find the NB associated with each of the three probabilities associated with D2. Do this by subtracting the fixed and retrofit costs associated with the D2 path sequence from the value of each of the three possible outcomes. Thus the probability distribution values for NB of D2 are as follows:

For probability of 0.6,

$$\$4{,}500{,}000 - \$3{,}000{,}000 - \$500{,}000 = \$1{,}000{,}000$$

For probability of 0.3,

$$\$3{,}000{,}000 - \$3{,}000{,}000 - \$500{,}000 = -\$500{,}000$$

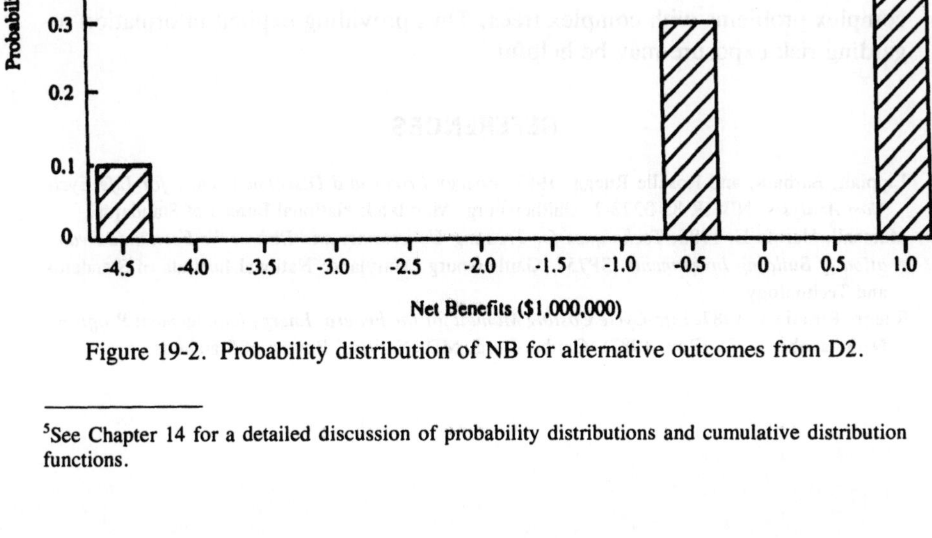

Figure 19-2. Probability distribution of NB for alternative outcomes from D2.

---

[5]See Chapter 14 for a detailed discussion of probability distributions and cumulative distribution functions.

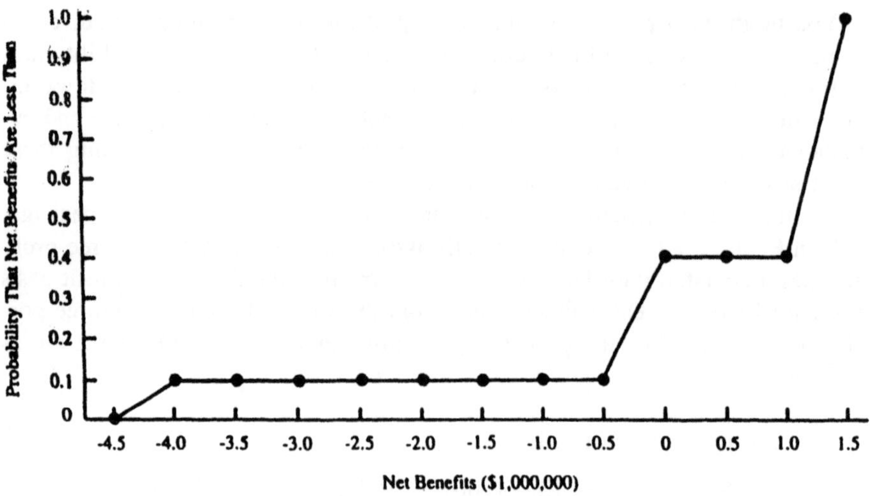

Figure 19-3. Cumulative distribution function of NB for alternative outcomes from D2.

For probability of 0.1,

$$-\$1{,}000{,}000 - \$3{,}000{,}000 - \$500{,}000 = -\$4{,}500{,}000$$

The discrete probability distribution in Figure 19-2 plots these values. By cumulating the probabilities, the cumulative distribution function of Figure 19-3 is derived. Note that these graphs give additional information to that provided by the decision tree, namely, that there is a 40% probability that D2 will not have positive NB. Whereas the perceptive reader might readily intuit such probability implications from the decision tree of Figure 19-1, it is less obvious for complex problems with complex trees. Thus providing explicit information regarding risk exposure may be helpful.

REFERENCES

Lippiatt, Barbara, and Rosalie Ruegg. 1987. *Energy Prices and Discount Factors for Life-Cycle Cost Analysis*. NBSIR 85-3273-2. Gaithersburg, Maryland: National Bureau of Standards.

Marshall, Harold E. 1988. *Techniques for Treating Uncertainty and Risk in the Economic Evaluation of Building Investments*. SP757. Gaithersburg, Maryland: National Institute of Standards and Technology.

Ruegg, Rosalie T. 1987. *Life-Cycle Costing Manual for the Federal Energy Management Program*. NBS handbook 135 (Rev. 1987). Gaithersburg, MD: National Bureau of Standards.

20

Decision to Accept or Reject a Project

Designers and managers of buildings frequently must decide whether to accept or reject discretionary investments in buildings—investments which may upgrade quality, extend life, lower costs, or increase revenue relative to the status quo. We call these "accept/reject decisions" because a possible alternative is to do nothing. Accept/reject decisions can be contrasted with design decisions, treated in Chapter 19, which require a selection from among alternative designs. For example, we call deciding whether to add storm windows an accept/reject decision and choosing the window style for a new building a design decision.

Five case studies illustrate how to make accept/reject decisions using the economic methods presented in Chapters 2 through 7, together with several uncertainty and risk analysis techniques. The first and second case studies show how to decide whether or not to add a discretionary system to a building on the basis of long-run costs. The first uses the LCC, NS, and SIR methods with breakeven analysis to evaluate the cost effectiveness of a waste heat recovery system. The second case study uses the LCC, NS, SIR, and payback methods to evaluate the cost effectiveness of a solar hot water system.

The third, fourth, and fifth case studies show how to decide whether or not to invest in various revenue-generating building projects. The third uses the BCR method to evaluate the profitability of investing in an apartment building. It provides a detailed treatment of income taxes and includes financing. The fourth case study uses NB, BCR, and ORR methods to evaluate the profitability of investing in a warehouse storage facility. Simulation analysis is applied to estimate the probitability that the investment will be profitable in the face of uncertainty. The fifth case study uses NB with the M/A technique to evaluate the profitability of a limited partnership investment in a company that buys and operates shopping centers.

20.1 WHETHER TO RETROFIT A COMPUTER FACILITY FOR WASTE-HEAT RECOVERY: EVALUATED WITH LCC, NS, SIR, AND BREAKEVEN ANALYSIS

This illustration shows how to evaluate a proposed project for retrofit to an existing office building.[1] The proposal is to reduce domestic water heating costs by using waste heat captured from a computer facility located within the building. We demonstrate how to use three methods—LCC, NS, and SIR—to determine if the project is economically feasible. We also show how to use the SIR to determine if the project is economically preferred when compared against alternative projects competing for limited retrofit funds. We use breakeven analysis (described in Chapter 15) to identify the maximum investment that can be made in the retrofit and still recover all costs. This is needed because of uncertainty over how much the heat recovery system will cost to purchase and install.

20.1.1 Problem

Suppose that you are the plant engineer for an office building owned by the county government.[2] A consultant proposes a waste-heat recovery project to save energy. The proposal is to install a heat exchanger to recover heat from waste condenser water from a computer room chiller, and then to use that recovered heat to preheat domestic hot water for the building. Other retrofit projects have also been proposed, and there is insufficient funding for all of them. It is up to you to decide whether or not to approve the proposal for the waste-heat recovery system.

20.1.2 Approach

By computing and comparing LCC without and with the retrofit, you can determine if it is cost effective. If the building's LCC diminishes with the retrofit, it yields positive NS. Since there is a limited budget for retrofit projects, you should also calculate the SIR to compare the waste-heat recovery project to other projects competing for the budget.

Calculating the maximum amount you can afford to pay for purchase and installation and still break even with expected energy savings gives a benchmark against which to compare the consultant's estimate of investment cost. Knowing the breakeven purchase and installation costs tells you how much leeway for error you have in deciding whether to accept or reject the heat exchanger.

---

[1]This case study is adapted from Ruegg and Marshall (1984, pp. 21-1 through 21-18).

[2]Analysis of this kind for a government agency requires no adjustments for tax effects.

20.1.3 Evaluation

20.1.3.1 Data and Assumptions

The office building is in the Maryland suburbs close to Washington, D.C. Purchased steam at $9.00 per thousand pounds (Mlb) is used to heat domestic hot water. The supplier of the steam uses coal to generate the steam. The analysis is taking place in 1987, and all costs are quoted in 1987 dollars.

The water chiller for the computer room currently delivers condenser water to a cooling tower, where thermal energy dissipates to the atmosphere. The heat exchanger retrofit will capture waste heat generated in the computer room and use it to augment domestic water heating. A consultant estimates the total installed cost of the heat exchanger, including all piping, insulation, and valves, to be about $7,000. This is a best-guess estimate, and the consultant has indicated several scenarios in which the cost might be higher or lower, depending on the final cost of labor and materials.

Estimated costs for maintaining the heat exchanger are $300 per year in 1987 dollars. The estimated cost of retubing the heat exchanger at the end of 15 years is $600 in 1987 dollars. Both maintenance and repair costs are expected to change at about the same rate as prices in general.

Records reveal that domestic hot water consumption averages 1 gallon per person per day. The building is occupied 252 days per year, and daily occupancy averages 3,000 persons. The water intake temperature averages 60°F, and the hot water supply temperature is 120°F. The consultant estimates that the heat exchanger can preheat the 60°F supply water to 80°F.

Agency policy dictates a 10% real discount rate. The system is expected to last for at least 25 years.

20.1.3.2 Estimate Present Value Energy Costs Without and With the Heat Exchanger

The first step is to find the quantity of purchased steam used currently to heat domestic hot water. This requires calculating the number of gallons of hot water used and the corresponding quantity of steam necessary to raise the water to the desired temperature. Calculate the gallons of hot water used per year as follows:

$$(1 \text{ gallon/person/day})\,(3{,}000 \text{ persons})\,(252 \text{ days/year})$$
$$= 756{,}000 \text{ gallons/year}$$

Enough energy must be purchased to raise the temperature 60°F, i.e., from the 60°F entry temperature to the 120°F consumption temperature. Given that 1 gallon of water weighs 8.34 pounds, and it takes 1 Btu of energy to raise the temperature of a pound of water by 1°F, calculate Btu requirements as follows:

$$(8.34 \text{ lb/gallon/°F}) (1 \text{ Btu/lb}) (60°\text{F}) (756{,}000 \text{ gallons/year})$$
$$= 378{,}302{,}400 \text{ Btu/year}$$

Since 1.05×10^6 Btu are equivalent to a thousand pounds (Mlb) of steam, you can convert Btu to Mlb of steam as follows:

$$\frac{378{,}302{,}400 \text{ Btu}}{1.05 \times 10^6 \text{ Btu/Mlb steam}} = 360.3 \text{ Mlb steam}$$

The next step is to calculate the energy required to heat domestic hot water with the proposed heat exchanger. With the heat exchanger, you need only purchase steam to raise domestic hot water by 40°F (i.e., from 80°F to 120°F) rather than by 60°F, as required without the heat exchanger. Thus the energy requirements with the heat exchanger will be only two-thirds (40/60) of the 360.3 Mlb of steam calculated earlier, or 240.2 Mlb of steam.

The final step is to calculate the present value costs of steam requirements without and with the heat exchanger. Table 20-1 shows how this is done. The annual quantity of steam multiplied by its price is the cost of energy per year. The cost of energy per year multiplied by the UPV** factor (a multiplier which combines the effects of discounting and anticipated price escalation for steam in Maryland in a single number[3]) gives present value energy costs. Without the

Table 20-1. Energy costs without and with heat exchanger

Alternative (1)	Annual Quantity Steam (Mlb) (2)	Price per Mlb Steam ($) (3)	Cost per Year ($) (4) = (2)(3)	UPV** (5)	Present Value Energy Costs ($) (6) = (4)(5)
Keep current setup, no heat exchanger	360.3	9.00	3,243	9.97	32,333
Add the heat exchanger	240.2	9.00	2,162	9.97	21,555

[3]The UPV** factor in this problem is for a 25-year study period and a 10% real discount rate. The factor comes from the 1987 edition of a federal report on energy prices prepared for DoE (Lippiatt and Ruegg, 1987, p. 18). In the source report the factors are designated "UPW*" factors, but here we refer to them as "UPV**" factors to distinguish them from factors with a constant rate of change which we desingate "UPV*." The factor is based on differential rates of energy escalation (excluding general price inflation), which is required when using constant dollar analyses. Since no factors were available for steam price escalation, the factor for coal is used. The assumption is that the price of coal-generated steam increases at the same rate as the price of coal. See Section 12.5.3 for a description of how to use the factors, and Appendix G for sample factors for 1988.

Table 20-2. Present values of investment, maintenance, and repair costs for the heat exchanger

Type of Cost (1)	Frequency of Occurrence (2)	Factor Name (3)	Factor Value (4)	Amount ($) (5)	Present Value ($) (6) = (4)(5)
Purchase and installation	Once	SPV	1.00	7,000	7,000
Maintenance	Yearly	UPV	9.08	300	2,724
Repair	Year 15	SPV	0.24	600	144

heat exchanger, the present value energy costs are $32,333, and with the heat exchanger, $21,555.

20.1.3.3 Calculate the LCC, NS, and SIR of the Heat Exchanger

LCC *without* the heat exchanger is composed solely of the $32,333 in present value energy costs. *With* the heat exchanger, present value investment costs, maintenance costs, and repair costs must be added to the $21,555 present value energy costs. Table 20-2 shows how present values are determined for non-energy costs. Table 20-3 combines all costs to show LCC without and with the heat exchanger. Present value NS with the heat exchanger is estimated at $910.

The SIR for the heat exchanger has in its numerator the savings in energy costs ($32,333 − $21,555 = $10,778) less any extra costs for maintenance ($2,724) and repairs ($144). The ratio has in its denominator the purchase and installation costs of the heat exchanger ($7,000). The SIR for the heat exchanger, computed using equation 4-3, is as follows:

$$\frac{\$10{,}778 - \$2{,}724 - \$144}{\$7{,}000} = 1.13$$

Table 20-3. LCC and NS of the heat exchanger

Alternative (1)	Present Value Costs ($): Energy (2)	Maintenance (3)	Repair (4)	Purchase and Installation (5)	LCC (6) = (2) + (3) + (4) + (5)
Keep current setup, no heat exchanger	32,333	0	0	0	32,333
Add the heat exchanger	21,555	2,724	144	7,000	31,423
		NS from heat exchanger			910

20.1.3.4 Find the Breakeven Value of the Purchase and Installation Costs for the Heat Exchanger

To find the breakeven purchase and installation costs for the heat exchanger, construct an equation with purchase and installation costs as the unknown value. Set the unknown equal to energy savings minus extra maintenance and repair costs. Then solve for the unknown. The solution value is the breakeven purchase and installation amount, i.e., the maximum amount you can afford to pay for the heat exchanger and still break even. Compute the breakeven value of purchase and installation costs as follows:

$$\begin{aligned} \text{Breakeven value} &= \$10{,}778 - \$2{,}724 - \$144 \\ &= \$7{,}910 \end{aligned}$$

20.1.4 Decide Whether to Accept or Reject the Heat Exchanger

Ignoring uncertainties, the project appears cost effective with estimated NS of $910 and an SIR of 1.13. These measures of worth support acceptance of the heat exchanger. But if alternative projects are available whose SIRs exceed the 1.13 estimated for the heat exchanger, and if these projects would exhaust your budget, the heat exchanger should be passed over in favor of the other projects. Under these circumstances, you should reconsider it next year when your new budget is allocated and you have a new group of building retrofits to evaluate.

Due to the uncertainty about the costs of purchasing and installing the system, you might in any case wish to consider the breakeven value before making a final recommendation. If purchase and installation costs were just another $1,000 over the estimate, NS would become a net loss. The more uncertain you feel about costs of obtaining the system, the less inclined you will be to recommend the heat exchanger.

This is an investment where considerations other than economics might sway your decision. For example, if you are trying to achieve target reductions in energy purchases, you will likely accept the project if funding is available. If, on the other hand, there are no management directives to save energy, you might pass up the retrofit in the hopes of finding a better return for your funds in alternative projects yet to be proposed.

20.2 WHETHER TO ADD A SOLAR HOT WATER SYSTEM TO A MILITARY BASE'S LAUNDERETTE: EVALUATED WITH LCC, NS, SIR, AND PAYBACK

The energy crisis of the 1970s prompted legislation and presidential executive orders requiring federal facilities to be made LCC effective, taking into account energy consumption. Guidelines for performing economic analysis of federal energy conservation and renewable energy projects are established in the U.S. Code of Federal Regulations (10 CFR Part 436) and amplified in the *Life-Cycle Costing Manual for the Federal Energy Measurement Program* (Ruegg, 1987).[4] Here we demonstrate how to evaluate the economic performance of a solar energy system according to federal guidelines.[5]

20.2.1 Problem

A Marine Corps air base in the southwest corner of Arizona is looking for ways to reduce its energy consumption. Suppose that you are an energy consultant hired to identify and evaluate potential retrofit projects. Favorable climatic conditions cause solar energy applications to look promising. Retrofitting the base launderette with a solar hot water system appears technically feasible. Your job is to evaluate the cost effectiveness of the proposed solar hot water system according to federal guidelines, assign priority for funding, and advise the base energy officer on your findings.

20.2.2 Approach

The first step is to determine the federal requirements for economic analysis. You can learn these from the latest edition of the *Manual.*

Federal criteria[6] call for a real discount rate of 7%,[7] performance of the analysis in constant dollars, a study period no longer than 25 years, and federal energy price projections. Measures of LCC and NS are required to determine a project's cost effectiveness; the SIR is used to assign funding priority; and a measure of payback is required as a supplementary measure for solar energy projects.

To provide a baseline for comparison, LCC is calculated based on the existing method of heating water. Then LCC is calculated based on the proposed system. NS is found by subtracting LCC with the solar energy system from LCC without it. The SIR is calculated according to the formulation used in federal analyses: present value energy savings, less any increased maintenance and repair costs, divided by the present value of investment-related costs (including replacements and residual value in the denominator). Payback may be given in either simple or discounted terms and is used as a supplementary measure only. (Main reliance should be on the former measures.)

---

[4]The *Life-Cycle Costing Manual* is hereafter referred to as the *Manual* and the Federal Energy Management Program as the FEMP.

[5]This case study is based in part on Ruegg (1987, pp. 86-101).

[6]Criteria are those in effect as of 1989.

[7]The Federal Energy Management Improvement Act of 1986 allowed the Secretary of Energy to specify the discount rate. The first implementation of that provision of the act is October 1990. Prior to that time the discount rate was set at 7% (real) by the Energy Security Act of 1980 and, before 1980, at 10% (real) by OMB Circular A-94.

20.2.3 Evaluation

20.2.3.1 Data and Assumptions

The proposed solar hot water system will supply 60% of the hot water load for the 5,000 ft^2 launderette. The entire load is now met by an electric water heater. The load exists 7 days a week and totals approximately 1,750 million Btu over the year. The daily load profile closely matches the solar profile, reducing the need for storage.

The closed-loop drainback system has 2,500 ft^2 of collector area and stores water in a single 750-gallon storage tank. An external plate heat exchanger transfers heat between the collector loop and the storage tank. With each draw of water by the launderette, solar-heated water enters the electric water heater, where its temperature is boosted. City water replaces the draw from the storage tank.

The collector array is mounted directly on the roof, benefiting from the insulating effect of the roof and reducing support structure costs from what they otherwise would be. The single-glazed, black chrome, flat-plate collectors have internal manifolds. The piping is run in the attic space to provide weatherproofing and to reduce heat losses. All lines have R-7 insulation. The mechanical room houses the drainback tank, heat exchanger, collector, pumps, storage tank, controls, and auxiliary water heater. The room is located directly below the collectors to reduce piping costs.

The 40% of load not met by the solar energy system will be met by the existing electric heater. It will boost the temperature of the solar-heated water to 135°F. The efficiency of the electric unit is 100%. Both the solar energy system and the existing system have expected lifetimes of about 20 years.

Table 20-4 gives costs by component for purchasing and installing the solar hot water system. Costs are divided between those fixed in the short run and those that vary with collector area.[8] Fixed costs total $61,000, and variable costs total $25.30/$ft^2$. Purchase and installation costs include freight, workers compensation, social security, unemployment tax, sales tax, contingency, profit, bonds, permits, and liability insurance. They do not include design costs, which are estimated at 10% of purchase and installation costs. Purchase and installation cost estimates are provided by a solar energy contractor and are considered reliable. Design costs are based on the standard charge and are also considered reliable.

Energy price data come from the 1988 edition of the federal report on energy

---

[8]It is necessary to obtain estimates of fixed and variable costs when part of the problem is to design and size the system. Here it is not essential because we are evaluating a system that has been predesigned and sized.

Table 20-4. Purchase and installation costs of the solar hot water system[a]

(1) Component	(2) Fixed Costs ($)	(3) Variable Costs ($/ft^2 Collector Area)
Collectors	$30,000	$12.50
Roof modifications	5,000	3.00
Tanks	4,000	2.00
Piping and pumps	15,000	5.00
Insulation	3,000	1.20
Controls and instrumentation	3,500	1.40
Electrical	500	0.20
Totals	$61,000	$25.30

[a]Includes all freight, overhead, sales tax, and profit. Excludes design costs, which are calculated as 10% of purchase and installation costs.

prices and discount factors (Lippiatt and Ruegg, 1988). Sample factors are provided in Appendix G, and instructions on how to use them are in Section 12.5.3.

The price for electricity at commercial rates is $19.62 per million Btu. The 1988 UPV** factor for converting annual costs to present value costs over 20 years is 10.93 for electricity in Arizona, as found in appendix Table G-25. (The UPV** is a multiplier which combines future energy price escalation with discounting.) Electricity to operate the solar energy system is estimated at 2% of the value of solar energy delivered annually. Table 20-5 summarizes the energy data for the case study.

For an annual charge equal to 2% of purchase and installation costs (constant dollars), the contractor guarantees coverage of any major repair and replacement costs required under normal operating conditions over the system's 20-

Table 20-5. Energy data for evaluating the solar hot water system

(1) Description	(2) Data
Initial price of electricity, 1988[a]	$19.62/10^6 Btu
UPV** factor[a]	10.93
Annual water heating load	1,750 × 10^6 Btu
Solar fraction	60% of load
Annual operating energy for pump electricity	2% of annual solar energy delivered
Efficiency of auxiliary heater	100%

[a]UPV** factor comes from Lippiatt and Ruegg (1988). See sample data in Appendix G, Tables G-22 through G-26, and discussion of use in Section 12.5.3.

year life. Maintenance costs for the electric hot water system are estimated to be about the same whether it is used alone or as an auxiliary to the solar energy system. Any residual value remaining at the end of the system's life is assumed to be offset by disposal costs. All future costs except energy costs are assumed to change at the same rate as prices in general.

20.2.3.2 Calculate and Compare the LCC with and without the Solar System

Compute the LCC associated with heating water for the facility *without* the solar energy system. Only one cost is relevant: electricity costs of the existing system. (Maintenance costs of the existing electric system need not be included because they are unaffected by the solar energy system.)

Use equation 20-1 to compute present value energy costs, which, in this case, equal LCC without solar.

$$LCC_{wo} = PVE_{wo} = (Q)(p)(UPV^{**}) \qquad (20\text{-}1)$$

where

LCC_{wo} = LCC without the solar hot water system,
PVE_{wo} = present value energy costs without the solar hot water system,
Q = quantity of electricity consumed without the solar hot water system,
P = price of electricity per unit, where units are the same as those in which quantity is expressed, and
UPV^{**} = UPV** factor based on the federal discount rate for evaluating energy conservation and renewable energy projects and federal projected rates of change in energy prices over the study period.

Substituting energy data from Table 20-5 into equation 20-1 gives LCC without the solar hot water system:

$$LCC_{wo} = PVE_{wo} = (1{,}750 \times 10^6 \text{ Btu})(19.62/10^6 \text{ Btu})(10.93)$$
$$= \$375{,}282$$

LCC *with* the solar energy system includes purchase and installation costs, as well as maintenance costs for the solar energy system, and electricity costs for the auxiliary system and the solar energy system. It can be calculated with equation 20-2.

$$LCC_{ws} = PVI_{ws} + PVM_{ws} + PVE_{ws} \qquad (20\text{-}2)$$

where

LCC_{ws} = LCC with the solar hot water system,
PVI_{ws} = present value purchase and installation costs with the solar hot water system,
PVM_{ws} = present value maintenance and repair costs with the solar hot water system, and
PVE_{ws} = present value of energy costs with the solar hot water system.

The PVI_{ws} term of equation 20-2 can be calculated as follows:

$$PVI_{ws} = \left[FC + (VC)(CA)\right](1 + dc) \tag{20-3}$$

where

FC = fixed purchase and installation costs,
VC = variable purchase and installation costs, i.e., costs per unit of collector area,
CA = the size of the collector area, and
dc = design costs expressed as a fraction of purchase and installation costs.

The fixed and variable cost breakdown is a common approach in solar energy evaluations and is required when evaluating alternative designs. Substituting values from Table 20-4, together with collector size and the design costs fraction, into equation 20-3 gives present value purchase and installation costs:

$$\begin{aligned} PVI_{ws} &= \left[\$61{,}000 + (\$25.30/\text{ft}^2)(2{,}500\ \text{ft}^2)\right](1 + 0.10) \\ &= \$136{,}675 \end{aligned}$$

The PVM_{ws} term of equation 20-2 can be calculated as follows:

$$PVM_{ws} = \left[(FC + (VC)(CA))(m)\right](UPV) \tag{20-4}$$

where

m = annually recurring maintenance costs expressed as a fraction of purchase and installation (exclusive of design) costs, and
UPV = UPV discount factor based on the federal discount rate for evaluating energy conservation and renewable energy projects, with no differential price change.

Substituting cost data from Table 20-4, together with the maintenance cost fraction and the UPV discount factor for 7% from appendix Table G-3, into equation 20-4 gives present value maintenance costs with the solar hot water system:

$$PVM_{ws} = \left[\left(\$61{,}000 + \left((\$25.30/\text{ft}^2)(2{,}500 \text{ ft}^2) \right) \right) (0.02) \right] (10.59)$$
$$= \$26{,}316$$

The PVE_{ws} term of equation 20-2 can be calculated as follows:

$$PVE_{ws} = L \left[\frac{(1 - sf)}{\eta} + (ope)(sf) \right] (p) UPV^{**} \qquad (20\text{-}5)$$

where

L = energy load,
sf = solar fraction,
η = efficiency of the auxiliary system,
ope = annual operating energy for the solar energy system to power pumps, expressed as a fraction of annual solar energy delivered,
p = price of energy per unit, and
UPV^{**} = UPV** discount factor based on federal projected rates of change in energy prices over the study period.

Substituting values from Table 20-5 into equation 20-5 gives present value energy costs with the solar hot water system:

$$PVE_{ws} = 1{,}750 \times 10^6 \text{ Btu} \left[\left(\frac{1 - 0.6}{1.00} \right) + (0.02 \times 0.6) \right]$$
$$\times (\$19.62/10^6 \text{ Btu})(10.93)$$
$$= \$154{,}616$$

Substituting present values calculated using equations 20-3 through 20-5 into equation 20-2 gives LCC with the solar hot water system:

$$LCC_{ws} = \$136{,}675 + \$26{,}316 + \$154{,}616$$
$$= \$317{,}607$$

Compare LCC *with* and *without* the solar energy system to see which is lower. The estimated cost without the system is \$375,282; with the system, \$317,607. NS is \$57,675.

20.2.3.3 Calculate the SIR

Calculate the SIR to provide a ranking for this project relative to other projects competing for funding. The SIR for a federal project is calculated with equation 20-6:

$$SIR = \frac{(PVE_{wo} - PVE_{ws}) - (PVM_{ws} - PVM_{wo})}{(PVI_{ws} - PVI_{wo}) + (PVR_{ws} - PVR_{wo}) - (PVS_{ws} - PVS_{wo})} \tag{20-6}$$

where

PVE = present value energy costs,
PVM = present value maintenance and repair costs,
PVI = present value purchase and installation costs,
PVR = present value replacement costs,
PVS = present value residual value, and the subscripts "wo" and "ws" indicate "without solar" and "with solar," respectively.

Substituting values previously calculated into equation 20-6 gives the SIR:

$$SIR = \frac{(\$375{,}282 - \$154{,}616) - (\$26{,}316 - 0)}{\$136{,}675 - 0}$$

$$= 1.42$$

20.2.3.4 Calculate Time Until Payback

The last step is to estimate the time until savings from the project fully offset investment costs, i.e., time to payback. Table 20-6 shows the calculation for discounted payback (DPB).[9] Estimated payback occurs in the 12th year. Or, more precisely, payback occurs in 11.3 years.[10] At the end of the 12th year, NS is estimated at $6,115.

Alternatively, an approximation approach can be used to calculate discounted

---

[9]Given the continuous cost and savings functions in this case, discounted payback could be found quickly without calculating cumulative costs for every year. By testing cumulative NS (or net losses) for selected years, one could quickly determine the year of payback.

[10]In the 12th year, NS of $8,597 are estimated, i.e., enough to offset the net loss of $2,482 remaining at the end of the 11th year and leave an NS balance of $6,115 at the end of the 12th year. Assuming roughly even monthly accruals of $716 ($8,597/12 = $716), it will take 3.47 months into the 12th year to offset the net loss of $2,482. Hence, the payback time is estimated more precisely at 11.3 years.

Table 20-6. Calculate DPB

(1) Year	(2) Cumulative Energy Savings[a] (PV $)	(3) Cumulative Net Increase in Maintenance, Repair, Replacement, and Residual[b] (PV $)	(4) Cumulative Savings (4) = (2) − (3)	(5) Increased Purchase and Installation Cost (PV $)	(6) Cumulative Saving Minus Purchase and Installation Cost (6) = (4) − (5) (PV $)	(7) Discounted Payback (Value of Col. (1) When the Sign of the Value in Col. (6) Changes from − to +
0	. . .	. . .	. . .	136,675		
1	18,978	2,311	16,667	. . .	−120,008	. . .
2	36,542	4,498	32,044	. . .	−104,631	. . .
3	52,895	6,511	46,384	. . .	−90,291	. . .
4	68,239	8,424	59,815	. . .	−76,860	. . .
5	82,573	10,189	72,384	. . .	−64,291	. . .
6	96,100	11,853	84,247	. . .	−52,428	. . .
7	108,617	13,394	95,223	. . .	−41,452	. . .
8	120,528	14,835	105,693	. . .	−30,982	. . .
9	131,834	16,202	115,632	. . .	−21,043	. . .
10	142,736	17,445	125,291	. . .	−11,384	. . .
11	152,831	18,638	134,193	. . .	−2,482	. . .
12	162,521	19,731	142,790	. . .	+6,115	12

[a]Cumulative energy savings are calculated as follows: $1,750 × 10^6 Btu [(0.6/1.00) − ((0.02)(0.6))] ($19.62/$10^6$ Btu)(UPV_t^{**}), where t is the year to which savings are cumulated.

[b]Cumulative increased maintenance and repair costs are calculated as follows: [$61,000 + (($25.30/ft^2)(2,500/ft^2))](0.02)(UPV_t), where t is the year to which costs are cumulated.

payback more quickly. (See Chapter 7.) Assume that yearly cash flows are more or less uniform. Then calculate simple payback (SPB) by dividing investment costs by net yearly cash flow (without discounting and based on initial costs). In this example, investment costs are $136,675. Net yearly cash flow is $17,704, calculated by subtracting $16,631 (annual costs to heat water with solar, power the system, and maintain it[11]) from $34,335 (annual costs to heat water without solar energy[12]). Therefore, SPB = $136,675/$17,704 = 7.7.

Now go to appendix Table G-25 for federal UPV** factors for Arizona, and locate in the column for commercial electricity the factor closest in value to 7.7. You find that 7.7 falls between 7.57 for year 11 and 8.05 for year 12.

[11]Computed as ((1,750)($19.62)(0.4)) + ((1,750)(0.6)(0.02)($19.62)) + (($124,250)(0.02) = $16,631.

[12]Computed as (1,750)($19.62) = $34,335.

Hence, you conclude that discounted payback occurs in a little more than 11 years or, using interpolation, 11.3 years.

20.2.4 Recommend Whether or Not to Add the Solar Hot Water System and Indicate Its Relative Priority

The solar energy project meets federal criteria for cost effectiveness and is acceptable if funding permits. The solar energy system is estimated to reduce building LCC by about $60,000 over the 20-year study period. This means that the project will realize a rate of return in excess of the required 7%. Project savings are expected to offset time-equivalent project costs in a little more than 11 years. Relative to competing cost-effective projects, the solar energy project receives a priority rating based on its SIR of 1.42. If sufficient funding remains after taking projects with SIRs greater than 1.42, you should recommend that the solar hot water system be added to the military base's launderette.

20.3 WHETHER TO INVEST IN AN APARTMENT BUILDING: EVALUATED WITH THE BCR METHOD AND ILLUSTRATING CASH-FLOW ANALYSIS AND TREATMENT OF FINANCING AND TAXES

In this case study, we demonstrate how to decide whether to accept or reject an investment in real estate, using the BCR method.[13] We also demonstrate how to calculate year-by-year cash flows both before and after taxes. Attention is given to cash flow because it is often critical to financial survival. Investments which would be economically efficient over the long run may nevertheless fail in the short run if they cause unsustainable negative cash flows.

Another special feature of this example is the inclusion of financing. As we point out in Chapter 13, financing normally should be included in economic evaluations of projects only if terms of financing are an integral part of the investment package. The case study demonstrates how to take into account financing because the apartment purchase has financing as an integral part of the investment package. This situation is typical of real estate investments.

20.3.1 Problem

Suppose that you have been hired by a realty partnership to seek profitable real estate investments. The objective is to identify investments which will more

[13]This case study is adapted from an example given in American Society for Testing and Materials (1989, pp. 7–9).

than compensate the partnership for its estimated opportunity cost of 12% after taxes without increasing the average risk of the investment portfolio. The partnership has $2 million on hand to invest. Its target holding period for property is 5 years. It carefully monitors year-to-year cash flows from investments in order to avoid liquidity crunches.

Suppose that you find an apartment building for $10 million. The seller will finance 80% of the price over 5 years at an interest rate of 10%, with uniform payments at the end of each year. The seller's offer of financing is more favorable than other sources of financing available to the partnership and is available only on the apartment building. In this case, financing is an attractive component of the investment package.

20.3.2 Approach

One of several approaches for measuring profitability is to compute a BCR, using equity funds for the denominator. The resulting ratio gives average discounted benefits (over and above future costs including opportunity costs) per dollar of equity funds invested. A ratio in excess of 1.0 indicates that the investment more than meets the required rate of return of 12% and, hence, is acceptable on economic grounds.

Because the partnership is subject to federal and state income taxes, an after-tax analysis is required. And, because the terms of financing are part of the investment opportunity, the calculations take financing into account. Performing the analysis in current dollars using the partnership's after-tax opportunity cost as the market discount rate facilitates the tax analysis.

Computing period-by-period cash flows is advisable in light of concern over short-run liquidity. Cash-flow analysis can be done on a before- and an after-tax basis, both of which may be of interest to management.

20.3.3 Evaluation

20.3.3.1 Data and Assumptions

Table 20-7 summarizes the data and assumptions you need to evaluate the profitability of the apartment house investment. The study period of 5 years matches the intended holding period. The after-tax market discount rate is 12%. The assumed inflation rate is 5% per year.

Investment cost data come next. The purchase price is divided between land and improvements to facilitate calculation of depreciation and estimation of resale value. You need loan terms to calculate the cash-flow effects of financing.

Yearly costs of maintenance, energy, trash removal, insurance, real estate taxes, and other costs of operating the building come under the heading "Operating Costs." Combining them reduces the number of calculations.

Table 20-7. Realty partnership investment in apartment building: data and assumptions

Study period (investor's holding period)	5 years
Discount rate, after taxes (includes estimated inflation rate)	12%
Inflation rate (annual rate of general price change)	5%
Investment cost data	
Purchase price	$10 million
Land	$2.5 million
Improvements	$7.5 million
Down payment (20% of purchase price)	$2 million
Loan (80% of purchase price)	$8 million
Loan interest rate	10%
Loan life	5 years
Yearly loan payment ($8 million loan amortized over 5 years at 10%)	$2,110,400
Depreciation period	27.5 years
Depreciation amount (S-L method)	$272,727/year
Income tax treatment of loan interest	Fully Deductible
Resale of building (estimated net of selling costs at the end of 5 years)	$12.1 million
Operating cost data	
Initial yearly costs, including maintenance, energy, trash removal, insurance, real estate taxes, etc.	$1.2 million
Rental revenues	
Initial yearly rent from residential tenants	$4.2 million
Initial yearly lease revenue from concessions	$0.5 million
Yearly rate of increase	8%
Marginal income tax rates	
Federal rate	28%
State rate	4%
Combined rate	30.9%[a]

[a]Assuming deductibility of state tax from federal tax liability, the combined tax rate is 0.28(1 − 0.04) + 0.04 = 0.309, or 30.9%.

Rental revenue consists of payments of rent by tenants and lease payments from business concessions located within the building.

The last three items in Table 20-7 are marginal income tax rates. First is the federal rate, then the state rate, and lastly the combined rate (explained in Chapter 13). Use the combined marginal rate for all income tax adjustments.

20.3.3.2 Calculating Present Values

Present value benefits and costs are calculated in a series of five tables, 20-8 through 20-12. Calculations are presented step by step in numbered columns, with operators indicated in column headings. Explanatory footnotes illuminate the operations.

Table 20-8 shows in eight columns how to calculate the present value of the

Table 20-8. Present value financed investment costs after tax deductions for interest

(1) Year	(2) Yearly Loan Payment in Current Dollars	(3) Interest Payments[a] in Current Dollars	(4) Income Tax Rate	(5) Income Tax Savings from Interest Deductions (5) = (3) × (4)	(6) After-Tax Loan Payment (6) = (2) − (5)	(7) SPV Factor (12% Discount Rate)	(8) PV of Financed Investment Costs After Taxes (8) = (6) × (7)
0	. . .	. . .	. . .	. . .	. . .	. . .	. . .
1	$2,110,400	$800,000	0.309	$247,200	$1,863,200	0.8929	$1,663,651
2	2,110,400	668,960	0.309	206,709	1,903,691	0.7972	1,517,622
3	2,110,400	524,816	0.309	169,168	1,948,232	0.7118	1,386,752
4	2,110,400	366,258	0.309	113,174	1,997,226	0.6355	1,269,237
5	2,110,400	191,843	0.309	59,279	2,051,121	0.5674	1,163,806
				PV of financed investment costs after deductions for loan interest:			7,001,068

[a]Interest payment_t = remaining principal_t × interest rate. Remaining principal_t = remaining principal_{t-1} − (loan payment − interest payment_{t-1}), where remaining principal_t is at the beginning of year t. See Section 13.2.6.

Table 20-9. Present value income tax savings due to depreciation write-off

(1) Year	(2) Yearly Depreciation in Current Dollars[a]	(3) Combined Income Tax Rate	(4) Yearly Income Tax Savings Due to Depreciation Write-Off, in Current Dollars (4) = (2) × (3)	(5) SPV Factor (12% Discount Rate)	(6) PV of Income Tax Savings Due to Depreciation Write-Off (6) = (4) × (5)
0	. . .	. . .	. . .	. . .	. . .
1	$272,727	0.309	$84,273	0.8929	$75,247
2	272,727	0.309	84,273	0.7972	67,182
3	272,727	0.309	84,273	0.7118	59,985
4	272,727	0.309	84,273	0.6355	53,555
5	272,727	0.309	84,273	0.5674	47,817
		PV of income tax savings due to depreciation write-off:			303,787

[a]Based on straight-line depreciation of $7.5 million in capital improvements over 27.5 years. The yearly depreciation is tied to historical costs and does not change with general price inflation. Because the amount is fixed in current dollars, inflation erodes the constant dollar value of the depreciation allowance. This is accomplished by use of SPV factors based on a market discount rate which includes general price inflation.

financed investment cost where the financed amount is 80% of the purchase cost. Present value investment costs are estimated at $7,001,068, after deductions of loan interest from taxable income. This amount is subtracted from present value revenue in the numerator of the BCR, because the denominator consists solely of the $2 million equity investment.

Table 20-9 shows how to calculate the present value of income tax savings due to depreciation write-offs. The amount, $303,787, is added to the present value revenue in the numerator of the BCR.

Table 20-10 shows how to calculate present value operating costs after taxes. The present value amount, $3,430,460, is subtracted from the present value revenue in the numerator of the BCR.

Table 20-11 shows how to calculate present value resale, net of capital gains tax. For this example, assume that you follow the convention of estimating the future value of the land and building, beginning with its cost today.[14] You might assume that land values will appreciate faster than general price inflation. This could happen if the area in question experiences economic growth over the study period, leading to a reduced supply of available land for building. Build-

[14]Alternative methods of estimating resale value include (1) discounting the estimated stream of cash flows accruing to the asset over its life remaining after the time of resale and (2) basing the future value of an asset on the current market value of comparable assets of similar age and, if working in current dollars, adjusting for projected inflation. (See Section 12.4.)

Table 20-10. Present value operating costs after taxes

(1) Year	(2) Operating Costs in Base-Year Prices	(3) Multiplier to Adjust for Yearly Rate of Price Increase	(4) Yearly Operating Cost in Current Dollars (4) = (2) × (3)	(5) Income Tax Rate	(6) Tax Reduction Due to Operating Cost Deductions, in Current Dollars (6) = (4) × (5)	(7) Yearly Operating Costs After Taxes, in Current Dollars (7) = (4) − (6)	(8) SPV Factor	(9) PV of Operating Costs After Taxes (9) = (7) × (8)
0	$1,200,000	. . .	. . .	. . .	. . .	. . .	. . .	. . .
1	1,200,000	$(1 + 0.05)^1$	$1,260,000	0.309	$389,340	$870,660	0.8929	$777,412
2	1,200,000	$(1 + 0.05)^2$	1,323,000	0.309	408,807	914,193	0.7972	728,795
3	1,200,000	$(1 + 0.05)^3$	1,389,150	0.309	429,247	959,903	0.7118	683,259
4	1,200,000	$(1 + 0.05)^4$	1,458,608	0.309	450,710	1,007,898	0.6355	640,519
5	1,200,000	$(1 + 0.05)^5$	1,531,538	0.309	473,245	1,058,293	0.5674	600,475
						PV of operating costs after taxes:		3,430,460

Table 20-11. Present value resale, net of capital gains tax

(1) Year	(2) Resale Value at End of 5 Years in Current Dollars[a]	(3) Book Value at End of 5 Years[b]	(4) Capital Gain (4) = (2) − (3)	(5) Capital Gains Tax Rate	(6) Capital Gains Tax (6) = (4) × (5)	(7) Resale Value Net of Capital Gains Tax (7) = (2) − (6)	(8) SPV Factor, 12% Discount Rate	(9) PV of Resale Value Net of Capital Gains Tax (9) = (7) × (8)
5	$12,100,000	$8,636,365	$3,463,635	$0.309	$1,070,263	$11,029,737	0.5674	$6,258,273

[a]Resale value has two components: land and building. Land resale is based on land costs appreciating 10% per year over 5 years. Building resale is based on the building's value deteriorating over 5 years at a compound rate of 3.33% of the initial cost per year to reflect the fact that an existing building under normal circumstances tends to be worth less than an identical new building. At the same time, the remaining value of the building is assumed to appreciate at the rate of general price inflation. Thus, after 5 years the estimated resale value of the land in current dollars is $4,026,275 [i.e., $\$2,500,000 \times (1 + 0.10)^5$], the estimated resale value of the building is $8,081,023 [i.e., $\$7,500,000 \times (1 - 0.0333)^5(1 + 0.05)^5$], and the estimated total resale is $12,100,000 (i.e., $4,026,275 + $8,081,023), rounded to the nearest $100,000.

[b]Original book value of $10.0 million less 5 years of straight-line depreciation of the $7.5 million in capital improvements (i.e., $10,000,000 − $1,363,636 = $8,636,365).

ing value, on the other hand, might be assumed to increase in value slower than general price inflation. This could happen as a result of two opposing forces: (1) deterioration, which causes existing buildings to be worth less than counterpart new buildings (all other factors being identical), and (2) inflation, which causes the price of buildings to rise.

Estimated resale value is reduced by estimated capital gains tax, calculated at the ordinary tax rate. Present value resale, net of capital gains tax, is estimated at $6,258,273. This amount is added to present value revenue in the numerator of the BCR.

Table 20-12 shows how to calculate present value revenue (after income taxes). It consists of apartment rental income and leasing revenue from several concessions located in the building. The amount, $14,579,368, goes in the numerator of the BCR.

20.3.3.3 Calculate the BCR

Calculating the BCR is a simple matter once the after-tax present values of all cash flows are determined. Table 20-13 shows how to calculate the ratio according to equation 4-1. All future revenue and costs are combined in the numerator. The down payment, which is to be made out of equity funds, is the denominator. Since the down payment occurs at the present, no discounting is required for it. The ratio is 5.36.

20.3.3.4 Calculate Year-to-Year Net Cash Flow

To see if the realty partnership should expect any liquidity problems, calculate the year-to-year net cash flow. Table 20-14 shows cash flow before taxes; Table 20-15, cash flow after taxes. After the initial outflow of $2 million, cash flow is expected to be positive over the entire holding period, both before and after taxes.

20.3.4 Analyze the Results and Advise the Client

The BCR of 5.36 means that after-tax proceeds are estimated to be $5.36 per dollar of equity funds invested, over and above the after-tax minimum accepted rate of return of 12%. Part of the return is due to favorable terms of financing and part to apartment rental and concessions income. Year-to-year cash flows are expected to improve the partnership's liquidity.

The economic performance of the apartment investment appears strong. You will probably want to recommend that the realty partnership accept it.

Table 20-12. Present value revenue after income taxes

(1) Year	(2) Initial Yearly Rent in Base-Year Prices	(3) Initial Yearly Lease Revenues from Concessions in Base-Year Prices	(4) Total Yearly Revenue in Base-Year Prices $(4_0) = (2_0) + (3_0)$	(5) Multiplier to Adjust for Yearly Rate of Price Increase	(6) Total Yearly Revenue in Current Dollars $(6) = (4_0) \times (5)$	(7) Combined Income Tax Rate	(8) Income Tax Increase Due to Revenue $(8) = (6) \times (7)$	(9) Total Yearly Revenue After Income Taxes, in Current Dollars $(9) = (6) - (8)$	(10) SPV Factor	(11) PV of Revenues After Income Taxes $(11) = (9) \times (10)$
0	$4,200,000	$500,000	$4,700,000	. . .	. . .	. . .	. . .	. . .	. . .	. . .
1	. . .	. . .	. . .	$(1 + 0.08)^1$	$5,076,000	0.309	$1,568,484	$3,507,516	0.8929	$3,131,861
2	. . .	. . .	. . .	$(1 + 0.08)^2$	5,482,080	0.309	1,693,963	3,788,117	0.7972	3,019,887
3	. . .	. . .	. . .	$(1 + 0.08)^3$	5,920,646	0.309	1,829,480	4,091,166	0.7118	2,912,092
4	. . .	. . .	. . .	$(1 + 0.08)^4$	6,394,298	0.309	1,975,838	4,418,460	0.6355	2,807,931
5	. . .	. . .	. . .	$(1 + 0.08)^5$	6,905,842	0.309	2,133,905	4,771,937	0.5674	2,707,597
								PV of total revenue:		14,579,368

Table 20-13. BCR computed from after-tax revenues and costs

(1) PV of Revenue	(2) PV of Resale Value Net of Capital Gains Tax	(3) PV of Financed Investment Costs	(4) PV of Income Tax Savings Due to Depreciation Write-Off	(5) PV of Operating Costs	(6) PV of Total Revenue Less Future Costs (Numerator of the BCR) (6) = (1) + (2) − (3) + (4) − (5)	(7) Down Payment (Investor's Equity)	(8) BCR (8) = (6)/(7)
$14,579,368	$6,258,273	$7,001,068	$303,787	$3,430,460	$10,709,900	$2,000,000	5.36

Table 20-14. Cash-flow analysis before taxes

(1) Year	(2) Down Payment	(3) Loan Payment	(4) Operating Costs	(5) Resale Proceeds	(6) Revenue	(7) Cash Flow Before Taxes (7) = (6) + (5) − ((2) + (3) + (4))
0	$2,000,000	. . .	. . .	. . .	. . .	−$2,000,000
1	. . .	$2,110,400	$1,260,000	. . .	$5,076,000	1,705,600
2	. . .	2,110,400	1,323,000	. . .	5,482,080	2,048,680
3	. . .	2,110,400	1,389,150	. . .	5,920,646	2,421,096
4	. . .	2,110,400	1,458,608	. . .	6,394,298	2,825,290
5	. . .	2,110,400	1,531,538	$12,100,000	6,905,842	15,363,904

20.4 WILL AN INVESTMENT IN A PUBLIC STORAGE FACILITY BE PROFITABLE? EVALUATED WITH NB, BCR, AND ORR METHODS, AND USING THE SIMULATION TECHNIQUE FOR RISK ANALYSIS

20.4.1 Problem

Suppose that you are a consultant who has been asked by an entrepreneur to evaluate the economic merits of constructing small-scale warehouse storage facilities for rent.[15] The client requests after-tax measures of long-run profitability and return on investment. You select NB, BCR, and ORR measures for a comprehensive economic perspective. The client also requests that you account for uncertainty in estimates of occupancy rates, operating costs, resale price, and land preparation and construction costs when computing the measures of project worth. Your job is to determine how likely it is that the storage facility is a profitable investment after taxes.

20.4.2 Approach

Construct probability density functions for the uncertain inputs that the client identified. Draw, at random, combinations of each of the variable inputs and combine them with constant inputs to compute a single measure of worth for each of the three methods listed earlier. Repeat this simulation until the distributions of project worth are well defined.

Construct a cumulative distribution function for each of the three measures of worth so that you can evaluate the likely economic performance and risk exposure of investing in the warehouse facility.

[15]This case study is adapted from Marshall (1988, pp. 44-52).

Table 20-15. Cash-flow analysis after taxes

(1) Year	(2) Cash Flow Before Taxes	(3) Income Tax Savings Due to Interest Deductions	(4) Income Tax Savings Due to Depreciation Write-Offs	(5) Income Tax Savings Due to Operating Cost Deductions	(6) Capital Gains Tax	(7) Income Tax Increase Due to Revenue	(8) Impact of Taxes on Cash Flow (8) = (3) + (4) + (5) − (6) − (7)	(9) Cash Flow After Taxes (9) = (2) + (8)
0	−$2,000,000	. . .	. . .	. . .	. . .	. . .	. . .	−$2,000,000
1	1,705,600	$247,200	$84,273	$389,340	. . .	$1,568,484	−$847,671	857,929
2	2,048,680	206,709	84,273	408,807	. . .	1,693,963	−994,174	1,054,506
3	2,421,096	169,168	84,273	429,247	. . .	1,829,480	−1,146,792	1,274,304
4	2,825,290	113,174	84,273	450,710	. . .	1,975,838	−1,327,681	1,497,609
5	15,363,904	59,279	84,273	473,245	$1,070,263	2,133,905	−2,587,371	12,776,533

Table 20-16. Input data assumed certain for simulation of investment in storage facility

1. $180,000 land purchase costs (LC)
2. 600 units with 300 ft.^{2}/unit
3. $1,200 rent/unit/year; $9,801 present value rent/unit/10 years (RNT)
4. 10-year holding period ($N = 10$)
5. 7.5% after-tax market discount rate (D)[a]
   7.5% after-tax MARR
   7.5% after-tax reinvestment rate (r)
6. 3.5% inflation rate (I)
7. 3.86% after-tax, real discount rate (d)[b]
8. 28% marginal tax rate on personal income (MTRPI)
9. 28% marginal tax rate on capital gains (MTRCG)
10. 31.5 years for depreciation via the straight-line method

[a]D was selected as follows: The best available return (7.5%) on a low-risk, tax-exempt bond was compared against the after-tax equivalent return (7.37%) on the best available taxable investment of similar risk to the bond. The after-tax, equivalent rate was computed for the taxable investment by multiplying the before-tax opportunity rate by the factor $1 - T$, where T is the composite federal/state tax rate at the margin. In this case, $D = 11\% \ (1 - 0.33) = 7.37\%$. The better opportunity was the tax-exempt bond (7.5% > 7.37%); therefore 7.5% is the after-tax discount rate used in the analysis.

[b]Using equation 11-2, $d = \dfrac{1.075}{1.035} - 1 = 0.0386$ or 3.86%.

Present the results of the analysis to the client in terms of probabilities of economic measures being less or greater than target values.

20.4.3 Evaluation

20.4.3.1 Data and Assumptions

Table 20-16 lists input data which are assumed certain for the analysis.[16] Your client has an option to purchase 9 acres at $20,000/acre for a total of $180,000.[17] The $1,200/unit annual rent projection is based on comparable space currently rented in the neighborhood. It seems reasonable to assume that the constant dollar value of rent will remain the same over the 10 years; that is, annual rents will escalate in current dollars at the rate of general price inflation. Thus you multiply the $1,200 figure by the UPV factor for 10 years at 3.86% (i.e., UPV = 8.1677) to find the present value of total rents/unit/10 years ($9,801).[18]

---

[16]The case illustration is based on estimated costs and revenues of constructing and operating the described facility in a lightly urbanized area of Virginia.

[17]The cost of the option is not pertinent to the analysis because it is a sunk cost whether or not the investment is made in the warehouse.

[18]The compounding frequency used throughout the analysis is 1 year; rates of discount and inflation are quoted per year; and the convention of end-of-year discounting is followed. The derivation of the after-tax, real discount rate of 3.86% appears in a footnote to Table 20-16.

Tax effects are included to illustrate the net effect on the investor after taxes.[19] Since the investment is to be financed at prevailing commercial rates, borrowing costs need not be included in the analysis. It is assumed that probability distributions are independent. A selection from one distribution, therefore, does not affect selections from others.

20.4.3.2 Identify Uncertain Inputs and Describe Them with Probability Density Functions

The uncertain input data are the number of units occupied, operating costs, resale value (the projected warehouse selling price in 10 years), and construction costs (including site preparation costs). The probability density functions in Figures 20-1 through 20-4 describe the uncertainty associated with each of the inputs. In each iteration of the simulation, a draw is made from among possible values in each probability density function and entered into the formulas for computing measures of project worth.

The probability density function in Figure 20-1 depicts the number of units expected to be rented in any given year. A gamma distribution seems best to describe anticipated rentals. The distribution has a mean of 564 units (94% occupancy) with a standard deviation of 12 units (2%). The left-hand tail lies close to the horizontal axis up to an occupancy level of about 520. This is the

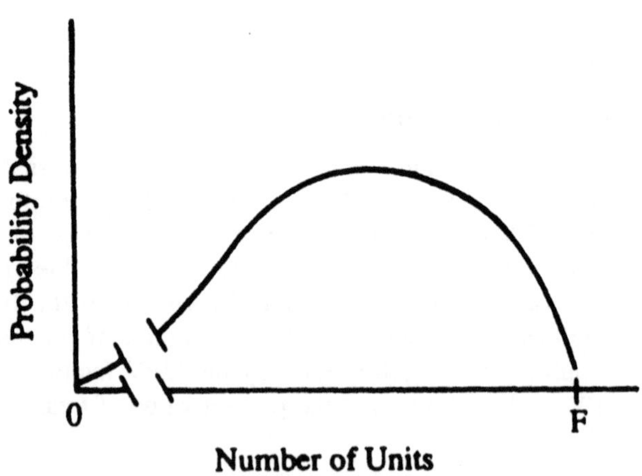

Figure 20-1. Probability density function of number of units rented.

[19]The effects of personal income taxes and capital gains taxes are computed in the case study. Property taxes do not enter into the calculation in this case because the local government has granted a 10-year exemption on property taxes to attract new business to a designated area for industrial development. If property taxes were levied, they should be accounted for in operating costs.

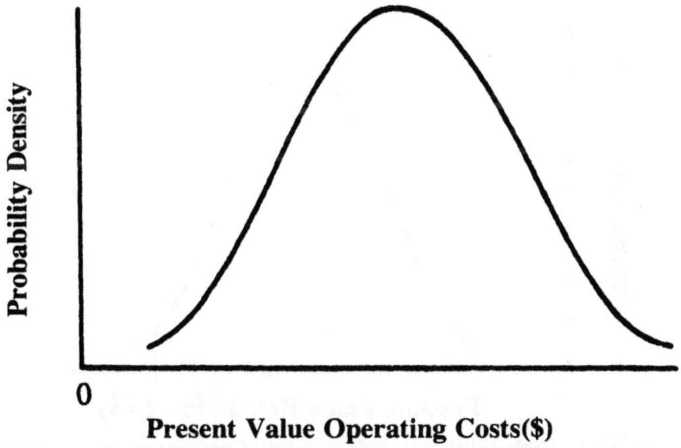

Figure 20-2. Probability density function of present value operating costs.

point at which the function in Figure 20-1 begins to rise significantly. (Gaps in the function and on the horizontal axis occur for a range of occupancy values less than 520. These segments are not shown because the occupancy values are relatively low-probability events and the tail would be very long.) The right-hand tail does not touch the horizontal axis because there is some likelihood that all 600 units (indicated by point F on the axis) could be rented. The facility is expected to fill up rapidly due to a shortage of storage in that location. Over the holding period for the facility, new renters are anticipated to replace quickly those that terminate their lease, making the probability density function about the same for each year. Figure 20-1, therefore, represents the average occupancy for every year in the 10-year period. In the simulation procedure, a value is drawn from this distribution on each iteration and multiplied by the present value rental rate ($9,801) for the 10 years to arrive at rental revenues in the measure of economic worth.

Annual operating costs are thought to be distributed normally, with a mean of $156,000 (base year dollars) and a standard deviation of $16,000 (about 10% variation from the mean).[20] Figure 20-2 shows the normal distribution of annual costs converted to a normal distribution of present value operating costs over the 10-year holding period. The mean of the distribution is $1,274,161.[21]

---

[20]Operating costs include the costs of gate attendants, liability insurance, administration, advertising, lighting, water and sewer service, and routine maintenance.

[21]Annual costs of operating the facility are expected to remain unchanged in constant dollars over the 10 years; that is, current dollar annual costs are assumed to increase at the same rate as general price inflation. Therefore you can convert annual operating costs to present values by multiplying any given uniform annual cost by the UPV factor for 10 years and 3.86%. The present value equivalent of the mean annual operating costs of $156,000 is $156,000 times the UPV factor 8.1677, or $1,274,161.

Figure 20-3. Probability density function of present value resale price.

The normal probability density function in Figure 20-3 depicts the present value resale price of the warehouse facility at the end of 10 years. It has a mean of $1,355,758 (in present value terms) and a standard deviation of $230,479 (about 17% variation from the mean). The resale price in constant dollars at the end of 10 years is projected to be $1,980,000.[22] This amount discounted at the after-tax, real rate of 3.86% over 10 years yields a present value of $1,355,758. This seemingly optimistic mean resale figure is based on the investor's perception that the property will likely hold its value as a storage facility. The higher values depicted by the right-hand tail might arise if the land were converted to a "higher and better use," thereby reflecting substantial "real" appreciation. Resale enters into the economic evaluation both as a benefit in terms of sale proceeds and as a component of capital gains on which taxes are levied.

Figure 20-4 depicts construction costs as a lognormal distribution. The mean is thought to be $1,800,000, with a standard deviation of $180,000 (about 10% variation from the mean).[23] The right-hand tail stretches out because unforeseen land preparation and construction problems always create the possibility of much higher costs then anticipated. Construction costs will be incurred over what is expected to be a relatively short construction period and are given in present value dollars.

---

[22]The projection is based on the supposition that the property will likely sell for about $1,980,000 in constant dollars, i.e., the equivalent of its initial total costs of land ($180,000), site preparation ($900,000), and building construction ($900,000).

[23]The estimated construction costs are based on a projected $900,000 in land preparation costs (concrete slab, paving, underground utilities, fence, lights, etc.) and $900,000 ($1,500/unit) in building construction costs.

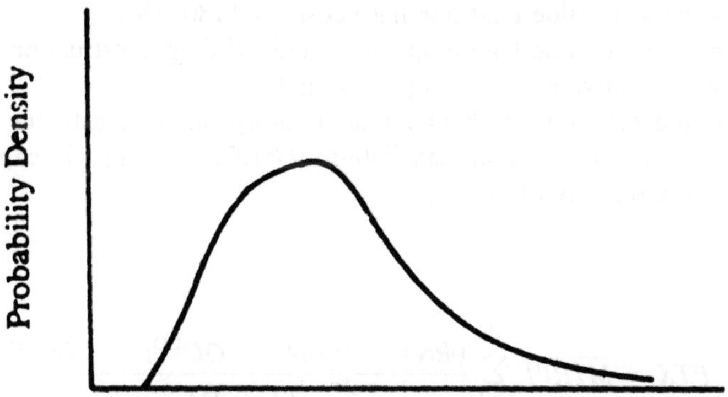

Figure 20-4. Probability density function of the present value of land preparation and construction costs.

20.4.3.3 Perform the Simulation and Compute Measures of Worth

In each iteration of the simulation, make a draw from among the possible values in each probability density function and enter the set of values into formulas for computing measures of project worth.

Three formulas (equations 20-7 to 20-9) based on method formulas described in Chapters 3, 4, and 6, and tax adjustment procedures described in Chapter 13, are needed to compute the economic measures of worth.[24] Because so many terms are in the calculations, simplified equation forms are helpful. Acronyms represent specific terms. Dollar values are in present value terms unless defined otherwise. Bars over parameters signify that they are constants. Their values are given in Table 20-16. Parameters without bars are uncertain variables represented by the distributions in Figures 20-1 to 20-4 or uncertain variables dependent on one of the distributions.

NET BENEFITS (NB)

$$NB = RS + (\overline{RNT} \cdot UN) - \overline{LC} - LPCC - OC - TX \qquad (20\text{-}7)$$

where

RS = present value resale proceeds,
$\overline{RNT}$ = present value rent rate/unit/10 years = \$9,801,
UN = number of units rented/year, on average, over 10 years,

[24]Note that in this case study current dollar amounts are used to account for tax effects and then reconciled with constant dollar amounts for nontax items.

$\overline{LC}$ = present value land purchase costs = \$180,000,
$LPCC$ = present value land preparation and building construction costs,
OC = present value operating costs, and
TX = present value of change in tax liability due to investment = change in personal income tax liability (PTX) + change in capital gains tax liability ($CGTX$),

where

$$PTX = \overline{MTRPI} \sum_{t=1}^{\overline{N}} \frac{(RNT^{CD}t)(UN) - OC^{CD}t - DEP^{CD}}{(1+\overline{D})^t},$$

$$CGTX = \overline{MTRCG}\, \frac{RS^{CD} - ((LPCC + \overline{LC}) - \overline{N}(DEP^{CD}))}{(1+\overline{D})^{\overline{N}}}$$

where

$RNT^{CD}t$ = rent rate/unit/year t in current dollars,
= \$1,200 $(1+\overline{I})^t$, or (\$1,200 · 1.035^t),
$\overline{MTRPI}$ = 28% marginal tax rate on personal income,
$OC^{CD}t$ = operating costs/year t in current dollars,
= constant dollar operating costs times 1.035^t,
DEP^{CD} = depreciation/year in current dollars,
= $LPCC/31.5$,
$\overline{MTRCG}$ = 28% marginal tax rate on capital gains,
RS^{CD} = resale price in current dollars,
= resale price in constant dollars times $1.035^{\overline{N}}$.

BENEFIT-COST RATIO (BCR)[25]

$$BCR = \frac{RS + (\overline{RNT} \cdot UN) - OC - TX}{\overline{LC} + LPCC} \qquad (20\text{-}8)$$

OVERALL RATE OF RETURN (ORR)

$$ORR = -1.0 + (1.0 + \bar{r})(BCR)^{1/\overline{N}} \qquad (20\text{-}9)$$

[25]As explained in Chapter 4, the choice of placing an item of cost or benefit in the numerator or denominator can affect the relative ranking of projects. Here we use equation 4-1 to compute BCR, placing only initial costs in the denominator in order to avoid biasing effects.

where

$\bar{r}$ = reinvestment rate, and
$\bar{r} = \overline{D}$ in this problem.

20.4.3.4 Construct Cumulative Distribution Functions for Measures of Worth and Evaluate Likely Economic Performance and Risk Exposure

We performed the simulations to show the kinds of cumulative distribution functions you might expect. Figures 20-5 through 20-7 give cumulative distribution functions for NB, BCR, and ORR, respectively. For example, looking first at Figure 20-5, you see a probability of approximately 5% that NB will be less than \$2 million, 50% that NB will be less than \$2.33 million, and about 98% that NB will be less than \$3 million. The cumulative distribution function gives a strong indication that the storage facility will be profitable, with a high likelihood of an NB between \$2 million and \$3 million.

Figure 20-6 provides supporting evidence of a profitable project, showing

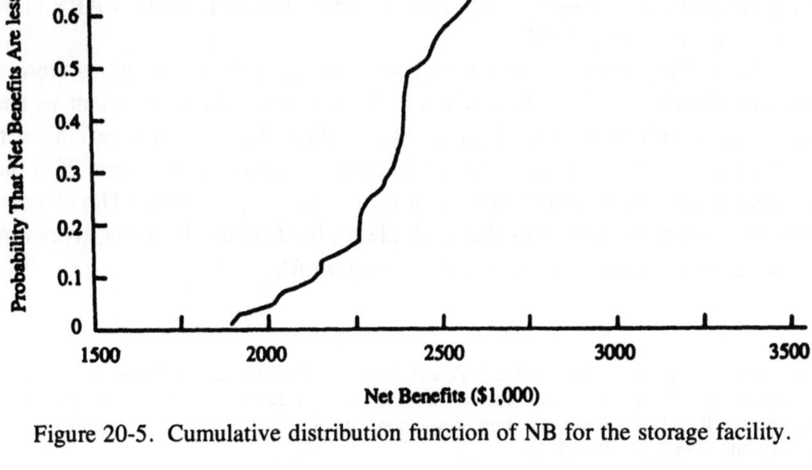

Figure 20-5. Cumulative distribution function of NB for the storage facility.

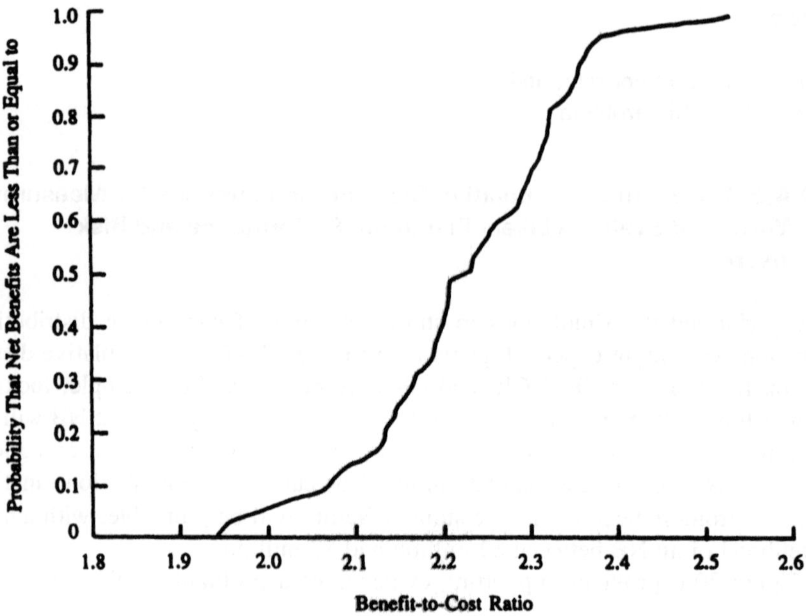

Figure 20-6. Cumulative distribution function of the BCR for the storage facility.

approximately a 5% probability that the BCR will be less than 1.97, 50% probability that the BCR will be less than 2.22, and 98% probability that the BCR will be less than 2.5.

Figure 20-7 gives similar results with the ORR—a 5% probability that it will be less than 15.2%, a 50% probability that it will be less than 16.4%, and a 98% probability that it will be less than 17.55%. Recall that the MARR shown in Table 20-16 is only 7.5%.

The Latin Hypercube sampling scheme was used to select values from the input distributions.[26] With this scheme, 50 simulations are sufficient to characterize distributions of project worth. To confirm this, two replications of the simulation, of 50 draws each, were performed to measure NB. Figure 20-8 plots the cumulative distribution functions for these two replications. The closeness of the two functions indicates that a single replication of 50 draws gives a relatively accurate picture of the distribution of worth.

[26]The name "Latin Hypercube" comes from a class of experimental designs known by statisticians as "Latin squares." This sampling scheme was selected because it requires fewer simulations than other schemes based on random sampling. For a more complete description of the sampling scheme, see Chapman (1988, pp. 360–371).

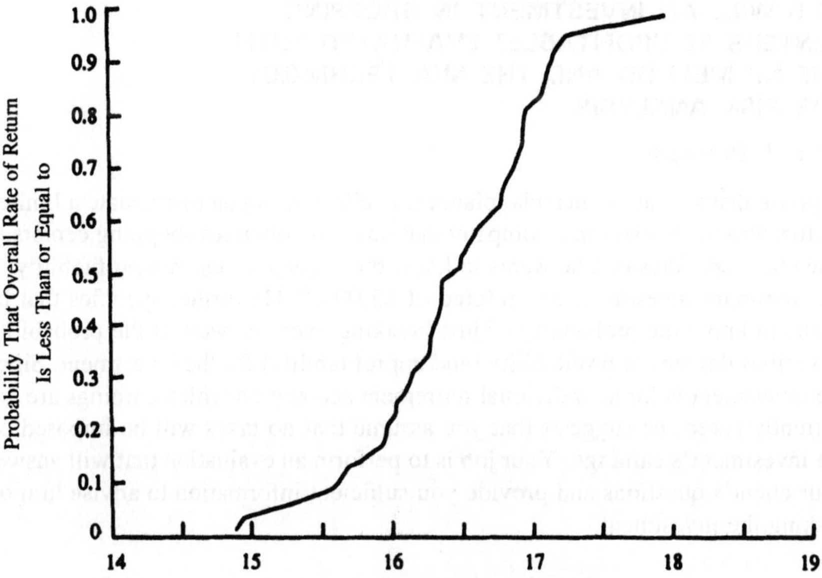

Figure 20-7. Cumulative distribution function of the ORR for the storage facility.

Figure 20-8. Two cumulative distribution functions of NB for the storage facility. Replication 1 = solid line and replication 2 = dotted line.

20.5 WILL AN INVESTMENT IN SHOPPING CENTERS BE PROFITABLE? EVALUATED WITH THE NB METHOD AND THE M/A TECHNIQUE FOR RISK ANALYSIS

20.5.1 Problem

Suppose that you are a financial planner. A client asks you to evaluate a limited partnership investment in a company that buys and operates shopping centers.[27] The client specifies that he wants to know the expected long-run profitability of the minimum investment unit offered of $5,000.[28] He further specifies that he wants to know the probability of just breaking even, as well as the probability of various degrees of profitability (and unprofitability) for the invc^tment. Since the investment is for an individual retirement account on which earnings are not currently taxed, he suggests that you assume that no taxes will be imposed on the investment's earnings. Your job is to perform an evaluation that will answer your client's questions and provide you sufficient information to advise him on making the investment.

20.5.2 Approach

Choose the NB method, which is described in Chapter 3, to provide a measure of profitability. Construct, on the basis of perceived uncertainty, the annual net cash flow distributions per $5,000 investment in the company that buys shopping centers. Then compute the mean, variance, and standard deviation of net cash flows for each of the 6 years in the holding period.

Use the M/A technique to calculate the mean and standard deviation of the probability density function of the present value of net cash flows over the 6 years.

Construct the probability density function and cumulative distribution function from that mean and standard deviation. Finally, compare the cumulative distribution function of net cash flows to the $5,000 certain investment cost to see how profitable the investment in shopping centers is likely to be.

20.5.3 Evaluation

20.5.3.1 Data and Assumptions[29]

A brokerage house is marketing investment shares in $5,000 units for a limited partnership that will purchase small shopping centers to be managed for both

---

[27]This case study is adapted from Marshall (1988, pp. 56–64).

[28]The client is considering the purchase of multiple $5,000 units. But since he prefers to keep that number confidential, he asks you to do the analysis for one unit only.

[29]We base this case illustration on data in part from a 1987 prospectus issued by a brokerage house sponsoring a limited partnership for buying shopping centers. The prospectus helped us to generate

income generation and capital appreciation. Management projects a 6-year holding period. Fewer or more than 6 years may pass before the shopping centers are sold and the proceeds distributed to investors. We know with relative certainty that the investment can be cashed in at the end of 6 years, however, because the brokerage house selling the partnerships buys and sells shares in the over-the-counter market.

Since the prospectus on which the data are based was published in 1987, we shall assume that is the year your client asked you to do the investment evaluation. You decide to use 1987 constant dollars in the analysis. Based on discussions with the client, you estimate that his real opportunity cost of capital is 8%, using Appendix B, Series A, Worksheet 2, Part I. You therefore set the real discount rate at 8%.

Assume that projections of before-tax net cash flows (annual revenues minus annual costs) are normally distributed in each period and independent among years.

Note that in this case study "net cash flows" does not mean *net* of the initial investment cost, but "net benefits" does. To calculate net benefits, you must subtract the $5,000 investment cost per share from the corresponding present value net cash flows.

20.5.3.2 Construct Annual Net Cash Flow Distributions on the Basis of Perceived Uncertainty

Table 20-17 provides probability distributions and means that describe potential earnings of the shopping centers. Five states (s) of business success are hypothesized for the group of shopping centers, ranging from the least profitable state, $s = 1$, to the most profitable, $s = 5$. Potential net cash flows (current earnings minus current expenses) are estimated for each of the 6 years ($t = 1$ through 6) for all five states of success. The net cash flow (C_{ts}) values are weighted by the probabilities (P_{ts}) that those respective business conditions might occur. The C_{ts} and P_{ts} values define the cash flow distributions for each year.

We used the following rationale in deriving the distributions for the cash illustration. The prospectus predicted an 8% return ($400 on the $5,000 investment) in the first period (i.e., year) based on current shopping center earnings. We based the dispersion of earnings about the mean (Table 20-18) on uncertainty about finding renters right away for shopping centers under new management. In the second period there is a higher probability associated with a larger mean and a smaller variance than in the first period. It is more likely that all of the space will be rented in the second year. For periods t_3 through

---

a realistic problem. We do not intend that the illustration provide economic advice about any particular investment opportunity in shopping centers.

Table 20-17. Probability distributions and means, by year, for net cash flows from shopping centers

States	Periods (t)																	
	1			2			3			4			5			6		
(s)	C_{1s}	P_{1s}	$C_{1s}P_{1s}$	C_{2s}	P_{2s}	$C_{2s}P_{2s}$	C_{3s}	P_{3s}	$C_{3s}P_{3s}$	C_{4s}	P_{4s}	$C_{4s}P_{4s}$	C_{5s}	P_{5s}	$C_{5s}P_{5s}$	C_{6s}	P_{6s}	$C_{6s}P_{6s}$
1	\$200	.15	\$30.0	\$250	.10	\$25.0	\$250	.05	\$12.5	\$300	.10	\$30	\$300	.10	\$30.0	\$6,900	.15	\$1035
2	300	.20	60.0	350	.20	70.0	350	.25	87.5	400	.25	100	450	.25	112.5	8,250	.20	1650
3	400	.30	120.0	450	.40	180.0	500	.40	200.0	550	.30	165	600	.30	180.0	9,600	.30	2880
4	500	.20	100.0	550	.20	110.0	650	.25	162.5	700	.25	175	750	.25	187.5	10,950	.20	2190
5	600	.15	90.0	650	.10	65.0	750	.05	37.5	800	.10	80	900	.10	90.0	12,300	.15	1845
	μ_1 = \$400			μ_2 = \$450			μ_3 = \$500			μ_4 = \$550			μ_5 = \$600			μ_6 = \$9,600		

Table 20-18. Standard deviations, by year, for net cash flows from shopping centers

States	Periods (t)								
	1			2			3		
(s)	$(C_{1s} - \mu_1)^2$	P_{1s}	$(C_{1s} - \mu_1)^2 P_{1s}$	$(C_{2s} - \mu_2)^2$	P_{2s}	$(C_{2s} - \mu_2)^2 P_{2s}$	$(C_{3s} - \mu_3)^2$	P_{3s}	$(C_{3s} - \mu_3)^2 P_{3s}$
1	$40,000	0.15	$6,000	$40,000	0.10	$4,000	$62,500	0.05	$3,125
2	10,000	0.20	2,000	10,000	0.20	2,000	22,500	0.25	5,625
3	0	0.30	0	0	0.40	0	0	0.40	0
4	10,000	0.20	2,000	10,000	0.20	2,000	22,500	0.25	5,625
5	40,000	0.15	6,000	40,000	0.10	4,000	62,500	0.05	3,125
			$\sigma_1^2 =$ $16,000			$\sigma_2^2 =$ $12,000			$\sigma_3^2 =$ $17,500
			$\sigma_1 =$ $126			$\sigma_2 =$ $110			$\sigma_3 =$ $132

States	Periods (t)								
	4			5			6		
(s)	$(C_{4s} - \mu_4)^2$	P_{4s}	$(C_{4s} - \mu_4)^2 P_{4s}$	$(C_{5s} - \mu_5)^2$	P_{5s}	$(C_{5s} - \mu_5)^2 P_{5s}$	$(C_{6s} - \mu_6)^2$	P_{6s}	$(C_{6s} - \mu_6)^2 P_{6s}$
1	$62,500	0.10	$6,250	$90,000	0.10	$9,000	$7,290,000	0.15	$1,093,500
2	22,500	0.25	5,625	22,500	0.25	5,625	1,822,500	0.20	364,500
3	0	0.30	0	0	0.30	0	0	0.30	0
4	22,500	0.25	5,625	22,500	0.25	5,625	1,822,500	0.20	364,500
5	62,500	0.10	6,250	90,000	0.10	9,000	7,290,000	0.15	1,093,500
			$\sigma_4^2 =$ $23,750			$\sigma_5^2 =$ $29,250			$\sigma_6^2 =$ $2,916,000
			$\sigma_4 =$ $154			$\sigma_5 =$ $171			$\sigma_6 =$ $1,708

t_5, the mean continues to rise because rent escalation contracts call for increasing rents (in constant dollar terms), while mortgage payments are fixed in current dollars (probably dropping in constant dollar terms). The variance about the mean increases, however, because estimates become less certain as the time horizon extends. Since we coupled resale proceeds with annual net cash flows in period 6, all values in the probability distribution for period 6 (column C_{6s} in Table 20-17) are larger than those in earlier periods. We estimate resale (in constant dollar terms) to be 1.2 to 2.4 times the \$5,000 purchase price. This is based on the company's past sales of similar properties.

20.5.3.3 Calculate the Mean, Variance, and Standard Deviation of Net Cash Flows for Each of the 6 Years in the Holding Period

Calculate the mean, shown as μ_t on the bottom line of Table 20-17, with equation 20-10. Calculate the variance, shown as σ_t^2 in Table 20-18, with equation 20-11. The standard deviation, shown as σ_t in Table 20-18, is simply the square root of σ_t^2, equation 20-12.

$$\mu_t = \sum_{s=1}^{K} (C_{ts} P_{ts}) \tag{20-10}$$

and

$$\sigma_t^2 = \sum_{s=1}^{K} (C_{ts} - \mu_t)^2 P_{ts} \tag{20-11}$$

$$\sigma_t = \left(\sum_{s=1}^{K} (C_{ts} - \mu_t)^2 P_{ts} \right)^{1/2} \tag{20-12}$$

where

μ_t = expected value of net cash flows in period t,
K = number of possible states of business success,
C_{ts} = net cash flow associated with state s in period t,
P_{ts} = probability of state s in period t,
σ_t^2 = variance of net cash flows in period t, and
σ_t = standard deviation of net cash flows in period t.

20.5.3.4 Calculate the Mean and Standard Deviation of the Probability Density Function of Present Value Net Cash Flows Over 6 Years

Use equations 20-13 and 20-14 to calculate the mean and standard deviation.

$$\mu_{PVC} = \sum_{t=1}^{N} \frac{\mu_t}{(1 + d)^t} \tag{20-13}$$

and

$$\sigma_{PVC} = \left(\sum_{t=1}^{N} \frac{\sigma_t^2}{(1+d)^{2t}} \right)^{1/2} \qquad (20\text{-}14)$$

where

μ_{PVC} = expected value of present value net cash flows over the project holding period,
σ_{PVC} = standard deviation of the present value of net cash flows over the project holding period, and
σ_t^2 = variance of net cash flows in period t.

Note that equations 20-13 and 20-14 differ slightly from equations 17-2 and 17-3. First, μ_t and σ_t^2 do not incorporate investment costs, whereas $E(B_t - \overline{C}_t)$ and Var $(B_t - \overline{C}_t)$ do. Second, σ_{PVC} is computed directly by equation 20-14 rather than computing the variance, as in equation 17-3.

Substituting into equations 20-13 and 20-14 yields a mean of the present value net cash flow of \$8,015 and a standard deviation of \$1,104, as shown below.

$$\mu_{PVC} = \frac{\$400}{(1.08)^1} + \frac{\$450}{(1.08)^2} + \frac{\$500}{(1.08)^3} + \frac{\$550}{(1.08)^4} + \frac{\$600}{(1.08)^5} + \frac{\$9{,}600}{(1.08)^6}$$

$$= \$370.37 + \$385.80 + \$396.92 + \$404.27 + \$408.35 + \$6{,}049.63$$

$$= \$8{,}015$$

$$\sigma_{PVC} = \left[\frac{\$16{,}000}{(1.08)^2} + \frac{\$12{,}000}{(1.08)^4} + \frac{\$17{,}500}{(1.08)^6} + \frac{\$23{,}750}{(1.08)^8} + \frac{\$29{,}250}{(1.08)^{10}} + \frac{\$2{,}916{,}000}{(1.08)^{12}} \right]^{1/2}$$

$$= [\$13{,}717.42 + \$8{,}820.36 + \$11{,}027.97 + \$12{,}831.39 + \$13{,}548.41 + \$1{,}157{,}983.70]^{1/2}$$

$$= [\$1{,}217{,}929.30]^{1/2}$$

$$= \$1{,}104$$

20.5.3.5 Construct the Probability Density Function and Cumulative Distribution Function for the Present Value of Net Cash Flows from the Shopping Centers, and Use Them to Evaluate the \$5,000 Investment

To construct a probability density function, use ordinates from a statistical table on normal curves to derive relative heights on the y-axis for corresponding present value net cash flows. Statistical tables, however, give ordinal values only

Table 20-19. Data for constructing the probability density function of present value net cash flows for shopping centers

	Net Cash Flow Values ($)		
z	$x = \mu - z\sigma$	$x = \mu + z\sigma$	Ordinate at z
(1)	(2)	(3)	(4)
0.0	8,015	8,015	0.3989
0.5	7,463	8,567	0.3521
1.0	6,911	9,119	0.2420
1.5	6,359	9,671	0.1295
2.0	5,807	10,223	0.0540
2.5	5,255	10,775	0.0175
3.0	4,703	11,327	0.0044

for standardized z values. Thus z values that correspond to specific dollar amounts must be available before corresponding ordinates on the y-axis can be found. Equation 20-15 shows the relationship between z and x, where x represents net cash flow. Each z value measures the number of standard deviations that any given net cash flow lies above or below the mean ($8,015).

$$z = \frac{x - \mu}{\sigma} \tag{20-15}$$

Table 20-19 illustrates the derivation of the probability density function with ordinate values. Select arbitrary z values (column 1) to establish the range of the distribution. Since 99.73% of all values in a normal distribution lie within three standard deviations of the distribution mean, z values from 0 to 3.0 seem appropriate. Compute the present value net cash flow (x values) corresponding to each z value for the range three standard deviations below (column 2) to three standard deviations above (column 3) the mean. Look up the ordinates (column 4) for corresponding z values in a statistical table.[30] Plot values for y in the same proportion to one another as the ordinates in column 4. Figure 20-9 shows a smooth function drawn through these y plots to construct the probability density function.

Note that the probability density function by itself gives considerable information about the economic worth of the shopping centers. For example, visual inspection indicates that there is little area under the left tail of the probability

[30]The ordinates here come from tables (Chemical Rubber Company, 1965) whose x and $f(x)$ values correspond, respectively, to z and ordinate values in this book.

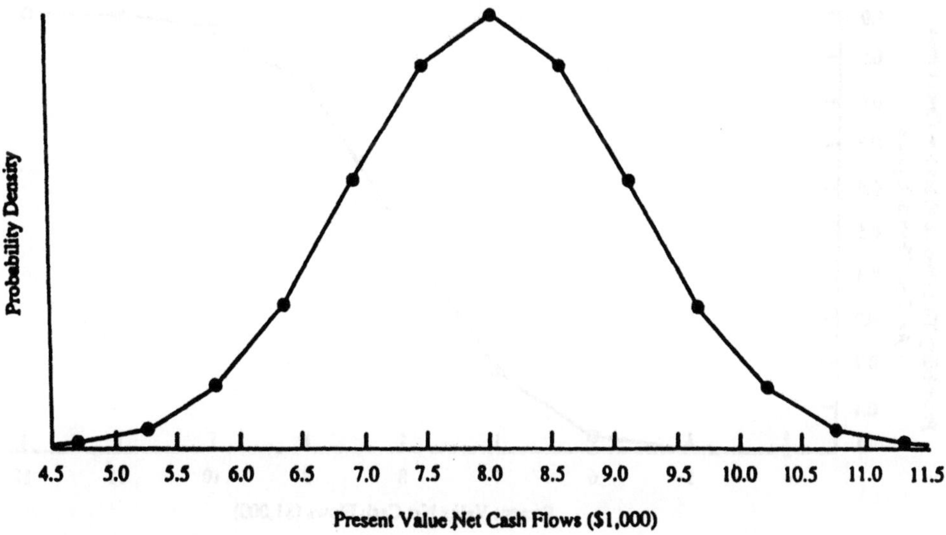

Figure 20-9. Probability density function of present value net cash flows for shopping centers.

density function for present value net cash flows less than $5,000, the amount that must be recovered to at least break even with the initial investment. Thus we know immediately that there is a low probability that the investment will not at least break even. Furthermore, we can see directly that the most likely value of the net cash flows is about $8,000.

The cumulative distribution function in Figure 20-10 is even more helpful, because it indicates the probability of being less than or just equal to alternative levels of present value net cash flows over the entire range of possible outcomes. Conversely, it can be used to determine the probability of exceeding alternative levels of present value net cash flows.

Table 20-20 shows how to obtain the numbers for plotting the cumulative distribution function. First, establish an arbitrary range of present value net cash flows (column 1) over which to plot the cumulative distribution function. Since 99.73% of all values lie within three standard deviations of the mean, a range of net cash flows between $\mu - 3\sigma$ and $\mu + 3\sigma$ seems reasonable. Looking back to Table 20-19, columns 2 and 3 show a corresponding range of net cash flow values from $4,703 to $11,327. For convenience, round to $4,000 and $12,000 with $1,000 intervals. Column 1 in Table 20-20 shows the present value net cash flow amounts for which the cumulative distribution function is constructed.

Compute a value for z (Table 20-20, column 2) for every net cash flow value (column 1), using equation 20-15. This is necessary because statistical tables give curve areas only for standardized z values. The associated normal curve

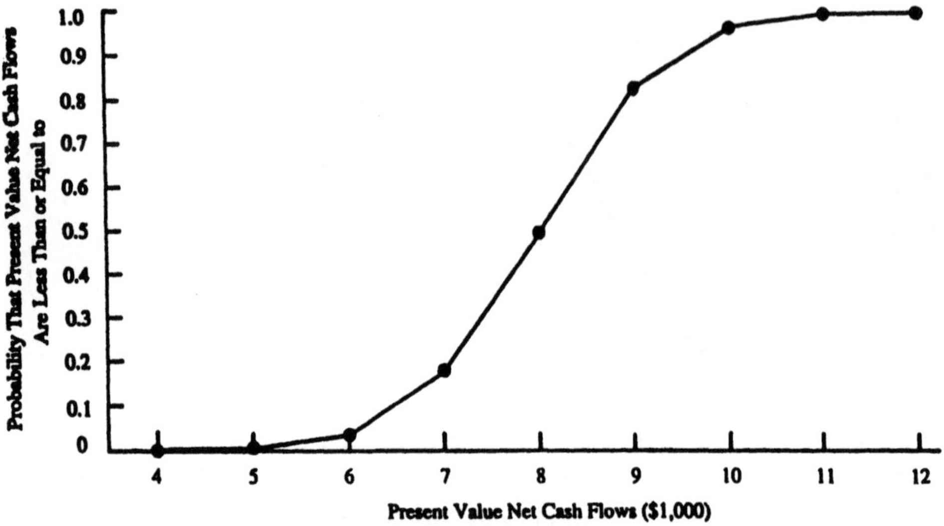

Figure 20-10. Cumulative distribution function of present value net cash flows for shopping centers.

areas for z values in column 2 are in column 3 of Table 20-20.[31] The area value describes what percent of the total area under the probability density function lies between the mean and some specified number of standard deviations (z value) to the right (positive area value) or left (negative area value). The probability that net cash flows will be less than or equal to any given value in column 1 is the area under the probability density function to the left of that value. To find the cumulative probability (column 4) at any given present value net cash flow, add the area of the corresponding z value to 0.5000.

You can use Table 20-20 to compute probabilities that the investment will achieve various degrees of profitability or loss, or you can read the probabilities directly off Figure 20-10. For example, use Table 20-20 to evaluate the likelihood that the $5,000 investment will at least break even. Find the area under the normal curve to the left of $5,000 as follows: The value of $z = -2.73$ [i.e., ($5,000 − $8,015)/$1,104]. The area under the normal curve between the mean and the z value of −2.73 is 0.4968. The difference between this area and the total area to the left of the mean is 0.0032 (i.e., 0.5000 − 0.4968). This suggests that there is less than a 1% chance that the shopping centers will not at least break even.

Now look at the cumulative distribution function to determine the likelihood of higher degrees of profitability. There is a 50% probability, for example, that

[31]The normal curve areas were found in Arkin and Colton (1950).

Table 20-20. Data for constructing the cumulative distribution function of present value net cash flows for shopping centers

Net Cash Flows ($1,000) (1)	*z* Value (2)	Area (3)	Cumulative Probability (4) = (3) + 0.5000
4.0	−3.64	−.4998	0.0002
5.0	−2.73	−.4968	0.0032
6.0	−1.83	−.4664	0.0336
7.0	−0.92	−.3212	0.1788
8.0	−0.01	−.0040	0.4960
9.0	+0.89	.3233	0.8233
10.0	+1.80	.4641	0.9641
11.0	+2.70	.4965	0.9965
12.0	+3.61	.4998	0.9998

returns will be less than $8,000 ($3,000 profit) and about a 97% probability that returns will be less than $10,000 ($5,000) profit.

20.5.4 Recommendation

The tables and functions present a profile of risk exposure that makes the investment in the company buying shopping centers appear attractive. You will probably want to recommend to your client that he accept this investment unless other higher-return alternatives with similar or less risk are competing for his limited budget or he is extremely risk averse.

REFERENCES

American Society for Testing and Materials (ASTM). 1989. *Standard Practice for Measuring Benefit-to-Cost and Savings-to-Investment Ratios for Buildings and Building Systems*. E964-89.

Arkin, H., and R. R. Colton. 1950. *Tables for Statisticians*. New York: Barnes and Noble.

Chapman, Robert E. 1988. The efficiency implications of safety standards. Ph.D. Dissertation for the George Washington University, Washington, DC.

Chemical Rubber Company. 1965. *Standard Mathematical Tables*. 14th edition. ed. Samuel M. Selby. Cleveland, Ohio.

Lippiatt, Barbara C., and Rosalie T. Ruegg. 1987. *Energy Prices and Discount Factors for Life-Cycle Cost Analysis*. NBSIR 85-3273-2. Gaithersburg, Maryland: National Bureau of Standards.

Lippiatt, Barbara C., and Rosalie T. Ruegg. 1988. *Energy Prices and Discount Factors for Life-Cycle Cost Analysis*. NBSIR 85-3273-3. Gaithersburg, Maryland: National Bureau of Standards.

Marshall, Harold E. 1988. *Techniques for Treating Uncertainty and Risk in the Economic Evaluation of Building Investments*. SP 757. Gaithersburg, Maryland: National Institute of Standards and Technology.

Ruegg, Rosalie T. 1987. *Life-Cycle Costing Manual for the Federal Energy Management Program*. NBS Handbook 132. Gaithersburg, Maryland: National Bureau of Standards.

Ruegg, Rosalie T., and Harold E. Marshall. 1984. *Economic Evaluation of Building Design, Construction, Operation and Maintenance: Instructor's Manual*, NBS Technical Note 1194. Gaithersburg, Maryland: National Bureau of Standards.

U.S. Code of Federal Regulations. 10 Part 436, Section 14.

U.S. Congress. *Federal Energy Management Improvement Act of 1986* (P.L. 100-615).

U.S. Congress. *Energy Security Act of 1980* (P.L. 96-294, 94 STAT 611, Sec. 405).

U.S. Office of Management and Budget. 1972. Circular A-94 (Revised). Discount rates to be used in evaluating time-distributed costs and benefits. Washington, DC, March 27.

21

Decisions on Building Location

Where to locate is a decision that is made for every building, and it can have an important economic impact. In the short run, location affects initial capital investment costs; in the long run, location affects ongoing operations and profitability. Choosing a location is a complex decision that requires consideration of economic and noneconomic factors.

No single analytical method will ensure the selection of an optimal location in light of complex circumstances and uncertainties, but the methods presented in Part 1 can help.[1] You can use economic evaluation methods to compare alternative locations with respect to costs, revenues, and profits. In Section 21.1, we show how to use an annual cost model to find the least-cost location for a factory; in Section 21.2, we show how to use the NB method to find the highest-profit location for a service facility. In the first example, we use sensitivity analysis to reflect uncertainty; in the second, we use probabilistic input data and sensitivity analysis.

21.1 WHERE TO LOCATE A GOODS-PRODUCING PLANT WITHOUT STRONG MARKET TIES: EVALUATED WITH AN ANNUAL COST MODEL AND SENSITIVITY ANALYSIS

21.1.1 Problem

Many goods-producing companies tend to focus on production and distribution costs in selecting a plant location. Sales revenue is often relatively insensitive to the choice. An example is a manufacturer of telephones which are shipped and marketed nationally. Sales volume, though uncertain, is not expected to be affected by the location of the manufacturing plant. Production and distribution costs, however, are expected to differ among alternative locations and according

[1] Locational choices are treated in urban economics, regional economics, spatial economics, and decision analysis. Among the techniques for analyzing economic aspects of location is mathematical programming. Distribution linear programming techniques, for example, are used to allocate production and distribution among locations for cost minimization.

to production volume. Production volume is a function of sales volume and inventory policy. In this case study, it is assumed that inventory is maintained at retail outlets, and all telephones produced are shipped. Because sales volume is uncertain, production volume is also uncertain.

Suppose that you have been hired as a consultant to advise the manufacturer on the location for a new plant. The objective is to find a location which maximizes company profits. You learn that sales revenue is not sensitive to location, but costs are. Thus you focus on identifying the candidate location which minimizes costs and which, in this case, also maximizes company profits. The company requests that you use no more than a 15-year study period.

21.1.2 Approach

The first step is to take into account constraints which might rule out certain locations on either economic or noneconomic grounds. Then identify promising locations for evaluation. Next, calculate and compare the costs of the alternatives. The final step is to select the location which appears most likely to minimize costs, taking into account uncertainty.

In this example, an annual LCC cost model is used. Sensitivity analysis is performed to show how annual costs in each location change as production volume changes.

21.1.3 Evaluation

21.1.3.1 Data and Assumptions

Locations A, B, and C in Table 21-1 have been identified as acceptable in terms of noneconomic considerations and promising in terms of economic considerations. The table summarizes for each location the yearly fixed cost and unit variable cost of producing and distributing the telephones. Fixed costs, such as

Table 21.1. Fixed and variable production and distribution costs associated with alternative plant locations

Location	Annualized Fixed Costs	Variable Cost/unit
A	$200,000	$50
B	400,000	25
C	800,000	10

the costs of acquiring the land and plant, are those which will be incurred in the short run regardless of the volume of output. Variable costs are those which vary directly with the volume of output, such as unit shipping costs. Note that fixed costs are annualized over the 15 year study period because an annual cost model is used to evaluate alternatives.

Location C has the highest fixed costs and the lowest variable costs. Location B has twice the fixed costs of location A but half the variable costs.

Based on the price of competitive products and other demand and supply data, telephones produced by the company are expected to sell for about $100 each in major markets, regardless of the location at which they are produced. Preliminary projections of annual sales volume range from 10,000 to 20,000 units. Since low inventory is a goal, the company wants to keep production volume about equal to projected sales volume.

21.1.3.2 Calculate Annual Costs for Each Location

Annual costs are calculated with equation 21-1.

$$AC = FC + VC(VOL) \tag{21-1}$$

where

AC = annual costs,
FC = fixed costs (annualized at zero production volume),
VC = annual variable costs (costs per unit as production volume is expanded past zero), and
VOL = annual production volume.

You can see from equation 21-1 that annual production volume is a critical variable in calculating annual costs. Because you are uncertain about what value to assign production volume, you will want to see how sensitive each location's annual costs are to it. And, since equation 21-1 is a straight-line function, you can easily prepare a sensitivity graph by calculating annual costs at each location based on two selected values of production volume and drawing the functions.

You might begin by calculating the annual costs at each location based on a zero production volume. At zero production, there are fixed costs but no variable costs. This is a convenient point to find because it simply equals fixed costs.

Since 20,000 units is the high end of the range of expected production volume, try substituting 20,000 into equation 21-1 as the second value of produc-

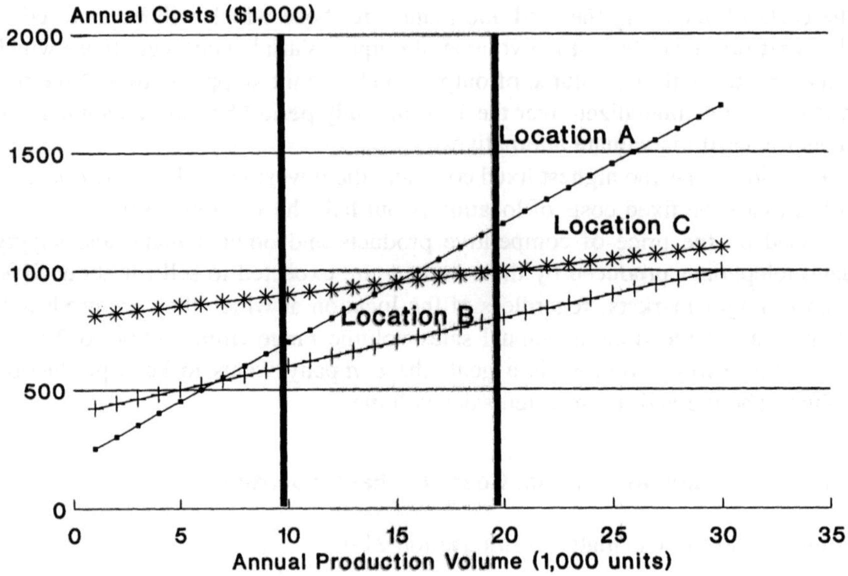

Figure 21-1. Annual costs at alternative plant locations as a function of production volume

tion volume. Compute annual costs for each location (AC_A, AC_B, and AC_C) as follows:

$$AC_A = \$200{,}000 + \$50(20{,}000) = \$1{,}200{,}000,$$

$$AC_B = \$400{,}000 + \$20(20{,}000) = \$800{,}000$$

$$AC_C = \$800{,}000 + \$10(20{,}000) = \$1{,}000{,}000.$$

Next, prepare the sensitivity graph which expresses annual costs as a function of production volume. The sensitivity graph will facilitate identification of the range of production volume over which a given location is preferred. Prepare the graph by plotting the two annual costs figures as a function of production volume for each location, as shown in Figure 21-1. Annual costs are on the vertical axis and production volume is on the horizontal axis.

Test the sensitivity of the three locations to production volume by comparing the cost functions. Within the entire range of expected production of 10,000 to 20,000 units, location B is preferred. For production volumes below about 7,000, location A is preferred. Sensitivity analysis shows that location B is the clear choice in the relevant range of production.

21.2 WHERE TO LOCATE A SERVICE BUSINESS WITH STRONG MARKET TIES: EVALUATED WITH THE NB METHOD AND PROBABILISTIC INPUT DATA

21.2.1 Problem

Service industries tend to focus on aspects of location which emphasize revenue potential. If either price per unit or sales volume is expected to vary by location, revenue as well as costs must be included in the comparison of locations. The most profitable location maximizes the excess of present or annual value revenue over costs. This may not be the location that minimizes costs.

Suppose that you are to select a location for an outlet of a shoe repair company, a highly market-oriented service. The problem is to determine a location that will maximize company profits in the face of uncertainty about sales volume. The company has a 10-year time horizon.

21.2.2. Approach

As in the previous case, eliminate from further consideration locations which are clearly not acceptable due to either economic or noneconomic constraints. Use market research to develop pricing policies and to estimate sales volumes based on proposed prices at each potential location. Estimate the costs of locating and operating an outlet at each location.

You can use NB to compare alternative locations. Because sales volume is the uncertain, critical factor to which profits are sensitive, express sales volumes at the various locations in probabilistic terms. By basing NB on expected value of sales volume, you will account for uncertainty in sales (see Section 17.3.1). Since most of the data are in annual terms, it is convenient to express dollar amounts as annual values.

21.2.3 Evaluation

21.2.3.1 Data and Assumptions

Locations X, Y, and Z are acceptable in terms of noneconomic considerations and are promising economically. Table 21-2 shows the prices which market research suggests can be charged per unit of service[2] at each location. Table 21-3 shows sales volume ranging from 25,000 to 300,000 units of service and their estimated probabilities of occurring at each location. Expressing sales vol-

[2]Principal repair services are broken down into units of service for the purpose of developing prices.

Table 21-2. Prices of shoe repair at alternative store locations

Location	Price/Unit of Service
X	$3.00
Y	1.50
Z	2.50

ume data in probabilistic terms helps you account for the uncertainty inherent in the estimates.

Table 21-4 gives the estimated annual costs of renting and operating the stores in each location. Fixed costs, such as rent and utilities, are listed first, and variable costs, such as materials and labor, are listed second.

21.2.3.2 Calculate AV NB (Profits) for Each Location

Equation 21-2 is used to calculate NB (profits):

$$\begin{aligned} \text{AV } NB &= TR - TC \\ &= [(p)(ESV)] - [FC + (VC)(ESV)] \end{aligned} \qquad (21\text{-}2)$$

where

AV NB = net benefits or profits, expressed in annual value dollars,
TR = total annual revenue, computed as unit price (p) times expected annual sales volume (ESV), and
TC = total annual costs, computed as annual fixed costs (FC) plus unit variable costs (VC) times expected annual sales volume (ESV)

Table 21.3. Probability of achieving different sales volumes at alternative store locations

	Probability of Achieving Sales Volume		
Annual Sales Volume	Location X	Location Y	Location Z
25,000	.20	.00	.00
50,000	.50	.00	.10
75,000	.20	.00	.25
100,000	.10	.10	.25
125,000	.00	.10	.25
150,000	.00	.20	.10
200,000	.00	.30	.05
250,000	.00	.20	.00
300,000	.00	.10	.00

Table 21.4. Yearly costs of renting and operating the stores in alternative locations

Costs	Locations		
	X	Y	Z
Fixed Costs			
Rent	$10,000	$15,000	$20,000
Utilities	30,000	20,000	25,000
Water	1,000	1,000	2,000
Equipment[a]	14,900	14,900	14,900
Total	55,900	50,900	61,900
Variable Costs			
Materials	$0.50/unit	$0.50/unit	$0.20/unit
Labor	1.00/unit	0.75/unit	0.75/unit
Total	1.50/unit	1.25/unit	0.95/unit

[a]$100,000 amortized at 8% over 10 years

First, use equation 21-3 to calculate the expected value of sales volume (*EVS*) for each location.

$$EVS = \sum_{j=1}^{q} V_j p_j \qquad (21\text{-}3)$$

where

EVS = expected value of annual sales volume,
V_j = the jth of q values for sales volume specified in Table 21-3, and
p_j = probability of the jth sales volume occurring.

Apply equation 21-3, using probability data for each location from Table 21-3, as follows:

$$EVS_X = (25{,}000 \times 0.20) + (50{,}000 \times 0.50) + (75{,}000 \times 0.20) + (100{,}000 \times 0.10) = 55{,}000$$

$$EVS_Y = (100{,}000 \times 0.10) + (125{,}000 \times 0.10) + (150{,}000 \times 0.20) + (200{,}000 \times 0.30) + (250{,}000 \times 0.20) + (300{,}000 \times 0.10) = 192{,}500$$

$$
\begin{aligned}
EVS_Z &= (50{,}000 \times 0.10) + (75{,}000 \times 0.25) + (100{,}000 \times 0.25) \\
&\quad + (125{,}000 \times 0.25) + (150{,}000 \times 0.10) + (200{,}000 \times 0.05) \\
&= 105{,}000
\end{aligned}
$$

Now use equation 21-2 to compute AV NB (or profits) for each location based on its expected value sales volume and its fixed and variable costs as follows:

$$
\begin{aligned}
\text{AV } NB_X &= (\$3.00 \times 55{,}000) - (\$55{,}900 + (\$1.50 \times 55{,}000)) \\
&= \$165{,}000 - \$138{,}400 \\
&= \$26{,}600. \\
\text{AV } NB_Y &= (\$1.50 \times 192{,}500) - (\$50{,}900 + (\$1.25 \times 192{,}500)) \\
&= \$288{,}750 - \$291{,}525 \\
&= -\$2{,}775. \\
\text{AV } NB_Z &= (\$2.50 \times 105{,}000) - (\$61{,}900 + (\$0.95 \times 105{,}000)) \\
&= \$262{,}500 - \$161{,}650 \\
&= \$100{,}850.
\end{aligned}
$$

Despite location Y's large expected value sales volume, it does not appear profitable because it services a low-price market. Location X, on the other hand, which services a low-volume, high-price market, does appear profitable. Location Z, however, which services a moderate-volume, moderate-price market and which has relatively low unit costs, appears to be the most profitable. Based on expected value sales volume, you would choose location Z for the shoe repair outlet.

Before recommending location Z, you may wish to think further about the uncertainty in the data and how it might affect the actual outcome. Estimated sales volume at each of the locations is the most uncertain input in the analysis. Profit estimates have been based on expected value sales volume. But the expected value of sales will rarely happen even with a discrete distribution such as the one you used here.[3] Given the concern of businesses about experiencing negative cash flow, you might wish to test the sensitivity of the net profit outcome to the lowest sales volume in the probability distribution.

For location X, there is a 20% probability of selling only 25,000 units. If

---

[3] See Section 17.3.1.

sales drop to 25,000, annual net losses will be $18,400 [i.e., ($3.00 × 25,000) − ($55,900 + ($1.50 × 25,000)) = −$18,400]. For location Y, there is a 10% probability of selling only 100,000 units. This level of sales means annual net losses of $25,900. For location Z, there is a 10% probability of selling only 50,000. But even with sales at the estimated lowest level, location Z is profitable with annual net benefits at $15,600 [i.e., ($2.50 × 50,000) − ($61,900 + ($0.95 × 50,000)) = $15,600]. This further supports recommending the selection of location Z.

22

Lease or Buy Decisions

When an organization requires a building, furnishings, or equipment, one question is whether to lease it or buy it. One approach may offer advantages over the other. The economic objective is to choose the alternative which will maximize net present value to the organization.

A complexity in analyzing which approach to choose is that the choice between buying and leasing may affect the overall financial position of the organization in ways that do not show up in a comparison of direct cash flows of alternatives. For example, a firm may use leasing to increase its financial leverage.[1] Leasing can increase financial leverage in two ways: (1) it enables a firm to obtain the use of more assets than it could under a secured loan agreement[2] and (2) it tends to have less impact on the future borrowing capacity of a firm than borrowing to purchase.

Another factor which may influence an organization to lease instead of buy is that it may be more successful administratively in securing regular replacements of assets within the terms of a lease than through purchase. Regular replacements may be important to avoid the costs of obsolescence.

Yet another factor which may influence choice is relative flexibility to cut losses. A lease may give a firm the ability to cancel quickly and avoid losses. Purchased assets may take longer to liquidate; in the meantime, obligations for loan repayment mount. Convenience is another factor which may be important to the decision but which may not be reflected in cash flows.

Some of the factors which influence the choice of leasing or buying affect cash flows directly. One factor is that relative tax positions may make it advantageous for another firm to own assets and for the user to obtain them through leasing. Firm A, for instance, may be better able to utilize investment tax credits, depreciation write-offs, and deduction of finance charges than firm B. This

[1]Financial leverage, usually measured as the ratio of total debt to total assets or to equity, enables an investor to gain the use of more assets for a given amount of equity. While this increases investor risk, it also increases earnings potential.

[2]A lessor tends to have less risk than a creditor who finances the purchase of an asset because legal ownership remains with the lessor. This often makes a lessor willing to provide leasing arrangements to a firm which cannot obtain a loan for purchasing assets due to a low credit rating.

may enable A to buy assets and lease them to B at favorable terms, thus lowering the amount of the lease payment. The lessee's ability to deduct lease payments fully from taxable income in the year they occur is another factor which is directly recorded in cash flow analysis. Another factor which affects cash flows directly is the presence of efficiencies in performing specialized services which may enable a lessor to provide services as part of a lease contract at a lower cost than the lessee would pay to obtain the services from other sources. Different expectations held by the lessor and lessee about residual value can also affect the choice of leasing or buying, as reflected directly in cash flows.

In the case study, we take into account only factors which show up directly in the cash flows of the lease and buy alternatives. We show how to use the LCC method, with a detailed treatment of taxes, to estimate whether it is more cost effective to lease or buy a building for an office supply company. We illustrate the RADR for taking into account uncertainty. We include a detailed cash-flow analysis to reflect concerns about solvency.

In the case study, a decision has been made to obtain use of an asset, and the decision centers on whether to obtain it by purchase or by lease. In this context, leasing versus buying is a financing decision, and the analysis compares the costs of leasing against the costs of borrowing to buy.[3]

22.1 PROBLEM

Suppose that you are a consultant to an office supply company that seeks space to accommodate an expanding business. The approximate location, size, and other features of the building have already been determined by the company. Your job is to advise whether the company should lease a building or buy one. The selection criterion is to minimize the long-run costs of meeting space requirements.

22.2 APPROACH

The approach is to (1) estimate the LCC of leasing, (2) estimate the LCC of buying, and (3) choose the one with the lower cost over the study period. Since the building is for a business, it is important that tax effects be taken into account. A detailed cash-flow analysis is advisable to help the company determine

[3]This is different from the case where the terms of the lease are important in deciding whether or not to obtain the asset at all. Then leasing becomes a capital budgeting as well as a financing decision. The first step in that case is to determine if the project is acceptable if purchased, generally excluding financing costs from the analysis. If net losses result, purchase is unacceptable. The next step is to determine if the project is acceptable if leased. See Clark et al. (1984) for treatment of leasing/buying as a joint capital budgeting and financing decision.

whether it can meet the short-run cash-flow requirements associated with the decision.

A pivotal factor in the decision to lease or buy is the residual or resale value[4] of the building at the end of the study period. Residual or resale value is important because a building tends to hold much of its value, or even to appreciate, over time. The lessor owns the property at the expiration of the lease, and the lessor's expectations about residual value are reflected in the lease charge. The buy alternative, on the other hand, reflects the office supply company's expectations about future residual or resale value. This is a value which will occur 10 years from now and is uncertain. A technique is needed to take into account the uncertainty. We treat uncertainty of resale by applying an RADR to resale value.

22.3 EVALUATION

22.3.1 Data and Assumptions

Table 22-1 gives data and assumptions for the leasing alternative. Table 22-2 gives data and assumptions for the purchase alternative.

A suitable building is available for lease. The lease payment is due at the beginning of each year, starting at $26,000.[5] The lease contract contains an escalation clause tied to the annual rate of change in the CPI. An up-front security deposit equal to half a year's lease payment is required. With only normal wear and tear on the building, the deposit will be refunded in full at the end of the lease, with an accrual of interest computed at a rate of 6% compounded annually.

A suitable building is for sale at a price of $180,000. The company does not have sufficient capital to purchase it outright. To purchase the building, it can

Table 22-1. Summary of data and assumptions for leasing

Yearly lease payment (in year 1)	$26,000
Payment due date	Beginning of year
Escalation clause[a]	5% compounded annually
Security deposit	$13,000
Refund of security deposit after 10 years (in current dollars, including 6% interest compounded annually)	$23,281

[a]Assumed to be 5% based on the CPI.

[4]For a discussion of the effect of residual values on lease/buy decisions, see Weston and Brigham (1981, p. 866).

[5]Lease and mortgage payments are typically charged monthly, but to reduce the number of entries in the cash-flow tables, it is assumed that they are made yearly.

Table 22-2. Summary of data and assumptions for purchase

Financing	
Purchase price	$180,000
Down payment	$27,000
Loan amount	$153,000
Loan interest rate	11.0%
Loan period	10 years
Closing costs	$5,000
Depreciation	
Depreciation period	31.5 years
Depreciation method	Straight-line
Adjusted cost basis	$140,000
Property Taxes	
Tax rate	2% of appraised value
Annual rate of increase in appraised value	3.0%
Resale value (in current dollars)	$360,000
Marginal income tax rates	
Federal	28.0%
State	5.0%
Combined	31.6%
Discount rate	
Average marginal cost of capital, after taxes	12.5%
Special risk-adjusted rate, after taxes	16.0%
Study period	10 years

make a down payment of 15% of the purchase price, or $27,000, and finance the remaining $153,000 with a bank loan at a rate of 11% over 10 years, repaid in 10 even installments.[6] The first payment is due at the beginning of the 2nd year and the last one at the beginning of the 11th year. Purchase of the building entails closing costs of $5,000 for lawyer's fees, transfer stamps, and title insurance.

Furnishings, equipment, supplies, and costs of operating, maintenance, and repair will be the responsibility of the office supply company under either ar-

[6]Note that in this example financing costs are included in evaluating the purchase alternative, even though the terms of financing are not assumed to be unique for the purchase. This would seem to violate the axiom of capital budget analysis explained in Section 13.3. That is, project evaluation is to exclude costs of financing unless the terms of financing are a unique, integral part of the project. The reason for including financing costs is the special nature of the lease/buy problem: The capital investment decision to acquire the building has already been made, independently of financing costs. The problem at hand is to determine the difference in the costs of leasing and borrowing to buy. One of these alternatives will be chosen because the decision has been made to acquire space and the company lacks funding to purchase outright. This requires an assessment of all cash flows which affect the decision. Interest costs and associated tax savings are two of the relevant cash flows.

rangement. Since these costs are expected to be about the same for the two buildings, you can ignore them for the analysis. Also, revenues are not expected to be affected by the lease/buy decision. So they too can be excluded. Either building will provide adequate space for the amount of expansion which is anticipated over the next 10 years. The lease payment is based on a 10-year lease contract; the loan period is also 10 years.

The building's adjusted cost basis may be depreciated over 31.5 years, using the straight-line method. The adjusted cost basis is calculated by deducting land costs from the purchase price and adding closing costs. Land costs are estimated at 25% of the unadjusted purchase price. Therefore, the adjusted cost basis is $140,000 [i.e., $180,000(1 − 0.25) + $5,000 = $140,000].

Property taxes on the two buildings are about the same. The property tax rate is 2% per $1,000 of appraised value. The appraised value for tax purposes over the coming 10 years is projected to increase at about the same rate as it did over the preceding 10 years: 3% compounded annually. If the company purchases the building, it will pay the property tax directly. If it leases, the company as lessee compensates the owner of the building for property taxes in the lease payment. Any change in the amount of property tax is taken into account in the lease by the CPI escalation clause.

Commercial real estate in the area is expected to double in price over the next 10 years. The projected resale value in current dollars is $360,000.

The company, a limited partnership, has a marginal federal income tax rate of 28%. The state marginal income tax rate is 5%. The combined marginal rate is 31.6% [i.e., 0.28(1 − 0.05) + 0.05 = 31.6%]. If the building is leased, the entire lease payment is deductible from taxable income in the year paid. If the building is purchased, interest payments, property taxes, and depreciation are deductible from taxable income.

On average, the company's after-tax cost of capital is 12.5% at the margin. This rate reflects the average riskiness of the company's operations. You could use it to discount the cash flows attributable to the project, except for the resale value of the building associated with the purchase alternative.[7] To reflect the greater uncertainty of the resale value, an RADR is applied to it. The rate suggested here is an after-tax rate of 16%, which is the rate the company uses to discount revenues from risky new product lines. Both rates are based on market interest rates (i.e., they include a projection of general price inflation).

---

[7]Rather than use the weighted marginal cost of capital as a discount rate, an analyst might use a lower rate, such as the after-tax cost of debt. The rationale is that the weighted marginal cost of capital reflects the average risk of the firm, while most cash flows associated with a lease/buy decision have a lower risk than the average company activity. We use instead the weighted marginal cost of capital as the discount rate for all cash flows, except for the uncertain resale value for which we use a higher rate reflecting the greater risk. Selection of a discount rate for lease/buy decisions is controversial. See Clark et al. (1984), pp. 385–386.

A higher discount rate for discounting resale value accounts for the greater uncertainty concerning its future amount compared with other amounts. There is no contractual arrangement fixing resale value. Unforeseen local developments in the real estate market could significantly affect the value. Lease payments and loan payments, on the other hand, are highly reliable estimates, since they are fixed by contract over the entire study period. Likewise, tax effects are highly predictable and subject to little variability. Even if there are changes in the tax codes, assets usually continue to be subject to the provisions under which they were acquired.[8]

The company has a 10-year planning horizon. If at the end of 10 years the company is still in business, it can renew the lease (or continue ownership of the building) or it may terminate the lease (or sell the building). For the evaluation, assume that the company uses a 10-year study period and terminates the lease or ownership at the end of the period.

22.3.2 Calculate the LCC of Leasing the Building

Table 22-3 shows in column 7 the detailed after-tax cash flow for the leasing alternative. The bottom figure in column 9 is the estimated LCC, after taxes, of the leasing alternative: about $137,000 in present value dollars over the 10 years. Computations are explained in column headings and footnotes to the table.

22.3.3 Calculate the LCC of Purchasing the Building

Analysis of the purchase alternative is shown in Tables 22-4 through 22-6. Table 22-4 analyzes the loan repayment schedule. It is necessary to divide the loan payment into principal and interest components in order to calculate the effect of deducting mortgage interest from taxable income. Column 4 of the table shows the amount of principal repaid each year, and column 7 shows the interest expense after tax deductions. These are two of the cash flows that go into the computations of LCC in Table 22-6.

Table 22-5 shows how to estimate property taxes taking into account their deduction from taxable income. The cash flow in column 5 goes into the computation of LCC in Table 22-6.

Table 22-6 combines cash outflows—down payment (column 2), principal

[8]A factor which introduces uncertainty into these other cash flows is the future rate of general price inflation. For example, the lease payment will vary in response to changes in the CPI, as required by the escalation clause in the contract. And, although the loan repayment is fixed in current dollars, its constant dollar amount will vary, depending on the inflation rate. Similarly, depreciation is fixed in current dollars but varies in constant dollars as the inflation rate varies.

Table 22-3. LCC of leasing the building (beginning-of-year cash flows)

(1) Year	(2) Deposit/ Refund	(3) Initial Lease Payment	(4) Escalation at 5%	(5) Yearly Lease Payment $(5) = (2_0) + ((3_0) \times (4))$	(6) Tax Savings Due to Deductions $(6) = (5) \times 0.316$	(7) After-Tax Cash Flow $(7) = (2) + (5) - (6)$	(8) SPV Factor for 12.5%	(9) PV Cost of Lease After Taxes $(9) = (7) \times (8)$
1	$13,000	$26,000	$(1 + 0.05)^0$	$39,000	$12,324	$26,676	1.0000	$26,676
2	. . .	. . .	$(1 + 0.05)^1$	27,300	8,627	18,673	0.8889	16,598
3	. . .	. . .	$(1 + 0.05)^2$	28,665	9,058	19,607	0.7901	15,492
4	. . .	. . .	$(1 + 0.05)^3$	30,098	9,511	20,587	0.7023	14,459
5	. . .	. . .	$(1 + 0.05)^4$	31,603	9,987	21,616	0.6243	13,495
6	. . .	. . .	$(1 + 0.05)^5$	33,183	10,486	22,697	0.5549	12,595
7	. . .	. . .	$(1 + 0.05)^6$	34,842	11,010	23,832	0.4933	11,756
8	. . .	. . .	$(1 + 0.05)^7$	36,585	11,561	25,024	0.4385	10,972
9	. . .	. . .	$(1 + 0.05)^8$	38,414	12,139	26,275	0.3897	10,241
10	. . .	. . .	$(1 + 0.05)^9$	40,335	12,746	27,589	0.3464	9,558
11[a]	−23,281	. . .	. . .	. . .	−7,357	−15,924	0.3079	−4,903
						LCC of leasing, after taxes		136,939

[a]The deposit is assumed to be refunded at the end of the 10th year, which is shown here as the beginning of the 11th year. Note that it receives a negative sign even though it is a cash inflow, because this is the way cash inflows are modeled in an LCC framework. Positive cash flows are subtracted from costs.

Table 22-4. Loan repayment schedule (beginning-of-year cash flows)

(1) Year	(2) Repayment[a]	(3) Interest Payment $(3)_t = 0.11 \times (5)_{t-1}$	(4) Principal Payment $(4) = (2) - (3)$	(5) Remaining Balance $(5)_t = (5)_{t-1} - (4)$	(6) Tax Savings due to Deductions of Interest $(6) = (3) \times 0.316$	(7) Interest After Tax Deductions $(7) = (3) - (6)$
1	. . .	. . .	. . .	$153,000	. . .	. . .
2	$25,979	$16,830	$9,149	143,851	$5,318	$11,512
3	25,979	15,824	10,155	133,696	5,000	10,824
4	25,979	14,707	11,272	122,424	4,647	10,060
5	25,979	13,467	12,512	109,912	4,256	9,211
6	25,979	12,090	13,889	96,023	3,820	8,270
7	25,979	10,563	15,416	80,607	3,338	7,225
8	25,979	8,867	17,112	63,495	2,802	6,065
9	25,979	6,984	18,995	44,508	2,207	4,777
10	25,979	4,895	21,084	23,424	1,547	3,348
11	25,979	2,577	23,402	22[b]	814	1,763

[a]The loan amount is amortized with the UCR at 11% over 10 years as follows: $153,000 × 0.1698 = $25,979.

[b]Rounding errors cause the remaining balance not to go exactly to zero.

Table 22-5. Property tax after income tax deductions (beginning-of-year cash flows[a])

(1) Year	(2) Appraised Value[b]	(3) Property Taxes (Rate = 2%) (3) = (2) × 0.02	(4) Tax Savings Due to Deductions of Property Taxes (4) = (3) × 0.316	(5) Property Tax After Income Tax Deductions (5) = (3) − (4)
1	$180,000	$3,600	$1,138	$2,462
2	185,400	3,708	1,172	2,536
3	190,962	3,819	1,207	2,612
4	196,691	3,934	1,243	2,691
5	202,592	4,052	1,280	2,772
6	208,669	4,173	1,319	2,854
7	214,929	4,299	1,358	2,941
8	221,377	4,428	1,399	3,029
9	228,019	4,560	1,441	3,119
10	234,859	4,697	1,484	3,213

[a]It is assumed that property taxes are collected in escrow at the beginning of the year in which they are due.
[b]The initial appraised value is assumed equal to market value. Subsequently, the appraised value is assumed to escalate at 3% compounded annually.

(column 3), interest after taxes (column 4), and property taxes after tax deductions (column 6)—with cash inflows—depreciation tax savings (column 5) and resale after taxes (column 7). Column 8 gives net cash flows, adjusted period by period for taxes, that are required to purchase the building. The last item in column 10 gives the LCC of the purchase alternative: about $93,000 in present value dollars over the 10-year period.

22.4 ANALYZE THE RESULTS AND DECIDE TO LEASE OR BUY

Comparing the LCCs of the two alternatives, as given in Tables 22-3 and 22-6, shows a cost advantage for purchasing over leasing of about $44,000 (i.e., $137,000 − $93,000). But by comparing the cash flows in column 7 of Table 22-3 and column 8 of Table 22-6, you can see that leasing generates smaller net cash outlays than buying over the first 7 years and is therefore preferable over that time period on a cash-flow basis. The LCC advantage of purchasing rests heavily on the resale value, which is the most uncertain of all the cash flows. The uncertainty, however, was taken into account by applying a RADR to estimated resale value.

An additional factor you may wish to consider is the sensitivity of the outcome to inflation, since it is another item in the analysis about which there is uncertainty. This can be done by intuitive reasoning.

Table 22-6. LCC of purchasing the building (beginning-of-year cash flows)

(1) Year	(2) Down Payment	(3) Principal Payment	(4) Interest Payment After Taxes	(5) Depreciation Tax Savings[a]	(6) Property Taxes After Tax Deductions	(7) Resale Value Less Capital Gains Tax	(8) Net Cash Flow from Purchase After Taxes (8) = (2) + (3) + (4) + (5) + (6) + (7)	(9) SPV	(10) PV of Purchase After Taxes (10) = (8) × (9)
1	$27,000	. . .	. . .	. . .	$2,462	. . .	$29,462	1.0000	$29,462
2	. . .	$9,149	$11,512	−$1,404	2,536	. . .	21,793	0.8889	19,372
3	. . .	10,155	10,824	−1,404	2,612	. . .	22,187	0.7901	17,530
4	. . .	11,272	10,060	−1,404	2,691	. . .	22,619	0.7023	15,885
5	. . .	12,512	9,211	−1,404	2,772	. . .	23,091	0.6243	14,416
6	. . .	13,889	8,270	−1,404	2,854	. . .	23,609	0.5549	13,101
7	. . .	15,416	7,225	−1,404	2,941	. . .	24,178	0.4933	11,927
8	. . .	17,112	6,065	−1,404	3,029	. . .	24,802	0.4385	10,876
9	. . .	18,995	4,777	−1,404	3,119	. . .	25,487	0.3897	9,932
10	. . .	21,084	3,348	−1,404	3,213	. . .	26,241	0.3464	9,090
11	. . .	23,402	1,763	−1,404	. . .	−$290,656[b]	−266,896	0.3079	−58,576[d]
								0.2267[c]	
							Total present value of purchase, after taxes		$93,015

[a]Savings from deductions of depreciation from taxable income are calculated as follows: ($140,000/31.5) × 0.316 = $1,404. This is given a negative sign because that is the way positive cash flows are treated in an LCC model.

[b]Resale value less capital gains tax is calculated as follows: $360,000 − [$360,000 − (($185,000 − $44,444) × 0.316)], where $185,000 is the original cost and $44,444 is 10 years' depreciation. (See Section 13.2.5.)

[c]The discount factor of 0.3079, based on a discount rate of 12.5%, is used to discount all items occurring at the beginning of year 11 except resale value; the factor 0.2267, based on a special risk-adjusted discount rate of 16%, is used to discount resale value.

[d](($23,402 + $1,763 − $1,404) × 0.3079) + (−$290,656 × 0.2267) = −$58,576.

A higher than expected rate of inflation reduces constant dollar values of principal and interest payments, thus improving the advantage of purchasing. Higher inflation reduces the tax benefits of deducting interest and depreciation because these are fixed dollar amounts, thus decreasing the advantage of purchasing. But this is a secondary effect that is much lower than the direct effect on principal and interest. A higher rate of inflation is reflected in future resale value, thus improving the advantage of purchasing. In contrast, a higher than anticipated inflation rate worsens the economics of leasing, because inflation is fully incorporated into the lease payment through the escalation clause.

It follows that a higher inflation rate will tend to improve the economics of owning over leasing. Conversely, a lower inflation rate will tend to benefit leasing over owning. Therefore, if you think that higher inflation is more likely than lower inflation, you have a stronger reason for recommending purchase.

You might also wish to review tax data and assumptions in the buy alternative. Reexamine, for instance, the percentage of purchase costs attributed to land. This can be important because land costs cannot be depreciated, whereas as a component of a lease payment, they are fully deductible.

Before making the final decision, it is a good idea to review other factors which have potential importance in deciding whether to lease or buy. One of these is whether the company's existing sources of credit are short and if leasing would conserve existing credit for other uses. Another factor is whether the difference in cash flows between leasing and buying is sufficiently large to be of any practical consequence to the company's solvency. Still another is whether the company will be able to utilize fully depreciation write-offs and other deductions from taxable income, as well as any tax credits if they are available. Based on the LCC analysis, you will probably recommend purchasing the building, unless one or more of these other factors is important to your client.

REFERENCES

Clark, John J., Thomas J. Hindelang, and Robert E. Pritchard, 1984. *Capital Budgeting*. Englewood Cliffs, New Jersey: Prentice-Hall, Inc.

Weston, J. Fred, and Eugene F. Brigham. 1981. *Managerial Finance*. Hinsdale, Illinois: The Dryden Press.

23

Economic Size of Building Components

In this chapter we demonstrate with two case studies how to determine the economic size of a component of a building or facility. The first case shows how to size attic insulation, i.e., how to choose the economically efficient level. NS is the evaluation measure, and sensitivity analysis is the technique for taking into account uncertainty.

The second case study takes us outside the building proper. It shows how to determine the economically efficient size, i.e., surface thickness, of a roadway and parking lot. LCC is the evaluation measure. To illustrate guidelines for sizing federal projects, we use a Veteran Administration hospital.

23.1 CHOOSING THE LEVEL OF ATTIC INSULATION: EVALUATED WITH NS AND SENSITIVITY ANALYSIS

23.1.1 Problem

Suppose that you are an energy consultant. The maintenance director of a church asks you to recommend how much insulation to put in the attic of one of the church buildings. The building committee will not act on your recommendation, however, until it next meets in 4 months. The delay between your analysis and the decision introduces uncertainty about the costs of materials and installation, since they have been rising rapidly. Your task is to determine what level of insulation, i.e., what resistance or R value, is likely to yield the largest long-run savings, net of costs, in light of the uncertainty.

23.1.2 Approach

Compute NS for several levels of insulation to see which is most cost effective. Given the uncertainty about purchase and installation costs, do a sensitivity analysis to see how variations in insulation costs might affect the efficient level of insulation.

23.1.3 Evaluation

23.1.3.1 Data and Assumptions

The uninsulated building is in a Washington, D.C., suburb. It is a parsonage heated by an electric resistance system. The analysis is taking place in 1987, and all costs are quoted in 1987 dollars. The current price of electricity is about \$0.07/kWh or \$20.03/10^6 Btu. The area of the attic to be insulated is 1,200 ft^2. You are to consider energy savings from heating only, since the building is not cooled.

The church plans to use the building for at least 25 more years. The financial officer of the church says that the best return the church can earn on surplus money in its "capital improvement fund" is 10% compounded annually on a AAA-rated bond. The financial officer asks you to assume an inflation rate of about 3% per year. Therefore the required rate of return or discount rate is about 7% in real terms. The appropriate UPV** factor is 10.97.[1]

A government study (Petersen, 1981) of a building with approximately the same heating system/ location/size characteristics as the church facility estimates reductions in Btu energy consumption for various increments of insulation. We used the study to estimate energy savings in the church building, as summarized in Table 23-1. The R value (e.g., R-11) represents the resistance

Table 23-1. Reductions in annual energy consumption for heating achieved with alternative levels of insulation in a 1,200 ft^2 house in Washington, D.C.

Batt Size (R Value)	Reductions in Energy Consumption (10^6 btu)
R-11	12.913
R-19	14.987
R-30	16.315
R-38	16.833

Source: Petersen (1981), p. 63.

[1]The 10.97 UPV** factor is based on a 25-year study period and a 7% real discount rate. The factor was taken from a report prepared for DoE on federal energy price projections for use in LCC studies (Lippiatt and Ruegg, 1987, p. 7). In the source report the factors are designated "UPW*" factors, but here we refer to them as "UPV**" factors to distinguish them from factors with a constant rate of change which we designate "UPV*." The factors are based on differential rates of energy escalation which are required when using constant dollar analysis. The initial energy price, \$20.03/10^6 Btu, is for 1987. It comes from a table of 1987 prices in the same report (p. 32). (See Section 12.5.3 for a discussion of how to use the factors, and Appendix G for sample UPV** factors for 1988.

Table 23-2. Insulation cost per square foot for Washington, D.C.

Batt Size (R Value)	Material ($/ft^2)	Labor ($/ft^2)	Overhead and Profit ($/ft^2)	Total ($/ft^2)
R-11	0.20	0.07	0.06	0.33
R-19	0.32	0.08	0.08	0.48
R-30	0.50	0.08	0.10	0.68
R-38	0.67	0.08	0.11	0.86

Source: Means (1987), p. 314.

value of an insulating batt. The greater the amount of insulation, the higher the R value achieved and the greater the reduction in energy consumption.

One approach to determing the cost of installing fiberglass batts in the attic is to use a cost estimating manual. We used the 1987 Means Company *Residential/Light Commercial Cost Data* guide to develop the data in Table 23-2. For each batt thickness, it shows material, labor, overhead and profit, and total estimated costs per square foot installed in the Washington, D.C., area. The total cost for 1.0 ft^2 of a batt of R-30, for example, is about \$0.68 (i.e., \$0.50 for materials + \$0.08 for labor + \$0.10 for overhead and profit = \$0.68).

23.1.3.2 Calculate NS and Perform Sensitivity Analysis

First, calculate present value dollar savings from reducing annual energy consumption for heating the house. For each R level, multiply corresponding reductions in energy consumption (shown in Table 23-1) by the current energy price(\$20.03/10^6 Btu) and the 10.97 UPV** factor. For example, calculate present value savings from installing R-30 insulation as follows:

$$(16.315 \times 10^6 \text{ Btu})\,(\$20.03/10^6 \text{ Btu})\,(10.97) = \$3{,}585$$

Table 23-3 shows present value savings for the four R levels. Note that present value savings increase as the R level increases.

Next, calculate present value costs of materials and installation for each of the four R levels considered. You can use the cost factors in Table 23-2 to compute a base-case or medium-cost projection, as shown in Table 23-4. The total installed costs (labor, materials, and overhead/markup) for each R level in the medium-cost column are the product of the insulation cost per square foot (Table 23-2) times the number of square feet of insulated space. The medium installed cost of R-38, for example, is as follows:

$$\$0.86/\text{ft}^2 \times 1{,}200 \text{ ft}^2 = \$1{,}032$$

Table 23-3. Present value savings from attic insulation in 1,200 ft² house in Washington D.C.

R Value	Present Value Savings ($)
R-11	2,837
R-19	3,293
R-30	3,585
R-38	3,699

To generate a range of costs for sensitivity analysis, you can estimate lower and upper range values of installed costs based on local advertisements. If you see 50% discounts on prices of installed insulation, for example, cut cost projections in half, as shown in the low-cost column in Table 23-4. If you think that material costs will not exceed those in Table 23-2, but that upswings in the local market could drive labor rates to possibly double those in the table, then base the high-cost column on doubled labor rates. For example, compute the installed cost of R-30 in the high-cost column of Table 23-4 as follows:

$$(\$0.50/\text{ft}^2\,(\text{material}) + \$0.16/\text{ft}^2\,(\text{labor}) + \$0.10/\text{ft}^2\,(\text{overhead and profit}))$$
$$\text{times } (1{,}200\ \text{ft}^2) = \$912.$$

Now perform sensitivity analysis by calculating present value NS for the low-, medium-, and high-cost scenarios. To do this, you must subtract present value costs from present value savings for each of the R levels for low, medium, and high insulation costs. Table 23-5 shows the sensitivity of present value *net* savings to the low-, medium-, and high-cost projections. The economically efficient level of insulation differs according to which projection you make about insulation costs. That is, under the low-cost scenario, R-38 is the economic choice. Under the medium- or high-cost scenario, R-30 is preferred.

Table 23-4. Present value costs of installing attic insulation in a 1,200 ft² house in Washington D.C.

	Present Value Costs ($)		
R Value	Low	Medium	High
R-11	198	396	480
R-19	288	576	672
R-30	408	816	912
R-38	516	1,032	1,128

Table 23-5. Sensitivity of present value NS to insulation cost for insulating a 1,200 ft^2 house in Washington, D.C.

	Present Value NS ($)		
R Value	Low Cost	Medium Cost	High Cost
R-11	2,639	2,441	2,357
R-19	3,005	2,717	2,621
R-30	3,177	2,769*	2,673*
R-38	3,183*	2,667	2,571

*designates the economic choice.

23.1.4 Recommendation

You probably want to recommend R-30 for insulating the attic. It is the level that maximizes NS for the medium and high projections of insulation cost. Furthermore, choosing R-38 provides little potential gain in comparison to potential losses. For example, suppose that you select R-38 instead of R-30, and the medium- or high-cost projection turns out to be correct. The loss is about $100 in present value terms. But if you select R-38 and the low-cost projection turns out to be correct, the additional savings compared to selecting R-30 is only a few dollars.

23.2 CHOOSING THE THICKNESS OF A ROADWAY/PARKING SURFACE FOR A VA HOSPITAL: EVALUATED WITH THE LCC METHOD ACCORDING TO FEDERAL GUIDELINES

This case study takes us outside the building proper. It demonstrates how to determine the economic thickness of a roadway and parking surface[2] for a Veteran Administration (VA) hospital according to prescribed federal guidelines. The case study also shows how to take into account a lag between the time the analysis is conducted and the investment occurs.[3]

---

[2]This type of problem has been considered by the Army Corps of Engineers, U.S. Department of Army (1986, pp. A-3–A-7).

[3]For a discussion of treating investment costs as a lump-sum amount at the beginning of the study period versus providing for a planning/construction period and delayed occupancy, see Ruegg (1987, pp. 38–40, 109, and 118–125).

23.2.1 Sizing Problem

Suppose that you work for an architectural and engineering (A&E) firm that has been awarded a federal contract for designing a VA hospital. The complete design includes a roadway and parking area. You have been assigned the job of developing specifications for the roadway and parking area. One of your decisions is whether to specify a 2-inch or 3-inch wearing surface for the asphalt. Either thickness of asphalt will meet basic performance requirements. The difference is that the 2-inch surface requires more replacement work.

All analyses must conform to applicable guidelines for performing economic evaluations of military construction projects. All analyses must be documented and submitted to the government design engineer in charge of the project.

23.2.2 Approach

The first step is to learn the applicable guidelines for evaluating projects in the military construction program. Guidelines are given in an army technical manual.[4]

For military design decisions, guidelines require that alternatives be compared in terms of their LCCs, first costs, and energy consumption. If alternatives differ conclusively in their LCCs, the one with the lowest LCC is selected. If the difference is statistically insignificant, selection is made on the basis of first cost or energy consumption if the alternatives differ significantly in either of these costs. Only alternatives which meet minimum performance criteria are considered.

Guidelines applying to most non-energy military construction projects require (1) the use of a real discount rate of 10%; (2) expressing all dollar amounts in constant dollars; (3) using actual calendar dates to describe the occurrence of cash flows; (4) charging initial construction costs at the projected midpoint of construction; (5) charging repair and replacement costs when they are actually expected to occur; (6) charging annually recurring costs at the midpoint of each year (end-of-year cash flows are used for most of the analyses in this book); and (7) using a study period that extends no farther than 25 years beyond the date of occupancy.

Analysts are encouraged to account for uncertainty in military construction projects if the results of the LCC analysis are not clear-cut and the decision might change with other plausible data and assumptions. Accounting for uncertainty is also encouraged for decisions likely to be controversial.

---

[4]U.S. Department of the Army (1986). Guidelines are not identical for federal energy and nonenergy conservation projects. We provide the roadway/parking surface case study to illustrate a federal nonenergy conservation analysis. It is performed according to Army guidelines in effect as of 1989. (We illustrate a federal energy analysis in Section 20.2.)

23.2.3 Evaluation

23.2.3.1 Data and Assumptions

Figures 23-1 and 23-2 are "Basic Input Data Summary Sheets" used to document data for analysis of military construction projects. They have been filled out for 2-inch and 3-inch asphalt paving alternatives.

Descriptive information in the upper-left-hand corner identifies the project and the design feature evaluated. The box in the upper-left-hand corner in the first line cites the criteria reference, that is, the source of the guidelines. In this example it is Headquarters, Department of the Army (HQDA). The next six lines record key dates which are necessary to describe cash flows according to the guidelines. The date of study (DOS) is whenever the study is performed. For the example, assume that you are doing the analysis in July 1989. The analysis base date (ABD) is the time to which all cash flows are discounted. ABD is normally the same as DOS. The analysis end date (AED) occurs 25 years after the beneficial occupancy date (BOD). To estimate this date, it is necessary to take into account how long it will be before construction begins and how long construction will last. (See Section 8.4.6 for discounting operations to incorporate delays.)

Suppose that you are now in the preliminary design phase, which usually occurs about 2 years prior to the beginning of construction. Assume that construction will last for about a year. That places the beginning of construction on July 1991, the end of construction and the BOD on July 1992, and the AED 25 years later, on July 2017. The midpoint of construction, the time construction costs are charged, is January 1992.

The next line designates the region in which the project is located. The region designation is for locating DoE projections to estimate energy costs. But there are no energy costs in this example.

The discount rate is listed next. It is a real rate of 10%. It is followed by a box for recording escalation rates which will be used in the analysis. The escalation rates must be differential rates, exclusive of the rate of general price inflation, to meet the requirement that the analysis be performed in constant dollars.[5] In the example, all costs are assumed to increase at about the same rate as prices in general. Therefore the differential escalation rates are zero.

The lower section of the Data Summary Sheet lists the cost elements, their amounts as of the ABD, projected dates of occurrence, and sources of data. A cash-flow diagram on the right-hand side completes the Data Summary sheet.

The example is quite simple, with four cost elements for the 2-inch alternative and three for the 3-inch alternative. Entering costs in dollars of the ABD means that all costs are in 1989 dollars.

Investment costs are \$40,500 for the thinner surface and \$47,000 for the

---

[5]See Section 12.5, for a discussion of differential escalation rates.

Project No. & Title 100 VA Hospital
Installation & Location Virginia
Design Feature Roadway / Parking Surface
Alt. No. 1 **Title** Asphalt w/2" Wearing Surface

LIFE CYCLE COST ANALYSIS
BASIC INPUT DATA SUMMARY

For use of this form, see TM 5-802-1; the proponent agency is USACE.

Criteria Reference		HQDA
Date of Study (DOS)		Jul 89
Analysis Base Date (ABD)		Jul 89
Analysis End Date (AED)		Jul 17
Midpoint of Construction		Jan 92
Beneficial Occupancy Date (BOD)	Actual Projected	Jul 92
	Assumed for Analysis	Jul 92
DOE Region		
Annual Discount Rate		10%

Type of Cost	Differential Escalation Rate per Year (%) Timeframe: 7/89 - 7/17		
All	0.00		

Principal Assumptions

Cash Flow Diagram

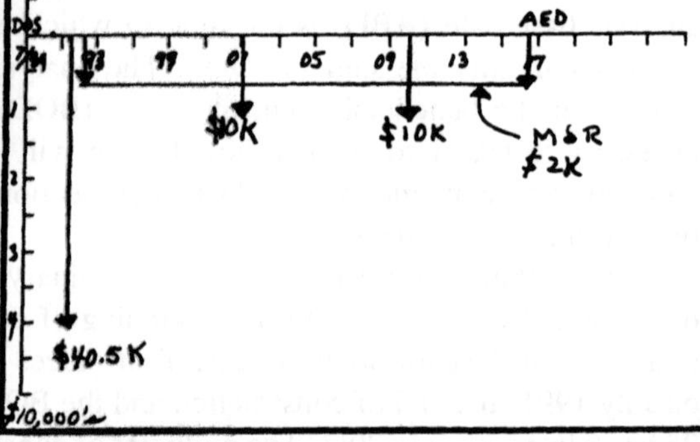

Cost Element	Cost on ABD ☒ $ x 10³ ☐ $ x 10⁶	Time Cost Incurred** Actual Projected Dates	Dates for Analysis (If Different)[b]	Source(s) of Data
Initial Investment	40.5	Jan 92		Contractor Quote
Replacement (1" top)[a]	10.0	July 01		Past Experience
Replacement (1" top)[a]	10.0	July 10		Past Experience
M + R	2.0	Jan 93[c] – Jan 17		Past Experience

DA FORM 5605-3-R, DEC 86 [Source: U.S. Department of the Army (1986)].

[a]The first replacement occurs 9 years after BOD, the second, 18 years after BOD.

[b]When 10 CFR436A criteria apply, i.e., for energy conservation and renewable energy projects.

[c]For recurring annual costs, show date of first and last costs only.

Sheet_______ of_______

Figure 23-1. Summary of data for 2-inch paving surface alternative

Project No. & Title 100 VA Hospital
Installation & Location Virginia
Design Feature Roadway/Parking Surface
Alt. No. 2 Title Asphalt w/3" Wearing Surface

LIFE CYCLE COST ANALYSIS
BASIC INPUT DATA SUMMARY

For use of this form, see TM 5-802-1; the proponent agency is USACE.

Criteria Reference		HQDA
Date of Study (DOS)		Jul 89
Analysis Base Date (ABD)		Jul 89
Analysis End Date (AED)		Jul 17
Midpoint of Construction		Jan 92
Beneficial Occupancy Date (BOD)	Actual Projected	Jul 92
	Assumed for Analysis	Jul 92
DOE Region		
Annual Discount Rate		10%

Type of Cost	Differential Escalation Rate per Year (%) Timeframe: 7/89 - 7/17		
All	0.00		

Principal Assumptions

Cash Flow Diagram

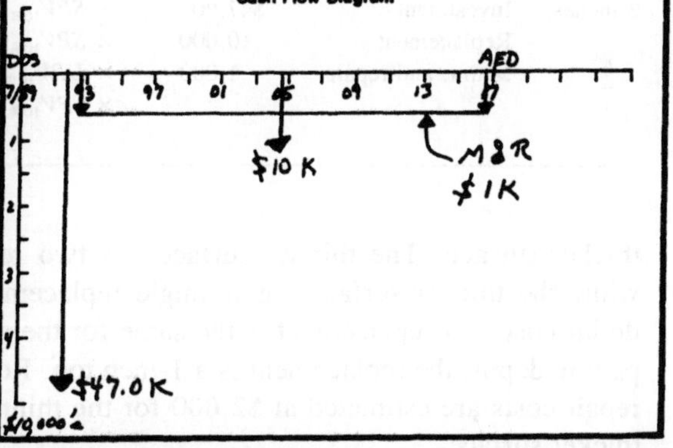

Cost Element	Cost on ABD ☑ $ x 10³ ☐ $ x 10⁶	Time Cost Incurred[c]: Actual Projected Dates	Dates for Analysis (If Different)[b]	Source(s) of Data
Initial Investment	47.0	Jan 92		Contractor Quote
Replacement (1" top)[a]	10.0	Jan 05		Past Experience
M&R	1.0	Jan 93 - Jan 17		Past Experience

DA FORM 5605-3-R, DEC 86 [Source: U.S. Department of the Army (1986)]

[a]The replacement occurs 12.5 years after BOD.

[b]When 10 CFR436A criteria apply, i.e., for energy conservation and renewable energy projects.

[c]For recurring annual costs, show date of first and last costs only.

Sheet ______ of ______

Figure 23-2. Summary of data for 3-inch paving surface alternative

Table 23-6. Calculating present values and LCCs of 2- and 3-inch paving depths

(1) Paving Depth	(2) Cost Element	(3) Amount (Constant $)	(4) Discount Operation	(5) Factor	(6) Present Value (6) = (3) × (5)
2 inches	Investment	$40,500	$\times SPV_{10\%, 2.5\,yr}$	0.7880	$31,913
	Replacement 1	10,000	$\times SPV_{10\%, 12\,yr}$	0.3186	3,186
	Replacement 2	10,000	$\times SPV_{10\%, 21\,yr}$	0.1351	1,351
	Maint. and repair	2,000	$\times UPV_{10\%, 25\,yr}$	9.0770	
			$\times SPV_{10\%, 2.5\,yr}$	0.7880	14,305
				LCC:	$50,755
3 inches	Investment	$47,000	$\times SPV_{10\%, 2.5\,yr}$	0.7880	$37,035
	Replacement	10,000	$\times SPV_{10\%, 15.5\,yr}$	0.2283	2,283
	Maint. and repair	1,000	$\times UPV_{10\%, 25\,yr}$	9.0770	
			$\times SPV_{10\%, 2.5\,yr}$	0.7880	7,153
				LCC:	$46,471

thicker surface. The thinner surface has two replacements of $10,000 each, while the thicker surface has a single replacement of $10,000. The constant dollar cost of a replacement is the same for the alternatives because, for either paving depth, the replacement is a 1-inch top. Routine annual maintenance and repair costs are estimated at $2,000 for the thinner surface and $1,000 for the thicker surface.

Recall the Army practice of charging initial costs at the midpoint of construction, repair and replacement costs whenever they are expected to occur, and annually recurring amounts at the midpoint of each year, beginning 6 months after the date of occupancy. In the example, initial investment costs are projected to occur in January 1992. The two replacements for the 2-inch alternative are projected to occur 9 and 18 years after occupancy, i.e., 12 and 21 years after the DOS. The single replacement for the 1-inch alternative is projected to occur 12.5 years after occupancy, i.e., 15.5 years after the DOS. Annual maintenance and repair costs are projected to occur first in January 1993 and last in January 2017.

23.2.3.2 Calculate LCCs and Compare Alternatives

Table 23-6 summarizes the operations to calculate present value cash flows and the LCC of each alternative. Note that it is necessary in this case to compute present value investment costs because they do not occur at the beginning of the study period. This is done by multiplying investment cost times the SPV discount formula for 10% and 2.5 years. Note also that in computing the present value of other costs, it is necessary to take into account the years prior to building occupancy to obtain the correct discount factors.

The 2-inch paving surface is estimated to have an LCC of about $50,800. For the 3-inch surface, the LCC is about $46,500.

23.2.4 Specify the Economic Roadway/Parking Surface and Submit Documentation of the Supporting Analysis

On the basis of the analysis, you would probably recommend the 3-inch paving surface. It probably is not necessary to perform uncertainty analysis in this case. There is high confidence in the data, the decision is routine, and there is a reasonably clear difference in the LCCs.

Attached documentation should include the Data Summary Sheets. Documentation should also include Table 23-6 or other documentation of calculations and copies of previous studies from which data are drawn.

REFERENCES

U.S. Department of the Army. 1986. *Economic Studies for Military Construction Design—Applications*. Technical Manual TM 5-802-1. Washington, DC: U.S. Department of the Army.

Lippiatt, Barbara, and Rosalie Ruegg. 1987. *Energy Prices and Discount Factors for Life-Cycle Cost Analysis*. NBSIR 85-3273-2. Gaithersburg, Maryland: National Bureau of Standards.

Petersen, Stephen R. 1981. *Economics and Energy Conservation in the Design of New Single-Family Housing*. NBSIR 81-2380. Gaithersburg, Maryland: National Bureau of Standards.

Residential/Light Commercial Cost Data. 1987. Kingston, Massachusetts: R. S. Means Company, Inc.

Ruegg, Rosalie T. 1987. *Life-Cycle Costing Manual for the Federal Energy Management Program*. NBS Handbook 135. Gaithersburg, MD: National Bureau of Standards.

24

Allocating Limited Budgets Among Competing Projects

The economic objective in allocating a budget is to maximize the total return on investment. In this chapter we illustrate practical approaches to budget allocation under two conditions. First, we illustrate allocation of a single budget, where the objective is to obtain the maximum return for that budget, without regard to future budgets. To do this, it is essential to make design/size increments of projects compete for funding.

The second illustration treats the condition where the investor wants to maximize the return on a series of annual budgets rather than on a single budget. The illustration points out that maximizing the return on each budget in the series may result in underdesigning and undersizing projects and, in the long run, fail to maximize the return on total investment. A practical approach is to design/size each project to minimum LCC or maximum NB *prior* to budget allocation.[1]

24.1 ALLOCATING A SINGLE BUDGET AMONG CANDIDATE PROJECTS FOR PLANT MODERNIZATION: EVALUATED WITH BCRS

24.1.1 Problem

A manufacturing company operates three plants. It earmarks $7.0 million of the current year's budget for modernizing plant and equipment. This is a one-time investment effort. The objective is to increase productivity and decrease operating costs as much as possible for the $7.0 million expenditure.

Each plant manager submits a list of proposed projects which appear economic. Some can be done in alternative designs or sizes. These projects are proposed first in their lowest-investment-cost version. Each expansion of the

[1]This chapter draws in part on Ruegg (1987, pp. 138–146).

project to a more costly design or size is offered as a separate investment decision.

Suppose that you, as chief financial officer, must allocate the $7.0 million budget among proposed projects and project expansions at the three plants. Since the proposed projects taken together exceed the budget, you must decide which to accept and which to reject.

24.1.2 Approach

Ranking projects and increments to projects according to their BCRs is one approach for guiding the allocation of limited funds. Make selections in descending order of ratios until the budget is exhausted. (See Section 4.2.2.)

24.1.3 Evaluation

24.1.3.1 Data and Assumptions

Table 24-1 lists in column 1 candidate projects submitted by each plant manager. The manager for plant 1 has proposed three projects, identified in the table as A, B, and C. Projects A and C may be done in their lowest first-cost versions (what we call the "base investment"), A(1) and C(1), or they may be expanded to more costly designs or larger sizes, A(2) and C(2). The notation A(1) → A(2) and C(1) → C(2) indicates that the increment is examined separately from the base investment. This allows an incremental project investment to compete for funding on its own merits. There are five investment possibilities for plant 1: project A in investment size (1) or (2), project B, and project C in investment size (1) or (2).

The manager for plant 2 has proposed four projects, identified in the table as D, E, F, and G. Project E may be done in base investment size E(1), or it may be expanded to E(2).

The manager for plant 3 has proposed two projects, H and I. Project H may be done in base investment size H(1), or it may be expanded to H(2).

Column 2 of Table 24-1 gives estimated investment costs. Note that they sum to $11.7 million, which is $4.7 million more than the $7.0 million budget. Column 3 gives estimated present value benefits, net of present value future costs (noninvestment costs). Column 4 gives NB. Column 5 gives incremental BCRs. The ratios on project expansions are clearly incremental ratios. But the ratios on the base investments are also incremental because the base size is the minimum feasible size for the projects (i.e., the increment is the difference between the base case size and no investment at all). Column 6 ranks the projects and their increments in descending order of BCRs.

Table 24-1. Allocating a limited budget among candidate projects for plant modernization

(1) Plant/Project	(2) Investment Costs ($1,000,000)	(3) Benefits (Net of Future Costs) ($1,000,000)	(4) NB ($1,000,000) (4) = (3) − (2)	(5) Incremental BCR (5) = (3)/(2)	(6) Ranking
Plant 1					
Project A(1)	1.00	5.00	4.00	5.0	5
A(1) → A(2)[a]	0.20	0.66	0.46	3.3	9
Project B	0.20	0.70	0.50	3.5	8
Project C(1)	0.70	2.00	1.30	2.9	10
C(1) → C(2)	0.25	0.45	0.20	1.8	12
Plant 2					
Project D	1.20	10.00	8.80	8.3	2
Project E(1)	0.50	3.25	2.75	6.5	4
E(1) → E(2)	0.10	0.45	0.35	4.5	6
Project F	0.30	3.50	3.20	11.7	1
Project G	2.00	14.00	12.00	7.0	3
Plant 3					
Project H(1)	1.50	6.00	4.50	4.0	7
H(1) → H(2)	0.25	0.65	0.40	2.6	11
Project I	3.50	4.50	1.00	1.3	13
Total	11.70				

[a]A(1) → A(2) indicates that the project is expandable from investment size A(1) to size A(2).

24.1.3.2 Allocating the Budget

Funds are allocated according to rankings. Project F, with the highest BCR of 11.7, is selected first, for an allocation of $0.30 million. Project D, with a BCR of 8.3, is selected second, for an additional allocation of $1.20 million, making a total allocation of $1.50 million. Project G is selected third, boosting the total allocation to $3.50 million. Project E in its base investment size is selected fourth, raising the total allocation to $4.0 million. The fifth selection is project A in its base investment size, increasing the total allocation to $5.0 million. The sixth choice is to expand project E to size E(2), increasing the total allocation to $5.10 million. The seventh choice is project H in its base investment size, raising the total budget allocation to $6.6 million. Project B is selected eighth, at a cost of $0.2 million, leaving $0.2 million unallocated. The ninth, and final, allocation of funds is to expand project A from its base investment size to size A(2), at a cost of $0.2 million, exhausting the $7.0 million budget.

NB for this allocation is estimated at $36.56 million. The budget has been allocated fully by following exactly the BCR rankings. NB is maximized, meaning that no other combination of projects and sizes within the budget would yield a larger NB.

24.2 ALLOCATING A SERIES OF ANNUAL BUDGETS FOR ENERGY CONSERVATION: EVALUATED WITH SIRS AND AGGREGATE NET SAVINGS

24.2.1 Problem

Suppose that you have taken the job of facilities manager for a large state university. You are told that you will receive ongoing annual budgets for retrofitting university buildings for energy conservation. State policy is to make public buildings LCC effective.

Candidate projects include some of variable design and size. One project, for example, is to increase the insulation in dormitories, classrooms, and administrative buildings. In allocating each year's budget, should you require project increments to compete for funding, or should you predesign and size projects as though there is no budget limitation, and then allocate the budget?

24.2.2 Approach

First compute SIRs for projects and project increments. Then select projects according to each allocation plan: (1) making project increments compete for funding and (2) predesigning and sizing projects prior to budget allocation. Next compute and compare aggregate present value NS for the two plans.

24.2.3 Evaluation

24.2.3.1 Data and Assumptions

Table 24-2 lists candidate projects. Suppose that project B, insulation, can be done only in two sizes: the base investment size, B(1), and a larger version, B(2). The increment of expanding project B from B(1) to B(2) is shown for two conditions. The first entry in the table is for the condition where the incremental level of insulation is installed at the same time that the base level of insulation is installed. For this condition, incremental investment costs are $10,000, incremental savings are $20,000, NS is $10,000, and the incremental

Table 24-2. Candidate projects for budget allocation

(1) Projects	(2) Investment Costs (PV $)	(3) Savings (PV $)	(4) NS (PV $)	(5) Incremental SIR[a]	(6) Total SIR
A	120,000	600,000	480,000	5.00	5.00
B(1)[b]	50,000	150,000	100,000	3.00	3.00
B(1) → B(2)[c]	10,000	20,000	10,000	2.00	. . .
B(1) → B(2)[d]	30,000	20,000	−10,000	0.67	. . .
B(2)[e]	60,000	170,000	110,000	. . .	2.83
D	30,000	120,000	90,000	4.00	4.00
E	80,000	120,000	40,000	1.50	1.50
F	50,000	145,000	95,000	2.90	2.90
G	10,000	12,000	2,000	1.20	1.20

[a]Incremental SIRs are based on the additional costs and savings of a separable investment increment. Because projects A, B(1), D, E, F, and G are in their lowest divisible investment size, there is no difference between their incremental and total SIRs.

[b]Size (1) is the smallest divisible level of project B.

[c]B(1) → B(2) represents the increment from expanding project B in size (1) to size (2).

[d]This second entry for expanding project B from size (1) to size (2) is based on the assumption that it would be done at a later time, thereby reincurring fixed installation costs.

[e]B(2) is project B in its larger size (2), done all at once.

SIR is 2.00. The second entry is for the condition where the incremental level of insulation is added in the second year, while the base level is installed in the first year. Incremental investment costs are higher under the second condition because fixed installation costs which are required for the base level are reincurred for the increment. Under this condition, the increment loses money.

In column 5, SIRs are incremental for all projects because projects A, D, E, F, and G are given in their smallest investment size, and the increment to B is given separately. In contrast, column 6 gives a total SIR for project B in size B(2), i.e., computed for total costs and savings of project B in size (2). There is no difference in incremental and total SIRs for the other projects.

The total package of projects requires investment of $350,000. This year's budget is $200,000. Next year's budget is expected to be $140,000.

24.2.3.2 Increments Compete for Funding

Table 24-3 shows allocation of the first and second year's budgets according to the first plan that makes project increments compete for funding. Take projects and project increments in descending order of their incremental SIRs, as given in Table 24-2. This results in an allocation of the first year's budget of $200,000 among projects A, D, and B(1). NS totals $670,000.

Following the same procedure to allocate the second year's budget results in

Table 24-3. Selecting projects according to budget allocation plan 1: making project increments compete for funding

	(1) Projects	(2) Investment Costs (PV $)	(3) Savings (PV $)	(4) = (3) − (2) NS (PV $)
		Project Selection in Year 1		
	A	120,000	600,000	480,000
	D	30,000	120,000	90,000
	B(1)	50,000	150,000	100,000
Aggregates		200,000	870,000	670,000
		Project Selection in Year 2		
	F	50,000	145,000	95,000
	E	80,000	120,000	40,000
	G	10,000	12,000	2,000
Aggregates		140,000	277,000	137,000

selection of projects F, E, and G. NS totals $137,000. Note that the increment to project B is not selected in the second year because its investment costs jump, causing it no longer to be cost effective.

24.2.3.3 Predesigned and Presized Projects Compete for Funding

Table 24-4 shows allocation of the first and second year's budgets according to the second plan that predesigns and presizes projects prior to their competition for funding. Presizing project B without regard to budget limitations means that project B is now considered only in size (2). This is because savings on the increment exceed its costs, meaning that size (2) is the cost-effective size for an unlimited budget. The total SIR on project B in size (2) is 2.83. This means that it is passed over in favor of project F in the first year's budget allocation. Projects A, D, and F are selected the first year. NS totals $665,000.

Project B in size (2) has the highest SIR of projects competing for the second year's budget. It takes $60,000 of available funds. Project E has the next highest SIR and takes the remaining $80,000 of the $140,000 budget.

24.2.3.4 Compute Aggregate Present Value NS for Alternative Budget Allocation Plans

The last step is to determine which of the two budget allocation plans will maximize NS for available funds. This is done by computing the present value

Table 24-4. Selecting projects according to budget allocation plan 2: predesigning and sizing projects prior to competition for funding

	(1) Projects	(2) Investment Costs (PV $)	(3) Savings (PV $)	(4) = (3) − (2) NS (PV $)
		Project Selection in Year 1		
	A	120,000	600,000	480,000
	D	30,000	120,000	90,000
	F	50,000	145,000	95,000
Aggregates		200,000	865,000	665,000
		Project Selection in Year 2		
	B(2)	60,000	170,000	110,000
	E	80,000	120,000	40,000
Aggregates		140,000	290,000	150,000

of aggregate NS from the combined budget over the two years for each of the budget allocation plans.

$$PV\,ANS(1) = \$670 + \$137\,\frac{1}{(1 + 0.08)^1} = \$797$$

$$PV\,ANS(2) = \$665 + \$150\,\frac{1}{(1 + 0.08)^1} = \$804$$

where

$PV\,ANS(1)$ = present value of aggregate NS for budget allocation plan 1, making project increments compete separately for funding, and

$PV\,ANS(2)$ = present value of aggregate NS for budget allocation plan 2, predesigning and presizing projects prior to their competition for funding.

Aggregate NS from budget allocation plan 1 is $670,000 in the first year and $137,000 in the second year. NS in the first year is in present value dollars, but NS in the second year is delayed by 1 year. An annual discount rate of 8% is applied to second-year NS to convert it to present value. Budget allocation plan 1 results in aggregate NS of $797,000.

Aggregate NS from budget allocation plan 2 is $665,000 in the first year and

$150,000 in the second year. Second-year NS is discounted to present value. Budget allocation plan 2 results in aggregate NS of $804,000.

24.2.4 Choose a Budget Allocation Procedure

Budget allocation plan 2 is estimated to maximize the return on the series of two budgets. The point of this case study is to illustrate that an overly short-run view can cause underdesigning and undersizing of projects when the problem is to allocate a series of ongoing budgets. Here, for example, taking a short view of selecting B(1) instead of B(2) means losing the opportunity to save $2.0, on the average, for every $1.0 invested in additional insulation.

A practical guideline when allocating one in a series of budgets is to design/size individual projects to minimize LCC or maximize NB as though funds are not limited, despite some tendency toward overdesigning and sizing. Then let the predesigned and presized projects compete for funding. To be sure of making design and size choices which maximize NB, however, requires a quantitative comparison of the benefits and costs of funding projects at higher levels and delaying them versus funding them at lower levels and doing them sooner.

REFERENCE

Ruegg, Rosalie T. 1987. *Life-Cycle Costing Manual for the Federal Management Program.* NBS Handbook 135. Gaithersburg, MD: National Bureau of Standards.

25

Decisions on Timing of Equipment Replacements

Replacement analysis is performed to ensure that equipment and facilities are not kept beyond their economic lives. Economic life is the time over which an item is the least-cost alternative for meeting its objective. If kept beyond its economic life, an item may continue to give service, but a replacement could do the job more cost effectively.

Equipment and components may need to be replaced because of physical deterioration, which causes maintenance expenses, energy consumption, or other operating costs to increase, productivity to fall, or physical appearance to suffer.

Replacement may also be prompted by technical obsolescence. Equipment and components become technically obsolete when new, improved substitutes offer lower operating costs, increased output capacity, enhanced appearance, or increased performance capabilities.

Replacement analysis determines *when* to replace and *with what* to replace. It can be performed with the basic tools of investment evaluation presented in Part 1.

Here we illustrate how to determine when to replace building equipment under two conditions. The first case study shows replacement of equipment with like equipment. The second shows replacement of equipment with unlike equipment.

25.1 WHEN TO REPLACE A SERIES OF LIKE EQUIPMENT: DETERMINED WITH AN ANNUAL COST MODEL

A simple replacement decision is that of setting a replacement schedule for a physical asset that is to be replaced by a series of like assets. When to replace a standard filter or compressor due to normal wear and tear are examples. Here we illustrate how to determine a replacement schedule for equipment when the technology is not expected to change.

25.1.1 Replacement Problem

Suppose that you wish to establish a replacement policy for an item of equipment used in a building. The item's potential useful life is 5 years. Your problem is to determine its economic life and to set up a replacement schedule based on the economic life.

25.1.2 Approach

You can solve this problem by finding the number of years in service for which the annualized LCC (i.e., annual cost) is minimized. Calculate the annual cost based on each number of years the equipment potentially could be kept in service. In each case include the purchase and operating costs associated with that time. In the example, there will be five annual cost numbers, because there are 5 years in which the equipment might be replaced. Find the number of years for which the annual cost is lowest. Then set the equipment replacement schedule to provide that service life.

25.1.3 Evaluation

25.1.3.1 Data and Assumptions

Equipment can be purchased for $20,000 and kept in service for 5 years if properly maintained and repaired. Table 25-1 shows how the after-tax costs of operating, maintaining, and repairing the equipment (OM&R costs) change over time. These costs are often called "operating costs" in replacement studies. Note that operating costs rise as the equipment ages.

Table 25-1 also gives after-tax salvage value remaining at the end of each year. Note that the $20,000 salvage value in year 0 is the purchase price of the

Table 25-1. How OM&R costs and salvage value change over time

(1) Years in Service	(2) OM&R Costs (After-Tax Dollars)	(3) Salvage Value (After-Tax Dollars)
0	0	20,000
1	2,000	12,000
2	3,000	10,000
3	4,000	8,000
4	5,000	6,000
5	6,000	2,000

equipment, which is assumed to be its value when put into service. The declining salvage value reflects increasing capital costs, often called "owning costs."

Assume that building performance is not affected by the age of the equipment; only owning and OM&R costs change. Assume further that no technological change in the equipment is anticipated in the foreseeable future.

Finally, assume that all OM&R costs, salvage values, and acquisition costs change at the same rate as prices in general, such that differential rates of change are zero. Assume that the real discount rate is 7% after taxes, compounded annually. All amounts are in constant dollars.

25.1.3.2 Determine the Optimal Replacement Schedule

Find the number of service years (N) for which the annualized LCC ($AC(N)$) is minimized. $AC(N)$ can be calculated using equation 25-1:

$$AC(N) = \left[\left[C - ((S(N))(SPV(N))) \right] + \sum_{j=1}^{N} \left[(OM\&R_j)(SPV_j) \right] \right] (UCR(N)) \qquad (25\text{-}1)$$

where

$AC(N)$ = annualized LCC of owning and operating the equipment when it is kept in service for N years,
C = initial acquisition costs,
$S(N)$ = salvage value of the equipment when it is disposed of after N years of service,
$SPV(N)$ = single present value discount factor for the time of disposal,
$OM\&R_j$ = operating, maintenance, and repair costs in year j,
SPV_j = single present value discount factor for year j, and
$UCR(N)$ = uniform capital recovery discount factor for the period N years.

Use equation 25-1 to calculate $AC(N)$ for a range of service years (N) from 1 up to 5, the latter being the useful life.

$$AC(1) = \left[\left[\$20{,}000 - ((\$12{,}000)(0.93)) \right] + \left[(\$2{,}000)(0.930) \right] \right] (1.07) = \$11{,}449$$

$$AC(2) = \left[\left[\$20{,}000 - ((\$10{,}000)(0.87)) \right] + \left[((\$2{,}000)(0.93)) + ((\$3{,}000)(0.87)) \right] \right] (0.553) = \$8{,}721$$

$$AC(3) = [[\$20{,}000 - ((\$8{,}000)(0.82))] + [((\$2{,}000)(0.93)) + ((\$3{,}000)(0.87)) + ((\$4{,}000)(0.82))]](0.381) = \$8{,}073$$

$$AC(4) = [[\$20{,}000 - ((\$6{,}000)(0.76))] + [((\$2{,}000)(0.93)) + ((\$3{,}000)(0.87)) + ((\$4{,}000)(0.82)) + ((\$5{,}000)(0.76))]](0.295) = \$7{,}962$$

$$AC(5) = [[\$20{,}000 - ((\$2{,}000)(0.71))] + [((\$2{,}000)(0.93)) + ((\$3{,}000)(0.87)) + ((\$4{,}000)(0.82)) + ((\$5{,}000)(0.76)) + ((\$6{,}000)(0.71))]](0.244) = \$8{,}391$$

Now compare the results of the calculations and identify the life for which the annual cost is lowest. Table 25-2 summarizes the annual cost for each potential service life. Since the annual cost is lowest when the equipment is kept in service for 4 years, the optimal replacement schedule is every 4 years. Figure 25-1 verifies this conclusion.

25.2 WHEN TO REPLACE IF A LOWER-COST SUBSTITUTE IS ANTICIPATED: DETERMINED WITH A PRESENT VALUE MODEL

In this section we build on the example of Section 25.1 to illustrate how to determine if and when to switch over from existing to improved equipment.

25.2.1 Replacement Problem

Suppose that you are asked to determine a replacement strategy for the equipment described in Section 25.1 when technological innovation is expected to

Table 25-2. Annual costs of owning and operating equipment over alternative service lives

Years in Service	Uniform Annual Cost
1	$11,449
2	8,721
3	8,073
4	7,962*
5	8,391

*denotes minimum annual costs.

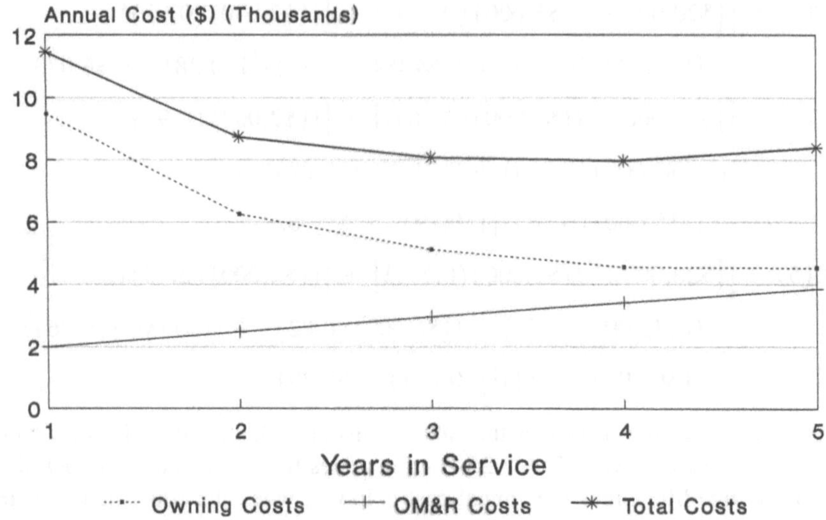

Figure 25-1. Annual costs by years in service. Note: Annualized owning costs plus annualized OM&R costs equals total annual costs.

bring into the marketplace a new, improved substitute for the existing equipment. The improved substitute will be available in about a year. It is expected to have a useful life of about 8 years. Its operating costs are expected to be less affected by aging than the existing equipment, and it is expected to retain its value somewhat better. Once on the market, its purchase price is expected to remain stable.

Here is your situation. It is the scheduled time for a replacement of the existing equipment which is on a 4-year replacement schedule. Improved equipment will not be available for a year. Therefore it is impossible to make a changeover now. Three alternative changeover strategies seem feasible. The objective is to choose the one that minimizes the LCC.

1. Replace the existing equipment now as scheduled, and wait for 4 years until the next regularly scheduled replacement period to switch to the improved equipment.
2. Replace the existing equipment now as scheduled, and switch to the improved equipment next year when it becomes available.
3. Do not replace existing equipment now. Instead, extend the service life for a fifth year, and then switch to the improved equipment.

25.2.2 Approach

To determine which changeover strategy is cost effective, calculate for each strategy the present value cost for the 4-year period from now until the next scheduled replacement of the existing equipment. There is no need to look beyond 4 years. If a changeover is appropriate, it should be made no later than the next scheduled replacement time.

25.2.3 Evaluation

25.2.3.1 Data and Assumptions

The improved equipment is expected to cost about $25,000. The annual cost of the improved equipment, estimated for each of eight potential service lives by equation 25-1, is summarized in Table 25-3. It is lowest if you replace the equipment every 6 years at an annual cost of $5,000.

Since $5,000 is less than $7,962 (the annual cost determined in Section 25.1 for the existing equipment), there is an advantage to changing from existing to improved equipment. Once the changeover is made, the improved equipment will be replaced every 6 years. The question is when to change over.

25.2.3.2 Calculate and Compare the Present Value Costs of Alternative Changeover Strategies

The present value of strategy 1 can be calculated with equation 25-2:

$$PV_1 = AC_E(4) \times UPV_{N=4,d=7\%}$$

$$= \$7{,}962 \times 3.39 = \$26{,}991 \qquad (25\text{-}2)$$

where

PV_1 = present value of replacement strategy 1, and
$AC_E(4)$ = the annual cost of the existing equipment when it is replaced once every 4 years.

Table 25-3. Service life and associated annual costs of owning and operating improved equipment

Years in Service	Uniform Annual Cost
1	$12,000
2	10,000
3	7,500
4	6,000
5	5,500
6	5,000*
7	6,200
8	7,500

*denotes minimum annual cost.

The present value of strategy 2 can be calculated with equation 25-3:

$$\begin{aligned} PV_2 &= \left[(AC_E(1))\,(SPV_{N=1,d=7\%})\right] \\ &\quad + \left[(AC_I(6))\,(UPV_{n=3,d=7\%})\,(SPV_{N=1,d=7\%})\right] \\ &= (\$11{,}449 \times 0.93) + (\$5{,}000 \times 2.62 \times 0.93) \\ &= \$22{,}831 \end{aligned} \tag{25-3}$$

where

PV_2 = present value of replacement strategy 2,
$AC_E(1)$ = annual cost of existing equipment if replaced once a year, and
$AC_I(6)$ = annual cost of improved equipment if replaced once every 6 years.

The present value of strategy 3 can be calculated with equation 25-4:

$$\begin{aligned} PV_3 &= \left[CSV_E - (OYS_E)(SPV_{N=1,d=7\%})\right] + \left[(OM\&R_E(1))\,(SPV_{N=1,d=7\%})\right] \\ &\quad + \left[(AC_I(6))\,(UPV_{N=3,d=7\%})(SPV_{n=1,d=7\%})\right] \\ &= \left[\$6{,}000 - (\$2{,}000 \times 0.93)\right] + (\$6{,}000 \times 0.93) + [\$5{,}000 \times 2.62 \\ &\quad \times 0.92] \\ &= \$21{,}903, \end{aligned} \tag{25-4}$$

where

CSV_E = current salvage value of the existing equipment.[1]
OYS_E = salvage value of the existing equipment one year from now, and
$OM\&R_E(1)$ = operating, maintenance, repair, and replacement costs associated with the existing system in its fifth year of operation.

25.2.3.3 Decide the Replacement Strategy

Strategy 3, "Keep existing equipment for one more year, and then switch to the improved equipment," appears to offer the lowest cost over the 4-year period. Once the changeover to the improved equipment is made 1 year hence, the new schedule of replacing the equipment every 6 years goes into effect.

---

[1]Note that the present value of extending the life in service of the existing equipment a fifth year is not calculated by simply plugging in the annual cost for 5 years from Table 25-2. This is because we are only interested in what costs will be in the fifth year, not the annualized costs based on a 5-year service life. Hence, we compute the fifth-year costs directly.

26

Selecting Combinations of Interdependent Systems

You should consider possible interdependencies among systems when selecting them for the same building or facility. Systems are interdependent if selecting one affects the economic efficiency of another. Failing to take into account interdependencies can result in selecting systems which are uneconomic or rejecting systems which are economic. In this chapter we explain how to take into account interdependencies among building systems.[1]

26.1 INTERDEPENDENCE

Systems are interdependent under at least three conditions: (1) when they are substitute means of accomplishing the same objective (but are not mutually exclusive[2]); (2) when they complement one another's performance; and (3) when one is a prerequisite to another.

Systems are *substitutes* when having more of one system means needing less of another. For example, increasing the insulation in a building envelope reduces the need for purchasing an HVAC system with a higher efficiency, because insulation lowers heating and cooling loads. Likewise, increasing the technical efficiency of the HVAC system lowers the requirement for purchased energy to meet existing loads. Increased insulation and HVAC efficiency both contribute to the objective of reducing energy consumption. Both may be selected. However, the more you have of one, the less energy savings you get by increasing the other.

Systems are *complements* when accepting one enhances the cost effectiveness of the other. In combination, complementary systems provide greater NB than the sum of their individual NB. For example, a passive solar energy system utilizing a massive wall might complement a noise reduction system. A power generation system and a solid waste disposal system might complement one

[1]This Chapter is based in part on Ruegg (1987, pp. 127–133).

[2]Mutually exclusive systems are not interdependent by definition, because choosing one means excluding the other, such that there can be no interdependence among the systems.

another. Evaluating complementary systems independently of synergistic effects understates their economic merit.

A system is *prerequisite* to another when the latter will not function as intended without the former. The cost effectiveness of the one is contingent on the presence of the other. For example, adding wall insulation in excess of a certain level is physically impossible unless the wall cavity is expanded. Expansion of the wall cavity is prerequisite to an increased level of insulation. Incorporating solar energy into a new building design is contingent on selecting a suitable building site. Further, installing a solar energy system directly on the roof is contingent on there being an adequate support system. Reinforcing the roof to provide stronger support may not be cost effective if evaluated independently of the solar energy system. But it may be cost effective when evaluated in conjunction with the solar energy system.

26.2 FIND THE COST-EFFECTIVE COMBINATION

We focus on finding cost-effective combinations of substitute systems, because that type of interdependence is most common in buildings. We illustrate two approaches: (1) the overall (or total) LCC approach (see Chapter 2), and (2) the incremental SIR approach applied sequentially (see Chapter 4). We focus only on the problem of interdependence and assume that the budget is sufficient to fund all options that are cost effective.

26.2.1 Overall LCC Approach

This approach calls for computing the overall LCCs of the building[3] for alternative combinations of interdependent systems and finding the combination for which costs are minimum. When there are many possible combinations which would require repetitious computations to identify the least-cost choice, using a computer program facilitates the search.[4]

Suppose that you want to find the economically efficient combination of energy-conserving modifications to the envelope of an existing building *and* to its heating system. Assume envelope retrofit options are (1) attic insulation, (2) storm windows, and (3) caulking existing windows. Attic insulation comes in batts that can be installed in any of four levels: R-11, R-19, R-30, and R-38.[5]

[3]The term "overall LCC of the building" includes only costs relevant to the decision but is sufficiently comprehensive to include all categories of costs affected by any of the candidate modifications. (see Chapter 2.)

[4]Recall that we assume that money is available to fund any combination of cost-effective projects.

[5]R = resistance level.

The degree of interdependence between storm windows and caulking is strong. In fact, storm windows substitute completely for caulking, such that energy savings from caulking are lost if storm windows are added. Storm windows, however, provide additional energy savings beyond those provided by caulking. This means that the energy savings of storm windows are reduced, but not lost, if windows are first caulked.

The only alternative considered for the existing electric resistance heating system is replacement with a heat pump. The heat pump has a coefficient of performance (COP) of 200%. The heat pump is interdependent with all of the envelope options.

Tables 26-1 and 26-2 summarize the modifications (including size options) and data needed for the analysis. Table 26-1 presents data for each modification as though it alone were installed. Column 2 of Table 26-1 in its first row shows purchased energy without modification to the building. Subsequent rows show purchased energy with each modification taken separately.

Table 26-2 presents data for combinations of modifications taking into ac-

Table 26-1. Overall LCCs of proposed modifications when examined separately

Proposed Modifications (1)	Annual Energy Purchased[a] (10^6 Btu) (2)	Investment Cost($)[b] (3)	Present Value Energy Costs($)[c] (4)	Overall LCC($)[d] (5) = (3) + (4)
None	50.0	0	10,000	10,000
Envelope				
R-11 attic insulation	37.0	300	7,400	7,700
R-19 attic insulation	35.0	450	7,000	7,450
R-30 attic insulation	34.0	630	6,800	7,430
R-38 attic insulation	33.5	780	6,700	7,480
Storm windows	47.0	200	9,400	9,600
Caulking existing windows	49.0	25	9,800	9,825
Heating System				
Heat pump (COP = 2.0)	25.0	3,500	5,000	8,500

[a](Annual space heating load)/(heating system efficiency) = annual energy purchased.
[b]The investment cost of the specified modification. Because R-11 is prerequisite to R-19, the total investment cost associated with R-19 is $450 [i.e., $300 (for R-11) + $150 (incremental cost for R-19)] = $450. LCC associated with R-19 = $7,000 + $450 = $7,450.
[c]Present value energy costs = annual energy purchased × price of energy × uniform present value discount factor modified for energy price escalation. For example, for the case of no modifications, present value energy costs = (50×10^6 Btu) × ($\$20/10^6$ Btu) × 10, where 10 is a factor to account for discounting and escalation over a 15-year period.
[d]The measure of "overall LCC" needs to include all of the cost categories affected by any of the proposed modifications, i.e., it must be comprehensive, but it should include only those building costs actually affected by one or more of the modifications.
Source: Adapted from Ruegg (1987, p. 130).

Table 26-2. Overall LCCs for combinations of proposed modifications

Proposed Modifications (1)	Annual Energy Purchased (10^6 Btu) (2)	Investment Cost($) (3)	Present Value Energy Costs($) (4)	Overall LCC($) (5) = (3) + (4)
R-30 attic insul./storm windows	31.0	830	6,200	7,030*
R-30 attic insul./caulking	33.0	655	6,600	7,255
R-30 attic insul./storm windows/caulking	31.0	855	6,200	7,055
Caulking/storm windows	47.0	225	9,400	9,625
R-11 attic insul./heat pump	18.5	3,800	3,700	7,500
R-19 attic insul./heat pump	17.5	3,950	3,500	7,450
R-30 attic insul./heat pump	17.0	4,130	3,400	7,530
R-11 attic insul./storm windows/heat pump	17.0	4,000	3,400	7,400
R-19 attic insul./storm windows/heat pump	16.0	4,150	3,200	7,350
R-30 attic insul./storm windows/heat pump	15.5	4,330	3,100	7,430

*designates most cost-effective combination.
Source: Ruegg (1987, p. 131).

count interdependence.[6] Column 2 of Table 26-2 shows purchased energy based on various combinations of modifications.

Column 3 of each table shows investment costs. Column 4 gives present value energy costs and column 5, overall LCCs.

Search through column 5 of the two tables to find the modification or combination of modifications for which overall LCC is lowest. It is lowest for the combination of insulating the attic to a level of R-30 and adding storm windows ($7,030 in the top row of Table 26-2).

26.2.2 SIR Approach Applied Sequentially

Now we demonstrate an approximation approach to finding the least-cost combination of independent building systems, which entails applying the SIRs sequentially. The first step is to calculate SIRs for each system and increment to systems. In this step, ignore interdependencies among systems. Table 26-3 shows the incremental SIR for each modification taken separately. Note that the incremental SIR for each increase in the level of insulation is computed on the basis of additional costs and savings for that increment alone.

---

[6]Not all combinations are given in Table 26-2. We eliminated several at the outset as having no possibility of producing a minimum LCC. For example, since we can see from Column 5 of Table 26-1 that R-38 is not cost effective when compared with other levels of insulation, we can deduce that it will not be cost effective used in combination with other modifications.

Table 26-3. Incremental SIRs computed for each modification

Proposed Modifications (1)	Incremental Investment Cost ($) (2)	Incremental Energy Savings ($) (3)	Incremental SIR[a] (4) = (3)/(2)
Envelope			
Insulation			
R-0 to R-11	300	2,600	8.7
R-11 to R-19	150	400	2.7
R-19 to R-30	180	200	1.1
R-30 to R-38	150	100	0.7
Storm windows	200	600	3.0
Window caulking	25	200	8.0
Heating System			
Heat pump	3,500	5,000	1.4

[a]The SIR must be based on incremental savings and costs, not totals, when used to guide project decisions regarding size/design choices, because SIRs based on totals cannot be relied upon to signal cost-effective project increments. The incremental SIR for increasing from R-11 to R-19, for example, is computed as follows:
$SIR = \Delta$Energy savings/Δinvestment costs = \$400/\$150 = 2.7.

Source: Adapted from Ruegg (1987, p. 130).

The second step is tentatively to select the system with the highest SIR. Then recalculate SIRs on the remaining options, based on acceptance of the first.

The third step is another selection round. Tentatively select the option with the next highest SIR and recalculate the SIRs on remaining options, based on acceptance of the second. In addition, recalculate the SIR on the first-selected system to make sure that it does not fall below the acceptance point as subsequent options are added. If the first option is no longer acceptable, the SIRs on the remaining options should be based only on acceptance of the second option.

The fourth step is another selection round. Tentatively select the option with the next highest SIR, and recalculate SIRs on remaining options and on previously selected options. Continue this process until there are no more options having SIRs greater than 1.0 or until the budget is exhausted. (You can verify that the resulting selections maximize NS using the aggregate NS method.)

Table 26-4 shows four rounds of selection with the sequential SIR approach. In the first selection round, choose R-11 attic insulation because its SIR is highest. Reduce the SIR of the heat pump accordingly. In the second round, select caulking. It has the highest SIR of modifications remaining after R-11 attic insulation is chosen. Reduce the SIRs of storm windows and heat pump accordingly. In round three, increase attic insulation to R-19 and drop the heat pump from consideration because the SIR falls below 1.0.

Table 26-4. Choosing among interdependent projects with SIRs applied sequentially

Sequential Selections (1)		Investment Cost (2)	Present Value Energy Savings (3)	SIR (4)	SIR After Prior Selection (5)
Round 1					
Tentative selection:	R-11 insulation	$300	$2,600	8.7	
Remaining projects:	Caulking	25	200		8.0
	Storm windows	200	600		3.0
	R-11 to R-19 insulation	150	400		2.7
	R-19 to R-30 insulation	180	200		1.1
	Heat pump	3,500	3,700		1.1
Round 2					
Tentative selections:	R-11 insulation	300	2,600	8.7	
	Caulking	25	200	8.0	
Remaining projects:	Storm windows	200	400		2.0
	R-11 to R-19 insulation	150	400		2.7
	R-19 to R-30 insulation	180	200		1.1
	Heat pump	3,500	3,600		1.0
Round 3					
Tentative selections:	R-11 insulation	300	2,600	8.7	
	R-11 to R-19 insulation[a]	150	400	2.7	
	Caulking	25	200	8.0	
Remaining projects:	Storm windows	200	400		2.0
	R-19 to R-30 insulation	180	200		1.1
	·····Heat pump·····	·····3,500·····	·····3,400·····		·····0.97·····
Round 4					
Final selections:	R-11 insulation	300	2,600	8.7	
	R-11 to R-19 insulation	150	400	2.7	
	R-19 to R-30 insulation	180	200	1.1	
	Storm windows	200	600	3.0	
	·····Caulking·····	·····25·····			·····0.04·····

Note: The dashed line indicates candidates deleted because their SIRs drop below 1.0.

[a]The breakdown of insulation modification by size increments is preserved even after tentative selection to allow for the possibility that part, but not all, of the modification might be later eliminated because of interdependencies with a subsequent selection.

Source: Ruegg (1987, p. 133).

In the final round, select storm windows. Also select the increment of R-19 to R-30 attic insulation, because its SIR exceeds 1.0 and its selection is not significantly affected by storm windows. Selection of an additional increment of attic insulation does not significantly affect preceding selections. But selection of storm windows causes the caulking SIR to fall below 1.0. Therefore, delete caulking.

As shown in round 4 of Table 26-4, the least-cost combination found by this approach is the same as that found using the overall LCC approach: storm windows and attic insulation to a level of R-30.

REFERENCE

Ruegg, Rosalie. 1987. *Life-Cycle Costing Manual for the Federal Energy Management Program.* NBS Handbook 135. Gaithersburg, MD: National Bureau of Standards.

APPENDIXES

Appendix A

Glossary

Adjusted Cost Basis—The depreciable costs of purchasing or constructing an asset, plus related depreciable expenses such as installation and site preparation.

Adjusted Internal Rate-of-Return (AIRR)—See Overall Rate of Return.

After-Tax Discount Rate—The discount rate reflecting the investor's minimum acceptable rate of return after tax effects are taken into account.

Aggregate Net Benefits (Aggregate Net Savings)—The sum of net benefits (or net savings) for a set of projects, either independent or interdependent.

Aggregate Net Benefits (Aggregate Net Savings) Method—A version of the net benefits or net savings method used to combine interdependent projects and to allocate budgets among competing projects.

Annual Compound Rate of Return (Annual Rate of Return; Annual Yield)—The annual percentage return on investment when earnings are compounded once a year.

Annually Recurring Costs—Costs incurred each year throughout the study period.

Annual Value (Annual Worth)—The time-equivalent value of past, present, or future cash flows expressed as an annually recurring uniform amount over the study period.

Annual Cost Model—Life-cycle costs expressed as a time-equivalent annually recurring uniform amount.

Annual Value (Annual Worth or Uniform Capital Recovery) Factor—A discount factor by which a present dollar amount may be multiplied to find its equivalent annual value, based on a given discount rate and a given period of time.

Base Case (Base-Case Alternative)—The situation against which an alternative design, system, or action requiring a higher investment cost is compared. For accept/reject decisions, the base case is not having the project in question. For design and sizing decisions, the base case is the lowest investment design or size.

Base Time—The date designated the present time for the purpose of expressing the time-equivalent present value of all benefits and costs (usually the date the study is performed).

Base Year—The first year of the study period, generally the year in which the life-cycle cost analysis is conducted.

Base-Year Energy Costs—The quantity of energy purchased in the base year, multiplied by the base-year price of fuel.

Base-Year Price—The price of a good or service as of the beginning of the first year of the study period.

Benefit-Cost Analysis—A method of evaluating projects or investments by comparing the present value or annual value of expected benefits to the present value or annual value of expected costs.

Benefit-to-Cost Ratio (BCR)—Benefits divided by costs, where both are discounted to a present value or annual value.

Book Value—The worth of property as shown in the accounting records of a company, usually its original cost adjusted for improvements, less depreciation expenses charged.

Breakeven Analysis—A technique for determining the minimum or maximum value that a variable can reach and still have a breakeven project, i.e., a project where benefits (savings) equal costs.

Capital Gains—The excess of net selling price of an asset over book value at the time of sale.

Capital Gains Tax—A tax on capital gains.

Capitalized Value—The present time-equivalent value of a uniform series recurring indefinitely.

Cash Flow—The stream of costs and benefits resulting from a project investment.

Capital Asset Pricing Model (CAPM)—A model that explains how assets should be priced according to their calculated risk and expected returns. It indicates what the return on an asset should be, based on the return on risk-free investments and the asset's risk relative to undiversifiable risk in the market.

Certainty Equivalent Factor (CEF)—The factor by which uncertain project returns are multiplied to determine the certain amount that a decision maker finds equally acceptable.

Certainty Equivalent Technique—A technique used to adjust economic measures of project worth to reflect risk exposure and risk attitude. Estimated project returns are multiplied by a certainty equivalent factor (CEF) to determine the certain amount that a decision maker finds equally acceptable.

Coefficient of Variation (CV)—The ratio of the standard deviation to the mean.

Complementary Systems—Systems for which the acceptance of one enhances the economic efficiency of the other. In combination, these systems provide greater NB than the sum of their individual NB.

Compounding—The frequency with which interest is credited. The more frequent the compounding, the higher the effective interest rate.

Compound Interest Factors or Formulas—See Discount Factors or Discount Formula.

Constant Dollars—Dollars of uniform purchasing power tied to a reference year (usually the base year) and exclusive of general price inflation or deflation.

Consumer Price Index—A series of price indices for items purchased by consumers, prepared by the U.S. Bureau of Labor Statistics. Prices in a selected base year are given an index value of 100. Subsequent increases above 100 indicate price increases.

Contingency—A cost element included in project cost estimation to cover costs that have some likelihood of occurrence but whose amounts cannot be predicated with certainty.

Continuous Cash Flow—Cash flows continuously flowing over time at a uniform rate, in contrast to discrete cash flow.

Contour Lines—Line segments that connect the various rays of a spider diagram.

Cost Basis—That part of an investment which can be depreciated.

Cost Effective—The condition whereby one building system has lower costs than competing systems and produces equal benefits.

Cumulative Distribution Function (cdf)—A function that shows on the *y*-axis the probability of a value being less than or equal to the corresponding value on the *x*-axis.

Current Dollars—Dollars of nonuniform purchasing power, including general price inflation or deflation, in which actual prices are stated. (With zero inflation or deflation, current dollars are identical to constant dollars.)

Decision Analysis—A technique for making economic decisions in an uncertain environment that allows a decision maker to include alternative outcomes, risk attitudes, and subjective impressions about uncertain events in an evaluation of investments. Decision analysis typically uses decision trees to represent decision problems.

Decision Tree—A decision flow diagram (in the shape of a tree) that is used to represent all possible outcomes, costs, and probabilities associated with a given decision problem. The tree serves as a road map to clarify possible alternatives and outcomes of sequential decisions.

Demand Charge—That portion of the charge for electric service based on the plant and equipment costs associated with supplying the electricity consumed.

Depreciation—The process of systematically recognizing the using up of an asset through wear, obsolescence, and the like, and of subtracting capital expenses from the revenue that the asset generates over time in computing income tax liability.

Depreciation Recapture Tax—A tax to collect ordinary income taxes previously avoided by deducting depreciation expenses from taxable income.

Differential Cost—The difference in the costs of an alternative with a higher investment cost and the base case.

Differential Price Escalation Rate—The difference between the projected rate of price increase assumed for a given item, such as energy, and the projected general rate of inflation.

Discounting—A technique for converting future cash flows to time-equivalent amounts as of a common time so that they can be correctly combined and compared.

Discount Factor—A multiplicative number that is used to compute the time-equivalent value of a cost or benefit occurring at a specified time. Discount factors are calculated from a discount formula for a given discount rate and time of occurrence.

Discount Formula—An expression of a mathematical relationship which enables the conversion of costs and benefits occurring at a given time to time-equivalent values at some other time.

Discount Rate—The minimum acceptable rate of return used in discounting benefits and costs occurring at different times to a common time. Discount rates reflect the investor's time value of money (or opportunity cost). Real discount rates reflect time value apart from changes in the purchasing power of the dollar (i.e., inflation or deflation) and are used to discount constant dollar cash flows. Market discount rates include changes in the purchasing power of the dollar and are used to discount current dollar cash flows.

Discounted Payback Period—The time required for cumulative savings from an investment to pay back the investment and other accrued costs, taking into account the time value of money.

Discrete Cash Flow—Cash flows discretely spaced within the periods they occur, in contrast to continuous cash flow.

Discretionary Project—A project that from a noneconomic standpoint is optional, such as adding a waste-heat recovery system.

Economic—economically efficient; cost effective.

Economic Efficiency—The condition reached when an investment maximizes profits or net benefits.

Economic Evaluation Methods—Methods for measuring and comparing the economic performance of alternatives. Methods include life-cycle cost, benefit-to-cost ratio, savings-to-investment ratio, internal rate of return, overall rate of return, discounted payback, and net savings.

Economic Life—The period of time over which a building or building system is considered to be the least-cost alternative for satisfying a particular need.

Effective Annual Interest Rate—the actual rate, taking into account the frequency of compounding.

Excise Tax—A tax on the manufacture, sale, or consumption of specified commodities.

Expected Value Analysis—(also known as "probability analysis"). One technique for estimating uncertain input data. The expected value of an input is the sum of the products of possible input values and their respective probabilities of occurrence.

Financial Leverage—The extent to which borrowing is used to control assets, usually measured as the ratio of total debt to total assets or of total debt to equity.

First Cost—See Initial Investment Costs.

Fixed Baseline—A single alternative against which all other alternatives are compared in designing or sizing a project.

Fixed Cost—Cost that does not vary with the level of output.

Functional Use Costs—The costs of labor and services required to carry out the intended functions of the building.

Future Value (Future Worth)—The time-equivalent value of past, present, or future cash flows expressed as of some time in the future.

Incremental Benefit-to-Cost (Savings-to-Investment) Ratio—A ratio computed by dividing the additional benefits (additional savings) by the additional investment costs associated with an expansion in investment.

Inflation—A rise in the general price level. Inflation can also be described as a decline in the general purchasing power of a currency.

Initial Investment Costs—Investment costs incurred at the beginning of the study period.

Internal Rate of Return (IRR)—The compound rate of interest that, when used to discount a project's cash flows, will equate costs and benefits.

Investment Costs—The costs associated with acquiring an asset, including such items as design, engineering, construction, purchase, and installation.

Investment Portfolio—A package of investments owned or managed as a unit.

Life-Cycle Cost (LCC)—The sum of all discounted costs of acquiring, owning, operating, and maintaining a building project over the study period.

Life-Cycle Costing Method—A method of economic evaluation that sums the discounted dollar costs of the initial investment (less the salvage value), replacements, operations (including energy usage), and maintenance and repair. Comparing life-cycle costs among mutually exclusive projects of equal performance is one way of determining relative cost effectiveness.

Mandatory Project—A project which is essential to the required performance of the building, such as replacing a deteriorated roof.

Mathematical/Analytical (M/A) Technique—A technique of obtaining probability functions for economic measures of project worth without the repeated trials of simulation.

Maintenance and Repair Cost—The total labor, material, and other related costs incurred in performing corrective and preventive maintenance and repair of a building, or of its system and components, or of both.

Market Discount Rate (Nominal Discount Rate)—The rate of interest reflecting the time value of money stemming from both inflation and the real earning power of money over time.

Mean (μ)—A statistical measure of central tendency. Specifically, the arithmetic mean is the sum of individual values for a group of items divided by the number of items in the group. It may also be referred to as the "average" value in a distribution.

Mean-Variance Criterion—A technique for evaluating the relative risk and return when choosing among competing projects. The mean-variance criterion is that the project with the higher mean (i.e., expected value of project worth) and lower standard deviation be chosen. This presumes that decision makers prefer less risk to more risk. It does not signal a clear choice when the project has both a higher mean *and* a higher standard deviation.

Methods of Analysis—The various ways in which project cash flows can be analyzed to measure project economic performance.

Minimum Acceptable Rate of Return (MARR)—The minimum percentage return required for an investment to be economically acceptable.

Modified Uniform Present Value Factor (UPV* and UPV)**—UPV* indicates a discount factor used to convert an annual amount escalating at a constant rate to a time-equivalent present value. UPV** indicates a discount factor from a special set published by the U.S. Government for computing present value energy costs based on variable energy price projections.

Mutually Exclusive—A condition whereby acceptance of one alternative precludes acceptance of others. Double glazing for a window is, for example, a mutually exclusive alternative to single glazing.

Net Benefits (NB) (Net Savings, NS)—The difference between benefits (or savings) and costs, where both are discounted to present or annual values. The NB and NS methods are used to measure project worth.

Net Present Value Method (Net Present Worth Method)—Net benefits or net savings method.

Nonrecurring Costs—Costs that are not incurred annually over the study period.

Nonfuel Operation and Maintenance Costs—Labor and material costs required for routine upkeep, repair, and operation, exclusive of energy costs.

Nominal Discount Rate—see Market Discount Rate.

Nonmutually Exclusive— A condition whereby the acceptance of one alternative does not preclude the acceptance of the others. Examples are wall insulation and ceiling insulation. (For contrast, see Mutually Exclusive.)

Operating Cost—Expenses incurred during the normal operation of a building, building system, or component, including labor, materials, utilities, and other related costs.

Opportunity Cost of Capital—The rate of return on the next best available use of funds of comparable risk.

Optimization—Economic optimization is the process by which economic analysis is applied to determine the most economically efficient or cost-effective choice among alternatives.

Overall Rate of Return (ORR) (also known as the Adjusted Internal Rate of Return, AIRR)—Annual yield from a project over the study period, taking into account reinvestment of interim receipts.

Overall (Total) Life-Cycle Costs—A comprehensive measure of life-cycle costs for a building or facility required to compare life-cycle costs of the alternative sets of candidate projects for budget allocation. Overall life-cycle costs must include costs affected by any of the projects within the sets being considered and must be consistently measured across sets.

Owning and Operating Costs—The combined costs of purchasing, installing, operating, maintaining, repairing, and replacing assets.

Payback Method—A technique of economic evaluation that determines the time required for the cumulative benefits from an investment to recover the investment cost and other accrued costs.

Payback Period—The time it takes for an investment's cumulative benefits or savings from a project to pay back the investment and other accrued costs.

Portfolio Analysis—A technique used to seek the combination of assets with the maximum return for any given degree of risk (i.e., variance of the return) or the minimum risk for any given rate of return.

Prerequisite System—A system which is required for another system to provide its intended function.

Present Value (Present Worth)—The time-equivalent value of past, present or future cash flows as of the beginning of the base year.

Present Value (Present Worth) Factor—A discount factor by which a future dollar amount may be multiplied to find its time-equivalent value as of the beginning of the base year. Single present value factors are used to convert single future amounts to present values. Uniform present value factors are used to convert annually recurring amounts to present values.

Probability Density Function—The derivative of a continuous, cumulative distribution function. The area under the probability density function must equal 1.0.

Producer Price Index (PPI)—A series of price indices for items purchased by producers prepared by the U.S. Bureau of Labor Statistics. Prices in a selected, base year are given an index value of 100. Subsequent increases above 100 indicate price increases.

Property (Ad Valorem) Tax—a tax levied on the appraised value of property.

Quoted Annual Interest Rate—The simple annual interest rate quoted without regard to compounding frequency.

Real Discount Rate—The rate of interest reflecting the portion of the time value of money attributable to the real earning power of money over time and not to general price inflation.

Real Rate of Return—A rate of return exclusive of the rate of general price inflation or deflation.

Reinvestment Rate—The assumed periodic rate of return on reinvested receipts from a project over the remainder of the study period.

Relevant Benefits—Positive cash flows such as revenues which are affected by an investment choice and, hence, influence the economic decision.

Relevant Costs—Negative cash flows such as maintenance costs which are affected by an investment choice and, hence, influence the economic decision.

Replacement Costs—Future costs, included in the capital budget, to replace a building system or a component during the study period.

Replacement Schedule—A planned frequency for replacing systems, components, or equipment.

Resale Value—The monetary sum expected from the sale of an asset.

Residual Value—The monetary value recaptured through resale, reuse, or scrap or the unused value remaining at the end of the study period when the asset continues in use.

Retrofit—The modification of an existing building or facility to include new systems or components.

Risk-Adjusted Discount Rate (RADR)—A discount rate that has been adjusted to account for risk. When using the RADR technique, projects with anticipated high variability in distributions of worth have their returns discounted at higher rates than projects with low variability.

Risk Analysis—The body of theory and practice that helps decision makers assess their risk exposures and risk attitudes so that the investment that is best for them can be selected. (Note that this definition is restricted to the types of analyses described in this book and is not necessarily consistent with how the term is used in reference to analyses in such areas as the environment or health.)

Risk Attitude—The willingness of decision makers to take chances or to gamble on investments of uncertain outcome. Risk attitudes are generally classified as risk averse, risk neutral, or risk taking. Risk-averse decision makers prefer a sure cash payment to a risky venture with a known expected value greater than the sure cash payment. Risk-neutral decision makers act on the basis of expected monetary value. They are indifferent between a sure cash payment and a risky venture with an expected value equal to the sure cash payment and therefore accept a fair gamble. Risk takers prefer a risky venture with a known expected value to a sure cash payment equal to the expected value.

Risk Averse (RA)—See Risk Attitude.

Risk Exposure—The probability of investing in a project whose economic outcome is less favorable than what is economically acceptable.

Risk Neutral (RN)—See Risk Attitude.

Risk Taking (RT)—See Risk Attitude.

Salvage Value—The residual value, net of any disposal costs, of any building system removed or replaced during or at the end of the study period.

Savings-to-Investment Ratio (SIR)—Discounted savings divided by discounted investment costs.

Sensitivity Analysis—A technique for measuring the impact on project outcomes of changing one or more key input values about which there is uncertainty.

Shifting Baseline—The alternative against which another alternative is compared in designing or sizing a project is the current best of the previous comparison.

Single Present Value (Worth) Factor (SPV or SPW)—The discount factor used to convert single future benefit and cost amounts to time-equivalent present values.

Simple Payback Period (SPB)—A measure of the length of time required for the cumulative savings from a project to recover the investment cost and other accrued costs, without taking into account the time value of money.

Simulation—A technique used to determine risk exposure from an investment decision by generating a probability distribution of the measure of economic performance. Input values are drawn randomly from probability distributions and are combined to compute a measure of economic performance. A distribution of the output is generated from repeated draws and computations.

Spider Diagram—A graph that presents a snapshot of the potential impact, taking one input at a time, of several uncertain input variables on project outcomes.

Standard Deviation (σ)—The square root of the variance. The standard deviation indicates the extent to which items in a distribution differ from the mean value.

Study Period—The period of time over which benefits and costs of a decision are analyzed.

Substitute Systems—Having more of one system means needing less of another.

Sunk Costs—Costs which have been incurred prior to making a current project decision and which therefore should not be considered in the current evaluation.

Systematic Risk—Nondiversifiable risk in an efficient market.

Tax Credit—A direct reduction in taxes owed, provided to stimulate the purchase of certain properties or services or investment in certain activities.

Tax Write-off—A deduction of expense (usually referring to depreciation) from taxable income.

Terminal Value—The value at the end of the study period (not including investment cost) of net cash flows and their earnings from reinvestment.

Time-Equivalent Values—Values occurring at different times which an investor is equally willing to pay (or receive).

Time Horizon—The time over which an investor is concerned about the benefits and costs of an investment.

Time Value of Money—The time-dependent value of money arising from the real earning potential of an investment over time and changes in the purchasing power of money. If project cash flows are stated in constant dollars, their adjustment to a common time basis is necessary to take into account the *real* earning potential of investments over time. If project cash flows are stated in current dollars, their adjustment to a common time basis is necessary to take into account not only the real earning potential over time but also price inflation or deflation.

Uncertainty—Uncertainty (or certainty), as used in this book, refers to a state of knowledge about the variable inputs to an economic analysis. If the analyst is unsure of input values, there is uncertainty. If the analyst is sure, there is certainty.

Uneconomic—Economically inefficient; cost ineffective.

Uniform Present Value (Worth) (UPV or UPW) Factor—The discount factor used to convert uniform annual values to a time-equivalent present value.

Useful Life—The period of time over which a building or building system continues to generate benefits or savings.

Utility Function—A function that shows how utility (i.e., satisfaction) varies with money or income. The shape of the utility function indicates the decision maker's risk attitude.

Variable Cost—Cost that varies with the level of output.

Variance (σ^2)—A statistical measure of dispersion about the mean. The square root of the variance is the standard deviation.

Weighted Average Cost of Capital—The weighted cost of funds on hand. (see Weighted Cost of Capital.)

Weighted Cost of Capital—The cost of funding to a firm computed by finding the after-tax cost of each source of funding, including equity and debt issue, and applying weights based on the proportion of funds obtained from each source.

Weighted Marginal Cost of Capital—The weighted cost of acquiring additional funds. (see Weighted Cost of Capital.)

Appendix B

Worksheets and Supporting Tables for Computing Life-Cycle Costs, Net Savings, Savings-to-Investment Ratio, and Payback

The following worksheets provide step-by-step instructions for evaluating decisions using the life-cycle cost, net savings, savings-to-investment ratio, and payback methods.[1] They are used to develop "ballpark" estimates of economic performance, not precise answers. The treatment is not intended to provide the in-depth knowledge of tax law needed to comply with IRS requirements. They do, however, accommodate the estimation of income tax credits, an approximation of depreciation write-offs, and deductions of current expenses from taxable income. (Not treated are property taxes, capital gains taxes, and depreciation recapture taxes.) The worksheets are designed to accommodate a variety of problems, and therefore allow for more entries than most project evaluations normally require.

There are three series of worksheets (series A, B, and C). Series A is for calculating present value net savings. Series B is for calculating simple and discounted payback under alternative conditions. Series C is for calculating savings-to-investment ratios.

Refer to Chapters 2, 3, 4, and 7 for descriptions of the methods and guidelines for applying them. Here, the emphasis is on calculations.

The worksheets require inputs from tables B-1 through B-9, which appear at the end of series C, as well as from tables in Appendix G. The tables are referenced in the worksheets as they are required. Tables B-1 through B-5 are indices for computing present value energy savings based on DoE-projected energy savings. Table B-6 contains factors for computing present value of annually recurring amounts, and Table B-7, factors for computing present value of non-annually recurring amounts. Tables B-8 and B-9 contain multipliers for estimating depreciation writeoffs.

[1]This appendix is based in part on Ruegg and Petersen (1987).

SERIES A—NET SAVINGS EVALUATION[a]

Worksheet 1. Project identification and basic assumptions

A **Description:** ______________________________

B **Objective: To determine** ☐ **If a Project is Cost Effective**

☐ **The Cost-Effective Design or Size of a Project**

C **Building Type:** ☐ **Owner-occupied Residence** ☐ **Rental Residence** ☐ **Commercial**

D **Study Period[b]:** _____ **Years.**

Planned Holding Period of the Building: _____ **Years.**

Expected Life of the Project: _____ **Years.**

E Market Discount Rate[c]: 0.___.

---

[a]These worksheets are intended for developing "ballpark" estimates of economic performance, not precise answers. The authors assume no liability resulting from decisions based on the use of any of the Worksheets in Series A, B, and C.

[b]For assistance, see Chapter 10, "Setting the Study Period." Note that the worksheet approach requires the assumptions that investment costs occur as a lump sum at the beginning of the study period and energy and other costs occur as end-of-year cash flows.

[c]For assistance, see Chapter 11, "Selecting a Discount Rate" and Worksheet 2.

SERIES A—NET SAVINGS EVALUATION

Worksheet 2. Guidance in estimating the discount rate[a]

Part I. ***Homeowner's Discount Rate***

A (1) If the project is to be financed, what is the interest rate at which the money can be borrowed? ___ %.

(2) The interest rate (highest) on debt outstanding is ____ %.

(3) The highest rate of taxable return, before taxes, available from other investments of comparable risk is ________ %.

(4) Enter here the highest rate from A(1), A(2), and A(3): ____ %.

B (1) Find in the Federal income tax schedule the maximum rate at which your income is taxed and write this as a decimal fraction: 0.___

(2) Find in your State income tax form (if applicable) the maximum tax rate at which your income is taxed and write this as a decimal fraction: 0.___

(3) Compute a "composite marginal income tax rate." Under tax laws in effect in 1986 this can be done by multiplying the State tax (from B(2)) times the remainder of one minus the Federal tax rate (from B(1)), and adding the product to the Federal tax rate:

0.___	×	(1.00−0.___)	+	0.___	=	0.___
marginal State tax rate		marginal Federal tax rate		marginal Federal tax rate		composite marginal income tax rate

C Multiply the rate from A(4) times the remainder of one minus the composite marginal income tax rate (from B(3)):

___ %	×	(1−0.___)	=	___ %.
Before tax rate from A(4)		Composite marginal tax rate from B(3)		After-tax rate

D Enter the result of C expressed as a decimal fraction: 0.___.

E Enter here the highest rate of *tax-exempt* return available from other investments of comparable risk, expressed as a decimal fraction: 0.___.

F Enter here the higher of the rates in D and E. 0.___.

(Enter this rate on Worksheet 1, line E. This is an estimate of your after-tax, market discount rate.)

[a]The concept is explained and illustrated in Chapter 11.

SERIES A—NET SAVINGS EVALUATION

Worksheet 2. (Continued) Guidance in estimating the discount rate

Part II. ***Firm's Discount Rate***

An approach to determining a proxy for the firm's appropriate discount rate is to compute it as a weighted average of the marginal costs associated with each source of funds to the firm.

A Complete the table below for the Applicable Sources of Funds:

Components of Capital for Investment (1)	Market Values ($ Amounts to be Raised) (2)	Market Weights (Market Values ÷ Total New Capital) (3)	After-Tax Cost of Capital (As Computed in Sec. B next page) (4)	Weighted After-Tax Cost of Capital (5)=(3)×(4)
(a) New Current Liabilities	_____	(a)(2)÷(f)(2)	B(1)	_____
(b) New Long-Term Debt	_____	(b)(2)÷(f)(2)	B(2)	_____
(c) New Preferred Stock	_____	(c)(2)÷(f)(2)	B(3)	_____
(d) New Common Stock	_____	(d)(2)÷(f)(2)	B(4)	_____
(e) Other Sources	_____	(e)(2)÷(f)(2)	B(5)	_____
(f) Total New Capital	_____			
(g) Marginal Cost of Capital, Sum col. (5) (This is a proxy for your after-tax, market discount rate. Enter it in Worksheet 1, line E)				_____

SERIES A—NET SAVINGS EVALUATION

Worksheet 2. (Continued) Guidance in estimating the discount rate

Part II. ***Firm's Discount Rate*** **(Continued)**

B Compute After-Tax Cost of Capital and enter results in Section A, Column (4), of preceding page[a]:

(1) Cost of New Current Liabilities[b]
(Enter in A(a)(4))

$$= \underbrace{0.\_\_\_}_{\text{interest rate to be paid}} \times (1-\underbrace{0.\_\_\_}_{\text{composite marginal income tax rate}^{c}}) = 0.\_\_\_$$

(2) Cost of New Long-Term Debt
(Enter in A(b)(4))

$$= \left[\underbrace{\$\_\_\_}_{\text{yearly interest to be paid}} \times (1-\underbrace{0.\_\_\_}_{\text{composite marginal income tax rate}^{c}})\right] \div \left[\underbrace{\$\_\_\_}_{\text{sale price of debt}} \times (1-\underbrace{0.\_\_\_}_{\text{flotation cost as a fraction of sales price}})\right] = 0.\_\_\_$$

(3) Cost of New Preferred Stock Equity
(Enter in A(c)(4))

$$= \left[\underbrace{\$\_\_\_}_{\text{stated dividend on new issue}}\right] \div \left[\underbrace{\$\_\_\_}_{\text{sale price of new preferred issue}} \times (1-\underbrace{0.\_\_\_}_{\text{flotation cost as a fraction of sales price}})\right] = 0.\_\_\_$$

(4) Cost of New Common Stock Equity[d]
(Enter in A(d)(4))

$$= \left[\underbrace{\$\_\_\_}_{\text{expected dividend in the next period}} \div \left[\underbrace{\$\_\_\_}_{\text{current market price of stock}} \times (1-\underbrace{0.\_\_\_}_{\text{flotation cost as a fraction of market price}})\right]\right] + \underbrace{0.\_\_\_}_{\text{annual growth rate of dividend}} = 0.\_\_\_$$

---

[a] Depreciation is often omitted from the cost of capital calculation on grounds that the calculated cost of capital *is* the cost of depreciation; hence separate treatment adds nothing.

[b] Assumes no flotation or service charge.

[c] Computed by multiplying the State tax rate times the remainder of one minus the Federal tax rate, and adding the product to the Federal tax rate, as shown on Worksheet 2, Part I "Homeowner's Discount Rate," B(3).

[d] Note that the value of common stock is assumed to reflect retained earnings and a separate calculation for retained earnings is not made. As an alternative to the estimation approach shown here, the cost of common-stock-equity capital is sometimes estimated according to the Capital Asset Pricing Model. (See Clark, et al., 1984, pp. 267–273.)

SERIES A—NET SAVINGS EVALUATION

Worksheet 3. Initial costs

		With the Project (1)	Without the Project[a] (Base Case) (2)	Differential Costs Attributable to the Project (3)=(1)−(2)
A	**Land**	$______	$______	
B	**Structure**	$______	$______	
C	**Equipment**	$______	$______	
D	**Other**	$______	$______	
E	**Initial Costs[b]** ($E=A+B+C+D$) [Proceed to Worksheet 4 if there are *no* tax credits or depreciation writeoffs to consider. If there *are* credits and/or depreciation writeoffs, proceed with E^1, E^2, or E^3, whichever is applicable.]	$______	$______	$______
E^1	**Initial Costs Adjusted for Tax Credits—*Owner-Occupied Residential*** ($E^1=E-F_j$) [First complete Section F, then E^1]	$______	$______	$______
E^2	**Initial Costs Adjusted for Tax Credits and Depreciation—*Rental Residential*** ($E^2=E-(F_j+Hd)$) [First complete Sections F&H, then E^2]	$______	$______	$______
E^3	**Initial Costs Adjusted for Tax Credits and Depreciation—*Nonresidential*** ($E^3=E-(G_j+Hd+Id)$) [First complete Sections G, H, & I, then E^3]	$______	$______	$______

[a] This col. is for specifying a base line against which to compare the project. For example, to evaluate the cost effectiveness of a new energy system to be retrofitted into an existing building, any initial capital costs associated with keeping the existing system would be needed for comparison with the new system. These might include current resale, reuse, or scrap value, net of disposal costs, plus any up-front renovation or repair costs necessary to keep the existing system in operation. To evaluate the cost effectiveness of an energy conserving design for a new building, the initial costs of a conventional building design would be needed for comparison.

[b] No construction scheduling is allowed in the Worksheet Approach; all acquisition costs are lumped together as of the beginning of the first year.

SERIES A—NET SAVINGS EVALUATION

Worksheet 3. (Continued) Adjustment of initial costs for tax credits Sec. F. *Residential Tax Credits*

F Residential Tax Credits[a]

	With the Project (1)	Without the Project (Base Case) (2)
Energy Conservation Tax Credit		
a Eligible Conservation Expenditure (a=Worksheet 3, eligible portion of B, C, and D)	$______	$______
b Percentage Credit Stated as a Decimal Fraction	$______	$______
c Amount of Energy Conservation Credit (c=a×b, not to exceed maximum allowable amount)	$______	$______
d SPV Factor for year 1[b] ______ [Go to Table G-2 and find the SPV factor for year 1 in the column for the discount rate closest to the rate specified in Worksheet 1, E.][c]		
e Present Value of Tax Savings from Energy Conservation Credit (e=c×d)	$______	$______
Renewable Energy Tax Credit		
f Eligible Renewable Energy Expenditure (f=Worksheet 3, eligible portion of B, C, and D)	$______	$______
g Percentage Credit Stated as a Decimal Fraction	$______	$______
h Amount of Renewable Energy Credit	$______	$______

[a]To supply the necessary data for tax-related calculations, the user requires familiarity with IRS rules.
[b]For an explanation of SPV factors, see Chapter 8.
[c]If greater accuracy is desired, interpolate in the table or calculate the factor directly.

SERIES A—NET SAVINGS EVALUATION

Worksheet 3. (Continued) Adjustment of initial costs for tax credits Sec. F. *Residential Tax Credits*

	With the Project (1)	Without the Project (Base Case) (2)
i Present Value of Tax Savings from Renewable Energy Credit (i=h×d)	$______	$______
j Present Value of Total Tax Savings from Residential Tax Credit (j=e+i)	$______	$______

[*If the building is owner-occupied residential,* subtract the amounts in j, cols. (1) and (2), from the amounts in Worksheet 3, E, cols. (1) and (2), and enter the results in Worksheet 3, E^1, cols. (1) and (2).

***If the building is rental residential* and**

- **if there are *no* depreciation expenses to be taken, subtract the amounts in j, cols. (1) and (2), from the amounts in Worksheet 3, E, cols. (1) and (2), and enter the results in Worksheet 3, E^2, cols. (1) and (2); or**
- **if there *are* depreciation expenses, do not yet subtract j from Worksheet 3, E; instead proceed to Section H.]**

SERIES A—NET SAVINGS EVALUATION

Worksheet 3. (Continued) Adjustment of initial costs for tax credits
Sec. G. *Nonresidential (Business) Tax Credits*

G Nonresidential (Business) Tax Credits

	With the Project (1)	Without the Project (Base Case) (2)
Investment Tax Credit		
a Qualifying Investment Expenditures (a=Worksheet 3, qualifying part of B, C, and D, less any deductions for old property traded in, adjusted for 1st year expensing if applicable, and taking into account the used property limitation and the "safe-harbor," at-risk rule (if applicable))	$______	$______
b Present Credit Stated as a Decimal Fraction	$______	$______
c Amount of Investment Credit (c=a×b, not to exceed maximum allowable amount in a given year)	$______	$______
d SPV Factor for year 1[a] ______ [Go to Table G-2 and find the SPV factor for year 1 in the column for discount rate closest to the rate you specified in Worksheet 1, E.][b]		
e Present Value of Tax Savings from Investment Tax Credit[c] (e=c×d)	$______	$______
Energy Tax Credit		
f Qualifying Energy Expenditures (f=Worksheet 3, qualifying part of B, C, and D)		
g Percentage Credit Stated as a Decimal Fraction	$______	$______
h Amount of Energy Credit (h=f×g)	$______	$______

[a] For explanation of SPV factor, see Chapter 8.

[b] If greater accuracy is desired, interpolate in the table or calculate the factor directly.

[c] If tax credits are carried back or forward, those amounts should be multiplied by the SPV for each year in which they are to be applied to tax liability.

SERIES A—NET SAVINGS EVALUATION

Worksheet 3. (Continued) Adjustment of initial costs for tax credits
Sec. G. *Nonresidential (Business) Tax Credits*

		With the Project (1)	Without the Project (Base Case) (2)
i	**Present Value of Tax Savings from Energy Tax Credit** **(i=h×d)**	$______	$______
j	**Present Value of Total Savings from Tax Credits** **(j=e+i)** **[*If the building is nonresidential* and** **• If there are *no* depreciation expenses to be taken, subtract the amounts in j, cols. (1) and (2), from the amounts in Worksheet 3, E, cols. (1) and (2), and enter the results in Worksheet 3, E^3, cols. (1) and (2);** **• if there *are* depreciation expenses, do not yet subtract j from Worksheet 3, E; instead proceed to Section H.]**	$______	$______

SERIES A—NET SAVINGS EVALUATION

Worksheet 3. (Continued) Adjustment of initial costs for depreciation

Sec. H. *Depreciation of Buildings and Building Components*

H Depreciation of Buildings and Building Components—Applies Only to Residential Rental and Nonresidential Buildings Used for Business
[Estimate depreciation using either of the two methods[a] provided below: 15-Year ACRS Method; S-L Method]

	With the Project (1)	Without the Project (Base Case) (2)
Accelerated Cost Recovery System (ACRS) Method		
a Depreciable Cost (a=Worksheet 3, eligible portion of B, C, and D)	$______	$______
b Multiplier for Computing Building Depreciation [Follow instructions for locating the appropriate multiplier in Table B-8.]	$______	$______
c Present Value of Accumulated Building Depreciation (c=a×b)	$______	$______
d Skip lines e–k and go to line 1, next page		
Straight-Line (S-L Method)		
e Depreciable Costs (e=Worksheet 3, eligible portion of B, C, and D)	$______	$______
f Estimated Salvage Value for Computing Depreciation[b]	$______	$______
g Net Depreciable Value (g=e−f)	$______	$______

[a] Depreciation regulations are part of the Federal income tax laws which are revised periodically. There are various methods of computing depreciation. Two methods are accommodated here for the purpose of estimating an approximate measure of the effect of depreciation on taxes paid; in practice you may find it advantageous to choose another method, or you may be required by law to choose another method. Use of the Accelerated Cost Recovery System (ACRS) method will provide a faster writeoff than use of the straight-line (S-L) method.

[b] Under 1986 revisions to the tax code, it is not required that salvage value be deducted from depreciable costs in order to compute net depreciable value, but that is the traditional way it was done. Under tax provisions in effect as of the publication date of the book, "f" would generally be set equal to zero.

SERIES A—NET SAVINGS EVALUATION

Worksheet 3. (Continued) Adjustment of initial costs for depreciation

Sec. H. *Depreciation of Buildings and Building Components*

		With the Project (1)	Without the Project (Base Case) (2)
h	**Depreciable Life** (Estimated useful life for computing depreciation; may be specified in tax regulations)	$______	$______
i	**Annual Depreciation Allowance** (i=g÷h)	$______	$______
j	UPV Multiplier for Computing Present Value [Go to Table G-3 and find the UPV factor for the discount rate (as given in Worksheet 1, E) and for the period corresponding to either the depreciable life or the study period, depending on whether (1) or (2) below is applicable: (1) Use the depreciable life (h above) if depreciable life < study period (as given in Worksheet 1, D); or (2) Use the study period if depreciable life > study period.]	$______	$______
k	**Present Value of Accumulated Building Depreciation** (k=i×j)	$______	$______
l	**Income Tax Savings from Building Depreciation**	$______	$______

If ACRS Method used,

l=c × composite tax rate (as computed in Worksheet 2, Part I, B(3))

If S-L Method used,

l=k × composite tax rate (as computed in Worksheet 2, Part I, B(3))

[***If the building is residential rental property***, and

- if there are ***no*** tax credits (section F), subtract the amounts in *l*, cols. (1) and (2), from the amounts in Worksheet 3, E, cols. (1) and (2), and enter the results in Worksheet 3, E^2, cols. (1) and (2);
- if there ***are*** tax credits (section F), add the amounts in *l*, cols. (1) and (2), to the amounts in Fj, cols. (1) and (2); subtract the totals from Worksheet 3, E, cols. (1) and (2), and enter the results in Worksheet 3, E^2, cols. (1) and (2).

If the building is nonresidential business property, and

- if there are ***no*** tax credits (section G) or ***no*** separate depreciation of equipment (section I), subtract the above amounts in *l*, cols. (1) and (2), from the amounts in Worksheet 3, E, cols. (1) and (2), and enter the results in Worksheet 3, E^3, cols. (1) and (2);
- if there ***are*** tax credits (section G) and/or separate depreciation of equipment (section I), add the above amounts in *l*, cols. (1) and (2), to the amounts in Gj and Id, cols. (1) and (2), subtract the totals from Worksheet 3, E, cols. (1) and (2); and enter the results in Worksheet 3, E^3, cols. (1) and (2).]

SERIES A—NET SAVINGS EVALUATION

Worksheet 3. (Continued) Adjustment of initial costs for depreciation
Sec. I. *Depreciation of Business Equipment*

I Depreciation of Business Equipment—Applies Only to Residential Rental and Nonresidential Buildings Used for Business [Estimate depreciation using either of the two methods provided below: 5-Year ACRS Method; S-L Method]

	With the Project (1)	Without the Project (Base Case) (2)
Accelerated Cost Recovery System (ACRS) Method		
a **Depreciation Costs** **(a=Worksheet 3, eligible portion of C and D)**	$_______	$_______
b Multiplier for Computing Equipment Depreciation [Follow instructions for locating the appropriate multiplier in Table B-9.]	$_______	$_______
c **Present Value of Accumulated Equipment Depreciation** **(c=a×b)**	$_______	$_______
d **Skip lines e–k and go to line *l*, next page**		
Straight-Line (S-L) Method		
e **Depreciable Costs** **(e=Worksheet 3, eligible portion of C and D)**	$_______	$_______
f **Estimated Salvage Value for Computing Depreciation**	$_______	$_______
g **Net Depreciable Value** **(g=e−f)**	$_______	$_______
h **Depreciable Life** **(Estimated useful life for computing depreciation; may be specified in tax regulations)**	$_______	$_______
i **Annual Depreciation Allowance** **(i=g÷h)**	$_______	$_______

SERIES A—NET SAVINGS EVALUATION

Worksheet 3. (Continued) Adjustment of initial costs for depreciation
Sec. I. *Depreciation of Business Equipment*

	With the Project (1)	Without the Project (Base Case) (2)
j UPV Multiplier for Computing Present Value [Go to Table G-3 and find the UPV factor for the discount rate (as given in Worksheet 1, E) and for the period corresponding to either the depreciable life or the study period depending on whether (1) or (2) below is applicable: (1) Use the depreciation life (h above) if depreciation life < study period (as given in Worksheet 1, D); or (2) Use the study period if depreciable life > study period.]	$______	$______
k Present Value of Accumulated Equipment Depreciation ($k=i\times j$)	$______	$______
l **Income Tax Savings from Equipment Depreciation** ***If ACRS Method used,*** **$l=c\times$ composite tax rate (as compared in Worksheet 2, B(3))** ***If S-L Method used,*** **$l=k\times$ composite tax rate (as computed in Worksheet 2, B(3))**	$______	$______

[*If the building is nonresidential*, and
- **if there are *no* tax credits (section G), or depreciation of real property (section H), subtract the above amounts in *l*, cols. (1) and (2), from the amounts in Worksheet 3, E, cols. (1) and (2), and enter the results in Worksheet 3, E^3, cols. (1) and (2);**
- **if there *are* tax credits (section G) and/or depreciation of real property (section H), add the above amounts in *l*, cols. (1) and (2), to the amounts in Gj and Hd, cols. (1) and (2); subtract the totals from Worksheet 3, E, cols. (1) and (2); and enter the results in Worksheet 3, E^3, cols. (1) and (2).]**

Worksheet 4. Energy costs & savings

Choose one of the following 5 Approaches to Estimating the Present Value of Energy Costs Over the Study Period:

- ☐ **Approach 1—Estimate future energy costs on the basis of U.S. Department of Energy Forecasts, and assume the annual quantity purchased to remain constant.**
- ☐ **Approach 2—Estimate future energy costs on the basis of U.S. Department of Energy Forecasts, and allow for a changing annual quantity to be purchased.**
- ☐ **Approach 3—Estimate future energy costs on the basis of some other forecast of future yearly energy costs and assume the annual quantity purchased to remain constant.**
- ☐ **Approach 4—Estimate future energy costs on the basis of some other forecast of future yearly costs and allow for a changing annual quantity to be purchased.**
- ☐ **Approach 5—Estimate future energy costs on the basis of a projected constant rate of change in annual energy costs (due either to price or quantity change or a combination thereof) over the study period.**

Complete the sections below applicable to the estimation approach chosen:

Section A. **Initial Cost of Energy—Applies to Approaches 1 through 5.**
For Approaches 1, 2, and 5 **complete lines a through e below:**
For Approaches 3 and 4**, complete only item b below:**

	With the Project (1)		**Without the Project (Base Case) (2)**		**Annual Savings Attributable to the Project (3)=(2)−(1)**	
	Energy Type 1 ( )	**Energy Type 2 ( )**	**Energy Type 1 ( )**	**Energy Type 2 ( )**	**Energy Type 1 ( )**	**Energy Type 2 ( )**
a Today's Price per Unit of Purchased Energy	$____	$____	$____	$____	$____	$____
b Annual Quantity to be purchased (In same units as price per unit is stated in (a) above)	____	____	____	____	____	____
c Cost of Annual Energy Purchased at Today's Price (c=a×b)	$____	$____	$____	$____		
d Other Annual Energy Charges at Today's Prices, e.g., Demand Charge	$____	$____	$____	$____		
e Total Annual Energy Cost at Today's Prices (e=c+d)	$____	$____	$____	$____	$____	$____

SERIES A—NET SAVINGS EVALUATION

Worksheet 4. Energy costs & savings (continued)

Section B**. Future Energy Savings—Applies to Approaches 1, 2, 3, and 4.**

For Approaches 1 and 2: Go to Tables B-1 through B-5 and follow instructions at the beginning of the tables to derive estimates of future dollar savings for each type of energy and enter below for each year of the study period:

For Approaches 3 and 4: Energy Price Forecasts provided by ____________. [Note: These forecasted prices should include estimated general price inflation if you are basing the discount rate on a market rate of interest as provided for in Worksheet 2.] Multiply the forecasted price for each year of your study period times the reduction in the annual quantity of energy from Section A(b)(3) of Worksheet 4, adjust for any demand or other charges not directly proportioned to quantity, and enter the results below:

Year Ending:

19 __

SAVINGS—ENERGY TYPE 1**:** __________

__ __

SAVINGS—ENERGY TYPE 2**:** __________

__ __

Section C**. Quantity Adjustment—Applies to Approaches 2 and 4.**

Enter a multiplier below for each year of the study period for which you wish to change the estimated quantity of energy saved from the initial estimate. Calculate the multiplier as the ratio of the projected quantity of energy used in the future year to the quantity used in the first year. (No entry is assumed equal to 1.)

Quantity Adjustment Multiplier—ENERGY TYPE 1**:** __________

__ __

Quantity Adjustment Multiplier—ENERGY TYPE 2**:** __________

__ __

SERIES A—NET SAVINGS EVALUATION

Worksheet 4. Energy costs & savings (Continued)

Section D SPV Discount Factors—Applies to *Approaches 1, 2, 3, and 4.*

Go to Table G-2 and find the column for the discount rate closest to the rate you specified in Worksheet 1, E. Enter an SPV discount factor corresponding to each year of the study period:

Year Ending:

19__ __

__ __

Section E **Present Value of Each Year's Energy Costs—Applies to Approaches 1, 2, 3, and 4.**

For Approaches 1 & 3, multiply each amount in Section B times the corresponding SPV discount factor in Section D and enter the results below for each year:

For Approaches 2 & 4, multiply each amount in Section B times the corresponding quantity multiplier in Section C and the product times the corresponding SPV discount factor in Section D and enter the results below for each year:

Year:

19__ __

Present Value Savings—ENERGY TYPE 1: ____________

$__ $__

Present Value Savings—ENERGY TYPE 2: ____________

$__ $__

Section F **Present Value Total Energy Savings Over the Study Period—Applies to Approaches 1, 2, 3, and 4.**

a Present Value Energy Savings for Each Type of Energy:

(Sum across years for each energy type in Section E above)	**ENERGY SAVINGS TYPE 1:** ____ **(1)**	**ENERGY SAVINGS TYPE 2:** ____ **(2)**
	$______	**$**______

b Present Value Energy Savings—*Owner-Occupied Residential:* **$**______
(b=a(1)+a(2))

c Present Value Energy Savings Adjusted for Income Tax Effects—*Rental Residential or Nonresidential*[a]: $ ______
(c=b×(1−Composite Marginal Income Tax Rate[b]))

[a] It is assumed that the after-tax value of energy savings to the building owner will be the same whether the owner or the tenant pays the energy costs. (This ignores the effect on consumption of who pays.) If the owner pays, the energy costs are tax deductible; hence, after-tax energy savings are reduced accordingly. If the tenant pays the energy costs, it is assumed that rents are reduced by a corresponding amount from what they would be if the owner paid these costs; hence a reduction in energy costs is assumed to translate into an equal increase in taxable rental income to the owner. With this assumption, the value to the owner can be estimated according to line c in either case.

[b] As computed in Worksheet 2, B(3).

SERIES A—NET SAVINGS EVALUATION

Worksheet 4. Energy costs & savings (continued)

Section G. **Present Value of Energy Savings Over the Study Period—Applies Only to Approach 5.**

		ENERGY TYPE 1 (1)	ENERGY TYPE 2 (2)
a	**Average Annual Compound Rate of Change in Energy Costs Forecasted Over the Study Period[a]:**	0.___	0.___
b	**Difference Between the Discount Rate and the Projected Annual Rate of Change in Energy Costs: (Gb=(Worksheet 1, E)−(Worksheet 4, Ga))**	0.___	0.___
c	Present Value Multiplier [Follow instructions for finding the appropriate multiplier for energy savings in Table B-6. The difference computed in Gb above is used to locate the multiplier.][b]	___	___
d	**Present Value Energy Savings Over the Study Period (Gd=(Worksheet 4, Ae(3) Energy Type)×(Worksheet 4, Gc))**	$___	$___

e **Present Value Energy Savings—*Owner-Occupied Residential*: $_____**
(Ge=Gd(1)+Gd(2))

f Present Value Energy Savings Adjusted for Income Tax Effects—*Rental Residential or Nonresidential*[c]: $_____
(Gf=Ge×(1−Composite Marginal Income Tax Rate[d]))

[a] To specify this rate, it is useful to start by estimating the average annual rate of general price change (inflation or deflation) which you expect to occur over the study period, and then consider the rate of change in energy prices relative to this general rate of change in prices. If you expect energy prices to change at about the same rate as most other prices, then enter your forecasted annual rate of change in general prices in Ga. If you expect energy prices to rise faster than most other prices, adjust the forecasted rate of change in general prices upward to reflect the excess price rise and enter the adjusted rate in Ga. If you expect energy prices to rise slower than most other prices, adjust the rate of change in general prices downward by the amount of the difference and enter the adjusted rate in Ga.

[b] Alternatively, if the annual rate of change falls within the range 1% to 10% find the present value multiplier in tables G-5–G-24.

[c] See footnote a, preceding page for assumptions.

[d] As computed in Worksheet 2, B(3).

SERIES A—NET SAVINGS EVALUATION

Worksheet 5. Annually recurring non-energy costs & savings (e.g., routine maintenance)

	With the Project (1)	Without the Project (Base Case) (2)	Savings(+) or Extra Costs(−) Attributable to the Project (3)=(2)−(1)
A Annual Costs **(Priced as though incurred at the beginning of the study period)**	$______	$______	$______

B Average Annual Compound Rate of Change in Annually Recurring Non-Energy Costs Forecasted Over the Study Period[a]: 0.____

C Difference Between the Discount Rate and the Projected Rate of Change in Annually Recurring Non-Energy Costs: 0.____
(C=Worksheet 1, E−Worksheet 5, B)

D Present Value Multiplier ________
[Follow instructions for finding the appropriate multiplier for annually recurring non-energy amounts in Table B-6. The difference computed in C above is used to locate the multiplier.][b]

E Present Value Annually Recurring Non-Energy Savings(+) or Extra Costs(−)—*Owner-Occupied Residential:* $______
(E=A(3)×D)

F Present Value Annually Recurring Non-Energy Savings(+) or Extra Costs(−) After Income Tax Effects—*Rental Residential or Nonresidential*[c]: $______
(F=E×(1−Composite Marginal Income Tax Rate[d]))

[a]To specify this rate (as in the case of energy in Worksheet 4, Ga), it is useful to start by estimating the rate of change in general prices (inflation or deflation). If you forecast annual recurring non-energy costs to change at about the same rate as prices in general, enter the rate of general inflation and deflation in B. If you expect annually recurring non-energy costs to increase faster than prices in general, adjust the forecasted rate of change in general prices upward and enter the adjusted rate in B. If you expect these costs to increase slower than prices in general, adjust the foreecasted rate of change in general prices downward and enter the adjusted rate in B.

[b]Alternatively, if the annual rate of change falls within the range 1% to 10%, find the present value multiplier in tables G-5–G-24.

[c]It is assumed that the after-tax value of annually recurring, non-energy savings or extra costs to the building owner will be the same whether the owner or the tenant pays these costs. (See the discussion in footnote a to Worksheet 4, F.)

[d]As computed in Worksheet 2, B(3).

SERIES A—NET SAVINGS EVALUATION

Worksheet 6. Non-annually recurring costs & savings (e.g., repairs and replacements)

A Projected Future Expenditures (Priced as though incurred at the beginning of the study period) **Year[a]:**	19____	____	____	____	____	____
a ***Without the Project***	$____	$____	$____	$____	$____	$____
b ***With the Project***	$____	$____	$____	$____	$____	$____
c **Savings(+) or Extra Costs(−) Attributable to the Project (c=a−b)**	$____	$____	$____	$____	$____	$____

B Average Annual Compound Rate of Change in Non-Annually Recurring Costs Forecasted Over the Study Period[b]: 0.____

C Difference Between the Discount Rate and the Projected Rate of Change in these Costs: 0.____
(C=Worksheet 1, E−Worksheet 6, B)

D Present Value Multipliers [Follow instructions for finding the appropriate multiplier in Table B-7 for each year in which savings or extra costs occur and enter them here. The difference computed in C above is used to locate the multiplier.][c]	____	____	____	____	____	____
E Present Value of Each Future Year's Amount of Savings(+) or Extra Costs(−) (E=Ac (by year)×D)	$____	$____	$____	$____	$____	$____

F Present Value Non-Annually Recurring Savings(+) or Extra Costs(−) Over the Study Period—*Owner-Occupied Residential*: $____
(sum E across)

G Present Value Non-Annually Recurring Savings(+) or Extra Costs(−) After Income Tax Effects—*Rental Residential or Nonresidential*[d]: $____
(G=F×(1−Composite Marginal Income Tax Rate[e]))

[a] If you wish to assume beginning-of-year, instead of end-of-year payments, enter the amount under the year preceding; e.g., enter a replacement cost expected to occur at the beginning of year 5 in year 4, because end-of-year 4 is the same as beginning of year 5.

[b] For assistance in estimating this rate, see Worksheet 5, footnote a.

[c] Alternatively, find the SCA factor in Table G-1 for the average annual compound rate of change, and the SPV factor in Table G-2 for the discount rate, and multiply the two together to find the present value multiplier with slightly greater accuracy.

[d] It is assumed that the after-tax value of non-annually recurring savings or extra costs to the building owner will be the same whether the owner or the tenant pays these costs. (See the discussion in footnote a to Worksheet 4, F.)

[e] As computed in Worksheet 2, B(3).

SERIES A—NET SAVINGS EVALUATION

Worksheet 7. Change in resale or scrap value

		With the Project (1)	Without the Project (Base Case) (2)	Increase(+) or Decrease(−) Attributable to the Project (3)=(1)−(2)
A	**Year of Resale or Disposal** ______			
B	**Resale/Scrap Amount** **(Valued as though received at the beginning of the study period)**	$______	$______	
C	**Amount of Selling/Disposal Costs** **(Valued as though incurred at the beginning of the study period)**	$______	$______	
D	**Net Resale/Scrap Proceeds** **(D=B−C)**	$______	$______	$______

E **Average Annual Compound Rate of Change in net resale/scrap value forecasted over the study period[a]: 0.____**

F **Difference between the discount rate and the projected rate of change in net resale/scrap value: 0.____**
(F=Worksheet 1, E−Worksheet 7, E)

G Present Value Multiplier ______
[Follow instructions for finding the appropriate multipliers(s) in Table B-7 for the year in which the resale or disposal occurs (A above) and enter it here. The difference computed in F above is used to locate the multipliers.][b]

H **Present Value Net Resale/Scrap Increase(+) or Decrease(−)**—***Owner-Occupied Residential/Rental Residential or*** *Nonresidential*[c]: $______
(H=D(3)×G)

[a] For assistance in estimating this rate, see Worksheet 5, footnote a.
[b] Alternatively, find the SCA factor in Table G-1 for the average annual compound rate of change, and the SPV factor in Table G-2 for the discount rate, and multiply the two together to find the present value multiplier with slightly greater accuracy.
[c] Capital gains and depreciation recapture taxes are not treated in this Worksheet approach.

SERIES A—NET SAVINGS EVALUATION

Worksheet 8. Total net savings in present value dollars

		Present Value Dollars
A	**Energy Savings** **(From Worksheet 4, F b or Ge for *Owner-Occupied Residential*)** **(From Worksheet 4, F c or Gf for *Rental Residential or Nonresidential*)**	$________
B	**Annually Recurring Non-Energy Savings(+) or Cost Increase(−)** **(From Worksheet 5, E for *Owner-Occupied Residential*)** **(From Worksheet 5, F for *Rental Residential or Nonresidential*)**	$________
C	Non-Annually Recurring Cost Savings(+) or Cost Increase(−) **(From Worksheet 6, F for *Owner-Occupied Residential*)** **(From Worksheet 6, G for *Rental Residential or Nonresidential*)**	$________
D	**Resale or Scrap Value Increase(+) or Decrease(−)** **(From Worksheet 7, H for *all types of buildings*)**	$________
E	**Total Savings** **(E=A+B+C+D, remembering to observe signs designating increases or decreases)**	$________
F	**Initial Costs** **(From Worksheet 3, E or E^1 for *Owner-Occupied Residential*)** **(From Worksheet 3, E^2 for *Rental Residential*)** **(From Worksheet 3, E^3 for *Nonresidential*)**	$________
G	**Net Savings (or Net Losses) of Project[a]** **(G=E−F)**	$________

[a]This completes the net savings evaluations. For guidance in interpreting the results, see Chapter 3.

At this point you may wish to perform sensitivity analysis by reworking the evaluation to see the results based on other data and assumptions. You may also wish to proceed to Series B, Payback Evaluation, and/or Series C, Savings-to-Investment Ratio Evaluation, to compute these supplementary measures.

SERIES B—PAYBACK EVALUATION
(COMPUTED ON A BEFORE-TAX BASIS)

[Use these worksheets to find approximately when the project will recover its costs]

Worksheet 1. Simple and discounted payback—Applies when future annual cash flows are constant

---

A Do you wish to assume that the quantity of energy saved will remain about constant over the life of the project or until the time to payback, whichever comes first? ☐ Yes ☐ No
If Yes, proceed to B; if No, go to Worksheet 3.

B Do you wish to assume that all future prices will remain unchanged over the life of the project or until the time to payback, whichever comes first? ☐ Yes ☐ No
If Yes, proceed to C; if No, go to Worksheet 3.

C Do you expect any significant future costs or savings that will not recur annually? ☐ Yes ☐ No
If No, proceed to D; if Yes, go to Worksheet 3.

Differential Costs or Savings Attributable to the Project

D Initial Costs (Note: No allowance is made for construction scheduling in the Worksheet Approach. All acquisition costs are lumped together as of the beginning of the first year.) $________

E Total Annual Energy Savings at Today's Prices (Series A-Worksheet 4, Ae(3), summation across energy types) $________

F Amount of Annually Recurring Non-Energy Extra Costs(−) or Savings(+) at Today's Prices (Series A-Worksheet 5, A(3)) $________

G Simple Payback (SPB) (G=D÷(E+F)) ________ years

H Discounted Payback (DPB) ________ years
[First specify a discount rate[a]: ________ %. Then go to Table G-3 and find the column for the specified discount rate. Search down that column for the UPV factor closest in value to the simple payback found in line G above. In the left hand column "Period," read off the corresponding number and enter it here.[b]]

---

[a] For assistance, see Chapter 11 and Worksheet 2 of Series A.
[b] Interpolation may be used to obtain a more precise estimate of DPB. For guidance in interpreting the results, see Chapter 7.

SERIES B—PAYBACK EVALUATION

Worksheet 2. Discounted payback/graphical solution—Applies when future cash flows are uniform or change at a constant rate
[To use this Worksheet, it is necessary first to find simple payback which may be done by completing the preceding Worksheet through line G.]

A **Simple Payback (SPB):** ________ **years**
(From Series B, Worksheet 1, G)

B **Discount Rate (d)[a]:** ________ **%**

C **Projected Average Annual Compound Rate of Change in Future Costs (e):** ________ **%**
[Note: Only a single average compound rate of change may be specified for all components of future cash flow.]

D **"k" Value** ________
[Go to Table C-1 and follow instructions for locating the "k" value based on the value of d given above in line B and the value of e given above in line B and the value of e given above in line C. Enter that "k" value here.]

E Appendix C Figure ________
[Go to Appendix C and follow instructions for locating the appropriate figure among figures C-1–C-4 and identify that figure by number here.]

F Discounted Payback (DPB) ________ years[b]
[Follow instructions for graphically determining DPB from SPB using the figure identified in E above and the "k" value in D above.]

---

[a]For assistance, see Chapter 11 and Worksheet 2 of Series A.
[b]For guidance in interpreting the results, see Chapter 7.

SERIES B—PAYBACK EVALUATION

Worksheet 3. Discounted payback—Applies when future cash flows are irregular. **[It is *not* necessary first to complete a preceding Worksheet to use this Worksheet.]**

A Discount Rate[a]: 0.____

B Complete the following Table:

Year (t)	Initial Investment Cost[b] (1)	Cumulative Present Value O&M Costs (From col. 6 of Section C following) (2)	Cumulative Present Value Repair and Replacement Costs (From col. 7 of Section D following) (3)	Cumulative Present Value Energy Savings (From col. 8 of Section E following) (4)	Cumulative Net Savings (5)=(4)−(1)−(2)−(3)	
0	$____	$0____	$0____	$0____	$____	
1		____	____	____	____	
2		____	____	____	____	
3		____	____	____	____	
4		____	____	____	____	
5		____	____	____	____	
6		____	____	____	____	
7		____	____	____	____	
8		____	____	____	____	Note: When this amount changes from negative to positive, discounted payback is indicated[c].
9		____	____	____	____	
10		____	____	____	____	
11		____	____	____	____	
12		____	____	____	____	
13		____	____	____	____	
14		____	____	____	____	
15		____	____	____	____	

[a] For assistance, see Chapter 11 and Worksheet 2 of Series A.

[b] Note: No allowance is made for construction scheduling in the Worksheet Approach. All acquisition costs are lumped together as of the beginning of the first year.

[c] For guidance in interpreting the results, see Chapter 7.

SERIES B—PAYBACK EVALUATION

Worksheet 3. (Continued) Discounted payback—Applies when future cash flows are irregular

C Calculation of Present Value Cumulative O&M Costs

	Today's Cost of O&M			Cumulative Present Value O&M Costs		Cumulative Increase(+) or Decrease(−) in O&M Costs Attributable to the Project
Year (t)	With the Project (1)	Without the Project (2)	Present Value Cumulative Multiplier[a] (3)	With the Project (4)=(1)×(3)	Without the Project (5)=(2)×(3)	(6)=(4)−(5)
0	$____	$____				
1			____	$____	$____	$____
2			____	____	____	____
3			____	____	____	____
4			____	____	____	____
5			____	____	____	____
6			____	____	____	____
7			____	____	____	____
8			____	____	____	____
9			____	____	____	____
10			____	____	____	____
11			____	____	____	____
12			____	____	____	____
13			____	____	____	____
14			____	____	____	____
15			____	____	____	____

[a] **To Find Present Value Cumulative Multipliers:**

- **First specify the projected average annual rate of change in O&M costs: 0.____.**
  (To specify this rate it is useful to start by estimating the rate of change in prices in general and to use that rate if you expect O&M costs to change about the same as the overall rate of price inflation or deflation. Adjust the rate upward or downward if you expect O&M costs to change at a faster or slower rate, respectively.)
- **Subtract the projected rate of change, specified above, from the discount rate specified in Series B, Worksheet 3, A and enter the difference here: 0.____.**
- Now go to Table B-6, find the column that corresponds to the difference calculated in the preceding step. Enter in this Worksheet, col. (3) above, a cumulative multiplier from the appropriate col. of Table B-6 for each year within the period over which payback is expected to occur.

SERIES B—PAYBACK EVALUATION

Worksheet 3. (Continued) Discounted payback—Applies when future cash flows are irregular

D Calculation of Present Value Repair and Replacement Costs

	Today's Cost of Repair and Replacement			Yearly Present Value Repair and Replacement Cost			
Year (t)	With the Project (1)	Without the Project (2)	Present Value Multiplier[a] (3)	With the Project (4)=(1)×(3)	Without the Project (5)=(2)×(3)	Change Attributable to the Project (6)=(4)−(5)	Cumulative Increase(+) or Decrease(−) Repair & Replacement Costs Attributable to the Project (7)=sum col. (6)
0	$____	$____	1	$____	$____	$____	____ (year 0, col. 6)
1	____	____	____	____	____	____	$____ (year 0+1, col. 6)
2	____	____	____	____	____	____	____ (years 0+1+2, col. 6)
3	____	____	____	____	____	____	____ (years 0+1+2+3, col. 6)
4	____	____	____	____	____	____	____
5	____	____	____	____	____	____	____
6	____	____	____	____	____	____	____
7	____	____	____	____	____	____	____
8	____	____	____	____	____	____	____
9	____	____	____	____	____	____	____
10	____	____	____	____	____	____	____
11	____	____	____	____	____	____	____
12	____	____	____	____	____	____	____
13	____	____	____	____	____	____	____
14	____	____	____	____	____	____	____
15	____	____	____	____	____	____	____

[a] Present Value Multiplier:

* First specify the projected average annual rate of change in repair and replacement costs: 0.____.
* Subtract the projected rate of change from the discount rate specified in Series B, Worksheet 3, A, and enter the difference here: 0.____.
* Now go to Table B-7, find the column that corresponds to the difference calculated in the preceding step. Enter in Worksheet 3, D(3) a multiplier from that column for each year in which there is an estimated repair or replacement cost.

SERIES B—PAYBACK EVALUATION

Worksheet 3. (Continued) Discounted payback—Applies when future cash flows are *not* uniform

E Calculation of Present Value Energy Savings

Year (t)	Yearly Quantity of Energy: With the Project (1)	Yearly Quantity of Energy: Without the Project (2)	Price of Energy[a]: With the Project (3)	Price of Energy[a]: Without the Project (4)	Yearly Energy Savings (5)=(2)×(4)−(1)×(3)	SPV[b] (6)	Present Value of Yearly Energy Savings (7)=(5)×(6)	Cumulative Present Value of Energy Savings (8)=sum col. (7)
0			$____	$____				
1	____	____	____	____	$____	____	$____	$____ (Year 1, col. 7)
2	____	____	____	____	____	____	____	____ (Years 1+2, col. 7)
3	____	____	____	____	____	____	____	____ (Years 1+2+3, col. 7)
4	____	____	____	____	____	____	____	____
5	____	____	____	____	____	____	____	____
6	____	____	____	____	____	____	____	____
7	____	____	____	____	____	____	____	____
8	____	____	____	____	____	____	____	____
9	____	____	____	____	____	____	____	____
10	____	____	____	____	____	____	____	____
11	____	____	____	____	____	____	____	____
12	____	____	____	____	____	____	____	____
13	____	____	____	____	____	____	____	____
14	____	____	____	____	____	____	____	____
15	____	____	____	____	____	____	____	____

[a] Energy Prices:

* For "Today's Price," use the price of energy in effect at the time of the analysis.
* To use U.S. Department of Energy Price Forecasts, go to Tables B-1 through B-5, and follow instructions for finding the set of price multipliers which applies most closely to your project. Multiply today's price (cols. (3) and (4), row 0 above) times the price multiplier from the appropriate Appendix B table for each year to obtain estimates of future prices and enter the results in cols. (3) and (4) for each future year.
* To use your own forecasted rates of change in energy prices, multiply the price of energy in the preceding year times (1 + forecasted rate of change) beginning with today's price. For example, if today's price is, say, $10 per unit and your projected rate of change over year 1 is 5 percent, the first year's price is entered as $10.50 per unit (i.e., $10 × 1.05). If the second year's projected rate of change is 7 percent, the price for year 2 will be entered as $11.03 (i.e., $10.50×1.07).

[b] Go to Appendix G, Table G-2 and find the col. of SPV multipliers for the discount rate specified above in Series B, Worksheet 3, A. Enter in col. (6) the SPV multiplier corresponding to each year.

SERIES C—SAVINGS-TO-INVESTMENT RATIO (SIR) EVALUATION

[Use this Worksheet to assist in developing project priorities when the budget is insufficient to fund all cost-effective projects]

A **Denominator—Initial Costs: $________**
(Bring forward from Series A, Worksheet 8, F)

B **Numerator—Total Savings: $________**
(Bring forward from Series A, Worksheet 8, E)

C **SIR Ratio[a]: ________**
(C=B÷A)

Note: Amounts required to complete this Worksheet are from Series A, Worksheet 8.
[a]For guidance in interpreting the results, see Chapter 4.

Supporting Tables to Use with the Worksheets

Tables B-1
through B-5: INDICES (MULTIPLIERS) FOR COMPUTING PRESENT VALUE ENERGY SAVINGS ACCORDING TO APPROACHES 1 AND 2 LISTED IN WORKSHEET 4 OF SERIES A

APPROACHES 1 and 2—use **Tables B-1** through **B-5**

[Note: These tables are updated in the fall of each year and published by the National Institute of Standards and Technology (Lippiatt and Ruegg, 1988). You should obtain and use updated tables. The tables given here are provided only to instruct in the method of using them.

INSTRUCTIONS:

* Find the table that applies to your geographical region as indicated by the table heading, e.g., Table B-1 applies to Census Region 1 which consists of Maine, New Hampshire, Vermont, Massachusetts, Connecticut, Rhode Island, New York, New Jersey, and Pennsylvania.

* Find the set of columns for the energy type in question.

* From this set of columns, select the column that most closely reflects your assumption regarding future general price inflation as indicated by the "General Price Inflation Rate" (GPIR). The choices in the sample tables are GPIR = 0, i.e., a stable purchasing power of the dollar; GPIR = 5%, i.e., an average annual inflation rate of 5%; GPIR = 10%, i.e., an average annual inflation rate of 10%; and GPIR = 15%, i.e., an average annual inflation rate of 15%. Interpolate between the columns to obtain indices for intermediate rates of price inflation. Note that the assumption regarding the GPIR should be consistently reflected in all parts of the evaluation, including the discount rate.

* Find the year marking the beginning of your study period in the first column, "year".

* If your study period begins with the first year listed in the table, multiply the annual dollar savings at today's prices from Worksheet 4, Section A(e) (3) times the index for each year of the study period beginning with the next index following 1.00 (assumes end-of-year cash flows), and en-

ter each product in Series A, Worksheet 4, Section B, under the appropriate year.

[Example: If your study period begins in 1988, the first year's savings would be calculated as $A(e)(3) × 1989 multiplier.]

* If your study period begins after the first year listed in the table, it is necessary to derive an estimated savings amount to correspond to the first year index of 1.00. This can be done by dividing the savings from Worksheet 4, Section A(e)(3), for the energy type and project alternative being considered, by the index for the first year of the study period. Then multiply the derived amount times the index for each year of the study period, beginning with the index following the base-year of the study period, and enter each product in Worksheet 4, Section B with the appropriate year.

[Example: If your study period begins in 1991, and the table begins in 1988, divide the 1991 dollar amount from A(e)(3) by the 1991 index to derive the 1988 amount. Then multiply the derived 1988 amount times the index for 1992 to estimate the first year's savings, multiply the derived 1988 amount times the index for 1993 to derive the second year's savings, etc.]

[Note: The indices in Tables B-1 through B-5 are related to the UPV** factors in Tables G-22 through G-26 as follows: They both reflect DoE's forecast of real changes in energy prices. But the indices are for estimating future energy costs or savings in current dollars, whereas the UPV** factors are for computing in a single step the present value equivalent (based on a 7% real discount rate) of future energy costs or savings estimated in constant dollars. The indices do not accomplish discounting. They must be used in combination with SPV factors to compute present value equivalents of future energy costs or savings.]

Table B-1. PROJECTED AVERAGE FUEL PRICE INDICES WITH ASSUMED GENERAL PRICE INFLATION RATES (GPIR) OF 0, 5, 10, AND 15 PERCENT BY END-USE SECTOR AND MAJOR FUEL

Census Region 1 (Maine, New Hampshire, Vermont, Massachusetts, Connecticut, Rhode Island, New York, New Jersey, Pennsylvania)

	RESIDENTIAL															
	ELECTRICITY (GPIR)				DISTILLATE FUEL (GPIR)				LIQUEFIED PETROLEUM GAS (GPIR)				NATURAL GAS (GPIR)			
YEAR	0%	5%	10%	15%	0%	5%	10%	15%	0%	5%	10%	15%	0%	5%	10%	15%
1988	1.00	1.00	1.00	1.00	1.00	1.00	1.00	1.00	1.00	1.00	1.00	1.00	1.00	1.00	1.00	1.00
1989	1.01	1.06	1.11	1.16	1.05	1.11	1.16	1.21	1.05	1.11	1.16	1.21	1.01	1.06	1.11	1.16
1990	0.99	1.09	1.20	1.31	1.07	1.18	1.29	1.41	1.07	1.17	1.29	1.41	1.02	1.12	1.23	1.35
1991	0.99	1.14	1.31	1.50	1.09	1.26	1.45	1.65	1.07	1.24	1.43	1.63	1.03	1.19	1.37	1.57
1992	0.98	1.19	1.43	1.71	1.11	1.35	1.62	1.94	1.08	1.31	1.58	1.89	1.04	1.27	1.53	1.82
1993	0.97	1.24	1.56	1.95	1.14	1.45	1.83	2.29	1.10	1.41	1.77	2.22	1.06	1.36	1.71	2.14
1994	0.97	1.30	1.71	2.24	1.18	1.58	2.09	2.73	1.14	1.52	2.01	2.63	1.09	1.46	1.93	2.52
1995	0.97	1.37	1.89	2.58	1.22	1.72	2.38	3.25	1.18	1.66	2.29	3.13	1.12	1.58	2.19	2.99
1996	0.98	1.45	2.10	2.99	1.28	1.89	2.74	3.91	1.23	1.82	2.64	3.76	1.15	1.70	2.47	3.52
1997	0.99	1.53	2.33	3.47	1.33	2.06	3.14	4.68	1.28	1.98	3.01	4.49	1.19	1.85	2.81	4.19
1998	1.00	1.63	2.59	4.04	1.38	2.25	3.59	5.60	1.32	2.15	3.42	5.33	1.23	2.00	3.18	4.96
1999	1.01	1.74	2.90	4.72	1.43	2.45	4.08	6.65	1.35	2.31	3.85	6.28	1.26	2.15	3.58	5.84
2000	1.02	1.82	3.19	5.43	1.47	2.63	4.60	7.84	1.37	2.46	4.30	7.33	1.28	2.30	4.02	6.86
2001	1.02	1.92	3.52	6.27	1.49	2.80	5.13	9.14	1.39	2.62	4.80	8.56	1.30	2.45	4.49	8.00
2002	1.02	2.02	3.88	7.23	1.51	2.98	5.72	10.65	1.41	2.79	5.35	9.96	1.32	2.61	5.00	9.31
2003	1.02	2.13	4.28	8.33	1.52	3.17	6.37	12.40	1.43	2.97	5.96	11.61	1.33	2.77	5.57	10.84
2004	1.03	2.24	4.72	9.61	1.55	3.38	7.12	14.49	1.45	3.16	6.65	13.55	1.35	2.95	6.22	12.66
2005	1.03	2.36	5.20	11.08	1.57	3.59	7.92	16.87	1.47	3.36	7.41	15.78	1.37	3.14	6.92	14.74
2006	1.03	2.48	5.74	12.77	1.58	3.79	8.76	19.50	1.48	3.55	8.20	18.26	1.38	3.32	7.66	17.06
2007	1.03	2.61	6.32	14.72	1.59	4.01	9.70	22.57	1.48	3.75	9.07	21.12	1.39	3.50	8.48	19.73
2008	1.04	2.75	6.97	16.96	1.60	4.23	10.74	26.12	1.49	3.96	10.04	24.43	1.39	3.70	9.38	22.83
2009	1.04	2.89	7.69	19.55	1.61	4.49	11.92	30.31	1.51	4.20	11.15	28.35	1.41	3.92	10.41	26.48
2010	1.04	3.05	8.47	22.53	1.62	4.74	13.18	35.04	1.52	4.43	12.34	32.80	1.42	4.14	11.52	30.64
2011	1.04	3.20	9.34	25.97	1.63	5.00	14.59	40.55	1.52	4.68	13.65	37.95	1.42	4.38	12.76	35.46
2012	1.05	3.37	10.30	29.93	1.64	5.29	16.14	46.92	1.53	4.94	15.10	43.88	1.43	4.62	14.11	41.00
2013	1.05	3.55	11.36	34.51	1.65	5.60	17.91	54.42	1.55	5.24	16.76	50.92	1.45	4.90	15.66	47.59

Source: Lippiatt and Ruegg (1988).

Table B-1, continued. PROJECTED AVERAGE FUEL PRICE INDICES WITH ASSUMED GENERAL PRICE INFLATION RATES (GPIR) OF 0, 5, 10, AND 15 PERCENT BY END-USE SECTOR AND MAJOR FUEL

Census Region 1 (Maine, New Hampshire, Vermont, Massachusetts, Connecticut, Rhode Island, New York, New Jersey, Pennsylvania)

	COMMERICAL																			
	ELECTRICITY (GPIR)				DISTILLATE FUEL (GPIR)				RESIDUAL FUEL (GPIR)				NATURAL GAS (GPIR)				STEAM COAL (GPIR)			
YEAR	0%	5%	10%	15%	0%	5%	10%	15%	0%	5%	10%	15%	0%	5%	10%	15%	0%	5%	10%	15%
1988	1.00	1.00	1.00	1.00	1.00	1.00	1.00	1.00	1.00	1.00	1.00	1.00	1.00	1.00	1.00	1.00	1.00	1.00	1.00	1.00
1989	1.01	1.06	1.11	1.17	1.08	1.13	1.18	1.24	1.05	1.10	1.15	1.20	1.01	1.06	1.11	1.16	1.01	1.06	1.11	1.16
1990	0.99	1.09	1.20	1.31	1.10	1.21	1.33	1.45	1.04	1.15	1.26	1.38	1.01	1.12	1.23	1.34	1.03	1.14	1.25	1.36
1991	0.99	1.14	1.31	1.50	1.13	1.30	1.50	1.71	1.05	1.22	1.40	1.60	1.02	1.18	1.36	1.55	1.05	1.22	1.40	1.60
1992	0.97	1.19	1.43	1.71	1.15	1.40	1.69	2.02	1.05	1.28	1.54	1.85	1.03	1.25	1.50	1.80	1.07	1.30	1.57	1.87
1993	0.97	1.24	1.56	1.95	1.20	1.53	1.93	2.41	1.11	1.41	1.78	2.23	1.04	1.33	1.68	2.10	1.09	1.39	1.75	2.19
1994	0.97	1.29	1.71	2.23	1.26	1.68	2.22	2.90	1.19	1.59	2.10	2.75	1.07	1.43	1.89	2.47	1.11	1.48	1.96	2.56
1995	0.97	1.36	1.89	2.58	1.32	1.85	2.57	3.51	1.29	1.81	2.51	3.43	1.10	1.55	2.14	2.92	1.13	1.58	2.19	2.99
1996	0.98	1.44	2.09	2.99	1.39	2.06	2.99	4.27	1.41	2.08	3.02	4.31	1.13	1.66	2.41	3.44	1.14	1.68	2.44	3.48
1997	0.99	1.53	2.33	3.47	1.47	2.28	3.47	5.17	1.53	2.38	3.61	5.39	1.16	1.81	2.74	4.09	1.16	1.80	2.73	4.07
1998	1.00	1.63	2.59	4.04	1.54	2.52	4.01	6.25	1.63	2.66	4.23	6.59	1.20	1.96	3.11	4.86	1.17	1.91	3.03	4.73
1999	1.02	1.74	2.90	4.73	1.61	2.75	4.59	7.49	1.70	2.91	4.85	7.91	1.23	2.10	3.50	5.71	1.19	2.04	3.41	5.56
2000	1.02	1.83	3.19	5.44	1.66	2.98	5.20	8.87	1.74	3.12	5.45	9.29	1.25	2.25	3.92	6.69	1.21	2.18	3.81	6.49
2001	1.02	1.92	3.52	6.27	1.68	3.17	5.81	10.35	1.76	3.32	6.08	10.84	1.27	2.39	4.38	7.81	1.23	2.31	4.23	7.55
2002	1.02	2.02	3.88	7.23	1.70	3.37	6.47	12.06	1.78	3.53	6.77	12.62	1.29	2.55	4.88	9.10	1.25	2.47	4.73	8.81
2003	1.03	2.13	4.28	8.34	1.72	3.59	7.20	14.03	1.81	3.75	7.54	14.69	1.30	2.70	5.43	10.58	1.26	2.63	5.28	10.29
2004	1.03	2.24	4.72	9.62	1.75	3.83	8.06	16.40	1.84	4.01	8.43	17.17	1.32	2.88	6.07	12.36	1.28	2.79	5.87	11.95
2005	1.03	2.36	5.21	11.09	1.77	4.06	8.96	19.08	1.86	4.26	9.39	19.99	1.34	3.06	6.76	14.39	1.30	2.97	6.55	13.94
2006	1.03	2.49	5.74	12.78	1.79	4.30	9.93	22.10	1.87	4.50	10.39	23.12	1.35	3.24	7.48	16.66	1.31	3.16	7.31	16.27
2007	1.03	2.62	6.33	14.73	1.80	4.54	10.98	25.55	1.88	4.75	11.49	26.74	1.35	3.42	8.28	19.26	1.33	3.37	8.15	18.98
2008	1.04	2.75	6.98	16.97	1.81	4.79	12.15	29.55	1.89	5.02	12.72	30.93	1.36	3.62	9.17	22.30	1.35	3.59	9.10	22.13
2009	1.04	2.90	7.69	19.57	1.82	5.08	13.48	34.29	1.91	5.32	14.12	35.92	1.37	3.83	10.17	25.86	1.37	3.82	10.15	25.81
2010	1.04	3.05	8.48	22.55	1.83	5.36	14.93	39.70	1.92	5.61	15.62	41.55	1.38	4.05	11.26	29.94	1.39	4.07	11.31	30.08
2011	1.04	3.21	9.35	25.99	1.84	5.66	16.51	45.90	1.93	5.93	17.29	48.05	1.39	4.27	12.45	34.61	1.41	4.33	12.61	35.07
2012	1.05	3.38	10.31	29.96	1.85	5.98	18.26	53.08	1.94	6.26	19.12	55.57	1.40	4.51	13.77	40.01	1.43	4.60	14.06	40.87
2013	1.05	3.55	11.37	34.54	1.87	6.33	20.27	61.57	1.96	6.64	21.23	64.51	1.41	4.78	15.28	46.44	1.45	4.90	15.67	47.62

Table B-2. PROJECTED AVERAGE FUEL PRICE INDICES WITH ASSUMED GENERAL PRICE INFLATION RATES (GPIR) OF 0, 5, 10, AND 15 PERCENT BY END-USE SECTOR AND MAJOR FUEL

Census Region 2 (Ohio, Indiana, Illinois, Michigan, Wisconsin, Minnesota, Iowa, Missouri, North Dakota, South Dakota, Nebraska, Kansas)

	RESIDENTIAL															
	ELECTRICITY (GPIR)				DISTILLATE FUEL (GPIR)				LIQUEFIED PETROLEUM GAS (GPIR)				NATURAL GAS (GPIR)			
YEAR	0%	5%	10%	15%	0%	5%	10%	15%	0%	5%	10%	15%	0%	5%	10%	15%
1988	1.00	1.00	1.00	1.00	1.00	1.00	1.00	1.00	1.00	1.00	1.00	1.00	1.00	1.00	1.00	1.00
1989	0.99	1.04	1.09	1.14	1.06	1.11	1.16	1.22	1.07	1.12	1.17	1.23	1.01	1.06	1.11	1.16
1990	0.99	1.09	1.20	1.31	1.07	1.18	1.30	1.42	1.08	1.20	1.31	1.43	1.03	1.13	1.24	1.36
1991	0.98	1.14	1.31	1.50	1.09	1.27	1.46	1.66	1.10	1.27	1.46	1.67	1.04	1.20	1.38	1.58
1992	0.98	1.19	1.43	1.71	1.11	1.35	1.63	1.95	1.10	1.34	1.62	1.93	1.06	1.28	1.55	1.85
1993	0.97	1.24	1.56	1.95	1.15	1.47	1.85	2.31	1.13	1.44	1.82	2.27	1.08	1.38	1.75	2.18
1994	0.96	1.29	1.71	2.23	1.19	1.60	2.11	2.76	1.17	1.57	2.08	2.71	1.12	1.50	1.98	2.59
1995	0.96	1.35	1.87	2.55	1.24	1.74	2.41	3.29	1.23	1.73	2.39	3.26	1.16	1.63	2.26	3.09
1996	0.96	1.41	2.05	2.93	1.30	1.92	2.78	3.97	1.29	1.91	2.77	3.96	1.20	1.77	2.57	3.66
1997	0.96	1.48	2.25	3.36	1.35	2.10	3.19	4.76	1.35	2.10	3.19	4.76	1.25	1.93	2.94	4.39
1998	0.96	1.56	2.48	3.87	1.41	2.30	3.66	5.70	1.41	2.29	3.65	5.69	1.30	2.11	3.36	5.24
1999	0.96	1.64	2.73	4.45	1.46	2.49	4.16	6.79	1.45	2.47	4.13	6.73	1.33	2.28	3.80	6.20
2000	0.97	1.74	3.04	5.19	1.50	2.69	4.70	8.01	1.47	2.65	4.63	7.89	1.36	2.45	4.28	7.30
2001	0.97	1.83	3.36	5.98	1.52	2.86	5.24	9.34	1.50	2.82	5.16	9.20	1.38	2.61	4.78	8.52
2002	0.97	1.93	3.70	6.90	1.54	3.04	5.84	10.87	1.51	3.00	5.75	10.72	1.40	2.77	5.32	9.92
2003	0.98	2.03	4.08	7.95	1.56	3.24	6.50	12.67	1.53	3.19	6.41	12.48	1.42	2.95	5.93	11.55
2004	0.98	2.14	4.50	9.17	1.58	3.45	7.26	14.79	1.56	3.40	7.16	14.57	1.44	3.15	6.63	13.49
2005	0.98	2.25	4.97	10.58	1.60	3.67	8.09	17.23	1.58	3.61	7.97	16.96	1.46	3.35	7.38	15.71
2006	0.98	2.37	5.48	12.19	1.61	3.87	8.95	19.93	1.59	3.82	8.82	19.63	1.47	3.54	8.17	18.18
2007	0.99	2.49	6.04	14.05	1.62	4.09	9.90	23.05	1.60	4.03	9.76	22.72	1.48	3.73	9.03	21.02
2008	0.99	2.62	6.65	16.19	1.63	4.32	10.97	26.68	1.61	4.26	10.80	26.28	1.49	3.94	10.00	24.33
2009	0.99	2.76	7.34	18.67	1.64	4.58	12.16	30.94	1.62	4.51	11.99	30.49	1.50	4.18	11.10	28.23
2010	0.99	2.91	8.09	21.51	1.65	4.84	13.47	35.81	1.63	4.77	13.26	35.26	1.51	4.41	12.27	32.63
2011	1.00	3.06	8.92	24.79	1.66	5.11	14.90	41.41	1.64	5.03	14.67	40.79	1.52	4.66	13.59	37.77
2012	1.00	3.22	9.83	28.58	1.67	5.39	16.48	47.88	1.65	5.32	16.24	47.18	1.53	4.92	15.02	43.66
2013	1.00	3.39	10.84	32.95	1.69	5.72	18.29	55.57	1.66	5.63	18.02	54.75	1.54	5.21	16.67	50.65

Table B2, continued. PROJECTED AVERAGE FUEL PRICE INDICES WITH ASSUMED GENERAL PRICE INFLATION RATES (GPIR) OF 0, 5, 10, AND 15 PERCENT BY END-USE SECTOR AND MAJOR FUEL

Census Region 2 (Ohio, Indiana, Illinois, Michigan, Wisconsin, Minnesota, Iowa, Missouri, North Dakota, South Dakota, Nebraska, Kansas)

	COMMERCIAL																			
	ELECTRICITY (GPIR)				DISTILLATE FUEL (GPIR)				RESIDUAL FUEL (GPIR)				NATURAL GAS (GPIR)				STEAM COAL (GPIR)			
YEAR	0%	5%	10%	15%	0%	5%	10%	15%	0%	5%	10%	15%	0%	5%	10%	15%	0%	5%	10%	15%
1988	1.00	1.00	1.00	1.00	1.00	1.00	1.00	1.00	1.00	1.00	1.00	1.00	1.00	1.00	1.00	1.00	1.00	1.00	1.00	1.00
1989	0.99	1.04	1.09	1.14	1.08	1.14	1.19	1.24	1.04	1.09	1.15	1.20	1.02	1.07	1.12	1.17	1.02	1.07	1.12	1.17
1990	0.99	1.09	1.19	1.31	1.10	1.22	1.34	1.46	1.04	1.15	1.26	1.37	1.02	1.12	1.23	1.35	1.04	1.15	1.26	1.37
1991	0.98	1.14	1.31	1.50	1.13	1.31	1.51	1.73	1.05	1.21	1.39	1.59	1.03	1.19	1.36	1.56	1.06	1.22	1.41	1.61
1992	0.97	1.18	1.43	1.70	1.17	1.42	1.71	2.04	1.05	1.27	1.54	1.83	1.03	1.26	1.51	1.81	1.08	1.31	1.58	1.89
1993	0.97	1.23	1.55	1.94	1.21	1.55	1.96	2.44	1.09	1.40	1.76	2.20	1.06	1.35	1.70	2.13	1.10	1.41	1.77	2.22
1994	0.96	1.29	1.70	2.22	1.28	1.71	2.26	2.95	1.17	1.56	2.06	2.70	1.09	1.46	1.93	2.52	1.12	1.50	1.98	2.59
1995	0.96	1.35	1.86	2.55	1.35	1.89	2.62	3.58	1.26	1.77	2.45	3.34	1.13	1.59	2.20	3.00	1.14	1.61	2.22	3.04
1996	0.95	1.41	2.04	2.92	1.43	2.11	3.06	4.37	1.36	2.01	2.92	4.17	1.16	1.71	2.48	3.54	1.16	1.71	2.48	3.54
1997	0.95	1.48	2.24	3.35	1.51	2.34	3.56	5.32	1.47	2.28	3.47	5.17	1.21	1.87	2.85	4.25	1.18	1.82	2.77	4.13
1998	0.95	1.55	2.47	3.85	1.59	2.59	4.13	6.44	1.56	2.54	4.04	6.30	1.25	2.04	3.25	5.08	1.20	1.95	3.11	4.85
1999	0.95	1.63	2.72	4.43	1.66	2.85	4.75	7.74	1.62	2.77	4.62	7.54	1.29	2.20	3.67	5.99	1.22	2.09	3.48	5.68
2000	0.97	1.74	3.03	5.17	1.72	3.09	5.40	9.20	1.65	2.97	5.19	8.84	1.32	2.37	4.13	7.05	1.24	2.23	3.90	6.65
2001	0.97	1.83	3.35	5.97	1.74	3.29	6.02	10.73	1.68	3.17	5.79	10.33	1.34	2.52	4.61	8.22	1.26	2.38	4.35	7.75
2002	0.97	1.92	3.69	6.88	1.77	3.50	6.70	12.49	1.70	3.36	6.45	12.01	1.35	2.68	5.13	9.57	1.28	2.53	4.85	9.03
2003	0.97	2.03	4.07	7.93	1.79	3.72	7.47	14.55	1.72	3.58	7.19	14.00	1.37	2.84	5.71	11.13	1.29	2.69	5.40	10.53
2004	0.98	2.13	4.49	9.15	1.82	3.96	8.35	17.00	1.75	3.81	8.03	16.35	1.39	3.04	6.39	13.02	1.31	2.86	6.02	12.27
2005	0.98	2.25	4.95	10.55	1.84	4.21	9.29	19.78	1.77	4.06	8.94	19.04	1.41	3.23	7.11	15.15	1.33	3.04	6.71	14.29
2006	0.98	2.36	5.46	12.15	1.85	4.45	10.28	22.89	1.78	4.28	9.89	22.02	1.42	3.41	7.88	17.53	1.35	3.25	7.51	16.71
2007	0.98	2.49	6.02	14.01	1.86	4.70	11.38	26.48	1.79	4.52	10.94	25.46	1.43	3.60	8.72	20.29	1.37	3.45	8.36	19.46
2008	0.99	2.62	6.64	16.14	1.87	4.97	12.59	30.63	1.80	4.78	12.12	29.49	1.43	3.80	9.64	23.45	1.38	3.67	9.31	22.65
2009	0.99	2.76	7.32	18.62	1.89	5.26	13.97	35.52	1.82	5.07	13.46	34.22	1.45	4.03	10.71	27.23	1.41	3.92	10.41	26.48
2010	0.99	2.90	8.07	21.45	1.90	5.55	15.45	41.09	1.83	5.35	14.88	39.56	1.45	4.25	11.83	31.47	1.42	4.16	11.59	30.82
2011	0.99	3.05	8.90	24.73	1.91	5.86	17.10	47.53	1.84	5.64	16.45	45.74	1.46	4.49	13.10	36.42	1.45	4.44	12.95	36.00
2012	1.00	3.21	9.81	28.50	1.92	6.20	18.94	55.03	1.85	5.96	18.20	52.88	1.47	4.75	14.50	42.14	1.46	4.72	14.41	41.89
2013	1.00	3.38	10.81	32.86	1.94	6.56	21.00	63.80	1.86	6.31	20.19	61.35	1.48	5.02	16.08	48.85	1.49	5.03	16.10	48.91

Table B-3. PROJECTED AVERAGE FUEL PRICE INDICES WITH ASSUMED GENERAL PRICE INFLATION RATES (GPIR) OF 0, 5, 10, AND 15 PERCENT BY END-USE SECTOR AND MAJOR FUEL

Census Region 3 (Delaware, Maryland, District of Columbia, Virginia, West Virginia, North Carolina, South Carolina, Georgia, Florida, Kentucky, Tennessee, Alabama, Mississippi, Arkansas, Louisiana, Oklahoma, Texas)

	RESIDENTIAL															
	ELECTRICITY (GPIR)				DISTILLATE FUEL (GPIR)				LIQUEFIED PETROLEUM GAS (GPIR)				NATURAL GAS (GPIR)			
YEAR	0%	5%	10%	15%	0%	5%	10%	15%	0%	5%	10%	15%	0%	5%	10%	15%
1988	1.00	1.00	1.00	1.00	1.00	1.00	1.00	1.00	1.00	1.00	1.00	1.00	1.00	1.00	1.00	1.00
1989	0.99	1.04	1.09	1.14	1.05	1.11	1.16	1.21	1.06	1.11	1.17	1.22	1.01	1.06	1.11	1.16
1990	0.98	1.08	1.19	1.30	1.07	1.18	1.29	1.41	1.07	1.18	1.30	1.42	1.02	1.13	1.24	1.35
1991	0.97	1.13	1.30	1.48	1.09	1.26	1.45	1.65	1.08	1.25	1.44	1.65	1.04	1.20	1.38	1.58
1992	0.97	1.17	1.41	1.69	1.11	1.35	1.62	1.94	1.09	1.33	1.60	1.91	1.05	1.28	1.54	1.84
1993	0.96	1.22	1.55	1.93	1.14	1.45	1.83	2.29	1.11	1.42	1.80	2.24	1.08	1.38	1.74	2.17
1994	0.96	1.28	1.70	2.22	1.18	1.58	2.09	2.73	1.15	1.55	2.04	2.67	1.12	1.49	1.98	2.58
1995	0.96	1.36	1.88	2.56	1.22	1.72	2.38	3.25	1.20	1.69	2.34	3.19	1.16	1.63	2.25	3.08
1996	0.97	1.43	2.07	2.95	1.28	1.89	2.74	3.91	1.26	1.86	2.70	3.85	1.19	1.76	2.55	3.64
1997	0.97	1.51	2.29	3.42	1.33	2.06	3.14	4.68	1.31	2.03	3.09	4.61	1.24	1.93	2.93	4.37
1998	0.98	1.59	2.54	3.96	1.38	2.25	3.59	5.60	1.36	2.21	3.52	5.50	1.29	2.10	3.34	5.21
1999	0.98	1.68	2.80	4.57	1.43	2.45	4.08	6.65	1.39	2.38	3.98	6.49	1.32	2.27	3.78	6.16
2000	1.00	1.79	3.13	5.33	1.47	2.63	4.60	7.84	1.42	2.55	4.45	7.59	1.36	2.44	4.26	7.26
2001	1.00	1.88	3.45	6.15	1.49	2.80	5.13	9.14	1.44	2.71	4.96	8.85	1.38	2.59	4.75	8.46
2002	1.00	1.98	3.81	7.09	1.50	2.98	5.71	10.64	1.46	2.88	5.53	10.31	1.39	2.76	5.29	9.85
2003	1.00	2.09	4.20	8.18	1.52	3.17	6.37	12.40	1.48	3.07	6.16	12.00	1.41	2.93	5.89	11.47
2004	1.01	2.20	4.63	9.43	1.55	3.38	7.11	14.48	1.50	3.27	6.89	14.02	1.43	3.13	6.58	13.40
2005	1.01	2.32	5.11	10.88	1.57	3.59	7.92	16.87	1.52	3.48	7.66	16.32	1.45	3.32	7.32	15.59
2006	1.01	2.44	5.63	12.54	1.58	3.79	8.76	19.50	1.53	3.67	8.48	18.88	1.46	3.51	8.11	18.05
2007	1.02	2.56	6.21	14.45	1.59	4.01	9.70	22.57	1.54	3.88	9.39	21.85	1.47	3.71	8.98	20.89
2008	1.02	2.70	6.84	16.65	1.60	4.23	10.74	26.12	1.54	4.10	10.39	25.28	1.48	3.92	9.94	24.18
2009	1.02	2.84	7.55	19.20	1.61	4.49	11.92	30.31	1.56	4.34	11.53	29.33	1.49	4.15	11.03	28.05
2010	1.02	2.99	8.32	22.13	1.62	4.74	13.18	35.04	1.57	4.59	12.76	33.93	1.50	4.38	12.20	32.43
2011	1.02	3.15	9.18	25.51	1.63	5.00	14.59	40.55	1.58	4.84	14.11	39.23	1.51	4.63	13.50	37.53
2012	1.03	3.31	10.11	29.39	1.64	5.29	16.14	46.92	1.59	5.11	15.62	45.38	1.52	4.89	14.94	43.42
2013	1.03	3.49	11.16	33.89	1.65	5.60	17.91	54.42	1.60	5.42	17.33	52.65	1.53	5.18	16.58	50.37

Table B-3, continued. PROJECTED AVERAGE FUEL PRICE INDICES WITH ASSUMED GENERAL PRICE INFLATION RATES (GPIR) OF 0, 5, 10, AND 15 PERCENT BY END-USE SECTOR AND MAJOR FUEL

Census Region 3 (Delaware, Maryland, District of Columbia, Virginia, West Virginia, North Carolina, South Carolina, Georgia, Florida, Kentucky, Tennessee, Alabama, Mississippi, Arkansas, Louisiana, Oklahoma, Texas)

	COMMERCIAL																			
	ELECTRICITY (GPIR)				DISTILLATE FUEL (GPIR)				RESIDUAL FUEL (GPIR)				NATURAL GAS (GPIR)				STEAM COAL (GPIR)			
YEAR	0%	5%	10%	15%	0%	5%	10%	15%	0%	5%	10%	15	0%	5%	10%	15%	0%	5%	10%	15%
1988	1.00	1.00	1.00	1.00	1.00	1.00	1.00	1.00	1.00	1.00	1.00	1.00	1.00	1.00	1.00	1.00	1.00	1.00	1.00	1.00
1989	0.99	1.04	1.09	1.14	1.08	1.13	1.19	1.24	1.05	1.10	1.15	1.21	1.02	1.07	1.12	1.17	1.02	1.07	1.12	1.17
1990	0.98	1.08	1.19	1.30	1.10	1.21	1.33	1.46	1.05	1.16	1.27	1.39	1.02	1.12	1.23	1.35	1.03	1.14	1.25	1.37
1991	0.97	1.13	1.29	1.48	1.13	1.31	1.51	1.72	1.06	1.22	1.41	1.61	1.03	1.19	1.36	1.56	1.06	1.22	1.41	1.61
1992	0.96	1.17	1.41	1.68	1.16	1.41	1.70	2.03	1.06	1.29	1.55	1.86	1.03	1.26	1.51	1.81	1.07	1.30	1.57	1.88
1993	0.96	1.22	1.54	1.92	1.21	1.54	1.95	2.43	1.12	1.42	1.80	2.24	1.06	1.35	1.70	2.13	1.09	1.39	1.75	2.19
1994	0.95	1.28	1.69	2.21	1.27	1.70	2.25	2.93	1.21	1.62	2.14	2.80	1.09	1.46	1.93	2.51	1.11	1.48	1.96	2.56
1995	0.96	1.35	1.87	2.55	1.34	1.88	2.60	3.55	1.33	1.86	2.58	3.52	1.13	1.58	2.19	2.99	1.13	1.59	2.20	3.01
1996	0.96	1.42	2.06	2.94	1.42	2.09	3.04	4.34	1.46	2.15	3.12	4.46	1.16	1.71	2.48	3.53	1.15	1.69	2.46	3.51
1997	0.97	1.50	2.28	3.40	1.50	2.32	3.53	5.27	1.59	2.47	3.76	5.60	1.21	1.87	2.84	4.24	1.15	1.79	2.72	4.06
1998	0.98	1.59	2.53	3.95	1.58	2.57	4.09	6.38	1.70	2.78	4.42	6.89	1.25	2.04	3.25	5.06	1.18	1.92	3.06	4.77
1999	0.98	1.68	2.80	4.56	1.65	2.81	4.69	7.65	1.78	3.05	5.08	8.29	1.28	2.19	3.66	5.97	1.20	2.04	3.41	5.56
2000	1.00	1.79	3.13	5.33	1.70	3.05	5.33	9.09	1.83	3.28	5.73	9.78	1.31	2.35	4.11	7.01	1.21	2.18	3.80	6.48
2001	1.00	1.88	3.45	6.15	1.72	3.25	5.95	10.60	1.85	3.49	6.39	11.39	1.33	2.51	4.60	8.19	1.23	2.31	4.24	7.55
2002	1.00	1.98	3.80	7.09	1.74	3.45	6.62	12.34	1.88	3.72	7.13	13.28	1.35	2.67	5.12	9.53	1.25	2.48	4.75	8.86
2003	1.00	2.09	4.20	8.17	1.77	3.68	7.38	14.38	1.90	3.95	7.94	15.47	1.37	2.84	5.70	11.11	1.27	2.64	5.30	10.32
2004	1.01	2.20	4.63	9.43	1.79	3.92	8.24	16.78	1.93	4.21	8.87	18.06	1.39	3.03	6.37	12.97	1.28	2.80	5.90	12.02
2005	1.01	2.32	5.11	10.87	1.82	4.17	9.19	19.56	1.95	4.48	9.88	21.04	1.40	3.21	7.09	15.09	1.30	2.98	6.57	14.00
2006	1.01	2.44	5.63	12.53	1.83	4.40	10.16	22.62	1.97	4.73	10.94	24.34	1.41	3.40	7.84	17.46	1.32	3.17	7.32	16.30
2007	1.01	2.56	6.20	14.44	1.84	4.65	11.25	26.17	1.98	4.99	12.08	28.11	1.42	3.59	8.69	20.21	1.33	3.37	8.15	18.98
2008	1.02	2.70	6.84	16.64	1.85	4.91	12.44	30.27	1.99	5.27	13.37	32.53	1.43	3.79	9.62	23.39	1.36	3.60	9.13	22.22
2009	1.02	2.84	7.54	19.18	1.87	5.20	13.81	35.13	2.01	5.59	14.86	37.80	1.44	4.01	10.66	27.12	1.37	3.83	10.17	25.86
2010	1.02	2.99	8.31	22.11	1.88	5.49	15.28	40.64	2.02	5.91	16.45	43.73	1.45	4.24	11.80	31.38	1.39	4.07	11.32	30.09
2011	1.02	3.15	9.17	25.49	1.89	5.80	16.91	47.00	2.03	6.23	18.17	50.50	1.46	4.48	13.06	36.31	1.41	4.35	12.67	35.21
2012	1.03	3.31	10.11	29.37	1.90	6.12	18.70	54.36	2.04	6.58	20.10	58.43	1.47	4.73	14.44	41.96	1.43	4.61	14.09	40.96
2013	1.03	3.48	11.15	33.86	1.92	6.49	20.76	63.08	2.06	6.98	22.34	67.87	1.48	5.01	16.03	48.70	1.46	4.93	15.77	47.91

Table B-4. PROJECTED AVERAGE FUEL PRICE INDICES WITH ASSUMED GENERAL PRICE INFLATION RATES (GPIR) OF 0, 5, 10, AND 15 PERCENT BY END-USE SECTOR AND MAJOR FUEL

Census Region 4 (Montana, Idaho, Wyoming, Colorado, New Mexico, Arizona, Utah, Nevada, Washington, Oregon, California, Alaska, Hawaii)

	RESIDENTIAL															
	ELECTRICITY (GPIR)				DISTILLATE FUEL (GPIR)				LIQUEFIED PETROLEUM GAS (GPIR)				NATURAL GAS (GPIR)			
YEAR	0%	5%	10%	15%	0%	5%	10%	15%	0%	5%	10%	15%	0%	5%	10%	15%
1988	1.00	1.00	1.00	1.00	1.00	1.00	1.00	1.00	1.00	1.00	1.00	1.00	1.00	1.00	1.00	1.00
1989	1.00	1.05	1.10	1.15	1.06	1.11	1.16	1.22	1.06	1.11	1.17	1.22	1.01	1.06	1.11	1.16
1990	1.00	1.10	1.21	1.32	1.07	1.18	1.30	1.42	1.07	1.18	1.30	1.42	1.03	1.13	1.24	1.36
1991	1.00	1.15	1.33	1.52	1.10	1.27	1.46	1.67	1.08	1.25	1.44	1.65	1.04	1.20	1.38	1.58
1992	0.99	1.21	1.45	1.74	1.12	1.36	1.63	1.95	1.09	1.33	1.60	1.91	1.06	1.28	1.55	1.85
1993	0.99	1.27	1.60	2.00	1.15	1.47	1.85	2.31	1.11	1.42	1.80	2.24	1.09	1.39	1.75	2.18
1994	1.00	1.35	1.78	2.32	1.19	1.60	2.11	2.76	1.15	1.54	2.04	2.66	1.12	1.50	1.98	2.59
1995	1.01	1.42	1.96	2.68	1.24	1.75	2.42	3.30	1.20	1.69	2.34	3.19	1.16	1.64	2.27	3.09
1996	1.02	1.50	2.18	3.11	1.30	1.92	2.79	3.98	1.26	1.86	2.70	3.85	1.20	1.77	2.57	3.67
1997	1.03	1.60	2.43	3.62	1.35	2.10	3.19	4.77	1.31	2.03	3.09	4.61	1.25	1.94	2.95	4.40
1998	1.05	1.71	2.72	4.25	1.41	2.30	3.66	5.71	1.36	2.21	3.52	5.49	1.30	2.12	3.37	5.26
1999	1.07	1.83	3.05	4.97	1.46	2.50	4.17	6.80	1.39	2.38	3.97	6.48	1.34	2.29	3.81	6.22
2000	1.08	1.93	3.38	5.76	1.50	2.69	4.71	8.02	1.42	2.54	4.45	7.58	1.37	2.46	4.30	7.33
2001	1.08	2.04	3.73	6.64	1.52	2.87	5.25	9.36	1.44	2.71	4.96	8.84	1.39	2.62	4.80	8.55
2002	1.08	2.14	4.11	7.66	1.54	3.05	5.85	10.90	1.46	2.88	5.53	10.30	1.41	2.79	5.34	9.96
2003	1.09	2.26	4.54	8.83	1.56	3.24	6.51	12.69	1.47	3.06	6.15	11.99	1.42	2.96	5.95	11.59
2004	1.09	2.38	5.00	10.19	1.58	3.46	7.28	14.82	1.50	3.27	6.88	14.01	1.45	3.16	6.65	13.53
2005	1.09	2.50	5.52	11.75	1.60	3.68	8.11	17.26	1.52	3.47	7.66	16.31	1.46	3.36	7.40	15.75
2006	1.09	2.63	6.08	13.54	1.61	3.88	8.97	19.96	1.52	3.67	8.48	18.87	1.47	3.55	8.19	18.24
2007	1.10	2.77	6.71	15.60	1.62	4.10	9.93	23.11	1.53	3.88	9.38	21.83	1.48	3.74	9.06	21.08
2008	1.10	2.92	7.39	17.99	1.63	4.33	10.99	26.72	1.54	4.09	10.38	25.24	1.49	3.96	10.03	24.41
2009	1.10	3.07	8.15	20.74	1.65	4.59	12.20	31.02	1.56	4.34	11.52	29.29	1.50	4.19	11.14	28.32
2010	1.10	3.23	8.99	23.91	1.66	4.85	13.49	35.88	1.57	4.58	12.75	33.89	1.51	4.43	12.31	32.74
2011	1.11	3.40	9.91	27.55	1.67	5.12	14.92	41.49	1.58	4.84	14.10	39.20	1.52	4.68	13.63	37.90
2012	1.11	3.58	10.93	31.75	1.68	5.41	16.52	48.02	1.58	5.11	15.61	45.35	1.53	4.94	15.09	43.86
2013	1.11	3.77	12.05	36.62	1.69	5.73	18.32	55.67	1.60	5.41	17.32	52.62	1.54	5.23	16.73	50.83

Table B-4, continued. PROJECTED AVERAGE FUEL PRICE INDICES WITH ASSUMED GENERAL PRICE INFLATION RATES (GPIR) OF 0, 5, 10, AND 15 PERCENT BY END-USE SECTOR AND MAJOR FUEL
Census Region 4 (Montana, Idaho, Wyoming, Colorado, New Mexico, Arizona, Utah, Nevada, Washington, Oregon, California, Alaska, Hawaii)

	COMMERCIAL																			
	ELECTRICITY (GPIR)				DISTILLATE FUEL (GPIR)				RESIDUAL FUEL (GPIR)				NATURAL GAS (GPIR)				STEAM COAL (GPIR)			
YEAR	0%	5%	10%	15%	0%	5%	10%	15%	0%	5%	10%	15%	0%	5%	10%	15%	0%	5%	10%	15%
1988	1.00	1.00	1.00	1.00	1.00	1.00	1.00	1.00	1.00	1.00	1.00	1.00	1.00	1.00	1.00	1.00	1.00	1.00	1.00	1.00
1989	1.00	1.05	1.10	1.15	1.08	1.13	1.19	1.24	1.04	1.09	1.14	1.20	1.01	1.06	1.12	1.17	1.03	1.08	1.13	1.19
1990	1.00	1.10	1.21	1.32	1.10	1.21	1.33	1.46	1.04	1.15	1.26	1.38	1.02	1.12	1.23	1.34	1.06	1.17	1.29	1.41
1991	1.00	1.15	1.33	1.51	1.13	1.31	1.50	1.72	1.05	1.21	1.39	1.59	1.02	1.18	1.36	1.55	1.09	1.27	1.46	1.66
1992	0.99	1.21	1.45	1.74	1.16	1.41	1.70	2.03	1.05	1.28	1.54	1.84	1.03	1.25	1.51	1.80	1.13	1.38	1.66	1.98
1993	0.99	1.27	1.60	2.00	1.21	1.54	1.95	2.43	1.09	1.40	1.76	2.20	1.05	1.34	1.69	2.11	1.16	1.48	1.87	2.34
1994	1.00	1.34	1.78	2.32	1.27	1.70	2.25	2.93	1.17	1.57	2.08	2.71	1.08	1.44	1.91	2.49	1.20	1.61	2.13	2.78
1995	1.01	1.42	1.96	2.68	1.33	1.88	2.60	3.55	1.27	1.78	2.47	3.37	1.11	1.56	2.16	2.95	1.23	1.73	2.40	3.28
1996	1.02	1.50	2.18	3.11	1.42	2.09	3.03	4.33	1.37	2.03	2.94	4.20	1.14	1.68	2.44	3.48	1.26	1.87	2.71	3.87
1997	1.03	1.60	2.43	3.62	1.49	2.32	3.52	5.26	1.48	2.30	3.50	5.22	1.18	1.83	2.78	4.15	1.30	2.01	3.05	4.56
1998	1.05	1.71	2.72	4.24	1.57	2.56	4.08	6.37	1.58	2.57	4.09	6.37	1.22	1.99	3.17	4.94	1.34	2.18	3.47	5.42
1999	1.07	1.82	3.04	4.96	1.64	2.81	4.69	7.64	1.64	2.80	4.68	7.63	1.25	2.14	3.56	5.81	1.38	2.37	3.95	6.44
2000	1.07	1.93	3.37	5.75	1.70	3.05	5.32	9.07	1.68	3.01	5.26	8.97	1.27	2.29	4.00	6.82	1.42	2.55	4.46	7.60
2001	1.08	2.03	3.72	6.63	1.72	3.24	5.94	10.58	1.70	3.21	5.87	10.46	1.29	2.44	4.46	7.95	1.44	2.72	4.97	8.86
2002	1.08	2.14	4.10	7.64	1.74	3.45	6.61	12.32	1.72	3.41	6.53	12.17	1.31	2.59	4.97	9.26	1.47	2.90	5.56	10.37
2003	1.08	2.25	4.52	8.81	1.76	3.67	7.37	14.36	1.74	3.63	7.29	14.19	1.32	2.75	5.53	10.78	1.48	3.09	6.20	12.08
2004	1.09	2.37	4.99	10.17	1.79	3.91	8.23	16.75	1.77	3.87	8.14	16.57	1.35	2.94	6.19	12.60	1.50	3.28	6.91	14.07
2005	1.09	2.50	5.51	11.72	1.81	4.16	9.17	19.52	1.79	4.11	9.05	19.28	1.36	3.12	6.89	14.66	1.52	3.49	7.69	16.38
2006	1.09	2.63	6.07	13.51	1.82	4.39	10.14	22.58	1.80	4.34	10.03	22.33	1.37	3.30	7.62	16.96	1.55	3.72	8.60	19.15
2007	1.09	2.76	6.69	15.57	1.84	4.64	11.22	26.12	1.81	4.59	11.10	25.83	1.38	3.48	8.43	19.62	1.57	3.96	9.58	22.29
2008	1.10	2.91	7.38	17.95	1.85	4.90	12.42	30.21	1.82	4.84	12.28	29.87	1.39	3.68	9.33	22.70	1.58	4.21	10.66	25.94
2009	1.10	3.06	8.13	20.69	1.86	5.19	13.79	35.06	1.84	5.13	13.63	34.66	1.40	3.90	10.35	26.33	1.61	4.49	11.91	30.30
2010	1.10	3.22	8.97	23.85	1.87	5.48	15.25	40.55	1.85	5.42	15.07	40.08	1.41	4.12	11.45	30.45	1.63	4.77	13.26	35.26
2011	1.10	3.39	9.89	27.49	1.88	5.79	16.87	46.90	1.86	5.72	16.67	46.35	1.42	4.35	12.69	35.27	1.65	5.08	14.81	41.17
2012	1.11	3.57	10.90	31.69	1.90	6.11	18.67	54.25	1.88	6.05	18.47	53.68	1.42	4.60	14.03	40.79	1.68	5.42	16.54	48.07
2013	1.11	3.76	12.03	36.55	1.91	6.48	20.72	62.95	1.89	6.40	20.47	62.18	1.44	4.87	15.57	47.30	1.70	5.75	18.40	55.90

Table B-5. PROJECTED AVERAGE FUEL PRICE INDICES WITH ASSUMED GENERAL PRICE INFLATION RATES (GPIR) OF 0, 5, 10, AND 15 PERCENT BY END-USE SECTOR AND MAJOR FUEL
United States Average

	RESIDENTIAL															
	ELECTRICITY (GPIR)				DISTILLATE FUEL (GPIR)				LIQUEFIED PETROLEUM GAS (GPIR)				NATURAL GAS (GPIR)			
YEAR	0%	5%	10%	15%	0%	5%	10%	15%	0%	5%	10%	15%	0%	5%	10%	15%
1988	1.00	1.00	1.00	1.00	1.00	1.00	1.00	1.00	1.00	1.00	1.00	1.00	1.00	1.00	1.00	1.00
1989	1.00	1.05	1.10	1.15	1.06	1.11	1.16	1.21	1.06	1.12	1.17	1.22	1.01	1.06	1.11	1.16
1990	0.99	1.09	1.19	1.31	1.07	1.18	1.29	1.41	1.08	1.19	1.30	1.42	1.02	1.13	1.24	1.35
1991	0.98	1.14	1.31	1.49	1.09	1.26	1.45	1.66	1.09	1.26	1.45	1.65	1.04	1.20	1.38	1.58
1992	0.97	1.18	1.42	1.70	1.11	1.35	1.63	1.94	1.09	1.33	1.60	1.92	1.05	1.28	1.54	1.84
1993	0.97	1.23	1.56	1.94	1.14	1.46	1.84	2.30	1.12	1.43	1.80	2.25	1.08	1.38	1.74	2.17
1994	0.97	1.29	1.71	2.23	1.18	1.59	2.10	2.74	1.16	1.55	2.05	2.68	1.11	1.49	1.97	2.57
1995	0.97	1.36	1.89	2.57	1.23	1.73	2.40	3.27	1.21	1.70	2.35	3.21	1.15	1.62	2.25	3.07
1996	0.97	1.43	2.08	2.97	1.29	1.90	2.76	3.93	1.27	1.87	2.72	3.88	1.19	1.76	2.55	3.64
1997	0.98	1.51	2.30	3.43	1.34	2.08	3.16	4.71	1.32	2.05	3.12	4.65	1.24	1.92	2.92	4.35
1998	0.98	1.60	2.55	3.98	1.39	2.27	3.61	5.64	1.37	2.23	3.56	5.55	1.28	2.09	3.33	5.19
1999	0.99	1.70	2.83	4.61	1.44	2.46	4.11	6.70	1.41	2.41	4.02	6.55	1.32	2.25	3.76	6.13
2000	1.00	1.80	3.15	5.36	1.48	2.65	4.63	7.90	1.43	2.57	4.50	7.67	1.35	2.43	4.24	7.23
2001	1.01	1.90	3.47	6.18	1.50	2.82	5.17	9.21	1.45	2.74	5.02	8.94	1.37	2.58	4.73	8.42
2002	1.01	2.00	3.83	7.13	1.52	3.00	5.76	10.73	1.47	2.91	5.59	10.41	1.39	2.75	5.27	9.82
2003	1.01	2.10	4.22	8.22	1.54	3.19	6.41	12.49	1.49	3.10	6.23	12.13	1.41	2.92	5.87	11.44
2004	1.01	2.21	4.66	9.49	1.56	3.40	7.16	14.59	1.51	3.31	6.96	14.17	1.43	3.12	6.56	13.36
2005	1.02	2.33	5.14	10.94	1.58	3.62	7.98	16.99	1.53	3.51	7.75	16.50	1.44	3.31	7.30	15.54
2006	1.02	2.45	5.66	12.61	1.59	3.82	8.83	19.65	1.54	3.71	8.58	19.09	1.45	3.50	8.08	17.98
2007	1.02	2.58	6.24	14.53	1.60	4.04	9.77	22.74	1.55	3.92	9.49	22.08	1.46	3.69	8.94	20.80
2008	1.02	2.71	6.88	16.74	1.61	4.26	10.81	26.30	1.56	4.14	10.50	25.55	1.47	3.90	9.90	24.07
2009	1.03	2.86	7.59	19.30	1.62	4.52	12.00	30.51	1.57	4.39	11.65	29.64	1.48	4.13	10.98	27.92
2010	1.03	3.01	8.37	22.25	1.63	4.77	13.28	35.31	1.58	4.63	12.90	34.29	1.49	4.37	12.15	32.31
2011	1.03	3.16	9.22	25.64	1.64	5.04	14.69	40.82	1.59	4.89	14.26	39.65	1.50	4.61	13.45	37.38
2012	1.03	3.33	10.17	29.55	1.65	5.32	16.25	47.24	1.60	5.17	15.79	45.88	1.51	4.87	14.88	43.25
2013	1.04	3.50	11.21	34.07	1.66	5.64	18.04	54.80	1.62	5.47	17.51	53.21	1.52	5.16	16.51	50.16

Table B-5, continued. PROJECTED AVERAGE FUEL PRICE INDICES WITH ASSUMED GENERAL PRICE INFLATION RATES (GPIR) OF 0, 5, 10, AND 15 PERCENT BY END-USE SECTOR AND MAJOR FUEL

United States Average

	COMMERCIAL																			
	ELECTRICITY (GPIR)				DISTILLATE FUEL (GPIR)				RESIDUAL FUEL (GPIR)				NATURAL GAS (GPIR)				STEAM COAL (GPIR)			
YEAR	0%	5%	10%	15%	0%	5%	10%	15%	0%	5%	10%	15%	0%	5%	10%	15%	0%	5%	10%	15%
1988	1.00	1.00	1.00	1.00	1.00	1.00	1.00	1.00	1.00	1.00	1.00	1.00	1.00	1.00	1.00	1.00	1.00	1.00	1.00	1.00
1989	1.00	1.05	1.10	1.15	1.08	1.13	1.19	1.24	1.04	1.09	1.15	1.20	1.01	1.06	1.11	1.16	1.02	1.07	1.12	1.17
1990	0.99	1.09	1.19	1.30	1.10	1.21	1.33	1.45	1.04	1.15	1.26	1.38	1.01	1.12	1.23	1.34	1.04	1.14	1.26	1.37
1991	0.98	1.13	1.30	1.49	1.13	1.31	1.50	1.72	1.05	1.21	1.40	1.59	1.02	1.18	1.36	1.55	1.06	1.22	1.41	1.61
1992	0.97	1.18	1.42	1.70	1.16	1.41	1.69	2.02	1.05	1.28	1.54	1.84	1.03	1.25	1.51	1.80	1.08	1.31	1.58	1.88
1993	0.96	1.23	1.55	1.94	1.21	1.54	1.94	2.42	1.10	1.41	1.77	2.21	1.05	1.34	1.69	2.11	1.10	1.41	1.77	2.22
1994	0.96	1.29	1.71	2.23	1.26	1.69	2.24	2.92	1.18	1.59	2.10	2.74	1.08	1.44	1.91	2.49	1.12	1.50	1.98	2.59
1995	0.97	1.36	1.88	2.57	1.33	1.87	2.59	3.53	1.29	1.82	2.52	3.44	1.12	1.57	2.17	2.97	1.14	1.60	2.22	3.03
1996	0.97	1.43	2.08	2.96	1.41	2.08	3.02	4.31	1.41	2.09	3.03	4.32	1.14	1.69	2.45	3.50	1.16	1.71	2.48	3.54
1997	0.97	1.51	2.30	3.43	1.49	2.31	3.51	5.23	1.54	2.38	3.62	5.40	1.19	1.84	2.80	4.18	1.18	1.83	2.78	4.14
1998	0.98	1.60	2.55	3.98	1.57	2.55	4.06	6.33	1.64	2.67	4.25	6.62	1.23	2.01	3.19	4.98	1.20	1.95	3.10	4.84
1999	0.99	1.70	2.83	4.62	1.63	2.79	4.66	7.60	1.71	2.92	4.87	7.95	1.26	2.16	3.60	5.88	1.22	2.09	3.49	5.68
2000	1.00	1.80	3.15	5.36	1.68	3.02	5.29	9.01	1.75	3.14	5.49	9.36	1.29	2.31	4.05	6.90	1.24	2.23	3.89	6.64
2001	1.01	1.90	3.47	6.19	1.71	3.22	5.90	10.51	1.77	3.34	6.12	10.90	1.31	2.46	4.51	8.04	1.26	2.37	4.35	7.75
2002	1.01	2.00	3.83	7.13	1.73	3.43	6.57	12.25	1.79	3.55	6.81	12.69	1.32	2.62	5.03	9.36	1.28	2.53	4.86	9.05
2003	1.01	2.10	4.22	8.23	1.75	3.64	7.32	14.26	1.82	3.78	7.59	14.78	1.34	2.79	5.60	10.91	1.30	2.70	5.42	10.56
2004	1.01	2.21	4.66	9.49	1.78	3.88	8.18	16.65	1.85	4.03	8.48	17.28	1.36	2.97	6.26	12.74	1.31	2.86	6.02	12.26
2005	1.02	2.33	5.14	10.94	1.80	4.13	9.10	19.38	1.87	4.28	9.45	20.11	1.38	3.16	6.97	14.84	1.33	3.05	6.72	14.30
2006	1.02	2.45	5.66	12.61	1.81	4.36	10.07	22.42	1.88	4.52	10.45	23.27	1.39	3.34	7.71	17.17	1.35	3.24	7.50	16.68
2007	1.02	2.58	6.24	14.53	1.82	4.61	11.16	25.96	1.89	4.78	11.57	26.92	1.40	3.53	8.54	19.87	1.37	3.45	8.36	19.46
2008	1.02	2.72	6.89	16.75	1.83	4.87	12.34	30.03	1.90	5.05	12.80	31.14	1.40	3.72	9.44	22.96	1.39	3.68	9.32	22.69
2009	1.03	2.86	7.59	19.31	1.85	5.16	13.70	34.85	1.92	5.35	14.22	36.16	1.42	3.94	10.47	26.64	1.41	3.91	10.40	26.45
2010	1.03	3.01	8.37	22.26	1.86	5.45	15.16	40.31	1.93	5.65	15.73	41.83	1.42	4.17	11.59	30.82	1.42	4.17	11.59	30.82
2011	1.03	3.17	9.23	25.65	1.87	5.75	16.77	46.62	1.94	5.97	17.41	48.38	1.43	4.40	12.83	35.66	1.44	4.43	12.92	35.92
2012	1.03	3.33	10.17	29.56	1.88	6.07	18.55	53.91	1.96	6.31	19.26	55.96	1.44	4.65	14.19	41.25	1.46	4.72	14.40	41.85
2013	1.04	3.51	11.22	34.08	1.90	6.44	20.59	62.56	1.97	6.67	21.34	64.85	1.45	4.92	15.75	47.86	1.49	5.04	16.11	48.96

Table B-6: UPV* MULTIPLIERS FOR COMPUTING PRESENT VALUE ENERGY SAVINGS ACCORDING TO APPROACH 5 LISTED IN WORKSHEET 4 OF SERIES A AND FOR COMPUTING PRESENT VALUE OF OTHER ANNUALLY RECURRING AMOUNTS SUCH AS OPERATING AND MAINTENANCE COSTS IN THE WORKSHEETS

INSTRUCTIONS for finding the UPV* multiplier(s) for computing present value energy savings according to Approach 5 (Series A, Worksheet 4):

* Find the column of Table B-6 which corresponds most closely to the difference between the discount rate and your projected average annual compound rate of change in energy prices as computed in Series A, Worksheet 4, Gb.

* Find the row which corresponds to the length of the study period as given in Series A, Worksheet 1, D.

* At the column/row intersection find the multiplier, and enter that multiplier in Series A, Worksheet 4, Gc.

* Repeat this process for each energy type.

INSTRUCTIONS for finding the multiplier(s) for computing present value of annually recurring non-energy amounts, e.g., operating and maintenance costs or savings:

* Use Table B-6 according to instructions provided above for energy savings; *except*, base the multiplier on the difference between the discount rate and your projected rate of change in the cost of the particular recurring cost in question (rather than on the change in energy prices), and enter the multiplier in Series A, Worksheet 5, D.

  [Note: For completing Series A Worksheets (Net Savings), only one multiplier is needed for each cash flow: the one corresponding to the length of the study period. For completing Series B Worksheet 3 (Payback), a multiplier is needed for each trial year until the payback year is identified.]

  [Note: Table B-6 reports, in a single table, factors that normally take many tables to present. For example, Table B-6 provides most of the factors given by Tables G-5 through G-21, and more. This is done by working off the difference between the discount rate and the projected rate of change, and rounding the results. The factors are rough approximations. If greater accuracy is required, it is recommended that you use Tables G-5 through G-21 instead, or calculate the factor directly from the formula given in Section 8.5.4]

Table B-6. Uniform Present Value Modified (UPV*) factors (cumulative multipliers)—for computing present value of annually recurring amounts (e.g., energy and O&M costs) projected to change over time at a constant rate

Study Period	Discount Rate Minus Projected Rate of Change in Cost of Energy (or Non-Energy Component)												
	−6	−5	−4	−3	−2	−1	0	1	2	3	4	5	6
1	1	1	1	1	1	1	1	1	1	1	1	1	.9
2	2	2	2	2	2	2	2	2	2	2	2	2	2
3	3	3	3	3	3	3	3	3	3	3	3	3	3
4	5	4	4	4	4	4	4	4	4	4	4	4	3
5	6	6	6	5	5	5	5	5	5	5	4	4	4
6	7	7	7	7	6	6	6	6	6	5	5	5	5
7	9	8	8	8	8	7	7	7	7	6	6	6	6
8	10	10	9	9	9	8	8	8	7	7	7	7	7
9	12	11	11	10	10	9	9	9	8	8	8	7	7
10	14	13	12	12	11	11	10	10	9	9	8	8	7
11	16	15	14	13	12	12	11	10	10	9	9	8	8
12	18	16	15	14	14	13	12	11	11	10	10	9	9
13	20	18	17	16	15	14	13	12	11	11	10	10	9
14	22	20	19	17	16	15	14	13	12	11	11	10	9
15	24	22	20	19	17	16	15	14	13	12	11	11	10
16	27	24	22	20	19	17	16	15	14	13	12	11	10
17	29	26	24	22	20	18	17	16	14	13	12	12	11
18	32	28	26	24	21	20	18	17	15	14	13	12	11
19	35	31	28	25	23	21	19	17	16	15	13	12	11
20	38	33	30	27	24	22	20	18	17	15	14	13	12
21	41	36	32	29	26	23	21	19	17	16	14	13	12
22	44	38	34	31	27	24	22	20	18	16	15	14	12
23	48	41	36	32	29	26	23	21	19	17	15	14	13
24	52	44	39	34	30	27	24	21	19	17	16	14	13
25	56	47	41	36	32	28	25	22	20	18	16	15	13
26	60	50	44	38	34	29	26	23	21	18	16	15	13
27	64	54	47	40	35	31	27	24	21	19	17	15	14
28	69	57	49	43	37	32	28	25	22	19	17	15	14
29	74	61	52	45	39	33	29	25	22	20	18	16	14
30	79	65	55	47	40	35	30	26	23	20	18	16	14

Ruegg and Petersen (1987).

Table B-7: SPV* MULTIPLIERS FOR COMPUTING PRESENT VALUE OF NON-ANNUALLY RECURRING AMOUNTS (E.G., REPAIR AND REPLACEMENT COSTS AND RESALE OR SCRAP VALUE) IN THE WORKSHEETS

TABLE B-7

INSTRUCTIONS: Find the column of Table B-7 which corresponds most closely to the difference between the discount rate and your projected rate of change in non-annually recurring amounts such as repair and replacement costs (as computed in Series A, Worksheet 6, C) or in salvage or scrap value, (as computed in Series A, Worksheet 7, F). Within that column, find the multiplier that corresponds to each year for which an amount has been entered on the respective Worksheet, and enter the multiplier(s) on the Worksheet.

[Note: Table B-7 reports, in a single table, factors that normally take many tables to present. The factors combine the estimation of a future amount with the discounting operation. Due to rounding, they are rough approximations. If greater accuracy is required, use Tables G-1 and G-2 in combination.]

Table B-7 Single Present Value Modified (SPV*) factors (multipliers)—for computing present value of non-annually recurring amounts (e.g., repair and replacement costs and resale or scrap value) projected to change over time at a constant rate

	Discount Rate Minus Projected Rate of Change in Non-Annually Recurring Amount												
Year	−6	−5	−4	−3	−2	−1	0	1	2	3	4	5	6
1	1.1	1.0	1.0	1.0	1.0	1.0	1.0	1.0	1.0	1.0	1.0	1.0	0.9
2	1.1	1.1	1.1	1.1	1.0	1.0	1.0	1.0	1.0	0.9	0.9	0.9	0.9
3	1.2	1.1	1.1	1.1	1.1	1.0	1.0	1.0	0.9	0.9	0.9	0.9	0.9
4	1.2	1.2	1.2	1.1	1.1	1.0	1.0	1.0	0.9	0.9	0.9	0.8	0.8
5	1.3	1.3	1.2	1.1	1.1	1.0	1.0	1.0	0.9	0.9	0.8	0.8	0.8
6	1.4	1.3	1.2	1.2	1.1	1.1	1.0	0.9	0.9	0.8	0.8	0.8	0.7
7	1.5	1.4	1.3	1.2	1.1	1.1	1.0	0.9	0.9	0.8	0.8	0.7	0.7
8	1.5	1.4	1.3	1.2	1.2	1.1	1.0	0.9	0.9	0.8	0.7	0.7	0.6
9	1.6	1.5	1.4	1.3	1.2	1.1	1.0	0.9	0.8	0.8	0.7	0.7	0.6
10	1.7	1.6	1.4	1.3	1.2	1.1	1.0	0.9	0.8	0.8	0.7	0.6	0.6
11	1.8	1.6	1.5	1.4	1.2	1.1	1.0	0.9	0.8	0.7	0.7	0.6	0.6
12	1.9	1.7	1.5	1.4	1.2	1.1	1.0	0.9	0.8	0.7	0.6	0.6	0.5
13	2.0	1.8	1.6	1.4	1.3	1.1	1.0	0.9	0.8	0.7	0.6	0.6	0.5
14	2.1	1.9	1.7	1.5	1.3	1.1	1.0	0.9	0.8	0.7	0.6	0.5	0.5
15	2.3	2.0	1.7	1.5	1.3	1.1	1.0	0.9	0.8	0.7	0.6	0.5	0.4
16	2.4	2.1	1.8	1.6	1.3	1.2	1.0	0.9	0.7	0.6	0.6	0.5	0.4
17	2.5	2.2	1.9	1.6	1.4	1.2	1.0	0.9	0.7	0.6	0.5	0.5	0.4
18	2.6	2.3	1.9	1.6	1.4	1.2	1.0	0.9	0.7	0.6	0.5	0.4	0.4
19	2.8	2.4	2.0	1.7	1.4	1.2	1.0	0.8	0.7	0.6	0.5	0.4	0.4
20	2.9	2.5	2.1	1.7	1.4	1.2	1.0	0.8	0.7	0.6	0.5	0.4	0.4
21	3.1	2.6	2.1	1.8	1.5	1.2	1.0	0.8	0.7	0.6	0.5	0.4	0.3
22	3.3	2.7	2.2	1.8	1.5	1.2	1.0	0.8	0.7	0.5	0.4	0.4	0.3
23	3.5	2.8	2.3	1.9	1.5	1.2	1.0	0.8	0.7	0.5	0.4	0.4	0.3
24	3.7	3.0	2.4	1.9	1.6	1.2	1.0	0.8	0.6	0.5	0.4	0.3	0.3
25	3.9	3.1	2.5	2.0	1.6	1.2	1.0	0.8	0.6	0.5	0.4	0.3	0.3
26	4.1	3.2	2.6	2.0	1.6	1.3	1.0	0.8	0.6	0.5	0.4	0.3	0.2
27	4.3	3.4	2.7	2.1	1.6	1.3	1.0	0.8	0.6	0.5	0.4	0.3	0.2
28	4.5	3.6	2.8	2.2	1.7	1.3	1.0	0.8	0.6	0.5	0.4	0.3	0.2
29	4.8	3.7	2.9	2.2	1.7	1.3	1.0	0.8	0.6	0.5	0.3	0.3	0.2
30	5.1	3.9	3.0	2.3	1.7	1.3	1.0	0.8	0.6	0.4	0.3	0.3	0.2

Source: Ruegg and Petersen (1987).

Tables B-8–B-9: FOR ESTIMATING PRESENT VALUE DEPRECIATION WRITEOFFS IN THE WORKSHEETS

Table B-8 Multipliers for estimating present value depreciation of buildings and building components based on 15-year ACRS and building or component placed in service the first month of a taxable year.

INSTRUCTIONS: Go to Table B-8. Find the number in column 1 closest to your intended study period as specified in Series A, Worksheet 1, D, and the number in column headings 2-9 closest to your after-tax discount rate as specified in Series A, Worksheet 1, E. Enter the corresponding multiplier in Series A, Worksheet 3, Hb. Interpolate for greater accuracy.

Table B-9 Multipliers for estimating present value depreciation of business equipment based on 5-year ACRS.

INSTRUCTIONS: Go to Table B-9. find the number in column 1 closest to your intended holding period, and the number in column headings 2-9 closest to your after-tax discount rate. Enter the corresponding multiplier in Worksheet 3, Ib. Interpolate for greater accuracy.

Table B-8 Multipliers for estimating present value depreciation of buildings and building components based on 15-year ACRS and building or component placed in service the first month of a taxable year

(1)	(2)	(3)	(4)	(5)	(6)	(7)	(8)	(9)
	After-Tax Discount Rate (From Worksheet 1, E)							
Study Period (Years)	6	8	10	12	14	16	18	20
1	.11	.11	.11	.11	.11	.10	.10	.10
2	.20	.20	.19	.19	.19	.17	.17	.17
3	.28	.27	.26	.25	.25	.23	.22	.22
4	.34	.33	.31	.30	.30	.27	.26	.26
5	.39	.38	.35	.34	.34	.30	.29	.29
6	.43	.42	.38	.37	.37	.32	.31	.31
7	.47	.46	.41	.40	.39	.34	.33	.33
8	.51	.49	.44	.42	.41	.36	.35	.34
9	.55	.52	.47	.44	.43	.38	.36	.35
10	.58	.54	.49	.46	.44	.39	.37	.36
11	.61	.56	.51	.47	.45	.40	.38	.37
12	.63	.58	.53	.48	.46	.41	.39	.38
13	.65	.60	.54	.49	.47	.42	.40	.38
14	.67	.62	.55	.50	.48	.43	.40	.39
15 and Beyond	.69	.64	.56	.51	.49	.44	.41	.39

Source: Ruegg and Petersen (1987).

Table B-9. Multipliers for estimating present value depreciation of business equipment based on 5-year ACRS

(1)	(2)	(3)	(4)	(5)	(6)	(7)	(8)	(9)
	After-Tax Discount Rate (From Worksheet 1, E)							
Study Period (Years)	6	8	10	12	14	16	18	20
1	.14	.14	.14	.13	.13	.13	.13	.12
2	.34	.33	.32	.31	.30	.29	.29	.27
3	.52	.50	.48	.46	.44	.42	.42	.39
4	.69	.65	.62	.59	.56	.54	.53	.49
5	.85	.79	.75	.71	.67	.64	.62	.57

Source: Ruegg and Petersen (1987).

REFERENCES

Clark, John J., Thomas J. Hindelang, and Robert E. Pritchard. 1984. *Capital Budgeting; Planning and Control of Capital Expenditures*. Englewood cliffs, New Jersey: Prentice-Hall, Inc.

Lippiatt, Barbara C. and Rosalie T. Ruegg. 1988. *Energy Prices and Discount Factors for Life-Cycle Cost Analysis*. NBSIR 88-3273. Gaithersburg, Maryland: National Bureau of Standards.

Ruegg, Rosalie T. and Stephen R. Petersen. 1987. *Comprehensive Guide for Least-Cost Energy Decisions*. NBS SP 709. Gaithersburg, Maryland: National Bureau of Standards.

Appendix C

Graphs for Converting SPB to DPB[1]

This appendix provides a set of graphs that help you approximate DPB quickly and easily from SPB without complex calculations.[2] SPB is easy to compute but misstates the true time to payback. These graphs allow you to convert SPB to DPB with little additional effort and thereby improve the measure of payback.

Table C-1 and Figures C-1 through C-4 are used to estimate DPB when SPB is known. The graphical procedure applies both when annual net cash flows are uniform and when they increase at a relatively constant rate. An example of a project with escalating net cash flows might be energy conservation, where the value of energy saved increases annually because energy prices increase more rapidly than prices of other goods in the economy.

C.1 HOW TO USE THE MATRIX AND GRAPHS

The first step is to compute SPB. You can do this using equation 7-3. The second step is to find the applicable k value from the matrix of k values in Table C-1. Go down the discount rate column headed "$d\%$" until you find the discount rate for the project. Then go across the top row of the matrix headed "$e\%$" until you find the escalation rate for net cash flows. Note that $e = 0$ if annual net cash flows are assumed to be uniform. Then find the k value in the matrix that corresponds to these two coordinate rates.

The third step is to find the graph among Figures C-1 through C-4 that corresponds both to the SPB calculated in the first step and to the k value found in the second step.

The final step is to determine the DPB. Find the SPB value on the horizontal axis. Draw a line upward, perpendicular to the horizontal axis, from the SPB value to the curve representing the k value already determined. Then draw a line to the left, perpendicular to the vertical axis, from the point of intersection on the k curve to the vertical axis labeled DPB. The value at the intersection with the vertical axis is the DPB.

C.2 PROBLEM ILLUSTRATION

Find the DPB for the following project. An initial investment of \$40,000 yields annual NS of \$8,000, with an escalation rate of 3% for NS and a discount rate of 12%. The

[1]See Chapter 7 for an in-depth treatment of SPB and DPB. See Chart 7-1 for a summary of key features of the payback method.

[2]This appendix is based in part on Marshall (1984).

Table C-1 Matrix of k values for combinations of e and d

d% \ e %	−4	−3	−2	−1	0	1	2	3	4	5	6	7	8	9	10	11	12	13	14	15
−2	0.98	0.99	1.00	1.01	1.02	1.03	1.04	1.05	1.06	1.07	1.08	1.09	1.10	1.11	1.12	1.13	1.14	1.15	1.16	1.17
−1	0.97	0.98	0.99	1.00	1.01	1.02	1.03	1.04	1.05	1.06	1.07	1.08	1.09	1.10	1.11	1.12	1.13	1.14	1.15	1.16
0	0.96	0.97	0.98	0.99	1.00	1.01	1.02	1.03	1.04	1.05	1.06	1.07	1.08	1.09	1.10	1.11	1.12	1.13	1.14	1.15
1	0.95	0.96	0.97	0.98	0.99	1.00	1.01	1.02	1.03	1.04	1.05	1.06	1.07	1.08	1.09	1.10	1.11	1.12	1.13	1.14
2	0.94	0.95	0.96	0.97	0.98	0.99	1.00	1.01	1.02	1.03	1.04	1.05	1.06	1.07	1.08	1.09	1.10	1.11	1.12	1.13
3	0.93	0.94	0.95	0.96	0.97	0.98	0.99	1.00	1.01	1.02	1.03	1.04	1.05	1.06	1.07	1.08	1.09	1.10	1.11	1.12
4	0.92	0.93	0.94	0.95	0.96	0.97	0.98	0.99	1.00	1.01	1.02	1.03	1.04	1.05	1.06	1.07	1.08	1.09	1.10	1.11
5	0.91	0.92	0.93	0.94	0.95	0.96	0.97	0.98	0.99	1.00	1.01	1.02	1.03	1.04	1.05	1.06	1.07	1.08	1.09	1.10
6	0.91	0.92	0.92	0.93	0.94	0.95	0.96	0.97	0.98	0.99	1.00	1.01	1.02	1.03	1.04	1.05	1.06	1.07	1.08	1.08
7	0.90	0.91	0.92	0.93	0.93	0.94	0.95	0.96	0.97	0.98	0.99	1.00	1.01	1.02	1.03	1.04	1.05	1.06	1.07	1.07
8	0.89	0.90	0.91	0.92	0.93	0.94	0.94	0.95	0.96	0.97	0.98	0.99	1.00	1.01	1.02	1.03	1.04	1.05	1.06	1.06
9	0.88	0.89	0.90	0.91	0.92	0.93	0.94	0.94	0.95	0.96	0.97	0.98	0.99	1.00	1.01	1.02	1.03	1.04	1.05	1.06
10	0.87	0.88	0.89	0.90	0.91	0.92	0.93	0.94	0.95	0.95	0.96	0.97	0.98	0.99	1.00	1.01	1.02	1.03	1.04	1.05
11	0.86	0.87	0.88	0.89	0.90	0.91	0.92	0.93	0.94	0.95	0.95	0.96	0.97	0.98	0.99	1.00	1.01	1.02	1.03	1.04
12	0.86	0.87	0.88	0.88	0.89	0.90	0.91	0.92	0.93	0.94	0.95	0.96	0.96	0.97	0.98	0.99	1.00	1.01	1.02	1.03
13	0.85	0.86	0.87	0.88	0.88	0.89	0.90	0.91	0.92	0.93	0.94	0.95	0.96	0.96	0.97	0.98	0.99	1.00	1.01	1.02
14	0.84	0.85	0.86	0.87	0.88	0.89	0.89	0.90	0.91	0.92	0.93	0.94	0.95	0.96	0.96	0.97	0.98	0.99	1.00	1.01
15	0.83	0.84	0.85	0.86	0.87	0.88	0.89	0.90	0.90	0.91	0.92	0.93	0.94	0.95	0.96	0.97	0.97	0.98	0.99	1.00
16	0.83	0.84	0.84	0.85	0.86	0.87	0.88	0.89	0.90	0.91	0.91	0.92	0.93	0.94	0.95	0.96	0.97	0.97	0.98	0.99
17	0.82	0.83	0.84	0.85	0.85	0.86	0.87	0.88	0.89	0.90	0.91	0.91	0.92	0.93	0.94	0.95	0.96	0.97	0.97	0.98
18	0.81	0.82	0.83	0.84	0.85	0.86	0.86	0.87	0.88	0.89	0.90	0.91	0.92	0.92	0.93	0.94	0.95	0.96	0.97	0.97
19	0.81	0.82	0.82	0.83	0.84	0.85	0.86	0.87	0.87	0.88	0.89	0.90	0.91	0.92	0.92	0.93	0.94	0.95	0.96	0.97
20	0.80	0.81	0.82	0.82	0.83	0.84	0.85	0.86	0.87	0.88	0.88	0.89	0.90	0.91	0.92	0.93	0.93	0.94	0.95	0.96
21	0.79	0.80	0.81	0.82	0.83	0.83	0.84	0.85	0.86	0.87	0.88	0.88	0.89	0.90	0.91	0.92	0.93	0.93	0.94	0.95
22	0.79	0.80	0.80	0.81	0.82	0.83	0.84	0.84	0.85	0.86	0.87	0.88	0.89	0.89	0.90	0.91	0.92	0.93	0.93	0.94
23	0.78	0.79	0.80	0.80	0.81	0.82	0.83	0.84	0.85	0.85	0.86	0.87	0.88	0.89	0.89	0.90	0.91	0.92	0.93	0.93
24	0.77	0.78	0.79	0.80	0.81	0.81	0.82	0.83	0.84	0.85	0.85	0.86	0.87	0.88	0.89	0.90	0.90	0.91	0.92	0.93
25	0.77	0.78	0.78	0.79	0.80	0.81	0.82	0.82	0.83	0.84	0.85	0.86	0.86	0.87	0.88	0.89	0.90	0.90	0.91	0.92

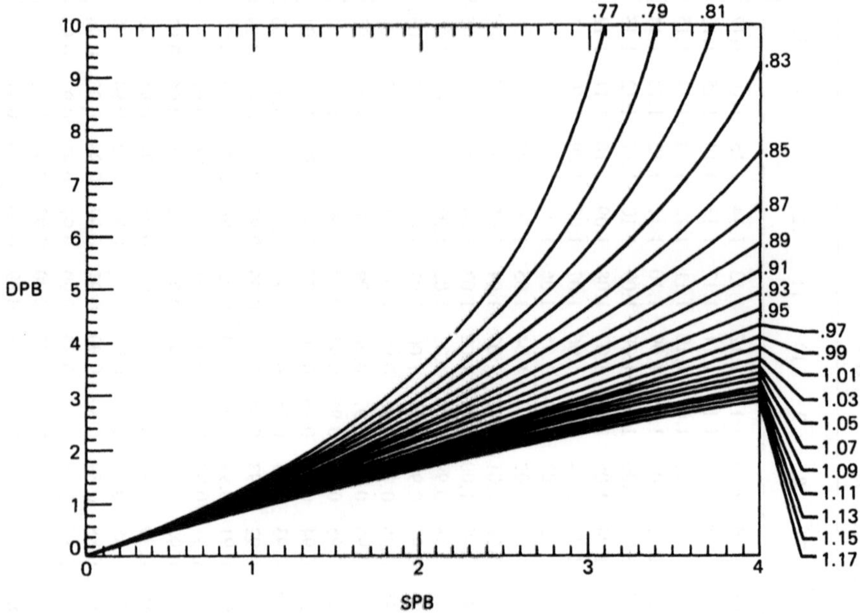

Figure C-1 Discounted payback with escalation for SPB = 0 to 4 years, k = .77 to 1.17 (odd)

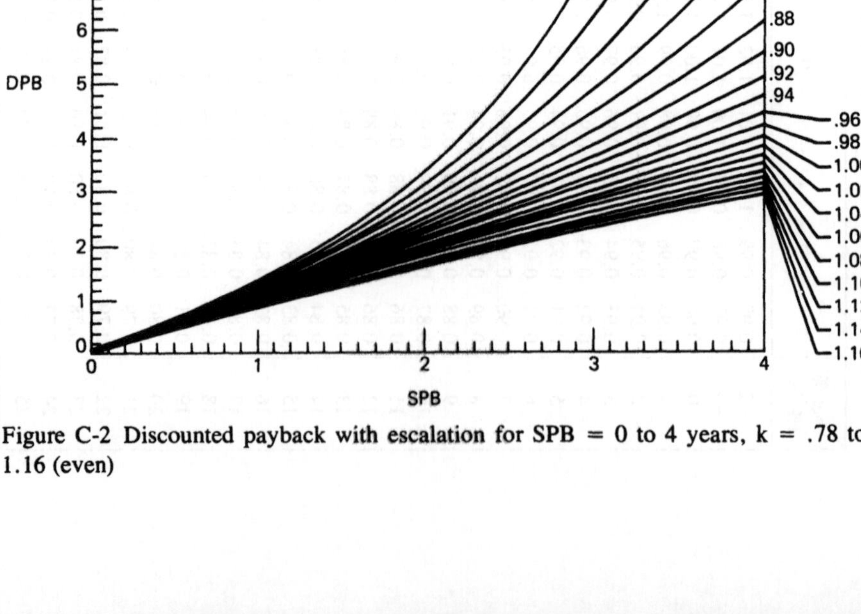

Figure C-2 Discounted payback with escalation for SPB = 0 to 4 years, k = .78 to 1.16 (even)

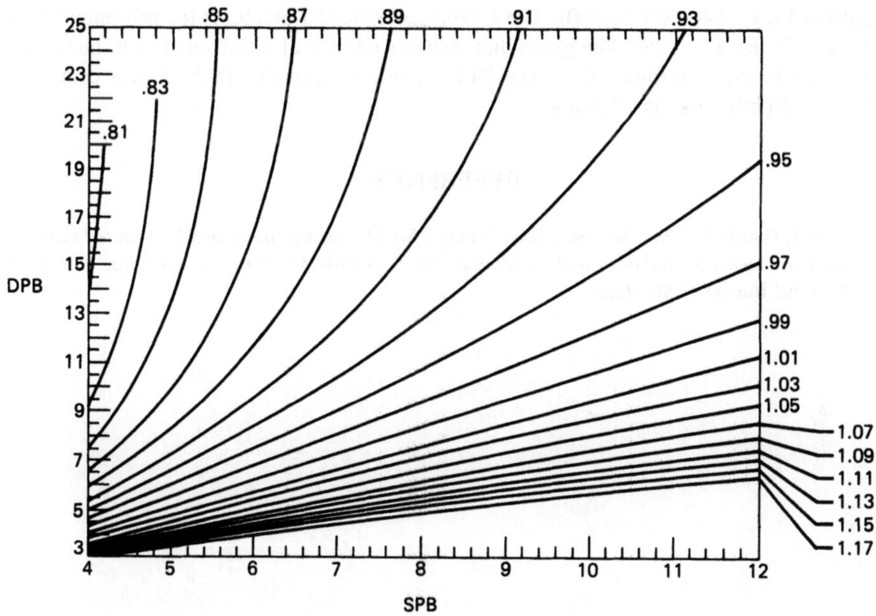

Figure C-3 Discounted payback with escalation for SPB = 4 to 12 years, k = .81 to 1.17 (odd)

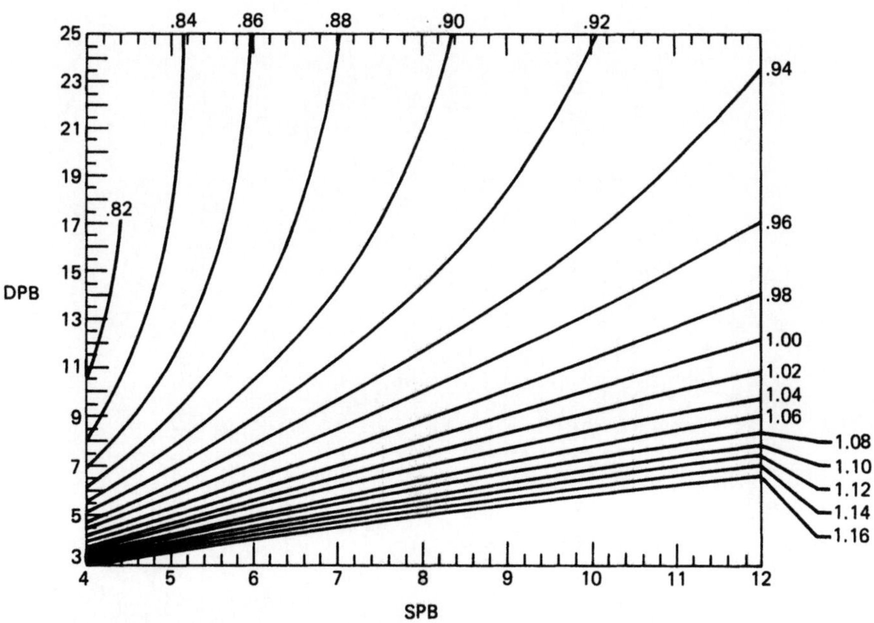

Figure C-4 Discounted payback with escalation for SPB = 4 to 12 years, k = .82 to 1.16 (even)

SPB is 5 (i.e., \$40,000/\$8,000). The k value in Table C-1 is 0.92 in the cell intersection for $e = 3$ and $d = 12\%$. The graph that includes an SPB of 5 years and even-numbered k values is found in Figure C-4. The DPB value corresponding to the 5-year computed SPB is slightly less than 7 years.

REFERENCE

Marshall, Harold E. 1984. *Recommended Practice for Measuring Simple and Discounted Payback for Investments in Buildings and Building Systems.* NBSIR 84-2850. Gaithersburg, Maryland: National Bureau of Standards.

Appendix D

Computer Programs

Computer programs can be invaluable in improving the speed and accuracy with which you apply the methods and techniques presented in this book. They are particularly helpful in tackling complex investment problems that would be difficult to evaluate manually without error and within a reasonable time. In addition, the graphics and spreadsheet capabilities of computer programs can enlarge and enhance display possibilities far beyond what we can do manually.

It is important to understand what computer programs can and cannot do for you. Computer programs can do the following:

1. Reduce tedious hand calculations, saving time and reducing mathematical errors.
2. Reduce the need to consult handbooks and tables by incorporating reference libraries of data within the program's data files.
3. Reduce the need to know all the methodological nuances in order to perform sophisticated analysis.

But most computer programs will not do the following:

1. Determine the objectives of the analysis.
2. Assemble all required input information for the analysis.
3. Interpret results and make decisions on that basis.

In short, despite the availability of elaborate programs, you need to be able to diagnose the problem, set specific objectives for the analysis, choose the appropriate software, compile at least part of the data and make assumptions, interpret the results of the analysis, and make decisions. This book provides the basics for performing economic analysis—both the background you need *before* using computer-aided tools and the knowledge of how to interpret economic measures and make decisions.

In selecting computer software for economic analysis, you should keep in mind that many methodologies and models are used by all sectors of the economy, and no single computer program will likely serve them all. You should ask for what specific purpose, customer, or application a computer program was developed; who the current users are; and whether the program is endorsed by a particular organization. Additionally, you will need to know if it will run on your computer; if the program is well documented and supported; if the information reported by the program meets your needs; and whether

Table D-1. List of computer programs reviewed

Section	Title
D.1	Range Estimating Program (REP/PC)
D.2	VENTURER
D.3	MAXRETURN
D.4	Risk Analysis Package (R-UAPAC)
D.5	FINFEAS2
D.6	National Bureau of Standards Life-Cycle Cost Program (NBSLCC)
D.7	Federal Building Life-Cycle Cost Program (FBLCC)
D.8	Financial Analysis Standard (FAST 123)
D.9	Lotus 1-2-3 LCC Templates
D.10	Supertree Professional Decision Analysis Software
D.11	ELSIE Expert System

the program is likely to increase productivity. It pays to select your computer program carefully. The wrong program can hinder rather than help you make good decisions.

We have selected for review a few of the many computer programs which might be useful for analyzing building decisions. Our purpose is to acquaint you with the possibilities for using computer programs. We make no guarantee of their current availability, accuracy, or the consequences of using them. In other words, you are on your own in obtaining and using them.

The programs reviewed are listed in Table D-1. They range from simple programs that perform basic discounting operations to elaborate programs that incorporate risk analysis techniques, multiple financial arrangements, and tax treatments. Some of the programs are proprietary, with concealed code and available only for a fee, whereas others are public-domain programs, with transparent code and available free or for a nominal charge.

D.1 RANGE-ESTIMATING PROGRAM (REP/PC)

REP is a risk analysis program designed and marketed by the Decision Sciences Corporation, Box 28848, St. Louis, Missouri. This program enables the decision maker to do the following:

1. Identify the mathematical probability of having a cost overrun.
2. Determine the maximum potential overrun and the maximum potential underrun.
3. Identify, quantify, and rank the reasons why the probability of overrun is high or low.

In the fields of engineering and construction, REP is used for reviewing cost estimates, setting contingencies, establishing competitive bids, negotiating prices with the owner, and managing ongoing projects.

REP/PC runs on IBM and compatible PCs. Minimal requirements are 256K RAM, two 362KB floppy drives or one floppy drive and a hard disk, and the Disk Operating System (DOS).

D.2 VENTURER

VENTURER is an interactive, menu-driven program for making investment decisions. Using Monte Carlo simulation, it performs risk analysis of a venture and measures the sensitivity of project objectives to influential factors. VENTURER enables decision makers to do the following:

1. Make economic assessments of business ventures.
2. Perform risk analyses in comparing project alternatives.
3. Design more economically efficient buildings.

VENTURER runs on an MS-DOS/PC-DOS operating system. For further information on the program, contact Dr. G. Singh, Civil Engineering Department, University of Leeds, LS2 9JT, United Kingdom.

D.3 MAXRETURN

MAXRETURN is a user-friendly, interactive microcomputer program for helping managers make resource allocation decisions when choosing among projects competing for a limited budget. By using dynamic programming, MAXRETURN enables the decision maker to evaluate projects which are lumpy (i.e., do not exhaust all of the available budget). Moreover, the technique can be used to determine simultaneously the efficient scale of investment in each project, as well as the optimal combination of projects.

MAXRETURN enables the decision maker to do the following:

1. Choose among investments competing for a limited budget.
2. Employ the dynamic programming technique without having any prior knowledge of it.
3. Take into account risk attitude by using the utility module.

MAXRETURN runs on an MS-DOS/PC-DOS operating system.

For further information on MAXRETURN, see G. Singh et al., Application of a microcomputer in investment decisions using dynamic programming, *Construction Management and Economics*, Vol. 4, 1986, pp. 189–199, or contact Dr. G. Singh, Civil Engineering Department, University of Leeds, LS2 9JT, United Kingdom.

D.4 RISK ANALYSIS PACKAGE (R-UAPAC)

R-UAPAC is a general-purpose microcomputer software package for risk analysis. It has the helpful feature of initially leading the user through a structured sequence of questions to translate qualitative uncertainty expressed by the user into a quantitative probability profile for each input variable.

This program enables decision makers to do the following:

1. Generate probability profiles of input data.
2. Investigate the likely outcome of any problem, process, or decision which is subject to uncertainty.

3. Combine in any mathematical form any number of variables at risk through Monte Carlo simulations.
4. Provide output graphically (cumulative distributions) or numerically.

R-UAPAC runs on an MS-DOS/PC-DOS operating system.

For further information on the program, contact Dr. Alan Wilson, Ishai and Wilson, Dov Hoz 35, Tel Aviv, Israel.

D.5 FINFEAS2

FINFEAS2 is a quantitative LCC program for the assessment of the financial feasibility of large building projects. It comes with complete documentation.

FINFEAS2 describes the uncertainty of input values with beta distributions. The program uses simulation techniques to select values randomly for each input from its respective distribution and iteratively builds up a cumulative probability distribution of the selected measure of feasibility.

The program enables the user to do the following:

1. Calculate net present value, internal rate of return, and overall rate of return, both before and after taxes.
2. Display tables and graphs of cash flows (monthly during construction and annually over project life).
3. Generate graphs of sensitivity analysis and cumulative probability distributions of each of the financial feasibility measures.

FINFEAS2 runs on a microcomputer such as an IBM-PC using a Lotus 1-2-3 Spreadsheet (Release 2, Lotus Development Corporation 1985). At least 640 bytes of memory are required.

For further information on the program, contact Dr. S. N. Tucker or Mr. M. Rahilly, CSIRO Division of Building Construction and Engineering, P.O. Box 56, Highett, Victoria 3190, Australia.

D.6 NATIONAL BUREAU OF STANDARDS LCC PROGRAM (NBSLCC)

The NBSLCC program and user's guide provide a convenient, quick, and accurate mechanism for making LCC-related calculations for a given building or component design. It is especially useful for evaluating design improvements to buildings that increase their first costs but reduce future energy costs.

NBSLCC enables the user to do the following:

1. Calculate LCC, NB, SIR, and ORR.
2. Account for financing, depreciation, and taxes.
3. Evaluate energy conservation investments, using long-term energy price projections by the DoE which are retrieved automatically from the disk.

Note that the economic methodologies incorporated in the program are completely consistent with the economic methods described in Part 1. Note also that a stand-alone program called DISCOUNT for computing discounted cash flows is a part of NBSLCC.

NBSLCC is designed to be used with a microcomputer having an MS-DOS operating system (IBM-PC/XT/AT and compatibles). The complete program is contained on a single $5\frac{1}{4}$-inch floppy disk.

For a description of the program and directions on how to order it, see Rosalie T. Ruegg and Stephen R. Petersen, "Comprehensive Guide for Least-Cost Energy Decisions," National Bureau of Standards Special Publication 709, January 1987. A single complimentary copy of Special Publication 709 can be ordered from the Applied Economics Group, Center for Computing and Applied Mathematics, National Institute of Standards and Technology, Building 101, Room A415, Gaithersburg, MD 20899.

D.7 FEDERAL BUILDING LCC (FBLCC) COMPUTER PROGRAM

This program and user's guide is virtually identical to the NBSBLCC program and guide described in Section D.6, except that it is directed to U.S. government building decisions. Thus it makes no provision for taxes and financing, and it incorporates federal criteria for performing evaluations.

For a description of FBLCC and directions on how to order it, see Stephen R. Petersen, *A User's Guide to the Federal Building Life-Cycle Cost (FBLCC) Computer Program*, National Bureau of Standards Technical Note 1222, May 1986. A single complimentary copy of Technical Note 1222 can be ordered from the Applied Economics Group, Center for Computing and Applied Mathematics, National Institute of Standards and Technology, Building 101, Room A415, Gaithersburg, MD 20899.

D.8 FINANCIAL ANALYSIS STANDARD (FAST 123)

FAST 123 is an integrated set of computer programs that permit sophisticated and detailed financial analysis of conventional and alternative energy system projects.

FAST 123 allows the user to do the following:

1. Perform financial analysis of most types of energy systems and power plants including coal, oil, nuclear, cogeneration, hydro, solar and wind electric, geothermal, and biomass.
2. Perform financial analysis for independent power producers or utility companies—regulated investor-owned, regulated public utility, or cooperative utility.

The program calculates detailed cash flows, return on investment, net present value, and payback for a given revenue stream. It determines the system's value for given revenues and required rate of return. It calculates levelized revenues in a rapid analysis. Finally, it matches the output of various renewable plants with specified utility seasonal peak, part-peak, and off-peak times. Both normalized and flow-through accounting are included for the utility models.

Corporate and limited partnership financing cash flow programs allow treatment of a variety of financing arrangements. Federal, state, and local tax effects are taken into account. The program is available for the metric system. Given sufficient customer interest, the developer will make program modifications to incorporate country-specific requirements, such as tax treatments.

The integrated programs are formulated as templates for Lotus 1-2-3. System requirements are for an IBM-PC or compatible computer, with at least 256K of memory, MS-DOS Version 2.0 or higher, Lotus 1-2-3 Version 1A or 2, and at least one double-sided diskette drive.

Information on the program, which is proprietary, can be obtained from Polydyne, Inc., 1900 South Norfolk Street, Suite 209, San Mateo, CA 94403; telephone (415) 570-7797.

D.9 LOTUS 1-2-3 LCC TEMPLATES

This program is available in conjunction with seminars for design professionals which emphasize how to use microcomputers and off-the-shelf software for performing LCC analysis of energy conservation investments. A diskette containing Lotus 1-2-3 LCC template examples for institutional and private business is provided to seminar participants or can be purchased directly from the supplier. Specialized templates can be requested.

System requirements are for an IBM-PC or compatible computer with at least one double-sided diskette drive, MS-DOS, and Lotus 1-2-3.

For more information, contact Mr. Barry Benator, Benatech, Inc., 1215 Hightower Trail, Suite D-220, Atlanta, GA 30350; telephone (404) 998-6090.

D.10 SUPERTREE PROFESSIONAL DECISION ANALYSIS SOFTWARE

Supertree is a decision analysis program designed and marketed by SDG Decision Systems, 3000 Sand Hill Road, Menlo Park, CA 94025-7127; telephone (415) 854-9000. The program enables the user to do the following:

1. Perform sensitivity analysis to find which variables have a significant effect on project outcome.
2. Construct decision trees, determine endpoint values, and roll back the tree.
3. Draw cumulative distribution functions for analyzing risk exposure.
4. Consider risk attitude.
5. Calculate the value of extra information.

A textbook and a training course on how to use the software are available from SDG Decision Systems. Supertree runs on an IBM-PC or compatible computer, as well as on the Macintosh. Suggested memory for the PC-DOS/MS-DOS machine is 640K. For the version that runs on the Macintosh, 1024K of memory is recommended. SDG Decision Systems recommends a hard disk and floppy disk drives. Supertree interfaces with Lotus 1-2-3, Multiplan, PC Excel, IFPS, and other spreadsheets.

D.11 ELSIE EXPERT SYSTEM

ELSIE is an expert system for the strategic planning of commercial construction projects at the inception stage. The knowledge-based system translates the client's requirements as expressed in the client's brief into a strategy which the design team can follow. ELSIE supports the quantity surveyor or building economist in the "lead consultant" role when advising clients in the early stages of a construction project.

ELSIE has four modules that are linked through a data base. The program does the following:

1. Provides a budget estimate of how much the project will cost and a statement of assumptions on which the estimate is based.
2. Analyzes the client's needs and constraints, and then recommends a suitable procurement approach.
3. Selects, sequences, and estimates the duration of the critical activities most likely to be required in the building process to satisfy project needs, based on the knowledge of the client and project particulars.
4. Evaluates the financial viability of the project using a conventional valuation model.

The user can perform sensitivity analysis with the majority of the variables identified. A 20-page printout describes the design strategy.

ELSIE runs on an IBM-PC and compatibles with at least 640K of memory and a hard disk. The knowledge base in the system relates to United Kingdom conditions, so it must be adapted for other countries.

For further information on ELSIE, contact Professor P. S. Brandon, Imaginor Systems, 41 Leslie Hough Way, Salford University Business Park, Salford M6 6AJ, England, United Kingdom.

Appendix E

Derivation of a Utility Function

This appendix[1] explains one approach[2] to constructing utility functions. The approach derives the utility function directly through questions to a decision maker about willingness to trade guaranteed sums for larger, uncertain sums. The appendix also presents the data on which the three utility functions in Figure 14-7 are based. Figure E-1 is used here to explain the derivation of the three functions in figure 14-7. Figure E-1 is exactly like Figure 14-7, except that the points through which the curves were fitted are shown in Figure E-1.

The function in Figure E-1 that bends over to the right indicates a risk-averse (RA) decision maker. The RA function is used to explain the process of constructing a utility function. The steps taken to construct it were as follows:

1. Two arbitrary incomes were selected. The decision maker chose \$0 and \$100,000. Then the decision maker was asked to assign arbitrary utility values (U) to these two incomes. They were as follows:

$$U(\$100{,}000) = 70 \qquad U(\$0) = 0$$

2. The decision maker was then asked to choose the amount of money (certainty equivalent) which he would be indifferent to receiving when compared to the following lottery:

$$0.50(\$100{,}000) + 0.50(\$0)$$

The decision maker chose \$25,000, half the \$50,000 expected value of the lottery. By substitution we solve for the corresponding utility value as follows:

$$U(\$25{,}000) = 0.50(70) + 0.50(0) = 35$$

3. Using the $U(\$25{,}000)$ as a known value, another point is determined experimentally by finding the certainty equivalent which the decision maker would be indif-

[1]This appendix is adapted from Marshall (1988, pp. 75–77).

[2]Examples of other approaches to constructing utility functions are to base utility measures (1) on observations of individual or firm behavior in practice and (2) on recorded behavior in games of chance designed to test risk attitude.

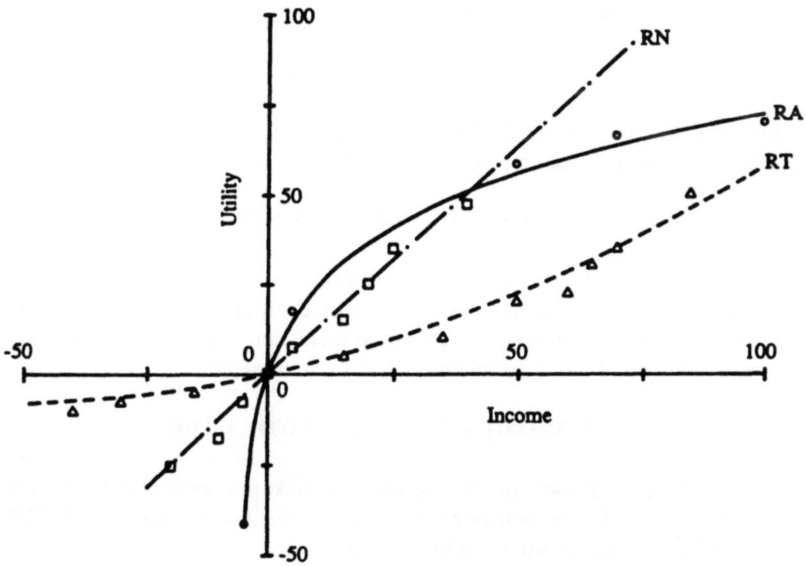

Figure E-1. Three types of risk attitude: data and fitted utility functions: RN = risk neutral; RA = risk averse; RT = risk taking.

ferent to receiving when compared to the following lottery:

$$0.50(\$25{,}000) + 0.50(\$0)$$

The choice was $5,000, considerably less than the $12,500 expected value of the lottery. By substitution we find the corresponding utility value of $5,000 as follows:

$$U(\$5{,}000) = 0.50(35) + 0.50(0) = 17.5$$

4. To find utility values for levels of income greater than $25,000, alternative lotteries are posed that use the known utility values of $25,000 and $100,000. For example, the decision maker was asked to choose the amount of money which he would be indifferent to receiving when compared to the following lottery:

$$0.90(\$100{,}000) + 0.10(\$25{,}000)$$

The decision maker chose $70,000, compared to an expected value of $92,500. By substitution we solve for the corresponding utility value as follows:

$$U(\$70{,}000) = 0.90(70) + 0.10(35) = 66.5$$

5. When asked to choose a value for the certainty equivalent of this lottery

$$0.75(\$70{,}000) + 0.25(\$25{,}000),$$

the decision maker chose \$50,000, compared to an expected value of \$58,750. By substitution we find the corresponding utility value as follows:

$$U(\$50{,}000) = 0.75(66.5) + 0.25(35) = 58.6$$

6. To find a point in the range where the decision maker could lose money, a loss situation was introduced into the lottery. The decision maker was asked to find the probabilities of outcomes which would make him indifferent between the following lottery and the certainty equivalent from doing nothing.

$$P(\$5{,}000) + (1 - P)(-\$5{,}000) = (\$0)$$

where P and $(1 - P)$ sum to 1.0. The decision maker assigned to P the probability 0.70 and to $(1 - P)$ the probability 0.30. Substituting known values for $U(\$5{,}000)$ and $U(\$0)$ yielded the utility of losing \$5,000, i.e.,

$$0.70(17.5) + (0.30) \cdot U(-\$5{,}000) = 0$$

$$U(-\$5{,}000) = \frac{-0.70(17.5)}{0.30}$$

$$= -40.8$$

To complete the process of constructing the RA utility function, a curve was statistically "fitted" through the combination of points selected arbitrarily (\$0, 0; \$100,000,

Table E-1. Data for plotting utility functions

Risk Taker		Risk Neutral	
(\$1,000)	Utility	(\$1,000)	Utility
−40	−10.0	−20	−25.0
−30	− 7.5	−10	−17.5
−15	− 5.0	− 5	− 7.5
0	0.0	0	0.0
15	5.0	5	7.5
35	10.0	15	15.0
50	20.0	20	25.0
60	22.5	25	35.0
65	30.0	40	47.5
70	35.0		
80	42.5		
85	50.0		

70) and experimentally ($25,000, 35; $5,000, 17.5; $70,000, 66.5; $50,000, 58.6; and −$5,000, −40.8), as shown in function RA in Figure E-1.[3] A curve-fitting routine from DATAPLOT, a computer program developed at the National Institute of Standards and Technology, was used to establish the utility function. The risk-taking (RT) and risk-neutral (RN) functions in Figure E-1 were constructed with the same routine, using the data shown in Table E-1.

REFERENCES

Berger, James O. 1985. *Statistical Decision Theory and Bayesian Analysis.* 2nd edition. New York: Springer-Verlag.

Howard, Ronald A. 1977. Risk preference. *Readings in Decision Analysis.* ed. Ronald Howard, James E. Matheson, and Katherine L. Miller, pp. 431–465. Menlo Park, California: SRI International.

Marshall, Harold E. 1988. *Techniques for Treating Uncertainty and Risk in the Economic Evaluation of Building Investments.* SP757. Gaithersburg, Maryland: National Institute of Standards and Technology.

[3]For additional information on how to construct utility functions, see Berger (1985, pp. 46–70) and Howard (1977, pp. 431–465).

Appendix F

Guide to Published Sources of Data

Lack of data is a prime obstacle to putting economic analysis into practice. Securing data at reasonable cost is essential to conducting economic analysis with an appropriate expenditure of resources.

This appendix identifies published sources of data. Though not an exhaustive treatment and limited to sources in the United States, it gives you an idea of the kinds of data which are available and helps you get started. Learning to assemble data from published sources is key to being able to perform analysis quickly and early in the decision process.

F.1 GENERAL ECONOMIC AND FINANCIAL DATA

Certain economic analyses of building projects require information about general economic conditions, both national and regional. For example, a decision to expand a plant may depend on sales forecasts, which in turn are a function of the outlook for general economic growth; plant location may depend on revenue forecasts for different locations, which are a function of regional economic growth; whether or not to replace an existing building with a new one may be influenced by general business conditions.

General financial data are also needed for design decisions. For example, data are often needed on prices, housing units, vacancy rates, construction contracts, market interest rates, and tax rates. The following publications contain general economic and financial data:

Business Conditions Digest, Bureau of Economic Analysis, U.S. Department of Commerce, Washington, D.C. (Published monthly).

Business Statistics: 1990, Bureau of Economic Analysis, U.S. Department of Commerce, Washington, D.C. (Published Biannually)

Directories, such as *Statistics Sources*, Gale Research, Detroit, Michigan; and *Business Trends and Forecasting Information Sources*, Gale Research, Detroit, Michigan.

Economic Almanac, National Industrial Conference Board, New York. (Published annually).

Economic Report of the President. (Published annually).

Historical Chart Book, Federal Reserve Board of Governors, Washington, D.C. (Published annually, updated monthly).

Monthly Labor Review (Consumer Price Index & Producer Price Index), U.S. Department of Labor, Washington, D.C. (Published monthly).
National Economic Projections, National Planning Association, Washington, D.C.
Statistical Abstract of the United States, U.S. Government Printing Office, Washington, D.C. (Published annually).
Survey of Current Business, Bureau of Economic Analysis, U.S. Department of Commerce, Washington, D.C. (Published monthly).
Tax Information on Depreciation, Internal Revenue Service Publication 534 (Issued annually).

F.2 GENERAL INDUSTRY DATA

Published industry sources, such as the following, include data on sales, assets, profit, labor costs, materials costs, investment in plant, and other statistics by industry group.

Census of Business, U.S. Bureau of the Census, Washington, D.C.
Survey of Manufacturers, U.S. Department of Commerce, Washington, D.C. (Published annually).
U.S. Industrial Outlook, U.S. Department of Commerce, Washington, D.C. (Published annually).

F.3 BUILDING-RELATED DATA

Published data on the costs of construction, operation, maintenance, repair, and replacement, such as the following, are invaluable in evaluating design alternatives early in the decision process.

ASHRAE Journal, American Society of Heating, Refrigerating and Air Conditioning Engineers, Commercial Air Conditioning Division, La Crosse, WI.
Boeckh Building Valuation Manual, The American Appraisal Company, Milwaukee, WI.
Construction Index, Architext, Chicago, Illinois
Cost Engineers Notebook, American Association of Cost Engineers, West VA. (Lists International Cost/Price Indexes).
Department of Commerce Composite Construction Cost Index, U.S. Bureau of the Census, Washington, D.C.
Design Cost and Data, Allan Thompson, Glendora, CA.
Downtown and Surburban Office Building Experience Report, Building Owners and Management Association International, Washington, DC.
Energy Prices and Discount Factors for Life-Cycle Cost Analysis, National Bureau of Standards, NBSIR 85-3273-2, Gaithersburg, MD. (Updated annually).
Engineering News-Record Building Cost Index, McGraw-Hill, Inc., New York.

Life-Cycle Cost Data, Alphonse J. Dell'isola and Stephen J. Kirk, McGraw-Hill Book Company, New York, NY (Published 1983).
Maintenance and Repair Database: Architectural, Electrical, Plumbing, and HVAC Reports, Construction Engineering Research Laboratory, Facility Systems Division, Champaign, IL.
Means Building Construction Cost Data Book, R. S. Means Company, Kingston, MA.
Means Construction Cost Indexes, R. S. Means Company, Kingston, MA.
Means Square Foot Costs, R. S. Means Company, Kingston, MA.
The Energy Review, Data Resources, Inc., Lexington, MA. (Published quarterly).

Appendix G

Discount Factor Tables

See chapter 8 for directions on using the following six sets of tables:

Table G-1. Single compound amount (SCA) factors
Table G-2. Single present value (SPV) factors
Table G-3. Uniform present value (UPV) factors
Table G-4. Uniform capital recovery (UCR) factors
Tables G-5 through G-21. Uniform present value factors modified (UPV*)
Tables G-22 through G-27. Uniform present value factors modified (UPV**) for DOE-projected fuel price escalation

Table G-1 Single Compound Amount (SCA) factors (multipliers)—for finding the time-equivalent future value of a present amount or for estimating a future amount based on a present amount and its projected rate of change

	Discount Rate or Rate of Change																
n	1%	2%	3%	4%	5%	6%	7%	8%	9%	10%	12%	14%	16%	18%	20%	24%	28%
1	1.01	1.02	1.03	1.04	1.05	1.06	1.07	1.08	1.09	1.10	1.12	1.14	1.16	1.18	1.20	1.24	1.28
2	1.02	1.04	1.06	1.08	1.10	1.12	1.14	1.17	1.19	1.21	1.25	1.30	1.35	1.39	1.44	1.54	1.64
3	1.03	1.06	1.09	1.12	1.16	1.19	1.23	1.26	1.30	1.33	1.40	1.48	1.56	1.64	1.73	1.91	2.10
4	1.04	1.08	1.13	1.17	1.22	1.26	1.31	1.36	1.41	1.46	1.57	1.69	1.81	1.94	2.07	2.36	2.68
5	1.05	1.10	1.16	1.22	1.28	1.34	1.40	1.47	1.54	1.61	1.76	1.93	2.10	2.29	2.49	2.93	3.44
6	1.06	1.13	1.19	1.27	1.34	1.42	1.50	1.59	1.68	1.77	1.97	2.19	2.44	2.70	2.99	3.64	4.40
7	1.07	1.15	1.23	1.32	1.41	1.50	1.61	1.71	1.83	1.95	2.21	2.50	2.83	3.19	3.58	4.51	5.63
8	1.08	1.17	1.27	1.37	1.48	1.59	1.72	1.85	1.99	2.14	2.48	2.85	3.28	3.76	4.30	5.59	7.21
9	1.09	1.20	1.30	1.42	1.55	1.69	1.84	2.00	2.17	2.36	2.77	3.25	3.80	4.44	5.16	6.93	9.22
10	1.10	1.22	1.34	1.48	1.63	1.79	1.97	2.16	2.37	2.59	3.11	3.71	4.41	5.23	6.19	8.59	11.81
11	1.12	1.24	1.38	1.54	1.71	1.90	2.10	2.33	2.58	2.85	3.48	4.23	5.12	6.18	7.43	10.66	15.11
12	1.13	1.27	1.43	1.60	1.80	2.01	2.25	2.52	2.81	3.14	3.90	4.82	5.94	7.29	8.92	13.21	19.34
13	1.14	1.29	1.47	1.67	1.89	2.13	2.41	2.72	3.07	3.45	4.36	5.49	6.89	8.60	10.70	16.39	24.76
14	1.15	1.32	1.51	1.73	1.98	2.26	2.58	2.94	3.34	3.80	4.89	6.26	7.99	10.15	12.84	20.32	31.69
15	1.16	1.35	1.56	1.80	2.08	2.40	2.76	3.17	3.64	4.18	5.47	7.14	9.27	11.97	15.41	25.20	40.56
16	1.17	1.37	1.60	1.87	2.18	2.54	2.95	3.43	3.97	4.59	6.13	8.14	10.75	14.13	18.49	31.24	51.92
17	1.18	1.40	1.65	1.95	2.29	2.69	3.16	3.70	4.33	5.05	6.87	9.28	12.47	16.67	22.19	38.74	66.46
18	1.20	1.43	1.70	2.03	2.41	2.85	3.38	4.00	4.72	5.56	7.69	10.58	14.46	19.67	26.62	48.04	85.07
19	1.21	1.46	1.75	2.11	2.53	3.03	3.62	4.32	5.14	6.12	8.61	12.06	16.78	23.21	31.95	59.57	108.89
20	1.22	1.49	1.81	2.19	2.65	3.21	3.87	4.66	5.60	6.73	9.65	13.74	19.46	27.39	38.34	73.86	139.38
21	1.23	1.52	1.86	2.28	2.79	3.40	4.14	5.03	6.11	7.40	10.80	15.67	22.57	32.32	46.01	91.59	178.41
22	1.24	1.55	1.92	2.37	2.93	3.60	4.43	5.44	6.66	8.14	12.10	17.86	26.19	38.14	55.21	113.57	228.36
23	1.26	1.58	1.97	2.46	3.07	3.82	4.74	5.87	7.26	8.95	13.55	20.36	30.38	45.01	66.25	140.83	292.30
24	1.27	1.61	2.03	2.56	3.23	4.05	5.07	6.34	7.91	9.85	15.18	23.21	35.24	53.11	79.50	174.63	374.14
25	1.28	1.64	2.09	2.67	3.39	4.29	5.43	6.85	8.62	10.83	17.00	26.46	40.87	62.67	95.40	216.54	478.90
26	1.30	1.67	2.16	2.77	3.56	4.55	5.81	7.40	9.40	11.92	19.04	30.17	47.41	73.95	114.48	268.51	613.00
27	1.31	1.71	2.22	2.88	3.73	4.82	6.21	7.99	10.25	13.11	21.32	34.39	55.00	87.26	137.37	332.95	784.64
28	1.32	1.74	2.29	3.00	3.92	5.11	6.65	8.63	11.17	14.42	23.88	39.20	63.80	102.97	164.84	412.86	1,004.34
29	1.33	1.78	2.36	3.12	4.12	5.42	7.11	9.32	12.17	15.86	26.75	44.69	74.01	121.50	197.81	511.95	1,285.55
30	1.35	1.81	2.43	3.24	4.32	5.74	7.61	10.06	13.27	17.45	29.96	50.95	85.85	143.37	237.38	634.82	1,645.50

Source: Ruegg and Petersen (1987).

Table G-2 Single Present Value (SPV) factors (multipliers)—for finding the present value of a future amount

Discount Rate

n	1%	2%	3%	4%	5%	6%	7%	8%	9%	10%	12%	14%	16%	18%	20%	24%	28%
1	0.99	0.98	0.97	0.96	0.95	0.94	0.93	0.93	0.92	0.91	0.89	0.88	0.86	0.85	0.83	0.81	0.78
2	0.98	0.96	0.94	0.92	0.91	0.89	0.87	0.86	0.84	0.83	0.80	0.77	0.74	0.72	0.69	0.65	0.61
3	0.97	0.94	0.92	0.89	0.86	0.84	0.82	0.79	0.77	0.75	0.71	0.67	0.64	0.61	0.58	0.52	0.48
4	0.96	0.92	0.89	0.85	0.82	0.79	0.76	0.74	0.71	0.68	0.64	0.59	0.55	0.52	0.48	0.42	0.37
5	0.95	0.91	0.86	0.82	0.78	0.75	0.71	0.68	0.65	0.62	0.57	0.52	0.48	0.44	0.40	0.34	0.29
6	0.94	0.89	0.84	0.79	0.75	0.70	0.67	0.63	0.60	0.56	0.51	0.46	0.41	0.37	0.33	0.28	0.23
7	0.93	0.87	0.81	0.76	0.71	0.67	0.62	0.58	0.55	0.51	0.45	0.40	0.35	0.31	0.28	0.22	0.18
8	0.92	0.85	0.79	0.73	0.68	0.63	0.58	0.54	0.50	0.47	0.40	0.35	0.31	0.27	0.23	0.18	0.14
9	0.91	0.84	0.77	0.70	0.64	0.59	0.54	0.50	0.46	0.42	0.36	0.31	0.26	0.23	0.19	0.14	0.11
10	0.91	0.82	0.74	0.68	0.61	0.56	0.51	0.46	0.42	0.39	0.32	0.27	0.23	0.19	0.16	0.12	0.08
11	0.90	0.80	0.72	0.65	0.58	0.53	0.48	0.43	0.39	0.35	0.29	0.24	0.20	0.16	0.13	0.09	0.07
12	0.89	0.79	0.70	0.62	0.56	0.50	0.44	0.40	0.36	0.32	0.26	0.21	0.17	0.14	0.11	0.08	0.05
13	0.88	0.77	0.68	0.60	0.53	0.47	0.41	0.37	0.33	0.29	0.23	0.18	0.15	0.12	0.09	0.06	0.04
14	0.87	0.76	0.66	0.58	0.51	0.44	0.39	0.34	0.30	0.26	0.20	0.16	0.13	0.10	0.08	0.05	0.03
15	0.86	0.74	0.64	0.56	0.48	0.42	0.36	0.32	0.27	0.24	0.18	0.14	0.11	0.08	0.06	0.04	0.02
16	0.85	0.73	0.62	0.53	0.46	0.39	0.34	0.29	0.25	0.22	0.16	0.12	0.09	0.07	0.05	0.03	0.02
17	0.84	0.71	0.61	0.51	0.44	0.37	0.32	0.27	0.23	0.20	0.15	0.11	0.08	0.06	0.05	0.03	0.02
18	0.84	0.70	0.59	0.49	0.42	0.35	0.30	0.25	0.21	0.18	0.13	0.09	0.07	0.05	0.04	0.02	0.01
19	0.83	0.69	0.57	0.47	0.40	0.33	0.28	0.23	0.19	0.16	0.12	0.08	0.06	0.04	0.03	0.02	0.01
20	0.82	0.67	0.55	0.46	0.38	0.31	0.26	0.21	0.18	0.15	0.10	0.07	0.05	0.04	0.03	0.01	0.01
21	0.81	0.66	0.54	0.44	0.36	0.29	0.24	0.20	0.16	0.14	0.09	0.06	0.04	0.03	0.02	0.01	0.01
22	0.80	0.65	0.52	0.42	0.34	0.28	0.23	0.18	0.15	0.12	0.08	0.06	0.04	0.03	0.02	0.01	0.00
23	0.80	0.63	0.51	0.41	0.33	0.26	0.21	0.17	0.14	0.11	0.07	0.05	0.03	0.02	0.02	0.01	0.00
24	0.79	0.62	0.49	0.39	0.31	0.25	0.20	0.16	0.13	0.10	0.07	0.04	0.03	0.02	0.01	0.01	0.00
25	0.78	0.61	0.48	0.38	0.30	0.23	0.18	0.15	0.12	0.09	0.06	0.04	0.02	0.02	0.01	0.00	0.00
26	0.77	0.60	0.46	0.36	0.28	0.22	0.17	0.14	0.11	0.08	0.05	0.03	0.02	0.01	0.01	0.00	0.00
27	0.76	0.59	0.45	0.35	0.27	0.21	0.16	0.13	0.10	0.08	0.05	0.03	0.02	0.01	0.01	0.00	0.00
28	0.76	0.57	0.44	0.33	0.26	0.20	0.15	0.12	0.09	0.07	0.04	0.03	0.02	0.01	0.01	0.00	0.00
29	0.75	0.56	0.42	0.32	0.24	0.18	0.14	0.11	0.08	0.06	0.04	0.02	0.01	0.01	0.01	0.00	0.00
30	0.74	0.55	0.41	0.31	0.23	0.17	0.13	0.10	0.08	0.06	0.03	0.02	0.01	0.01	0.00	0.00	0.00

Table G-3 Uniform Present Value (UPV) factors (multipliers)—for finding the present value of a uniformly recurring future amount

	Discount Rate																
n	1%	2%	3%	4%	5%	6%	7%	8%	9%	10%	12%	14%	16%	18%	20%	24%	28%
1	0.99	0.98	0.97	0.96	0.95	0.94	0.93	0.93	0.92	0.91	0.89	0.88	0.86	0.85	0.83	0.81	0.78
2	1.97	1.94	1.91	1.89	1.86	1.83	1.81	1.78	1.76	1.74	1.69	1.65	1.61	1.57	1.53	1.46	1.39
3	2.94	2.88	2.83	2.78	2.72	2.67	2.62	2.58	2.53	2.49	2.40	2.32	2.25	2.17	2.11	1.98	1.87
4	3.90	3.81	3.72	3.63	3.55	3.47	3.39	3.31	3.24	3.17	3.04	2.91	2.80	2.69	2.59	2.40	2.24
5	4.85	4.71	4.58	4.45	4.33	4.21	4.10	3.99	3.89	3.79	3.60	3.43	3.27	3.13	2.99	2.75	2.53
6	5.80	5.60	5.42	5.24	5.08	4.92	4.77	4.62	4.49	4.36	4.11	3.89	3.68	3.50	3.33	3.02	2.76
7	6.73	6.47	6.23	6.00	5.79	5.58	5.39	5.21	5.03	4.87	4.56	4.29	4.04	3.81	3.60	3.24	2.94
8	7.65	7.33	7.02	6.73	6.46	6.21	5.97	5.75	5.53	5.33	4.97	4.64	4.34	4.08	3.84	3.42	3.08
9	8.57	8.16	7.79	7.44	7.11	6.80	6.52	6.25	6.00	5.76	5.33	4.95	4.61	4.30	4.03	3.57	3.18
10	9.47	8.98	8.53	8.11	7.72	7.36	7.02	6.71	6.42	6.14	5.65	5.22	4.83	4.49	4.19	3.68	3.27
11	10.37	9.79	9.25	8.76	8.31	7.89	7.50	7.14	6.81	6.50	5.94	5.45	5.03	4.66	4.33	3.78	3.34
12	11.26	10.58	9.95	9.39	8.86	8.38	7.94	7.54	7.16	6.81	6.19	5.66	5.20	4.79	4.44	3.85	3.39
13	12.13	11.35	10.63	9.99	9.39	8.85	8.36	7.90	7.49	7.10	6.42	5.84	5.34	4.91	4.53	3.91	3.43
14	13.00	12.11	11.30	10.56	9.90	9.29	8.75	8.24	7.79	7.37	6.63	6.00	5.47	5.01	4.61	3.96	3.46
15	13.87	12.85	11.94	11.12	10.38	9.71	9.11	8.56	8.06	7.61	6.81	6.14	5.58	5.09	4.68	4.00	3.48
16	14.72	13.58	12.56	11.65	10.84	10.11	9.45	8.85	8.31	7.82	6.97	6.27	5.67	5.16	4.73	4.03	3.50
17	15.56	14.29	13.17	12.17	11.27	10.48	9.76	9.12	8.54	8.02	7.12	6.37	5.75	5.22	4.77	4.06	3.52
18	16.40	14.99	13.75	12.66	11.69	10.83	10.06	9.37	8.76	8.20	7.25	6.47	5.82	5.27	4.81	4.08	3.53
19	17.23	15.68	14.32	13.13	12.09	11.16	10.34	9.60	8.95	8.36	7.37	6.55	5.88	5.32	4.84	4.10	3.54
20	18.05	16.35	14.88	13.59	12.46	11.47	10.59	9.82	9.13	8.51	7.47	6.62	5.93	5.35	4.87	4.11	3.55
21	18.86	17.01	15.42	14.03	12.82	11.76	10.84	10.02	9.29	8.65	7.56	6.69	5.97	5.38	4.89	4.12	3.55
22	19.66	17.66	15.94	14.45	13.16	12.04	11.06	10.20	9.44	8.77	7.64	6.74	6.01	5.41	4.91	4.13	3.56
23	20.46	18.29	16.44	14.86	13.49	12.30	11.27	10.37	9.58	8.88	7.72	6.79	6.04	5.43	4.92	4.14	3.56
24	21.24	18.91	16.94	15.25	13.80	12.55	11.47	10.53	9.71	8.98	7.78	6.84	6.07	5.45	4.94	4.14	3.56
25	22.02	19.52	17.41	15.62	14.09	12.78	11.65	10.67	9.82	9.08	7.84	6.87	6.10	5.47	4.95	4.15	3.56
26	22.80	20.12	17.88	15.98	14.38	13.00	11.83	10.81	9.93	9.16	7.90	6.91	6.12	5.48	4.96	4.15	3.57
27	23.56	20.71	18.33	16.33	14.64	13.21	11.99	10.94	10.03	9.24	7.94	6.94	6.14	5.49	4.96	4.15	3.57
28	24.32	21.28	18.76	16.66	14.90	13.41	12.14	11.05	10.12	9.31	7.98	6.96	6.15	5.50	4.97	4.16	3.57
29	25.07	21.84	19.19	16.98	15.14	13.59	12.28	11.16	10.20	9.37	8.02	6.98	6.17	5.51	4.97	4.16	3.57
30	25.81	22.40	19.60	17.29	15.37	13.76	12.41	11.26	10.27	9.43	8.06	7.00	6.18	5.52	4.98	4.16	3.57

Table G-4 Uniform Capital Recovery (UCR) factors (multipliers)—for finding the uniform annual equivalent of a present amount

	Discount Rate																
n	1%	2%	3%	4%	5%	6%	7%	8%	9%	10%	12%	14%	16%	18%	20%	24%	28%
1	1.01	1.02	1.03	1.04	1.05	1.06	1.07	1.08	1.09	1.10	1.12	1.14	1.16	1.18	1.20	1.24	1.28
2	0.51	0.52	0.52	0.53	0.54	0.55	0.55	0.56	0.57	0.58	0.59	0.61	0.62	0.64	0.65	0.69	0.72
3	0.34	0.35	0.35	0.36	0.37	0.37	0.38	0.39	0.40	0.40	0.42	0.43	0.45	0.46	0.47	0.50	0.54
4	0.26	0.26	0.27	0.28	0.28	0.29	0.30	0.30	0.31	0.32	0.33	0.34	0.36	0.37	0.39	0.42	0.45
5	0.21	0.21	0.22	0.22	0.23	0.24	0.24	0.25	0.26	0.26	0.28	0.29	0.31	0.32	0.33	0.36	0.39
6	0.17	0.18	0.18	0.19	0.20	0.20	0.21	0.22	0.22	0.23	0.24	0.26	0.27	0.29	0.30	0.33	0.36
7	0.15	0.15	0.16	0.17	0.17	0.18	0.19	0.19	0.20	0.21	0.22	0.23	0.25	0.26	0.28	0.31	0.34
8	0.13	0.14	0.14	0.15	0.15	0.16	0.17	0.17	0.18	0.19	0.20	0.22	0.23	0.25	0.26	0.29	0.33
9	0.12	0.12	0.13	0.13	0.14	0.15	0.15	0.16	0.17	0.17	0.19	0.20	0.22	0.23	0.25	0.28	0.31
10	0.11	0.11	0.12	0.12	0.13	0.14	0.14	0.15	0.16	0.16	0.18	0.19	0.21	0.22	0.24	0.27	0.31
11	0.10	0.10	0.11	0.11	0.12	0.13	0.13	0.14	0.15	0.15	0.17	0.18	0.20	0.21	0.23	0.26	0.30
12	0.09	0.09	0.10	0.11	0.11	0.12	0.13	0.13	0.14	0.15	0.16	0.18	0.19	0.21	0.23	0.26	0.30
13	0.08	0.09	0.09	0.10	0.11	0.11	0.12	0.13	0.13	0.14	0.16	0.17	0.19	0.20	0.22	0.26	0.29
14	0.08	0.08	0.09	0.09	0.10	0.11	0.11	0.12	0.13	0.14	0.15	0.17	0.18	0.20	0.22	0.25	0.29
15	0.07	0.08	0.08	0.09	0.10	0.10	0.11	0.12	0.12	0.13	0.15	0.16	0.18	0.20	0.21	0.25	0.29
16	0.07	0.07	0.08	0.09	0.09	0.10	0.11	0.11	0.12	0.13	0.14	0.16	0.18	0.19	0.21	0.25	0.29
17	0.06	0.07	0.08	0.08	0.09	0.10	0.10	0.11	0.12	0.12	0.14	0.16	0.17	0.19	0.21	0.25	0.28
18	0.06	0.07	0.07	0.08	0.09	0.09	0.10	0.11	0.11	0.12	0.14	0.15	0.17	0.19	0.21	0.25	0.28
19	0.06	0.06	0.07	0.08	0.08	0.09	0.10	0.10	0.11	0.12	0.14	0.15	0.17	0.19	0.21	0.24	0.28
20	0.06	0.06	0.07	0.07	0.08	0.09	0.09	0.10	0.11	0.12	0.13	0.15	0.17	0.19	0.21	0.24	0.28
21	0.05	0.06	0.06	0.07	0.08	0.09	0.09	0.10	0.11	0.12	0.13	0.15	0.17	0.19	0.20	0.24	0.28
22	0.05	0.06	0.06	0.07	0.08	0.08	0.09	0.10	0.11	0.11	0.13	0.15	0.17	0.18	0.20	0.24	0.28
23	0.05	0.05	0.06	0.07	0.07	0.08	0.09	0.10	0.10	0.11	0.13	0.15	0.17	0.18	0.20	0.24	0.28
24	0.05	0.05	0.06	0.07	0.07	0.08	0.09	0.09	0.10	0.11	0.13	0.15	0.16	0.18	0.20	0.24	0.28
25	0.05	0.05	0.06	0.06	0.07	0.08	0.09	0.09	0.10	0.11	0.13	0.15	0.16	0.18	0.20	0.24	0.28
26	0.04	0.05	0.06	0.06	0.07	0.08	0.08	0.09	0.10	0.11	0.13	0.14	0.16	0.18	0.20	0.24	0.28
27	0.04	0.05	0.05	0.06	0.07	0.08	0.08	0.09	0.10	0.11	0.13	0.14	0.16	0.18	0.20	0.24	0.28
28	0.04	0.05	0.05	0.06	0.07	0.07	0.08	0.09	0.10	0.11	0.13	0.14	0.16	0.18	0.20	0.24	0.28
29	0.04	0.05	0.05	0.06	0.07	0.07	0.08	0.09	0.10	0.11	0.12	0.14	0.16	0.18	0.20	0.24	0.28
30	0.04	0.04	0.05	0.06	0.07	0.07	0.08	0.09	0.10	0.11	0.12	0.14	0.16	0.18	0.20	0.24	0.28

Table G-5 Uniform Present Value Factors Modified (UPV*)
Discount rate = 1%

	Rate of Price Increase per Period														
Period	1%	2%	3%	4%	5%	6%	7%	8%	9%	10%	12%	14%	16%	18%	20%
1	1.000	1.010	1.020	1.030	1.040	1.050	1.059	1.069	1.079	1.089	1.109	1.129	1.149	1.168	1.188
2	2.000	2.030	2.060	2.090	2.120	2.151	2.182	2.213	2.244	2.275	2.339	2.403	2.468	2.533	2.600
3	3.000	3.060	3.120	3.182	3.244	3.307	3.371	3.435	3.501	3.567	3.702	3.841	3.983	4.128	4.277
4	4.000	4.100	4.202	4.306	4.412	4.520	4.630	4.743	4.857	4.974	5.214	5.464	5.723	5.991	6.270
5	5.000	5.150	5.305	5.464	5.626	5.793	5.965	6.141	6.321	6.506	6.891	7.296	7.721	8.168	8.637
6	6.000	6.211	6.430	6.656	6.889	7.130	7.379	7.636	7.901	8.175	8.751	9.363	10.02	10.71	11.45
7	7.000	7.283	7.577	7.883	8.201	8.532	8.876	9.234	9.606	9.993	10.81	11.70	12.65	13.68	14.79
8	8.000	8.365	8.747	9.147	9.566	10.00	10.46	10.94	11.45	11.97	13.10	14.33	15.68	17.15	18.76
9	9.000	9.457	9.940	10.45	10.98	11.55	12.14	12.77	13.43	14.13	15.63	17.31	19.16	21.21	23.48
10	10.00	10.56	11.16	11.79	12.46	13.17	13.92	14.73	15.58	16.48	18.45	20.66	23.15	25.95	29.09
11	11.00	11.68	12.40	13.17	13.99	14.87	15.81	16.82	17.89	19.03	21.56	24.45	27.74	31.48	5.75
12	12.00	12.80	13.66	14.59	15.59	16.66	17.81	19.05	20.38	21.82	25.02	28.72	33.01	37.95	43.66
13	13.00	13.94	14.95	16.05	17.24	18.53	19.93	21.44	23.08	24.85	28.86	33.55	39.06	45.51	53.06
14	14.00	15.09	16.27	17.56	18.96	20.50	22.17	24.00	25.99	28.16	33.11	39.00	46.00	54.33	64.23
15	15.00	16.24	17.61	19.11	20.76	22.56	24.55	26.73	29.12	31.75	37.82	45.15	53.99	64.65	77.50
16	16.00	17.42	18.98	20.71	22.62	24.73	27.07	29.65	32.51	35.67	43.05	52.09	63.15	76.70	93.27
17	17.00	18.60	20.37	22.35	24.55	27.00	29.73	32.77	36.16	39.94	48.85	59.92	73.68	90.77	112.0
18	18.00	19.79	21.80	24.05	26.56	29.39	32.56	36.11	40.11	44.59	55.28	68.76	85.77	107.2	134.3
19	19.00	21.00	23.25	25.79	28.66	31.89	35.55	39.69	44.36	49.65	62.41	78.74	99.66	126.4	160.7
20	20.00	22.22	24.73	27.58	30.83	34.52	38.72	43.51	48.96	55.16	70.31	90.00	115.6	148.9	192.1
21	21.00	23.45	26.24	29.43	33.09	37.28	42.08	47.59	53.91	61.17	79.08	102.7	133.9	175.1	229.5
22	22.00	24.69	27.78	31.34	35.44	40.18	45.64	51.96	59.26	67.71	88.80	117.1	155.0	205.8	273.8
23	23.00	25.94	29.35	33.30	37.88	43.21	49.41	56.63	65.04	74.83	99.58	133.3	179.1	241.6	326.5
24	24.00	27.21	30.95	35.32	40.42	46.40	53.41	61.62	71.27	82.59	111.5	151.5	206.9	283.4	389.1
25	25.00	28.49	32.58	37.40	43.07	49.75	57.64	66.96	77.99	91.04	124.8	172.2	238.7	332.3	463.5
26	26.00	29.78	34.25	39.54	45.81	53.26	62.12	72.67	85.25	100.2	139.5	195.5	275.4	389.3	551.9
27	27.00	31.08	35.95	41.74	48.66	56.95	66.87	78.78	93.08	110.3	155.8	221.8	317.4	456.0	656.9
28	28.00	32.40	37.68	44.01	51.63	60.82	71.91	85.31	101.5	121.2	173.9	251.4	365.7	534.0	781.6
29	29.00	33.73	39.44	46.35	54.72	64.88	77.24	92.29	110.7	133.1	193.9	284.9	421.1	625.0	929.9
30	30.00	35.08	41.24	48.75	57.92	69.14	82.88	99.76	120.5	146.0	216.1	322.7	484.8	731.4	1106
31	31.00	36.43	43.08	51.23	61.26	73.61	88.87	107.7	131.1	160.1	240.8	365.4	558.0	855.7	1315
32	32.00	37.80	44.95	53.78	64.72	78.30	95.21	116.3	142.6	175.5	268.1	413.5	642.0	1001	1564
33	33.00	39.19	46.86	56.41	68.32	83.23	101.9	125.4	155.0	192.2	298.4	467.9	738.5	1170	1859
34	34.00	40.59	48.81	59.11	72.07	88.40	109.0	135.2	168.3	210.4	332.0	529.3	849.3	1369	2210
35	35.00	42.00	50.80	61.90	75.96	93.82	116.6	145.6	182.7	230.2	369.3	598.5	976.6	1600	2627
36	36.00	43.42	52.82	64.77	80.01	99.52	124.6	156.8	198.3	251.8	410.6	676.7	1123	1871	3123
37	37.00	44.86	54.89	67.72	84.22	105.5	133.0	168.7	215.1	275.4	456.5	764.9	1291	2187	3711
38	38.00	46.32	56.99	70.76	88.59	111.8	142.0	181.5	233.2	301.0	507.3	864.5	1484	2556	4410
39	39.00	47.79	59.14	73.90	93.14	118.4	151.5	195.1	252.7	328.9	563.7	976.9	1705	2987	5241
40	40.00	49.27	61.33	77.12	97.87	125.3	161.5	209.7	273.8	359.3	626.2	1104	1959	3491	6228

Source: Petersen (1989).

Table G-6 Uniform Present Value Factors Modified (UPV*) Discount rate = 2%

Period	Rate of Price Increase per Period 1%	2%	3%	4%	5%	6%	7%	8%	9%	10%	12%	14%	16%	18%	20%
1	0.9902	1.000	1.010	1.020	1.029	1.039	1.049	1.059	1.069	1.078	1.098	1.118	1.137	1.157	1.176
2	1.971	2.000	2.029	2.059	2.089	2.119	2.149	2.180	2.211	2.241	2.304	2.367	2.431	2.495	2.561
3	2.942	3.000	3.059	3.119	3.180	3.242	3.304	3.367	3.431	3.496	3.628	3.763	3.901	4.043	4.189
4	3.903	4.000	4.099	4.200	4.303	4.408	4.515	4.624	4.735	4.848	5.081	5.323	5.574	5.835	6.105
5	4.855	5.000	5.149	5.302	5.459	5.620	5.785	5.955	6.129	6.307	6.678	7.067	7.477	7.907	8.358
6	5.797	6.000	6.209	6.425	6.649	6.880	7.118	7.364	7.618	7.880	8.430	9.016	9.640	10.30	11.01
7	6.731	7.000	7.280	7.571	7.874	8.189	8.516	8.856	9.209	9.577	10.35	11.19	12.10	13.08	14.13
8	7.655	8.000	8.361	8.739	9.135	9.549	9.982	10.44	10.91	11.41	12.47	13.63	14.90	16.29	17.80
9	8.570	9.000	9.453	9.930	10.43	10.96	11.52	12.11	12.73	13.38	14.79	16.35	18.08	20.00	22.12
10	9.476	10.00	10.56	11.14	11.77	12.43	13.13	13.88	14.67	15.51	17.34	19.39	21.70	24.29	27.20
11	10.37	11.00	11.67	12.38	13.14	13.96	14.83	15.75	16.74	17.80	20.13	22.79	25.82	29.26	33.17
12	11.26	12.00	12.79	13.64	14.56	15.54	16.60	17.74	18.96	20.28	23.21	26.59	30.50	35.00	40.20
13	12.14	13.00	13.93	14.93	16.02	17.19	18.47	19.84	21.33	22.94	26.58	30.84	35.82	41.65	48.47
14	13.01	14.00	15.07	16.24	17.52	18.91	20.42	22.07	23.86	25.82	30.28	35.58	41.87	49.34	58.20
15	13.88	15.00	16.23	17.58	19.06	20.69	22.47	24.43	26.57	28.93	34.35	40.88	48.76	58.24	69.65
16	14.73	16.00	17.40	18.95	20.65	22.54	24.62	26.92	29.46	32.27	38.82	46.81	56.59	68.53	83.12
17	15.58	17.00	18.58	20.34	22.29	24.46	26.88	29.56	32.55	35.88	43.72	53.44	65.49	80.44	98.96
18	16.41	18.00	19.77	21.76	23.98	26.46	29.24	32.36	35.86	39.78	49.10	60.84	75.62	94.21	117.6
19	17.24	19.00	20.98	23.20	25.71	28.54	31.73	35.32	39.39	43.97	55.02	69.12	87.13	110.1	139.5
20	18.06	20.00	22.19	24.68	27.50	30.70	34.33	38.46	43.16	48.50	61.51	78.37	100.2	128.6	165.3
21	18.88	21.00	23.42	26.18	29.33	32.94	37.06	41.78	47.19	53.38	68.64	88.70	115.1	149.9	195.7
22	19.68	22.00	24.66	27.71	31.23	35.27	39.93	45.30	51.50	58.65	76.46	100.3	132.1	174.6	231.4
23	20.48	23.00	25.91	29.28	33.17	37.69	42.93	49.02	56.10	64.33	85.06	113.2	151.3	203.1	273.4
24	21.27	24.00	27.17	30.87	35.18	40.21	46.09	52.96	61.02	70.45	94.49	127.6	173.2	236.1	322.8
25	22.05	25.00	28.45	32.50	37.24	42.82	49.40	57.14	66.27	77.06	104.9	143.7	198.1	274.3	381.0
26	22.82	26.00	29.74	34.15	39.37	45.54	52.87	61.56	71.89	84.18	116.2	161.8	226.5	318.5	449.4
27	23.59	27.00	31.04	35.84	41.56	48.37	56.51	66.24	77.89	91.86	128.7	181.9	258.7	369.7	529.9
28	24.35	28.00	32.36	37.56	43.81	51.30	60.33	71.19	84.31	100.1	142.4	204.4	295.3	428.8	624.6
29	25.10	29.00	33.68	39.32	46.12	54.36	64.33	76.44	91.16	109.1	157.5	229.6	337.0	497.2	735.9
30	25.85	30.00	35.02	41.11	48.51	57.53	68.53	82.00	98.48	118.7	174.0	257.7	384.4	576.4	867.0
31	26.58	31.00	36.38	42.94	50.97	60.82	72.94	87.88	106.3	129.1	192.2	289.2	438.3	667.9	1021
32	27.31	32.00	37.74	44.80	53.50	64.25	77.57	94.11	114.7	140.3	212.2	324.3	499.6	773.9	1203
33	28.03	33.00	39.12	46.70	56.10	67.80	82.42	100.7	123.6	152.4	234.0	363.6	569.3	896.4	1416
34	28.75	34.00	40.51	48.63	58.78	71.50	87.51	107.7	133.2	165.4	258.1	407.4	648.6	1038	1667
35	29.46	35.00	41.92	50.60	61.54	75.35	92.85	115.1	143.4	179.5	284.5	456.5	738.8	1202	1962
36	30.16	36.00	43.34	52.62	64.37	79.34	98.45	122.9	154.3	194.6	313.5	511.3	841.3	1392	2310
37	30.85	37.00	44.78	54.67	67.30	83.49	104.3	131.2	165.9	211.0	345.3	572.6	957.9	1611	2719
38	31.54	38.00	46.23	56.76	70.31	87.80	110.5	140.0	178.4	228.6	380.3	641.1	1091	1865	3200
39	32.22	39.00	47.69	58.89	73.40	92.29	116.9	149.3	191.7	247.6	418.6	717.6	1241	2159	3765
40	32.90	40.00	49.17	61.07	76.59	96.95	123.7	159.1	205.9	268.1	460.8	803.2	1413	2499	4431

Source: Petersen (1989).

Table G-7 Uniform Present Value Factors Modified (UPV*)
Discount rate = 3%

Period	Rate of Price Increase per Period														
	1%	2%	3%	4%	5%	6%	7%	8%	9%	10%	12%	14%	16%	18%	20%
1	0.9806	0.9903	1.000	1.010	1.019	1.029	1.039	1.049	1.058	1.068	1.087	1.107	1.126	1.146	1.165
2	1.942	1.971	2.000	2.029	2.059	2.088	2.118	2.148	2.178	2.209	2.270	2.332	2.395	2.458	2.522
3	2.885	2.942	3.000	3.059	3.118	3.178	3.239	3.301	3.363	3.427	3.555	3.688	3.823	3.962	4.104
4	3.810	3.904	4.000	4.098	4.198	4.300	4.404	4.510	4.617	4.727	4.954	5.188	5.432	5.684	5.946
5	4.716	4.856	5.000	5.148	5.299	5.454	5.614	5.777	5.945	6.117	6.474	6.849	7.244	7.658	8.093
6	5.605	5.799	6.000	6.207	6.421	6.642	6.870	7.106	7.349	7.600	8.127	8.687	9.284	9.919	10.59
7	6.477	6.733	7.000	7.277	7.565	7.865	8.176	8.500	8.836	9.185	9.924	10.72	11.58	12.51	13.51
8	7.332	7.658	8.000	8.358	8.732	9.123	9.532	9.961	10.41	10.88	11.88	12.97	14.17	15.48	16.90
9	8.170	8.574	9.000	9.448	9.921	10.42	10.94	11.49	12.07	12.68	14.00	15.47	17.08	18.88	20.86
10	8.992	9.481	10.00	10.55	11.13	11.75	12.41	13.10	13.83	14.61	16.32	18.22	20.37	22.77	25.46
11	9.798	10.38	11.00	11.66	12.37	13.12	13.93	14.78	15.70	16.68	18.83	21.28	24.06	27.23	30.83
12	10.59	11.27	12.00	12.78	13.63	14.53	15.51	16.55	17.67	18.88	21.56	24.66	28.23	32.34	37.08
13	11.36	12.15	13.00	13.92	14.91	15.99	17.15	18.40	19.76	21.23	24.53	28.40	32.92	38.20	44.37
14	12.12	13.02	14.00	15.06	16.22	17.48	18.85	20.34	21.97	23.74	27.76	32.54	38.20	44.91	52.86
15	12.87	13.89	15.00	16.22	17.56	19.02	20.62	22.38	24.31	26.42	31.28	37.12	44.14	52.59	62.75
16	13.60	14.74	16.00	17.39	18.92	20.60	22.46	24.51	26.78	29.28	35.10	42.19	50.84	61.40	74.27
17	14.32	15.59	17.00	18.57	20.30	22.23	24.37	26.75	29.40	32.34	39.25	47.80	58.39	71.48	87.69
18	15.02	16.43	18.00	19.76	21.72	23.91	26.36	29.10	32.17	35.61	43.77	54.01	66.88	83.04	103.3
19	15.71	17.26	19.00	20.96	23.16	25.63	28.42	31.56	35.10	39.09	48.68	60.89	76.45	96.28	121.5
20	16.38	18.08	20.00	22.17	24.63	27.41	30.56	34.14	38.21	42.82	54.02	68.50	87.22	111.4	142.8
21	17.05	18.90	21.00	23.39	26.12	29.24	32.79	36.85	41.49	46.80	59.83	76.92	99.36	128.8	167.5
22	17.69	19.70	22.00	24.63	27.65	31.12	35.10	39.69	44.96	51.05	66.14	86.24	113.0	148.7	196.3
23	18.33	20.50	23.00	25.88	29.21	33.05	37.50	42.66	48.64	55.58	73.01	96.56	128.4	171.5	229.9
24	18.96	21.29	24.00	27.14	30.79	35.04	40.00	45.78	52.53	60.43	80.48	108.0	145.7	197.7	269.0
25	19.57	22.08	25.00	28.41	32.41	37.09	42.59	49.05	56.65	65.60	88.60	120.6	165.3	227.6	[illegible]
26	20.17	22.85	26.00	29.70	34.06	39.20	45.28	52.48	61.01	71.13	97.42	134.6	187.3	261.9	367.6
27	20.76	23.62	27.00	31.00	35.74	41.37	48.08	56.08	65.62	77.03	107.0	150.1	212.0	301.2	429.5
28	21.34	24.38	28.00	32.31	37.45	43.61	50.99	59.85	70.50	83.33	117.5	167.2	239.9	346.2	501.5
29	21.90	25.14	29.00	33.63	39.20	45.91	54.01	63.80	75.67	90.07	128.8	186.2	271.3	397.7	585.5
30	22.46	25.88	30.00	34.97	40.98	48.27	57.14	67.95	81.13	97.25	141.2	207.2	306.7	456.8	683.3
31	23.00	26.62	31.00	36.32	42.80	50.71	60.40	72.29	86.92	104.9	154.6	230.4	346.5	524.5	797.2
32	23.54	27.35	32.00	37.68	44.65	53.22	63.78	76.85	93.04	113.1	169.2	256.1	391.4	602.0	929.9
33	24.06	28.08	33.00	39.06	46.53	55.79	67.30	81.63	99.52	121.9	185.0	284.6	441.9	690.8	1085
34	24.57	28.79	34.00	40.44	48.46	58.45	70.95	86.64	106.4	131.2	202.3	316.1	498.8	792.5	1265
35	25.08	29.51	35.00	41.85	50.42	61.18	74.75	91.90	113.6	141.2	221.1	350.9	562.9	909.1	1475
36	25.57	30.21	36.00	43.26	52.41	63.99	78.69	97.41	121.3	151.9	241.5	389.5	635.0	1043	1719
37	26.05	30.91	37.00	44.69	54.45	66.88	82.78	103.2	129.4	163.3	263.7	432.2	716.3	1196	2004
38	26.53	31.60	38.00	46.14	56.53	69.86	87.04	109.2	138.0	175.4	287.8	479.5	807.9	1371	2336
39	26.99	32.28	39.00	47.59	58.64	72.92	91.46	115.6	147.1	188.4	314.0	531.8	910.9	1572	2723
40	27.45	32.96	40.00	49.07	60.80	76.08	96.05	122.3	156.8	202.3	342.5	589.7	1027	1802	3173

Table G-8 Uniform Present Value Factors Modified (UPV*)
Discount rate = 4%

	Rate of Price Increase per Period													
Period	1%	2%	3%	4%	5%	6%	7%	8%	9%	10%	12%	14%	16%	18%	20%
1	0.9712	0.9808	0.9904	1.000	1.010	1.019	1.029	1.038	1.048	1.058	1.077	1.096	1.115	1.135	1.154
2	1.914	1.943	1.971	2.000	2.029	2.058	2.087	2.117	2.147	2.176	2.237	2.298	2.359	2.422	2.485
3	2.830	2.886	2.943	3.000	3.058	3.117	3.176	3.237	3.298	3.360	3.486	3.615	3.747	3.883	4.021
4	3.720	3.811	3.905	4.000	4.097	4.196	4.297	4.400	4.504	4.611	4.831	5.059	5.295	5.540	5.794
5	4.584	4.719	4.858	5.000	5.146	5.296	5.450	5.607	5.769	5.935	6.279	6.641	7.021	7.420	7.839
6	5.423	5.609	5.801	6.000	6.205	6.417	6.636	6.862	7.095	7.335	7.839	8.376	8.947	9.554	10.20
7	6.237	6.482	6.736	7.000	7.275	7.560	7.856	8.164	8.484	8.816	9.519	10.28	11.09	11.97	12.92
8	7.028	7.338	7.661	8.000	8.354	8.724	9.112	9.516	9.940	10.38	11.33	12.36	13.49	14.72	16.06
9	7.797	8.178	8.578	9.000	9.444	9.911	10.40	10.92	11.47	12.04	13.28	14.65	16.16	17.84	19.69
10	8.543	9.001	9.486	10.00	10.54	11.12	11.73	12.38	13.06	13.79	15.37	17.15	19.14	21.37	23.87
11	9.268	9.809	10.39	11.00	11.66	12.35	13.10	13.89	14.74	15.64	17.63	19.90	22.47	25.38	28.70
12	9.972	10.60	11.28	12.00	12.78	13.61	14.51	15.47	16.50	17.60	20.07	22.91	26.17	29.94	34.27
13	10.66	11.38	12.16	13.00	13.91	14.89	15.95	17.10	18.34	19.68	22.69	26.20	30.31	35.10	40.69
14	11.32	12.14	13.03	14.00	15.05	16.20	17.44	18.80	20.27	21.87	25.51	29.82	34.92	40.96	48.11
15	11.96	12.89	13.90	15.00	16.21	17.53	18.97	20.56	22.29	24.19	28.55	33.78	40.07	47.61	56.66
16	12.59	13.62	14.75	16.00	17.37	18.88	20.55	22.39	24.41	26.64	31.82	38.13	45.80	55.15	66.53
17	13.20	14.34	15.60	17.00	18.55	20.27	22.17	24.29	26.63	29.24	35.35	42.89	52.21	63.71	77.92
18	13.79	15.04	16.44	18.00	19.74	21.68	23.84	26.26	28.96	31.98	39.14	48.11	59.34	73.42	91.07
19	14.36	15.74	17.27	19.00	20.94	23.11	25.56	28.31	31.40	34.89	43.23	53.83	67.31	84.44	106.2
20	14.92	16.41	18.10	20.00	22.15	24.58	27.32	30.43	33.96	37.96	47.63	60.10	76.19	96.94	123.7
21	15.46	17.08	18.91	21.00	23.37	26.07	29.14	32.64	36.64	41.20	52.38	66.98	86.10	111.1	143.9
22	15.98	17.73	19.72	22.00	24.60	27.59	31.01	34.94	39.45	44.64	57.48	74.52	97.15	127.2	167.2
23	16.49	18.37	20.52	23.00	25.85	29.14	32.93	37.32	42.39	48.27	62.98	82.78	109.5	145.5	194.1
24	16.99	19.00	21.32	24.00	27.11	30.72	34.91	39.79	45.48	52.11	68.90	91.83	123.2	166.2	225.1
25	17.47	19.61	22.10	25.00	28.38	32.33	36.95	42.36	48.72	56.18	75.28	101.8	138.5	189.7	260.9
26	17.94	20.22	22.88	26.00	29.66	33.97	39.04	45.03	52.11	60.48	82.15	112.6	155.7	216.4	302.2
27	18.39	20.81	23.65	27.00	30.96	35.64	41.20	47.80	55.66	65.02	89.54	124.6	174.7	246.6	349.8
28	18.83	21.39	24.41	28.00	32.26	37.35	43.42	50.68	59.38	69.83	97.51	137.6	196.0	281.0	404.8
29	19.26	21.96	25.17	29.00	33.58	39.08	45.70	53.67	63.29	74.92	106.1	152.0	219.7	319.9	468.2
30	19.68	22.52	25.92	30.00	34.92	40.85	48.04	56.77	67.38	80.30	115.3	167.7	246.2	364.1	541.4
31	20.08	23.07	26.66	31.00	36.26	42.66	50.46	59.99	71.66	85.99	125.3	184.9	275.7	414.3	625.9
32	20.47	23.60	27.39	32.00	37.62	44.50	52.94	63.34	76.16	92.01	136.0	203.8	308.7	471.2	723.3
33	20.85	24.13	28.12	33.00	38.99	46.37	55.50	66.81	80.87	98.37	147.5	224.5	345.4	535.8	835.7
34	21.22	24.65	28.84	34.00	40.38	48.28	58.13	70.42	85.80	105.1	159.9	247.1	386.4	609.0	965.5
35	21.58	25.15	29.55	35.00	41.77	50.23	60.83	74.16	90.98	112.2	173.3	272.0	432.0	692.2	1115
36	21.93	25.65	30.26	36.00	43.19	52.22	63.62	78.06	96.40	119.8	187.7	299.3	483.0	786.5	1288
37	22.27	26.14	30.96	37.00	44.61	54.24	66.48	82.10	102.1	127.7	203.3	329.1	539.9	893.5	1487
38	22.60	26.62	31.65	38.00	46.05	56.30	69.43	86.29	108.0	136.2	220.0	361.9	603.3	1015	1717
39	22.92	27.08	32.34	39.00	47.50	58.40	72.46	90.65	114.3	145.1	238.0	397.8	674.0	1153	1982
40	23.23	27.54	33.02	40.00	48.97	60.55	75.58	95.17	120.8	154.5	257.3	437.1	752.9	1309	2289

Table G-9 Uniform Present Value Factors Modified (UPV*)
Discount rate = 5%

Period	Rate of Price Increase per Period														
	1%	2%	3%	4%	5%	6%	7%	8%	9%	10%	12%	14%	16%	18%	20%
1	0.9619	0.9714	0.9810	0.9905	1.000	1.010	1.019	1.029	1.038	1.048	1.067	1.086	1.105	1.124	1.143
2	1.887	1.915	1.943	1.972	2.000	2.029	2.058	2.087	2.116	2.145	2.204	2.264	2.325	2.387	2.449
3	2.777	2.832	2.887	2.943	3.000	3.057	3.116	3.175	3.234	3.295	3.418	3.544	3.674	3.806	3.942
4	3.633	3.722	3.813	3.906	4.000	4.096	4.194	4.294	4.396	4.499	4.713	4.934	5.163	5.401	5.648
5	4.457	4.587	4.721	4.859	5.000	5.145	5.293	5.445	5.601	5.761	6.093	6.442	6.809	7.194	7.597
6	5.249	5.428	5.612	5.803	6.000	6.203	6.413	6.629	6.853	7.083	7.566	8.080	8.627	9.208	9.826
7	6.011	6.244	6.487	6.738	7.000	7.272	7.554	7.847	8.152	8.468	9.137	9.859	10.64	11.47	12.37
8	6.744	7.037	7.344	7.665	8.000	8.351	8.717	9.100	9.501	9.919	10.81	11.79	12.85	14.02	15.28
9	7.449	7.808	8.185	8.582	9.000	9.440	9.902	10.39	10.90	11.44	12.60	13.89	15.31	16.88	18.61
10	8.127	8.556	9.010	9.491	10.00	10.54	11.11	11.71	12.35	13.03	14.51	16.16	18.01	20.09	22.41
11	8.779	9.283	9.819	10.39	11.00	11.65	12.34	13.08	13.86	14.70	16.54	18.63	21.01	23.70	26.75
12	9.407	9.989	10.61	11.28	12.00	12.77	13.59	14.48	15.43	16.45	18.71	21.32	24.31	27.76	31.72
13	10.01	10.68	11.39	12.17	13.00	13.90	14.87	15.92	17.05	18.28	21.02	24.23	27.96	32.32	37.39
14	10.59	11.34	12.16	13.04	14.00	15.04	16.18	17.41	18.74	20.20	23.49	27.39	32.00	37.44	43.88
15	11.15	11.99	12.91	13.91	15.00	16.20	17.50	18.93	20.49	22.21	26.13	30.82	36.45	43.20	51.29
16	11.69	12.62	13.64	14.76	16.00	17.36	18.85	20.50	22.31	24.31	28.93	34.55	41.38	49.68	59.76
17	12.20	13.23	14.36	15.61	17.00	18.53	20.23	22.12	24.20	26.52	31.93	38.60	46.82	56.95	69.44
18	12.70	13.82	15.07	16.46	18.00	19.72	21.64	23.78	26.16	28.83	35.13	42.99	52.83	65.12	80.50
19	13.18	14.40	15.76	17.29	19.00	20.92	23.07	25.48	28.20	31.25	38.53	47.76	59.47	74.31	93.14
20	13.64	14.96	16.44	18.12	20.00	22.13	24.53	27.24	30.31	33.78	42.17	52.94	66.80	84.63	107.6
21	14.08	15.50	17.11	18.93	21.00	23.35	26.01	29.05	32.50	36.44	46.05	58.57	74.90	96.24	124.1
22	14.51	16.03	17.77	19.74	22.00	24.58	27.53	30.91	34.78	39.22	50.18	64.67	83.86	109.3	143.0
23	14.92	16.54	18.41	20.55	23.00	25.82	29.07	32.82	37.14	42.14	54.60	71.30	93.75	123.9	164.5
24	15.31	17.04	19.04	21.34	24.00	27.08	30.64	34.78	39.59	45.19	59.30	78.50	104.7	140.4	189.2
25	15.69	17.53	19.66	22.13	25.00	28.34	32.25	36.81	42.14	48.39	64.32	86.31	116.7	158.9	217.4
26	16.05	18.00	20.26	22.91	26.00	29.62	33.88	38.89	44.78	51.74	69.68	94.80	130.1	179.7	249.6
27	16.40	18.46	20.86	23.68	27.00	30.92	35.54	41.03	47.53	55.25	75.39	104.0	144.8	203.1	286.4
28	16.74	18.90	21.44	24.45	28.00	32.22	37.24	43.23	50.38	58.93	81.48	114.0	161.1	229.3	328.4
29	17.06	19.33	22.02	25.20	29.00	33.54	38.97	45.49	53.33	62.79	87.98	124.9	179.1	258.9	376.5
30	17.38	19.75	22.58	25.95	30.00	34.86	40.73	47.82	56.40	66.82	94.91	136.7	198.9	292.0	431.4
31	17.68	20.16	23.13	26.70	31.00	36.21	42.52	50.21	59.59	71.05	102.3	149.5	220.9	329.3	494.2
32	17.96	20.55	23.67	27.43	32.00	37.56	44.35	52.68	62.90	75.48	110.2	163.4	245.1	371.2	565.9
33	18.24	20.94	24.20	28.16	33.00	38.93	46.22	55.21	66.33	80.13	118.6	178.4	271.9	418.3	647.9
34	18.51	21.31	24.72	28.88	34.00	40.31	48.12	57.82	69.90	84.99	127.6	194.8	301.5	471.2	741.6
35	18.77	21.67	25.23	29.60	35.00	41.70	50.05	60.50	73.60	90.08	137.2	212.6	334.2	530.7	848.7
36	19.01	22.03	25.73	30.31	36.00	43.11	52.03	63.25	77.44	95.42	147.4	231.9	370.3	597.5	971.0
37	19.25	22.37	26.22	31.01	37.00	44.53	54.04	66.09	81.43	101.0	158.3	252.9	410.2	672.6	1111
38	19.48	22.70	26.70	31.71	38.00	45.96	56.08	69.01	85.57	106.9	169.9	275.6	454.3	757.0	1271
39	19.70	23.02	27.17	32.39	39.00	47.41	58.17	72.01	89.87	113.0	182.3	300.4	503.0	851.8	1453
40	19.91	23.34	27.64	33.08	40.00	48.87	60.30	75.09	94.33	119.4	195.5	327.2	556.8	958.4	1662

Table G-10 Uniform Present Value Factors Modified (UPV*)
Discount rate = 6%

	Rate of Price Increase per Period														
Period	1%	2%	3%	4%	5%	6%	7%	8%	9%	10%	12%	14%	16%	18%	20%
1	0.9528	0.9623	0.9717	0.9811	0.9906	1.000	1.009	1.019	1.028	1.038	1.057	1.075	1.094	1.113	1.132
2	1.861	1.888	1.916	1.944	1.972	2.000	2.028	2.057	2.086	2.115	2.173	2.232	2.292	2.352	2.414
3	2.726	2.779	2.833	2.888	2.944	3.000	3.057	3.115	3.173	3.232	3.353	3.476	3.602	3.732	3.865
4	3.550	3.637	3.725	3.815	3.907	4.000	4.095	4.192	4.291	4.392	4.599	4.814	5.037	5.268	5.507
5	4.335	4.462	4.591	4.724	4.860	5.000	5.143	5.290	5.441	5.595	5.916	6.253	6.606	6.977	7.366
6	5.084	5.256	5.433	5.616	5.805	6.000	6.201	6.409	6.623	6.844	7.307	7.800	8.324	8.880	9.471
7	5.797	6.019	6.251	6.491	6.741	7.000	7.269	7.549	7.839	8.140	8.778	9.464	10.20	11.00	11.85
8	6.476	6.755	7.046	7.350	7.668	8.000	8.347	8.710	9.089	9.485	10.33	11.25	12.26	13.36	14.55
9	7.124	7.462	7.818	8.192	8.586	9.000	9.436	9.893	10.37	10.88	11.97	13.18	14.51	15.98	17.61
10	7.740	8.143	8.568	9.019	9.496	10.00	10.53	11.10	11.70	12.33	13.71	15.25	16.97	18.91	21.06
11	8.328	8.798	9.298	9.830	10.40	11.00	11.64	12.33	13.06	13.83	15.54	17.48	19.67	22.16	24.98
12	8.888	9.428	10.01	10.63	11.29	12.00	12.76	13.58	14.45	15.39	17.48	19.87	22.62	25.78	29.41
13	9.422	10.03	10.69	11.41	12.17	13.00	13.89	14.85	15.89	17.01	19.52	22.44	25.85	29.81	34.43
14	9.930	10.62	11.36	12.17	13.05	14.00	15.03	16.15	17.37	18.69	21.68	25.21	29.38	34.30	40.10
15	10.41	11.18	12.01	12.92	13.92	15.00	16.18	17.48	18.89	20.43	23.97	28.19	33.25	39.30	46.53
16	10.88	11.72	12.65	13.66	14.78	16.00	17.35	18.82	20.45	22.24	26.38	31.40	37.48	44.86	53.81
17	11.32	12.24	13.26	14.38	15.63	17.00	18.52	20.20	22.06	24.12	28.93	34.84	42.11	51.05	62.05
18	11.73	12.74	13.86	15.09	16.47	18.00	19.70	21.60	23.71	26.07	31.62	38.55	47.18	57.94	71.38
19	12.13	13.22	14.44	15.79	17.30	19.00	20.90	23.03	25.41	28.09	34.47	42.53	52.72	65.61	81.94
20	12.51	13.69	15.00	16.47	18.13	20.00	22.11	24.48	27.16	30.19	37.48	46.82	58.79	74.16	93.89
21	12.88	14.13	15.55	17.14	18.95	21.00	23.32	25.96	28.96	32.36	40.66	51.42	65.43	83.66	107.4
22	13.22	14.56	16.08	17.80	19.76	22.00	24.55	27.47	30.80	34.62	44.01	56.38	72.70	94.25	122.7
23	13.55	14.97	16.59	18.45	20.57	23.00	25.79	29.01	32.70	36.97	47.56	61.71	80.65	106.0	140.1
24	13.87	15.37	17.10	19.08	21.36	24.00	27.05	30.57	34.66	39.40	51.31	67.44	89.35	119.1	159.7
25	14.16	15.75	17.58	19.70	22.15	25.00	28.31	32.17	36.67	41.92	55.27	73.61	98.87	133.8	181.9
26	14.45	16.12	18.06	20.31	22.93	26.00	29.59	33.79	38.73	44.54	59.46	80.24	109.3	150.0	207.1
27	14.72	16.47	18.52	20.91	23.71	27.00	30.88	35.45	40.86	47.26	63.88	87.37	120.7	168.1	235.6
28	14.98	16.81	18.97	21.49	24.48	28.00	32.18	37.14	43.04	50.08	68.55	95.04	133.2	188.2	267.8
29	15.22	17.14	19.40	22.07	25.24	29.00	33.49	38.86	45.29	53.01	73.49	103.3	146.8	210.7	304.4
30	15.46	17.46	19.82	22.64	25.99	30.00	34.82	40.61	47.60	56.05	78.70	112.2	161.8	235.6	345.7
31	15.68	17.76	20.23	23.19	26.73	31.00	36.15	42.39	49.97	59.20	84.22	121.7	178.1	263.4	392.5
32	15.90	18.05	20.63	23.73	27.47	32.00	37.50	44.21	52.42	62.47	90.04	132.0	196.0	294.4	445.4
33	16.10	18.33	21.02	24.27	28.20	33.00	38.87	46.07	54.93	65.87	96.19	143.0	215.6	328.8	505.4
34	16.29	18.60	21.40	24.79	28.93	34.00	40.24	47.95	57.51	69.39	102.7	154.9	237.1	367.1	573.3
35	16.48	18.86	21.76	25.30	29.65	35.00	41.63	49.88	60.17	73.05	109.6	167.6	260.5	409.8	650.1
36	16.65	19.12	22.12	25.81	30.36	36.00	43.03	51.84	62.90	76.84	116.8	181.4	286.2	457.3	737.1
37	16.82	19.36	22.47	26.30	31.06	37.00	44.45	53.83	65.71	80.78	124.5	196.1	314.3	510.2	835.6
38	16.98	19.59	22.80	26.79	31.76	38.00	45.88	55.87	68.59	84.87	132.6	212.0	345.0	569.1	947.1
39	17.13	19.81	23.13	27.26	32.45	39.00	47.32	57.94	71.56	89.11	141.2	229.1	378.7	634.6	1073
40	17.28	20.03	23.44	27.73	33.13	40.00	48.78	60.05	74.62	93.51	150.2	247.4	415.5	707.5	1216

Table G-11 Uniform Present Value Factors Modified (UPV*) Discount rate = 7%

	Rate of Price Increase per Period														
Period	1%	2%	3%	4%	5%	6%	7%	8%	9%	10%	12%	14%	16%	18%	20%
1	0.9439	0.9533	0.9626	0.9720	0.9813	0.9907	1.000	1.009	1.019	1.028	1.047	1.065	1.084	1.103	1.121
2	1.835	1.862	1.889	1.917	1.944	1.972	2.000	2.028	2.056	2.085	2.142	2.201	2.259	2.319	2.379
3	2.676	2.728	2.781	2.835	2.889	2.944	3.000	3.056	3.114	3.171	3.289	3.410	3.534	3.660	3.790
4	3.470	3.554	3.640	3.727	3.817	3.907	4.000	4.094	4.190	4.288	4.490	4.698	4.915	5.139	5.372
5	4.219	4.341	4.466	4.595	4.727	4.862	5.000	5.142	5.287	5.437	5.746	6.071	6.412	6.770	7.146
6	4.927	5.092	5.262	5.438	5.619	5.807	6.000	6.199	6.405	6.617	7.061	7.534	8.036	8.569	9.136
7	5.594	5.807	6.028	6.257	6.496	6.743	7.000	7.267	7.543	7.831	8.438	9.092	9.796	10.55	11.37
8	6.224	6.489	6.765	7.054	7.356	7.671	8.000	8.344	8.703	9.078	9.879	10.75	11.70	12.74	13.87
9	6.819	7.139	7.475	7.828	8.199	8.590	9.000	9.431	9.884	10.36	11.39	12.52	13.77	15.15	16.68
10	7.381	7.759	8.158	8.581	9.028	9.500	10.00	10.53	11.09	11.68	12.97	14.41	16.02	17.81	19.82
11	7.911	8.349	8.816	9.312	9.840	10.40	11.00	11.64	12.31	13.03	14.62	16.41	18.45	20.75	23.35
12	8.411	8.912	9.449	10.02	10.64	11.30	12.00	12.75	13.56	14.43	16.35	18.55	21.08	23.98	27.31
13	8.883	9.449	10.06	10.71	11.42	12.18	13.00	13.88	14.83	15.86	18.16	20.83	23.94	27.55	31.75
14	9.329	9.961	10.64	11.39	12.19	13.06	14.00	15.02	16.13	17.33	20.05	23.26	27.04	31.49	36.73
15	9.750	10.45	11.21	12.04	12.94	13.93	15.00	16.17	17.45	18.85	22.04	25.85	30.40	35.83	42.32
16	10.15	10.91	11.75	12.67	13.68	14.79	16.00	17.33	18.80	20.40	24.12	28.60	34.04	40.61	48.58
17	10.52	11.36	12.28	13.29	14.41	15.64	17.00	18.50	20.17	22.00	26.29	31.54	37.98	45.89	55.60
18	10.88	11.78	12.78	13.89	15.12	16.48	18.00	19.69	21.56	23.65	28.56	34.67	42.26	51.71	63.48
19	11.21	12.18	13.26	14.47	15.82	17.32	19.00	20.88	22.98	25.34	30.95	38.00	46.90	58.13	72.31
20	11.53	12.57	13.73	15.04	16.50	18.15	20.00	22.08	24.43	27.08	33.44	41.55	51.93	65.21	82.22
21	11.82	12.93	14.18	15.59	17.18	18.97	21.00	23.30	25.91	28.87	36.05	45.34	57.38	73.02	93.33
22	12.10	13.28	14.61	16.12	17.84	19.78	22.00	24.53	27.41	30.70	38.78	49.37	63.29	81.63	105.8
23	12.37	13.61	15.03	16.64	18.48	20.59	23.00	25.77	28.94	32.59	41.64	53.66	69.70	91.12	119.8
24	12.62	13.93	15.43	17.15	19.12	21.39	24.00	27.02	30.50	34.53	44.63	58.24	76.65	101.6	135.4
25	12.86	14.23	15.82	17.64	19.74	22.18	25.00	28.28	32.09	36.53	47.76	63.12	84.18	113.1	153.0
26	13.08	14.52	16.19	18.12	20.36	22.96	26.00	29.55	33.71	38.58	51.04	68.31	92.34	125.9	172.7
27	13.29	14.80	16.55	18.58	20.96	23.74	27.00	30.84	35.36	40.69	54.47	73.85	101.2	139.9	194.8
28	13.49	15.06	16.89	19.03	21.55	24.51	28.00	32.13	37.04	42.86	58.06	79.74	110.8	155.4	219.6
29	13.68	15.31	17.22	19.47	22.12	25.27	29.00	33.44	38.75	45.09	61.83	86.02	121.2	172.5	247.4
30	13.85	15.55	17.54	19.90	22.69	26.02	30.00	34.77	40.49	47.38	65.76	92.72	132.5	191.3	278.6
31	14.02	15.77	17.85	20.31	23.25	26.77	31.00	36.10	42.27	49.74	69.88	99.85	144.7	212.1	313.6
32	14.18	15.99	18.14	20.71	23.80	27.51	32.00	37.45	44.07	52.16	74.19	107.4	158.0	235.0	352.8
33	14.33	16.19	18.43	21.10	24.33	28.24	33.00	38.81	45.92	54.65	78.71	115.5	172.3	260.2	396.8
34	14.47	16.39	18.70	21.48	24.86	28.97	34.00	40.18	47.79	57.21	83.43	124.2	187.9	288.1	446.1
35	14.60	16.58	18.96	21.85	25.38	29.69	35.00	41.56	49.71	59.85	88.38	133.4	204.8	318.8	501.5
36	14.73	16.76	19.22	22.21	25.88	30.40	36.00	42.96	51.65	62.55	93.55	143.1	223.1	352.7	563.5
37	14.84	16.93	19.46	22.56	26.38	31.11	37.00	44.37	53.64	65.33	98.97	153.6	242.9	390.1	633.1
38	14.95	17.09	19.70	22.90	26.87	31.81	38.00	45.80	55.66	68.19	104.6	164.7	264.5	431.3	711.1
39	15.06	17.24	19.92	23.23	27.35	32.50	39.00	47.23	57.72	71.13	110.6	176.5	287.8	476.7	798.6
40	15.16	17.39	20.14	23.55	27.82	33.19	40.00	48.68	59.82	74.16	116.8	189.1	313.1	526.8	896.8

Table G-12 Uniform Present Value Factors Modified (UPV*)
Discount rate = 8%

	Rate of Price Increase per Period														
Period	1%	2%	3%	4%	5%	6%	7%	8%	9%	10%	12%	14%	16%	18%	20%
1	0.9352	0.9444	0.9537	0.9630	0.9722	0.9815	0.9907	1.000	1.009	1.019	1.037	1.056	1.074	1.093	1.111
2	1.810	1.836	1.863	1.890	1.917	1.945	1.972	2.000	2.028	2.056	2.112	2.170	2.228	2.286	2.346
3	2.628	2.679	2.731	2.783	2.836	2.890	2.945	3.000	3.056	3.112	3.228	3.346	3.467	3.591	3.717
4	3.393	3.474	3.558	3.643	3.730	3.818	3.908	4.000	4.093	4.189	4.384	4.587	4.798	5.016	5.242
5	4.108	4.226	4.347	4.471	4.598	4.729	4.863	5.000	5.141	5.285	5.584	5.898	6.227	6.573	6.935
6	4.777	4.936	5.099	5.269	5.443	5.623	5.809	6.000	6.197	6.401	6.828	7.281	7.762	8.274	8.817
7	5.402	5.606	5.817	6.036	6.264	6.500	6.745	7.000	7.264	7.538	8.118	8.741	9.412	10.13	10.91
8	5.987	6.239	6.501	6.776	7.062	7.361	7.674	8.000	8.341	8.696	9.455	10.28	11.18	12.16	13.23
9	6.534	6.837	7.154	7.488	7.838	8.207	8.593	9.000	9.427	9.876	10.84	11.91	13.09	14.38	15.81
10	7.046	7.401	7.777	8.173	8.593	9.036	9.505	10.00	10.52	11.08	12.28	13.63	15.13	16.81	18.68
11	7.525	7.935	8.370	8.834	9.326	9.850	10.41	11.00	11.63	12.30	13.77	15.44	17.32	19.46	21.87
12	7.972	8.438	8.937	9.469	10.04	10.65	11.30	12.00	12.75	13.55	15.32	17.35	19.68	22.35	25.41
13	8.391	8.914	9.476	10.08	10.73	11.43	12.19	13.00	13.87	14.82	16.92	19.37	22.21	25.51	29.34
14	8.782	9.363	9.991	10.67	11.41	12.20	13.07	14.00	15.01	16.11	18.59	21.50	24.93	28.97	33.71
15	9.148	9.787	10.48	11.24	12.06	12.96	13.94	15.00	16.16	17.43	20.31	23.75	27.85	32.74	38.57
16	9.490	10.19	10.95	11.79	12.70	13.70	14.80	16.00	17.32	18.77	22.10	26.13	30.99	36.86	43.97
17	9.810	10.57	11.40	12.31	13.32	14.43	15.65	17.00	18.49	20.13	23.96	28.64	34.36	41.37	49.96
18	10.11	10.92	11.82	12.82	13.92	15.14	16.50	18.00	19.67	21.52	25.88	31.28	37.98	46.29	56.62
19	10.39	11.26	12.23	13.31	14.51	15.84	17.33	19.00	20.86	22.94	27.88	34.08	41.87	51.67	64.03
20	10.65	11.58	12.62	13.78	15.08	16.53	18.16	20.00	22.06	24.39	29.95	37.02	46.04	57.55	72.25
21	10.90	11.88	12.99	14.23	15.63	17.21	18.99	21.00	23.28	25.86	32.09	40.14	50.53	63.97	81.39
22	11.13	12.17	13.34	14.67	16.17	17.87	19.80	22.00	24.50	27.35	34.32	43.42	55.34	70.99	91.55
23	11.34	12.43	13.68	15.09	16.69	18.52	20.61	23.00	25.74	28.88	36.63	46.89	60.52	78.65	102.8
24	11.54	12.69	14.00	15.49	17.20	19.16	21.41	24.00	26.99	30.43	39.02	50.55	66.07	87.03	115.4
25	11.73	12.93	14.30	15.88	17.69	19.79	22.20	25.00	28.24	32.01	41.50	54.41	72.04	96.18	129.3
26	11.90	13.15	14.59	16.25	18.17	20.40	22.99	26.00	29.52	33.62	44.08	58.49	78.45	106.2	144.8
27	12.07	13.37	14.87	16.62	18.64	21.00	23.77	27.00	30.80	35.27	46.75	62.80	85.34	117.1	162.0
28	12.22	13.57	15.14	16.96	19.10	21.60	24.54	28.00	32.09	36.94	49.52	67.34	92.73	129.0	181.1
29	12.36	13.76	15.39	17.30	19.54	22.18	25.30	29.00	33.40	38.64	52.39	72.14	100.7	142.1	202.3
30	12.50	13.94	15.63	17.62	19.97	22.75	26.06	30.00	34.72	40.37	55.37	77.20	109.2	156.3	225.9
31	12.62	14.11	15.86	17.93	20.38	23.31	26.81	31.00	36.05	42.14	58.45	82.55	118.4	171.9	252.1
32	12.74	14.27	16.08	18.23	20.79	23.86	27.55	32.00	37.39	43.94	61.65	88.19	128.2	188.9	281.2
33	12.85	14.42	16.29	18.52	21.19	24.40	28.28	33.00	38.75	45.77	64.98	94.14	138.8	207.5	313.6
34	12.95	14.57	16.49	18.79	21.57	24.93	29.01	34.00	40.11	47.64	68.42	100.4	150.1	227.8	349.5
35	13.05	14.70	16.68	19.06	21.94	25.45	29.73	35.00	41.49	49.54	71.99	107.1	162.3	250.0	389.5
36	13.14	14.83	16.86	19.32	22.31	25.96	30.45	36.00	42.89	51.47	75.69	114.1	175.4	274.2	433.9
37	13.22	14.95	17.03	19.57	22.66	26.46	31.16	37.00	44.29	53.45	79.53	121.5	189.5	300.7	483.2
38	13.30	15.06	17.20	19.80	23.00	26.95	31.86	38.00	45.71	55.45	83.52	129.3	204.6	329.6	538.0
39	13.37	15.17	17.36	20.03	23.33	27.43	32.56	39.00	47.15	57.50	87.65	137.5	220.8	361.2	598.9
40	13.44	15.27	17.51	20.25	23.66	27.91	33.25	40.00	48.59	59.58	91.93	146.2	238.3	395.8	666.6

Table G-13 Uniform Present Value Factors Modified (UPV*)
Discount rate = 9%

	Rate of Price Increase per Period														
Period	1%	2%	3%	4%	5%	6%	7%	8%	9%	10%	12%	14%	16%	18%	20%
1	0.9266	0.9358	0.9450	0.9541	0.9633	0.9725	0.9817	0.9908	1.000	1.009	1.028	1.046	1.064	1.083	1.101
2	1.785	1.811	1.838	1.864	1.891	1.918	1.945	1.973	2.000	2.028	2.083	2.140	2.197	2.255	2.313
3	2.581	2.631	2.682	2.733	2.785	2.838	2.891	2.945	3.000	3.055	3.168	3.284	3.402	3.523	3.647
4	3.318	3.398	3.479	3.562	3.646	3.732	3.820	3.909	4.000	4.093	4.283	4.480	4.685	4.897	5.116
5	4.001	4.115	4.232	4.353	4.476	4.602	4.731	4.864	5.000	5.139	5.428	5.732	6.050	6.384	6.733
6	4.634	4.787	4.944	5.107	5.275	5.448	5.626	5.810	6.000	6.196	6.605	7.040	7.503	7.993	8.514
7	5.221	5.415	5.617	5.827	6.045	6.270	6.505	6.748	7.000	7.262	7.815	8.409	9.049	9.736	10.47
8	5.764	6.003	6.253	6.514	6.786	7.070	7.367	7.677	8.000	8.337	9.057	9.841	10.69	11.62	12.63
9	6.268	6.553	6.854	7.169	7.500	7.848	8.213	8.597	9.000	9.423	10.33	11.34	12.44	13.66	15.01
10	6.734	7.068	7.421	7.794	8.188	8.605	9.044	9.509	10.00	10.52	11.65	12.90	14.31	15.88	17.62
11	7.166	7.550	7.958	8.391	8.851	9.340	9.860	10.41	11.00	11.62	12.99	14.54	16.29	18.27	20.50
12	7.567	8.001	8.465	8.960	9.490	10.06	10.66	11.31	12.00	12.74	14.38	16.25	18.40	20.86	23.67
13	7.938	8.423	8.944	9.503	10.10	10.75	11.45	12.19	13.00	13.87	15.80	18.05	20.65	23.66	27.16
14	8.282	8.818	9.396	10.02	10.70	11.43	12.22	13.07	14.00	15.00	17.26	19.92	23.04	26.70	31.00
15	8.601	9.187	9.824	10.52	11.27	12.09	12.98	13.94	15.00	16.15	18.77	21.88	25.58	29.99	35.23
16	8.896	9.533	10.23	10.99	11.82	12.73	13.72	14.81	16.00	17.31	20.31	23.93	28.29	33.55	39.89
17	9.170	9.857	10.61	11.44	12.35	13.35	14.45	15.66	17.00	18.47	21.90	26.07	31.17	37.40	45.02
18	9.424	10.16	10.97	11.87	12.86	13.95	15.17	16.51	18.00	19.65	23.53	28.31	34.24	41.57	50.66
19	9.659	10.44	11.31	12.28	13.35	14.54	15.87	17.35	19.00	20.84	25.20	30.66	37.50	46.09	56.88
20	9.876	10.71	11.63	12.67	13.82	15.11	16.56	18.18	20.00	22.04	26.92	33.11	40.97	50.97	63.72
21	10.08	10.96	11.94	13.04	14.28	15.67	17.24	19.00	21.00	23.25	28.69	35.68	44.67	56.26	71.25
22	10.26	11.19	12.23	13.40	14.72	16.21	17.90	19.82	22.00	24.48	30.51	38.36	48.60	61.99	79.54
23	10.44	11.41	12.50	13.74	15.14	16.74	18.56	20.63	23.00	25.71	32.38	41.16	52.78	68.19	88.67
24	10.60	11.61	12.76	14.06	15.55	17.25	19.20	21.43	24.00	26.96	34.30	44.10	57.24	74.91	98.71
25	10.75	11.80	13.00	14.37	15.94	17.75	19.83	22.23	25.00	28.21	36.27	47.17	61.98	82.17	109.8
26	10.89	11.98	13.23	14.66	16.32	18.23	20.44	23.01	26.00	29.48	38.29	50.38	67.02	90.04	122.0
27	11.01	12.14	13.44	14.95	16.68	18.70	21.05	23.79	27.00	30.76	40.38	53.73	72.39	98.56	135.4
28	11.13	12.30	13.65	15.21	17.04	19.16	21.65	24.57	28.00	32.05	42.51	57.24	78.10	107.8	150.1
29	11.24	12.45	13.84	15.47	17.37	19.60	22.23	25.33	29.00	33.35	44.71	60.92	84.18	117.8	166.4
30	11.34	12.58	14.03	15.72	17.70	20.04	22.80	26.09	30.00	34.67	46.97	64.76	90.66	128.6	184.3
31	11.44	12.71	14.20	15.95	18.01	20.46	23.37	26.84	31.00	36.00	49.29	68.77	97.54	140.3	204.0
32	11.52	12.83	14.36	16.17	18.32	20.87	23.92	27.59	32.00	37.34	51.67	72.97	104.9	152.9	225.7
33	11.60	12.94	14.52	16.38	18.61	21.27	24.46	28.32	33.00	38.69	54.12	77.37	112.7	166.6	249.5
34	11.68	13.05	14.66	16.59	18.89	21.65	25.00	29.05	34.00	40.05	56.64	81.96	121.0	181.5	275.8
35	11.75	13.14	14.80	16.78	19.16	22.03	25.52	29.78	35.00	41.43	59.23	86.77	129.8	197.5	304.7
36	11.81	13.24	14.93	16.96	19.42	22.40	26.03	30.50	36.00	42.82	61.88	91.79	139.2	214.9	336.6
37	11.87	13.32	15.05	17.14	19.67	22.75	26.54	31.21	37.00	44.22	64.62	97.05	149.2	233.8	371.7
38	11.93	13.40	15.17	17.31	19.91	23.10	27.03	31.91	38.00	45.64	67.42	102.5	159.9	254.2	410.3
39	11.98	13.48	15.28	17.47	20.14	23.44	27.52	32.61	39.00	47.06	70.30	108.3	171.2	276.2	452.8
40	12.03	13.55	15.38	17.62	20.37	23.76	27.99	33.30	40.00	48.50	73.27	114.3	183.2	300.1	499.6

Table G-14 Uniform Present Value Factors Modified (UPV*)
Discount rate = 10%

	Rate of Price Increase per Period														
Period	1%	2%	3%	4%	5%	6%	7%	8%	9%	10%	12%	14%	16%	18%	20%
1	0.9182	0.9273	0.9364	0.9455	0.9545	0.9636	0.9727	0.9818	0.9909	1.000	1.018	1.036	1.055	1.073	1.091
2	1.761	1.787	1.813	1.839	1.866	1.892	1.919	1.946	1.973	2.000	2.055	2.110	2.167	2.223	2.281
3	2.535	2.584	2.634	2.684	2.735	2.787	2.839	2.892	2.946	3.000	3.110	3.224	3.339	3.458	3.579
4	3.246	3.324	3.403	3.483	3.566	3.649	3.735	3.821	3.910	4.000	4.185	4.377	4.576	4.782	4.996
5	3.899	4.009	4.123	4.239	4.358	4.480	4.605	4.734	4.865	5.000	5.279	5.573	5.880	6.203	6.541
6	4.498	4.645	4.797	4.953	5.115	5.281	5.453	5.630	5.812	6.000	6.394	6.812	7.255	7.726	8.226
7	5.048	5.234	5.428	5.628	5.837	6.053	6.277	6.509	6.750	7.000	7.528	8.096	8.706	9.361	10.06
8	5.553	5.781	6.019	6.267	6.526	6.796	7.078	7.372	7.680	8.000	8.683	9.426	10.24	11.11	12.07
9	6.017	6.288	6.572	6.871	7.184	7.513	7.858	8.220	8.601	9.000	9.859	10.81	11.85	13.00	14.26
10	6.443	6.758	7.090	7.441	7.812	8.203	8.616	9.053	9.513	10.00	11.06	12.23	13.55	15.01	16.65
11	6.834	7.194	7.575	7.981	8.411	8.868	9.354	9.870	10.42	11.00	12.28	13.72	15.34	17.18	19.25
12	7.193	7.598	8.030	8.491	8.983	9.510	10.07	10.67	11.31	12.00	13.52	15.25	17.23	19.50	22.09
13	7.523	7.972	8.455	8.973	9.530	10.13	10.77	11.46	12.20	13.00	14.78	16.84	19.23	21.99	25.19
14	7.825	8.320	8.853	9.429	10.05	10.72	11.45	12.23	13.08	14.00	16.07	18.49	21.33	24.66	28.57
15	8.103	8.642	9.226	9.860	10.55	11.30	12.11	12.99	13.95	15.00	17.38	20.20	23.55	27.53	32.26
16	8.358	8.941	9.576	10.27	11.02	11.85	12.75	13.74	14.82	16.00	18.71	21.97	25.89	30.60	36.28
17	8.593	9.218	9.903	10.65	11.48	12.38	13.38	14.47	15.67	17.00	20.07	23.81	28.36	33.90	40.67
18	8.808	9.475	10.21	11.02	11.91	12.90	13.98	15.19	16.52	18.00	21.45	25.71	30.96	37.44	45.46
19	9.005	9.713	10.50	11.36	12.32	13.39	14.58	15.89	17.36	19.00	22.86	27.68	33.70	41.24	50.69
20	9.187	9.934	10.76	11.69	12.72	13.87	15.15	16.59	18.20	20.00	24.30	29.72	36.59	45.31	56.38
21	9.353	10.14	11.02	12.00	13.09	14.33	15.71	17.27	19.02	21.00	25.76	31.84	39.64	49.68	62.60
22	9.506	10.33	11.25	12.29	13.45	14.77	16.25	17.94	19.84	22.00	27.24	34.03	42.86	54.36	69.38
23	9.647	10.50	11.47	12.56	13.80	15.20	16.78	18.59	20.65	23.00	28.76	36.31	46.25	59.39	76.78
24	9.776	10.67	11.68	12.82	14.12	15.61	17.30	19.24	21.45	24.00	30.30	38.66	49.83	64.78	84.85
25	9.894	10.82	11.87	13.07	14.44	16.00	17.80	19.87	22.25	25.00	31.87	41.11	53.60	70.57	93.66
26	10.00	10.96	12.05	13.30	14.73	16.38	18.29	20.49	23.04	26.00	33.46	43.64	57.58	76.77	103.3
27	10.10	11.09	12.22	13.52	15.02	16.75	18.76	21.10	23.82	27.00	35.09	46.26	61.78	83.43	113.7
28	10.19	11.21	12.38	13.73	15.29	17.11	19.22	21.70	24.59	28.00	36.75	48.98	66.20	90.57	125.2
29	10.28	11.32	12.53	13.93	15.55	17.45	19.67	22.28	25.36	29.00	38.43	51.80	70.87	98.23	137.6
30	10.36	11.43	12.67	14.11	15.80	17.78	20.11	22.86	26.12	30.00	40.15	54.72	75.79	106.4	151.2
31	10.43	11.52	12.80	14.29	16.04	18.09	20.53	23.43	26.88	31.00	41.90	57.74	80.97	115.3	166.1
32	10.49	11.61	12.92	14.45	16.26	18.40	20.94	23.98	27.62	32.00	43.68	60.88	86.44	124.7	182.3
33	10.55	11.69	13.03	14.61	16.48	18.69	21.35	24.53	28.36	33.00	45.49	64.13	92.21	134.9	199.9
34	10.61	11.77	13.14	14.76	16.68	18.98	21.74	25.06	29.09	34.00	47.34	67.50	98.30	145.7	219.2
35	10.66	11.84	13.24	14.90	16.88	19.25	22.12	25.59	29.82	35.00	49.21	70.99	104.7	157.4	240.2
36	10.70	11.91	13.33	15.03	17.07	19.52	22.49	26.11	30.54	36.00	51.13	74.61	111.5	169.9	263.2
37	10.75	11.97	13.42	15.16	17.24	19.77	22.85	26.61	31.25	37.00	53.08	78.35	118.6	183.4	288.2
38	10.78	12.03	13.50	15.28	17.41	20.01	23.19	27.11	31.96	38.00	55.06	82.24	126.1	197.8	315.5
39	10.82	12.08	13.58	15.39	17.58	20.25	23.54	27.60	32.66	39.00	57.08	86.27	134.1	213.2	345.2
40	10.85	12.13	13.65	15.49	17.73	20.48	23.87	28.08	33.36	40.00	59.13	90.44	142.4	229.8	377.7

Table G-15 Uniform Present Value Factors Modified (UPV*) Discount rate = 12%

Period	Rate of Price Increase per Period 1%	2%	3%	4%	5%	6%	7%	8%	9%	10%	12%	14%	16%	18%	20%
1	0.9018	0.9107	0.9196	0.9286	0.9375	0.9464	0.9554	0.9643	0.9732	0.9821	1.000	1.018	1.036	1.054	1.071
2	1.715	1.740	1.765	1.791	1.816	1.842	1.868	1.894	1.920	1.947	2.000	2.054	2.108	2.164	2.219
3	2.448	2.495	2.543	2.591	2.640	2.690	2.740	2.791	2.842	2.894	3.000	3.108	3.219	3.333	3.449
4	3.110	3.183	3.258	3.335	3.413	3.492	3.573	3.655	3.739	3.825	4.000	4.182	4.370	4.565	4.767
5	3.706	3.810	3.916	4.025	4.137	4.252	4.369	4.489	4.612	4.738	5.000	5.274	5.562	5.863	6.179
6	4.244	4.380	4.521	4.666	4.816	4.970	5.129	5.293	5.462	5.636	6.000	6.386	6.796	7.231	7.692
7	4.729	4.900	5.078	5.262	5.452	5.650	5.856	6.068	6.289	6.517	7.000	7.518	8.075	8.672	9.313
8	5.166	5.373	5.589	5.814	6.049	6.294	6.550	6.816	7.094	7.383	8.000	8.670	9.399	10.19	11.05
9	5.561	5.804	6.060	6.328	6.609	6.903	7.213	7.537	7.877	8.234	9.000	9.843	10.77	11.79	12.91
10	5.916	6.197	6.492	6.804	7.133	7.480	7.846	8.232	8.639	9.069	10.00	11.04	12.19	13.47	14.90
11	6.237	6.554	6.890	7.247	7.625	8.026	8.451	8.902	9.381	9.889	11.00	12.25	13.66	15.25	17.04
12	6.526	6.880	7.256	7.658	8.086	8.542	9.029	9.549	10.10	10.69	12.00	13.49	15.19	17.12	19.33
13	6.787	7.176	7.593	8.039	8.518	9.031	9.581	10.17	10.81	11.49	13.00	14.75	16.76	19.09	21.78
14	7.022	7.446	7.902	8.394	8.923	9.494	10.11	10.77	11.49	12.26	14.00	16.03	18.40	21.17	24.41
15	7.234	7.692	8.187	8.723	9.303	9.931	10.61	11.35	12.15	13.03	15.00	17.33	20.09	23.36	27.22
16	7.426	7.916	8.449	9.028	9.659	10.35	11.09	11.91	12.80	13.78	16.00	18.66	21.84	25.66	30.24
17	7.598	8.120	8.689	9.312	9.993	10.74	11.55	12.45	13.43	14.51	17.00	20.01	23.66	28.09	33.47
18	7.754	8.306	8.911	9.575	10.31	11.11	11.99	12.97	14.05	15.23	18.00	21.39	25.54	30.65	36.93
19	7.894	8.475	9.114	9.820	10.60	11.46	12.41	13.47	14.64	15.94	19.00	22.79	27.49	33.34	40.64
20	8.020	8.629	9.302	10.05	10.87	11.79	12.82	13.95	15.22	16.64	20.00	24.21	29.51	36.18	44.62
21	8.134	8.769	9.474	10.26	11.13	12.11	13.20	14.42	15.79	17.33	21.00	25.66	31.60	39.17	48.87
22	8.237	8.897	9.632	10.45	11.37	12.41	13.56	14.87	16.34	18.00	22.00	27.14	33.76	42.33	53.44
23	8.330	9.013	9.778	10.64	11.60	12.69	13.91	15.30	16.88	18.66	23.00	28.64	36.00	45.65	58.32
24	8.414	9.119	9.912	10.80	11.81	12.95	14.25	15.72	17.40	19.31	24.00	30.17	38.32	49.15	63.56
25	8.489	9.216	10.03	10.96	12.01	13.21	14.57	16.12	17.90	19.95	25.00	31.72	40.73	52.83	69.17
26	8.557	9.304	10.15	11.11	12.20	13.45	14.87	16.51	18.40	20.57	26.00	33.31	43.22	56.72	75.19
27	8.619	9.384	10.25	11.24	12.37	13.67	15.16	16.89	18.88	21.19	27.00	34.92	45.80	60.81	81.63
28	8.674	9.456	10.35	11.37	12.54	13.89	15.44	17.25	19.35	21.79	28.00	36.56	48.47	65.12	88.53
29	8.724	9.523	10.44	11.48	12.69	14.09	15.71	17.60	19.80	22.38	29.00	38.23	51.23	69.66	95.92
30	8.769	9.583	10.52	11.59	12.84	14.28	15.96	17.93	20.24	22.97	30.00	39.93	54.10	74.45	103.8
31	8.809	9.638	10.59	11.69	12.97	14.46	16.21	18.26	20.67	23.54	31.00	41.67	57.07	79.49	112.3
32	8.846	9.689	10.66	11.79	13.10	14.63	16.44	18.57	21.09	24.10	32.00	43.43	60.14	84.80	121.4
33	8.879	9.734	10.72	11.87	13.22	14.80	16.66	18.87	21.50	24.65	33.00	45.22	63.32	90.40	131.2
34	8.909	9.776	10.78	11.95	13.33	14.95	16.87	19.16	21.90	25.19	34.00	47.05	66.62	96.29	141.6
35	8.935	9.814	10.83	12.03	13.43	15.09	17.07	19.44	22.29	25.73	35.00	48.90	70.04	102.5	152.8
36	8.960	9.848	10.88	12.10	13.53	15.23	17.27	19.71	22.66	26.25	36.00	50.80	73.57	109.0	164.8
37	8.981	9.880	10.93	12.16	13.62	15.36	17.45	19.97	23.03	26.76	37.00	52.72	77.24	115.9	177.6
38	9.001	9.908	10.97	12.22	13.71	15.49	17.63	20.22	23.38	27.27	38.00	54.68	81.03	123.2	191.4
39	9.019	9.934	11.01	12.28	13.79	15.60	17.80	20.46	23.73	27.76	39.00	56.67	84.96	130.9	206.1
40	9.035	9.958	11.04	12.33	13.87	15.71	17.98	20.70	24.07	28.25	40.00	58.70	89.03	138.9	221.9

Table G-16 Uniform Present Value Factors Modified (UPV*)
Discount rate = 14%

	Rate of Price Increase per Period														
Period	1%	2%	3%	4%	5%	6%	7%	8%	9%	10%	12%	14%	16%	18%	20%
1	0.8860	0.8947	0.9035	0.9123	0.9211	0.9298	0.9386	0.9474	0.9561	0.9649	0.9825	1.000	1.018	1.035	1.053
2	1.671	1.695	1.720	1.745	1.769	1.794	1.820	1.845	1.870	1.896	1.948	2.000	2.053	2.107	2.161
3	2.366	2.412	2.457	2.504	2.551	2.598	2.646	2.695	2.744	2.794	2.896	3.000	3.106	3.216	3.327
4	2.982	3.052	3.124	3.196	3.270	3.346	3.423	3.501	3.580	3.661	3.828	4.000	4.179	4.363	4.555
5	3.528	3.626	3.726	3.828	3.933	4.041	4.151	4.264	4.379	4.498	4.743	5.000	5.269	5.552	5.847
6	4.012	4.139	4.270	4.405	4.544	4.687	4.835	4.987	5.143	5.305	5.642	6.000	6.379	6.781	7.207
7	4.440	4.598	4.761	4.931	5.106	5.288	5.476	5.672	5.874	6.084	6.526	7.000	7.509	8.055	8.639
8	4.820	5.009	5.205	5.410	5.624	5.847	6.079	6.321	6.572	6.835	7.394	8.000	8.658	9.372	10.15
9	5.156	5.376	5.607	5.848	6.101	6.366	6.644	6.935	7.240	7.560	8.246	9.000	9.828	10.74	11.73
10	5.454	5.705	5.969	6.247	6.541	6.849	7.175	7.518	7.879	8.260	9.084	10.00	11.02	12.15	13.40
11	5.718	5.999	6.297	6.612	6.945	7.298	7.673	8.069	8.489	8.935	9.907	11.00	12.23	13.61	15.16
12	5.952	6.263	6.593	6.944	7.318	7.716	8.140	8.592	9.073	9.586	10.72	12.00	13.46	15.12	17.01
13	6.159	6.498	6.860	7.247	7.661	8.104	8.579	9.087	9.631	10.21	11.51	13.00	14.71	16.69	18.96
14	6.343	6.709	7.102	7.524	7.977	8.466	8.991	9.556	10.17	10.82	12.29	14.00	15.99	18.31	21.01
15	6.506	6.897	7.320	7.776	8.269	8.801	9.377	10.00	10.68	11.41	13.06	15.00	17.29	19.99	23.17
16	6.650	7.066	7.517	8.006	8.537	9.114	9.740	10.42	11.16	11.97	13.81	16.00	18.61	21.72	25.44
17	6.777	7.217	7.695	8.216	8.784	9.404	10.08	10.82	11.63	12.52	14.55	17.00	19.95	23.52	27.83
18	6.890	7.352	7.856	8.408	9.012	9.674	10.40	11.20	12.08	13.04	15.28	18.00	21.32	25.38	30.35
19	6.991	7.473	8.002	8.582	9.221	9.925	10.70	11.56	12.50	13.55	15.99	19.00	22.71	27.31	33.00
20	7.079	7.581	8.133	8.742	9.414	10.16	10.98	11.90	12.91	14.04	16.69	20.00	24.13	29.30	35.79
21	7.158	7.678	8.252	8.887	9.592	10.38	11.25	12.22	13.30	14.51	17.38	21.00	25.57	31.36	38.73
22	7.228	7.764	8.359	9.020	9.756	10.58	11.49	12.52	13.67	14.97	18.06	22.00	27.03	33.50	41.82
23	7.290	7.842	8.456	9.141	9.907	10.76	11.73	12.81	14.03	15.41	18.73	23.00	28.53	35.71	45.07
24	7.344	7.911	8.544	9.252	10.05	10.94	11.95	13.08	14.37	15.83	19.38	24.00	30.04	38.00	48.50
25	7.393	7.973	8.623	9.352	10.17	11.10	12.15	13.34	14.70	16.24	20.02	25.00	31.59	40.36	52.10
26	7.436	8.028	8.694	9.444	10.29	11.25	12.34	13.59	15.01	16.64	20.65	26.00	33.16	42.82	55.90
27	7.474	8.078	8.759	9.528	10.40	11.39	12.52	13.82	15.31	17.02	21.27	27.00	34.76	45.35	59.89
28	7.507	8.123	8.817	9.605	10.50	11.52	12.69	14.04	15.59	17.38	21.88	28.00	36.39	47.98	64.09
29	7.537	8.162	8.870	9.674	10.59	11.64	12.85	14.25	15.86	17.74	22.48	29.00	38.04	50.70	68.52
30	7.564	8.198	8.918	9.738	10.68	11.76	13.00	14.44	16.12	18.08	23.07	30.00	39.73	53.51	73.18
31	7.587	8.230	8.961	9.796	10.76	11.86	13.14	14.63	16.37	18.41	23.65	31.00	41.44	56.42	78.08
32	7.608	8.258	8.999	9.849	10.83	11.96	13.27	14.81	16.61	18.73	24.22	32.00	43.19	59.44	83.25
33	7.626	8.284	9.035	9.897	10.89	12.05	13.40	14.98	16.84	19.04	24.77	33.00	44.96	62.56	88.68
34	7.643	8.306	9.066	9.941	10.95	12.13	13.51	15.14	17.06	19.34	25.32	34.00	46.77	65.79	94.40
35	7.657	8.327	9.095	9.982	11.01	12.21	13.62	15.29	17.26	19.62	25.86	35.00	48.61	69.13	100.4
36	7.670	8.345	9.121	10.02	11.06	12.28	13.72	15.43	17.46	19.90	26.39	36.00	50.48	72.59	106.8
37	7.681	8.361	9.144	10.05	11.11	12.35	13.82	15.57	17.65	20.17	26.91	37.00	52.38	76.18	113.4
38	7.691	8.376	9.166	10.08	11.15	12.42	13.91	15.69	17.83	20.42	27.42	38.00	54.32	79.88	120.5
39	7.700	8.389	9.185	10.11	11.19	12.47	13.99	15.81	18.01	20.67	27.92	39.00	56.29	83.72	127.8
40	7.708	8.401	9.202	10.14	11.23	12.53	14.07	15.93	18.17	20.91	28.41	40.00	58.29	87.70	135.6

Table G-17 Uniform Present Value Factors Modified (UPV*)
Discount rate = 15%

	Rate of Price Increase per Period														
Period	1%	2%	3%	4%	5%	6%	7%	8%	9%	10%	12%	14%	16%	18%	20%
1	0.8783	0.8870	0.8957	0.9043	0.9130	0.9217	0.9304	0.9391	0.9478	0.9565	0.9739	0.9913	1.009	1.026	1.043
2	1.650	1.674	1.698	1.722	1.747	1.771	1.796	1.821	1.846	1.871	1.922	1.974	2.026	2.079	2.132
3	2.327	2.371	2.416	2.462	2.508	2.554	2.602	2.649	2.698	2.747	2.846	2.948	3.052	3.159	3.269
4	2.922	2.990	3.060	3.131	3.203	3.276	3.351	3.427	3.505	3.584	3.746	3.914	4.088	4.268	4.454
5	3.445	3.539	3.636	3.736	3.837	3.942	4.048	4.158	4.270	4.384	4.622	4.871	5.132	5.405	5.691
6	3.903	4.026	4.152	4.283	4.417	4.555	4.697	4.844	4.995	5.150	5.475	5.820	6.185	6.572	6.982
7	4.307	4.458	4.615	4.777	4.946	5.120	5.301	5.488	5.682	5.883	6.306	6.761	7.248	7.770	8.329
8	4.661	4.841	5.029	5.225	5.429	5.641	5.863	6.093	6.333	6.584	7.116	7.693	8.319	8.999	9.735
9	4.971	5.181	5.400	5.629	5.870	6.121	6.385	6.661	6.951	7.254	7.904	8.618	9.400	10.26	11.20
10	5.244	5.482	5.732	5.995	6.272	6.564	6.871	7.195	7.536	7.895	8.672	9.534	10.49	11.55	12.73
11	5.484	5.749	6.030	6.326	6.640	6.972	7.324	7.696	8.091	8.508	9.420	10.44	11.59	12.88	14.33
12	5.695	5.986	6.296	6.625	6.976	7.348	7.745	8.167	8.616	9.095	10.15	11.34	12.70	14.24	16.00
13	5.880	6.197	6.535	6.896	7.282	7.695	8.136	8.609	9.115	9.656	10.86	12.24	13.82	15.64	17.73
14	6.042	6.383	6.748	7.141	7.562	8.014	8.501	9.024	9.587	10.19	11.55	13.12	14.95	17.07	19.55
15	6.185	6.548	6.940	7.362	7.817	8.309	8.840	9.414	10.03	10.71	12.22	14.00	16.09	18.55	21.44
16	6.310	6.695	7.111	7.562	8.051	8.580	9.155	9.780	10.46	11.20	12.88	14.87	17.24	20.06	23.42
17	6.420	6.825	7.265	7.743	8.264	8.831	9.449	10.12	10.86	11.67	13.51	15.73	18.39	21.61	25.48
18	6.517	6.941	7.403	7.907	8.458	9.061	9.722	10.45	11.24	12.12	14.13	16.58	19.56	23.19	27.63
19	6.602	7.043	7.526	8.055	8.636	9.274	9.976	10.75	11.60	12.55	14.74	17.43	20.74	24.83	29.88
20	6.676	7.134	7.636	8.189	8.798	9.470	10.21	11.03	11.95	12.96	15.33	18.27	21.93	26.50	32.22
21	6.742	7.214	7.735	8.310	8.946	9.650	10.43	11.30	12.27	13.35	15.90	19.10	23.13	28.22	34.66
22	6.799	7.286	7.823	8.419	9.081	9.817	10.64	11.55	12.58	13.73	16.46	19.93	24.34	29.98	37.21
23	6.850	7.349	7.903	8.518	9.204	9.970	10.83	11.79	12.87	14.09	17.01	20.75	25.56	31.79	39.87
24	6.894	7.405	7.974	8.608	9.317	10.11	11.00	12.01	13.15	14.43	17.54	21.56	26.79	33.64	42.65
25	6.933	7.455	8.037	8.689	9.420	10.24	11.17	12.22	13.41	14.76	18.05	22.36	28.03	35.55	45.55
26	6.967	7.499	8.094	8.762	9.514	10.36	11.32	12.41	13.66	15.07	18.56	23.16	29.29	37.50	48.57
27	6.998	7.539	8.145	8.828	9.600	10.47	11.47	12.60	13.89	15.38	19.05	23.95	30.55	39.50	51.73
28	7.024	7.573	8.191	8.888	9.678	10.58	11.60	12.77	14.11	15.66	19.52	24.73	31.82	41.56	55.02
29	7.047	7.604	8.232	8.942	9.749	10.67	11.72	12.93	14.33	15.94	19.99	25.51	33.11	43.67	58.46
30	7.067	7.632	8.269	8.991	9.815	10.76	11.84	13.08	14.53	16.20	20.44	26.28	34.40	45.84	62.04
31	7.085	7.656	8.302	9.036	9.874	10.84	11.94	13.23	14.72	16.45	20.88	27.04	35.71	48.06	65.78
32	7.101	7.677	8.331	9.076	9.929	10.91	12.04	13.36	14.90	16.70	21.31	27.80	37.03	50.34	69.69
33	7.115	7.696	8.357	9.112	9.978	10.98	12.14	13.49	15.07	16.93	21.73	28.55	38.36	52.68	73.76
34	7.127	7.713	8.381	9.145	10.02	11.04	12.22	13.60	15.23	17.15	22.14	29.29	39.70	55.08	78.01
35	7.138	7.728	8.402	9.174	10.07	11.10	12.30	13.72	15.38	17.36	22.53	30.02	41.06	57.54	82.45
36	7.147	7.742	8.421	9.201	10.10	11.15	12.38	13.82	15.53	17.56	22.92	30.76	42.42	60.07	87.07
37	7.155	7.753	8.438	9.225	10.14	11.20	12.45	13.92	15.66	17.75	23.29	31.48	43.80	62.66	91.90
38	7.162	7.764	8.453	9.247	10.17	11.25	12.51	14.01	15.80	17.94	23.66	32.20	45.19	65.32	96.94
39	7.169	7.773	8.467	9.267	10.20	11.29	12.57	14.10	15.92	18.11	24.02	32.91	46.59	68.05	102.2
40	7.174	7.781	8.479	9.285	10.22	11.33	12.63	14.18	16.04	18.28	24.36	33.61	48.01	70.85	107.7

Table G-18 Uniform Present Value Factors Modified (UPV*)
Discount rate = 16%

Period	Rate of Price Increase per Period 1%	2%	3%	4%	5%	6%	7%	8%	9%	10%	12%	14%	16%	18%	20%
1	0.8707	0.8793	0.8879	0.8966	0.9052	0.9138	0.9224	0.9310	0.9397	0.9483	0.9655	0.9828	1.000	1.017	1.034
2	1.629	1.652	1.676	1.700	1.725	1.749	1.773	1.798	1.823	1.848	1.898	1.949	2.000	2.052	2.105
3	2.289	2.332	2.376	2.421	2.466	2.512	2.558	2.605	2.652	2.700	2.798	2.898	3.000	3.105	3.212
4	2.864	2.930	2.998	3.067	3.137	3.209	3.282	3.356	3.432	3.509	3.667	3.831	4.000	4.175	4.357
5	3.364	3.456	3.550	3.646	3.745	3.846	3.950	4.056	4.164	4.276	4.506	4.747	5.000	5.265	5.542
6	3.800	3.918	4.040	4.166	4.295	4.428	4.566	4.707	4.853	5.003	5.316	5.648	6.000	6.373	6.767
7	4.179	4.325	4.475	4.631	4.793	4.960	5.134	5.314	5.500	5.692	6.098	6.534	7.000	7.500	8.035
8	4.509	4.682	4.862	5.049	5.244	5.447	5.658	5.878	6.107	6.346	6.854	7.404	8.000	8.646	9.347
9	4.797	4.996	5.205	5.423	5.652	5.891	6.141	6.404	6.678	6.966	7.583	8.259	9.000	9.813	10.70
10	5.047	5.272	5.509	5.759	6.021	6.297	6.587	6.893	7.215	7.554	8.287	9.099	10.00	11.00	12.11
11	5.265	5.515	5.780	6.059	6.355	6.668	6.999	7.349	7.719	8.112	8.967	9.925	11.00	12.21	13.56
12	5.455	5.729	6.020	6.329	6.658	7.007	7.378	7.773	8.193	8.640	9.623	10.74	12.00	13.43	15.06
13	5.620	5.917	6.233	6.571	6.931	7.317	7.728	8.168	8.638	9.142	10.26	11.53	13.00	14.68	16.61
14	5.764	6.082	6.423	6.788	7.179	7.600	8.051	8.536	9.057	9.617	10.87	12.32	14.00	15.95	18.22
15	5.890	6.227	6.591	6.982	7.404	7.858	8.349	8.878	9.450	10.07	11.46	13.09	15.00	17.25	19.89
16	5.999	6.355	6.740	7.156	7.607	8.095	8.623	9.197	9.819	10.50	12.03	13.85	16.00	18.56	21.61
17	6.094	6.467	6.873	7.313	7.791	8.311	8.877	9.494	10.17	10.90	12.58	14.59	17.00	19.90	23.38
18	6.176	6.566	6.990	7.453	7.957	8.508	9.110	9.770	10.49	11.29	13.11	15.32	18.00	21.26	25.23
19	6.248	6.653	7.095	7.578	8.108	8.688	9.326	10.03	10.80	11.65	13.63	16.04	19.00	22.64	27.13
20	6.311	6.729	7.188	7.691	8.244	8.853	9.525	10.27	11.09	12.00	14.12	16.75	20.00	24.05	29.10
21	6.366	6.797	7.270	7.792	8.367	9.004	9.708	10.49	11.36	12.32	14.60	17.44	21.00	25.48	31.14
22	6.413	6.856	7.343	7.882	8.479	9.141	9.877	10.70	11.61	12.63	15.06	18.12	22.00	26.94	33.25
23	6.455	6.907	7.408	7.963	8.580	9.267	10.03	10.89	11.85	12.93	15.51	18.79	23.00	28.42	35.43
24	6.491	6.953	7.466	8.036	8.672	9.382	10.18	11.07	12.08	13.21	15.94	19.45	24.00	29.93	37.68
25	6.522	6.993	7.517	8.101	8.755	9.487	10.31	11.24	12.29	13.47	16.35	20.10	25.00	31.46	40.02
26	6.549	7.029	7.563	8.160	8.830	9.583	10.43	11.39	12.48	13.72	16.76	20.73	26.00	33.02	42.43
27	6.573	7.060	7.603	8.212	8.898	9.671	10.55	11.54	12.67	13.96	17.14	21.36	27.00	34.61	44.93
28	6.594	7.087	7.639	8.259	8.959	9.751	10.65	11.67	12.85	14.19	17.52	21.97	28.00	36.22	47.51
29	6.612	7.111	7.671	8.301	9.015	9.824	10.75	11.80	13.01	14.40	17.88	22.58	29.00	37.86	50.19
30	6.628	7.132	7.699	8.339	9.065	9.891	10.83	11.92	13.16	14.61	18.23	23.17	30.00	39.53	52.95
31	6.641	7.151	7.724	8.373	9.110	9.952	10.92	12.03	13.31	14.80	18.57	23.75	31.00	41.23	55.81
32	6.653	7.167	7.746	8.403	9.152	10.01	10.99	12.13	13.45	14.98	18.89	24.33	32.00	42.96	58.77
33	6.664	7.181	7.766	8.431	9.189	10.06	11.06	12.22	13.57	15.16	19.20	24.89	33.00	44.72	61.83
34	6.673	7.194	7.784	8.455	9.223	10.11	11.13	12.31	13.70	15.32	19.51	25.44	34.00	46.50	65.00
35	6.680	7.205	7.799	8.477	9.253	10.15	11.18	12.39	13.81	15.48	19.80	25.99	35.00	48.32	68.27
36	6.687	7.215	7.813	8.497	9.281	10.19	11.24	12.47	13.91	15.62	20.08	26.52	36.00	50.17	71.66
37	6.693	7.223	7.826	8.514	9.306	10.22	11.29	12.54	14.01	15.76	20.36	27.05	37.00	52.06	75.17
38	6.698	7.231	7.837	8.530	9.329	10.26	11.34	12.61	14.11	15.90	20.62	27.57	38.00	53.97	78.79
39	6.703	7.237	7.846	8.544	9.349	10.28	11.38	12.67	14.20	16.02	20.87	28.07	39.00	55.92	82.55
40	6.707	7.243	7.855	8.557	9.368	10.31	11.42	12.73	14.28	16.14	21.12	28.57	40.00	57.90	86.43

Table G-19 Uniform Present Value Factors Modified (UPV*)
Discount rate = 18%

Period	Rate of Price Increase per Period														
	1%	2%	3%	4%	5%	6%	7%	8%	9%	10%	12%	14%	16%	18%	20%
1	0.8559	0.8644	0.8729	0.8814	0.8898	0.8983	0.9068	0.9153	0.9237	0.9322	0.9492	0.9661	0.9831	1.000	1.017
2	1.589	1.612	1.635	1.658	1.682	1.705	1.729	1.753	1.777	1.801	1.850	1.899	1.949	2.000	2.051
3	2.216	2.257	2.300	2.343	2.386	2.430	2.475	2.520	2.565	2.611	2.705	2.801	2.899	3.000	3.103
4	2.752	2.816	2.880	2.946	3.013	3.081	3.151	3.221	3.293	3.366	3.517	3.672	3.833	4.000	4.172
5	3.212	3.298	3.387	3.478	3.571	3.666	3.764	3.864	3.966	4.070	4.287	4.514	4.751	5.000	5.260
6	3.605	3.716	3.829	3.947	4.067	4.192	4.320	4.451	4.587	4.727	5.018	5.327	5.654	6.000	6.366
7	3.942	4.076	4.216	4.360	4.509	4.664	4.824	4.989	5.161	5.338	5.712	6.113	6.541	7.000	7.491
8	4.230	4.388	4.553	4.724	4.902	5.088	5.281	5.482	5.691	5.909	6.371	6.871	7.413	8.000	8.635
9	4.476	4.657	4.847	5.045	5.252	5.469	5.695	5.933	6.181	6.440	6.996	7.605	8.271	9.000	9.798
10	4.687	4.890	5.103	5.328	5.563	5.811	6.071	6.345	6.633	6.936	7.590	8.313	9.114	10.00	10.98
11	4.868	5.092	5.328	5.577	5.840	6.118	6.412	6.723	7.051	7.398	8.153	8.997	9.942	11.00	12.18
12	5.023	5.266	5.523	5.797	6.087	6.394	6.721	7.068	7.437	7.829	8.687	9.658	10.76	12.00	13.41
13	5.155	5.416	5.694	5.990	6.306	6.642	7.001	7.384	7.793	8.230	9.195	10.30	11.56	13.00	14.65
14	5.268	5.546	5.843	6.161	6.501	6.865	7.255	7.674	8.123	8.604	9.676	10.91	12.34	14.00	15.92
15	5.365	5.658	5.973	6.311	6.675	7.065	7.486	7.939	8.427	8.953	10.13	11.51	13.12	15.00	17.20
16	5.448	5.756	6.087	6.444	6.829	7.245	7.695	8.181	8.708	9.278	10.57	12.09	13.88	16.00	18.51
17	5.519	5.840	6.186	6.561	6.967	7.407	7.884	8.403	8.967	9.581	10.98	12.64	14.63	17.00	19.84
18	5.580	5.912	6.272	6.664	7.089	7.552	8.056	8.606	9.207	9.864	11.37	13.18	15.36	18.00	21.20
19	5.632	5.975	6.348	6.754	7.198	7.682	8.212	8.792	9.429	10.13	11.74	13.70	16.08	19.00	22.57
20	5.677	6.029	6.414	6.834	7.295	7.799	8.353	8.962	9.633	10.37	12.09	14.20	16.80	20.00	23.97
21	5.715	6.076	6.471	6.905	7.381	7.904	8.481	9.118	9.822	10.60	12.43	14.69	17.49	21.00	25.40
22	5.747	6.117	6.522	6.967	7.457	7.999	8.597	9.261	9.997	10.82	12.74	15.15	18.18	22.00	26.84
23	5.775	6.152	6.566	7.022	7.526	8.084	8.703	9.391	10.16	11.01	13.05	15.61	18.86	23.00	28.31
24	5.799	6.182	6.604	7.070	7.586	8.160	8.798	9.510	10.31	11.20	13.33	16.04	19.52	24.00	29.81
25	5.820	6.208	6.637	7.113	7.640	8.228	8.885	9.620	10.44	11.37	13.60	16.47	20.17	25.00	31.33
26	5.837	6.231	6.666	7.150	7.689	8.290	8.963	9.720	10.57	11.53	13.86	16.87	20.81	26.00	32.88
27	5.852	6.250	6.692	7.183	7.731	8.345	9.035	9.811	10.69	11.68	14.10	17.27	21.44	27.00	34.46
28	5.865	6.267	6.714	7.212	7.769	8.395	9.099	9.895	10.80	11.82	14.34	17.65	22.06	28.00	36.06
29	5.876	6.282	6.733	7.238	7.803	8.439	9.158	9.972	10.90	11.95	14.56	18.02	22.67	29.00	37.69
30	5.885	6.294	6.750	7.261	7.833	8.479	9.211	10.04	10.99	12.08	14.77	18.37	23.27	30.00	39.34
31	5.893	6.305	6.765	7.280	7.860	8.515	9.259	10.11	11.08	12.19	14.96	18.72	23.86	31.00	41.02
32	5.900	6.315	6.778	7.298	7.884	8.548	9.303	10.17	11.15	12.30	15.15	19.05	24.44	32.00	42.74
33	5.906	6.323	6.789	7.314	7.905	8.577	9.342	10.22	11.23	12.39	15.33	19.37	25.01	33.00	44.48
34	5.911	6.330	6.799	7.327	7.924	8.603	9.378	10.27	11.30	12.49	15.50	19.68	25.57	34.00	46.25
35	5.916	6.336	6.808	7.339	7.941	8.626	9.411	10.31	11.36	12.57	15.66	19.98	26.12	35.00	48.05
36	5.919	6.341	6.815	7.350	7.956	8.647	9.440	10.35	11.41	12.65	15.81	20.26	26.66	36.00	49.88
37	5.922	6.346	6.822	7.359	7.969	8.666	9.467	10.39	11.47	12.73	15.96	20.54	27.19	37.00	51.74
38	5.925	6.350	6.827	7.367	7.981	8.683	9.491	10.43	11.52	12.80	16.10	20.81	27.71	38.00	53.64
39	5.927	6.353	6.832	7.375	7.992	8.699	9.513	10.46	11.56	12.86	16.23	21.07	28.22	39.00	55.56
40	5.929	6.356	6.837	7.381	8.001	8.712	9.533	10.49	11.60	12.92	16.35	21.33	28.73	40.00	57.52

Table G-20 Uniform Present Value Factors Modified (UPV*)
Discount rate = 20%

Period	Rate of Price Increase per Period														
	1%	2%	3%	4%	5%	6%	7%	8%	9%	10%	12%	14%	16%	18%	20%
1	0.8417	0.8500	0.8583	0.8667	0.8750	0.8833	0.8917	0.9000	0.9083	0.9167	0.9333	0.9500	0.9667	0.9833	1.000
2	1.550	1.573	1.595	1.618	1.641	1.664	1.687	1.710	1.733	1.757	1.804	1.853	1.901	1.950	2.000
3	2.146	2.187	2.227	2.269	2.311	2.353	2.396	2.439	2.483	2.527	2.617	2.710	2.804	2.901	3.000
4	2.648	2.709	2.770	2.833	2.897	2.962	3.028	3.095	3.164	3.233	3.376	3.524	3.678	3.836	4.000
5	3.071	3.152	3.236	3.322	3.410	3.499	3.591	3.686	3.782	3.880	4.085	4.298	4.522	4.755	5.000
6	3.426	3.529	3.636	3.746	3.858	3.975	4.094	4.217	4.344	4.474	4.746	5.033	5.338	5.660	6.000
7	3.725	3.850	3.979	4.113	4.251	4.394	4.542	4.695	4.854	5.018	5.363	5.732	6.126	6.549	7.000
8	3.977	4.123	4.274	4.431	4.595	4.765	4.942	5.126	5.317	5.516	5.938	6.395	6.889	7.423	8.000
9	4.189	4.354	4.527	4.707	4.895	5.092	5.298	5.513	5.738	5.973	6.476	7.025	7.626	8.282	9.000
10	4.367	4.551	4.744	4.946	5.158	5.382	5.616	5.862	6.120	6.392	6.977	7.624	8.338	9.128	10.00
11	4.518	4.718	4.930	5.153	5.389	5.637	5.899	6.176	6.468	6.776	7.446	8.193	9.027	9.959	11.00
12	4.644	4.861	5.090	5.333	5.590	5.863	6.152	6.458	6.783	7.128	7.883	8.733	9.693	10.78	12.00
13	4.750	4.982	5.227	5.488	5.766	6.062	6.377	6.712	7.070	7.451	8.290	9.247	10.34	11.58	13.00
14	4.840	5.084	5.345	5.623	5.921	6.238	6.578	6.941	7.330	7.746	8.671	9.734	10.96	12.37	14.00
15	4.915	5.172	5.446	5.740	6.055	6.394	6.757	7.147	7.566	8.018	9.026	10.20	11.56	13.15	15.00
16	4.979	5.246	5.533	5.842	6.174	6.531	6.917	7.332	7.781	8.266	9.358	10.64	12.14	13.91	16.00
17	5.032	5.309	5.607	5.929	6.277	6.652	7.059	7.499	7.976	8.494	9.667	11.06	12.70	14.66	17.00
18	5.077	5.363	5.671	6.005	6.367	6.760	7.186	7.649	8.153	8.703	9.956	11.45	13.25	15.40	18.00
19	5.115	5.408	5.726	6.071	6.446	6.854	7.299	7.784	8.314	8.894	10.23	11.83	13.77	16.13	19.00
20	5.147	5.447	5.773	6.129	6.516	6.938	7.400	7.906	8.461	9.070	10.48	12.19	14.28	16.84	20.00
21	5.173	5.480	5.814	6.178	6.576	7.012	7.490	8.015	8.593	9.231	10.71	12.53	14.77	17.55	21.00
22	5.196	5.508	5.849	6.221	6.629	7.077	7.570	8.114	8.714	9.378	10.93	12.85	15.24	18.24	22.00
23	5.215	5.532	5.878	6.258	6.675	7.135	7.642	8.202	8.823	9.513	11.14	13.16	15.70	18.92	23.00
24	5.231	5.552	5.904	6.290	6.716	7.186	7.706	8.282	8.923	9.637	11.33	13.45	16.15	19.58	24.00
25	5.244	5.569	5.926	6.318	6.752	7.231	7.762	8.354	9.013	9.751	11.51	13.73	16.57	20.24	25.00
26	5.256	5.584	5.945	6.343	6.783	7.271	7.813	8.419	9.095	9.855	11.67	13.99	16.99	20.89	26.00
27	5.265	5.596	5.961	6.364	6.810	7.306	7.858	8.477	9.170	9.950	11.83	14.24	17.39	21.52	27.00
28	5.273	5.607	5.975	6.382	6.834	7.337	7.899	8.529	9.238	10.04	11.97	14.48	17.78	22.15	28.00
29	5.280	5.616	5.987	6.398	6.854	7.364	7.935	8.576	9.299	10.12	12.11	14.71	18.15	22.76	29.00
30	5.286	5.623	5.997	6.411	6.873	7.388	7.967	8.618	9.355	10.19	12.23	14.92	18.51	23.37	30.00
31	5.290	5.630	6.006	6.423	6.888	7.410	7.995	8.657	9.406	10.26	12.35	15.13	18.86	23.96	31.00
32	5.294	5.635	6.013	6.433	6.902	7.428	8.021	8.691	9.452	10.32	12.46	15.32	19.20	24.54	32.00
33	5.298	5.640	6.020	6.442	6.915	7.445	8.044	8.722	9.494	10.38	12.56	15.50	19.53	25.12	33.00
34	5.301	5.644	6.025	6.450	6.925	7.460	8.064	8.750	9.532	10.43	12.66	15.68	19.84	25.68	34.00
35	5.303	5.647	6.030	6.457	6.935	7.473	8.082	8.775	9.567	10.48	12.75	15.84	20.15	26.24	35.00
36	5.305	5.650	6.034	6.462	6.943	7.484	8.098	8.797	9.598	10.52	12.83	16.00	20.44	26.78	36.00
37	5.307	5.653	6.038	6.467	6.950	7.495	8.112	8.818	9.627	10.56	12.91	16.15	20.73	27.32	37.00
38	5.308	5.655	6.041	6.472	6.956	7.504	8.125	8.836	9.652	10.60	12.98	16.29	21.00	27.85	38.00
39	5.309	5.657	6.043	6.476	6.962	7.511	8.137	8.852	9.676	10.63	13.05	16.43	21.27	28.37	39.00
40	5.310	5.658	6.045	6.479	6.966	7.518	8.147	8.867	9.697	10.66	13.11	16.56	21.53	28.88	40.00

Table G-21 Uniform Present Value Factors Modified (UPV*)
Discount rate = 24%

Period	Rate of Price Increase per Period														
	1%	2%	3%	4%	5%	6%	7%	8%	9%	10%	12%	14%	16%	18%	20%
1	0.8145	0.8226	0.8306	0.8387	0.8468	0.8548	0.8629	0.8710	0.8790	0.8871	0.9032	0.9194	0.9355	0.9516	0.9677
2	1.478	1.499	1.521	1.542	1.564	1.586	1.608	1.630	1.652	1.674	1.719	1.765	1.811	1.857	1.904
3	2.018	2.056	2.094	2.132	2.171	2.210	2.250	2.290	2.331	2.372	2.456	2.542	2.629	2.719	2.811
4	2.458	2.514	2.570	2.627	2.685	2.744	2.804	2.866	2.928	2.991	3.121	3.256	3.395	3.539	3.688
5	2.817	2.890	2.965	3.042	3.120	3.201	3.283	3.367	3.453	3.541	3.723	3.913	4.112	4.319	4.536
6	3.109	3.200	3.294	3.390	3.489	3.591	3.696	3.803	3.914	4.028	4.266	4.517	4.782	5.062	5.358
7	3.347	3.455	3.567	3.682	3.801	3.925	4.052	4.184	4.320	4.460	4.756	5.072	5.409	5.769	6.153
8	3.541	3.664	3.793	3.927	4.066	4.210	4.359	4.515	4.676	4.844	5.199	5.582	5.995	6.441	6.922
9	3.698	3.837	3.981	4.132	4.289	4.453	4.625	4.803	4.990	5.184	5.599	6.051	6.544	7.081	7.666
10	3.827	3.979	4.138	4.304	4.479	4.662	4.853	5.054	5.265	5.486	5.960	6.483	7.057	7.690	8.387
11	3.932	4.095	4.268	4.449	4.639	4.840	5.051	5.273	5.507	5.754	6.287	6.879	7.537	8.270	9.084
12	4.017	4.191	4.376	4.570	4.775	4.992	5.221	5.464	5.720	5.991	6.582	7.244	7.987	8.821	9.759
13	4.086	4.270	4.465	4.672	4.890	5.122	5.368	5.630	5.907	6.202	6.848	7.573	8.407	9.346	10.41
14	4.143	4.335	4.540	4.757	4.988	5.234	5.495	5.774	6.072	6.389	7.088	7.887	8.800	9.845	11.04
15	4.189	4.389	4.601	4.828	5.070	5.329	5.605	5.900	6.216	6.554	7.306	8.170	9.168	10.32	11.66
16	4.226	4.433	4.653	4.888	5.140	5.410	5.699	6.010	6.343	6.702	7.502	8.431	9.512	10.77	12.25
17	4.257	4.469	4.696	4.939	5.199	5.480	5.781	6.105	6.455	6.832	7.679	8.670	9.834	11.20	12.82
18	4.282	4.499	4.731	4.981	5.249	5.539	5.851	6.189	6.553	6.948	7.839	8.890	10.13	11.61	13.37
19	4.302	4.523	4.760	5.016	5.292	5.590	5.912	6.261	6.639	7.050	7.984	9.093	10.42	12.00	13.91
20	4.319	4.543	4.785	5.046	5.328	5.633	5.964	6.324	6.715	7.142	8.114	9.279	10.68	12.37	14.43
21	4.332	4.560	4.805	5.071	5.358	5.670	6.010	6.379	6.782	7.222	8.232	9.450	10.93	12.73	14.93
22	4.343	4.573	4.822	5.091	5.384	5.702	6.049	6.427	6.841	7.294	8.339	9.607	11.16	13.06	15.42
23	4.352	4.584	4.836	5.109	5.406	5.729	6.082	6.469	6.892	7.358	8.435	9.752	11.37	13.38	15.89
24	4.359	4.594	4.848	5.124	5.424	5.752	6.111	6.505	6.937	7.414	8.522	9.885	11.57	13.69	16.34
25	4.365	4.601	4.857	5.136	5.440	5.772	6.136	6.537	6.977	7.464	8.601	10.01	11.76	13.98	16.78
26	4.370	4.607	4.865	5.146	5.453	5.789	6.158	6.564	7.012	7.508	8.672	10.12	11.94	14.25	17.21
27	4.374	4.613	4.872	5.155	5.464	5.804	6.177	6.588	7.043	7.548	8.736	10.22	12.10	14.51	17.62
28	4.377	4.617	4.878	5.162	5.474	5.816	6.193	6.609	7.070	7.583	8.793	10.32	12.26	14.76	18.02
29	4.380	4.620	4.882	5.168	5.482	5.827	6.207	6.627	7.094	7.614	8.846	10.40	12.40	15.00	18.41
30	4.382	4.623	4.886	5.173	5.489	5.836	6.219	6.643	7.115	7.641	8.893	10.49	12.54	15.23	18.78
31	4.384	4.625	4.889	5.178	5.494	5.843	6.229	6.657	7.133	7.666	8.935	10.56	12.67	15.44	19.14
32	4.385	4.627	4.892	5.181	5.499	5.850	6.238	6.669	7.149	7.687	8.974	10.63	12.78	15.64	19.49
33	4.386	4.629	4.894	5.184	5.503	5.856	6.246	6.679	7.164	7.706	9.009	10.69	12.89	15.84	19.83
34	4.387	4.630	4.896	5.187	5.507	5.860	6.252	6.688	7.176	7.723	9.040	10.75	13.00	16.02	20.16
35	4.388	4.631	4.897	5.189	5.510	5.865	6.258	6.696	7.187	7.738	9.069	10.80	13.10	16.20	20.48
36	4.389	4.632	4.899	5.191	5.512	5.868	6.263	6.703	7.197	7.752	9.094	10.85	13.19	16.37	20.79
37	4.389	4.633	4.900	5.192	5.515	5.871	6.267	6.709	7.205	7.764	9.117	10.89	13.27	16.53	21.08
38	4.389	4.634	4.901	5.193	5.516	5.874	6.271	6.715	7.213	7.774	9.138	10.93	13.35	16.68	21.37
39	4.390	4.634	4.901	5.195	5.518	5.876	6.274	6.719	7.219	7.784	9.157	10.97	13.42	16.82	21.65
40	4.390	4.634	4.902	5.195	5.519	5.878	6.277	6.723	7.225	7.792	9.174	11.01	13.49	16.96	21.92

Table G-22 Uniform Present Value Factors Modified for DOE-Projected Average Fuel Price Escalation (UPV) by End-Use Sector and Major Fuel[1] Discount Rate = 7 Percent Census Region 1 (Maine, New Hampshire, Vermont, Massachusetts, Connecticut, Rhode Island, New York, New Jersey, Pennsylvania)**

	RESIDENTIAL				COMMERCIAL					INDUSTRIAL					TRANSPORTATION	
N	ELEC	DIST	LPG	NATGAS	ELEC	DIST	RESID	NATGAS	COAL	ELEC	DIST	RESID	NATGAS	COAL	GASLNE	N
1	0.95	0.98	0.98	0.94	0.95	1.01	0.98	0.94	0.95	0.95	1.01	0.98	0.95	0.95	0.95	1
2	1.81	1.92	1.91	1.83	1.81	1.96	1.89	1.83	1.85	1.81	1.97	1.89	1.85	1.85	1.86	2
3	2.62	2.80	2.79	2.67	2.62	2.88	2.75	2.66	2.70	2.61	2.90	2.75	2.71	2.71	2.71	3
4	3.36	3.65	3.62	3.47	3.36	3.76	3.55	3.45	3.52	3.34	3.78	3.55	3.54	3.52	3.51	4
5	4.05	4.46	4.40	4.23	4.05	4.62	4.34	4.19	4.30	4.02	4.65	4.34	4.34	4.29	4.29	5
6	4.70	5.25	5.16	4.96	4.70	5.45	5.13	4.90	5.03	4.65	5.50	5.12	5.12	5.02	5.04	6
7	5.30	6.01	5.89	5.66	5.30	6.28	5.94	5.59	5.73	5.24	6.33	5.92	5.89	5.71	5.79	7
8	5.87	6.75	6.61	6.32	5.87	7.09	6.76	6.24	6.40	5.80	7.16	6.73	6.63	6.36	6.53	8
9	6.41	7.48	7.30	6.97	6.41	7.89	7.59	6.88	7.03	6.34	7.97	7.56	7.37	6.98	7.26	9
10	6.92	8.18	7.97	7.60	6.91	8.67	8.42	7.49	7.62	6.84	8.78	8.38	8.10	7.56	7.98	10
11	7.40	8.86	8.61	8.19	7.40	9.44	9.23	8.07	8.19	7.33	9.56	9.17	8.80	8.11	8.66	11
12	7.85	9.51	9.22	8.76	7.85	10.17	10.00	8.63	8.73	7.79	10.32	9.94	9.48	8.64	9.31	12
13	8.27	10.13	9.80	9.30	8.27	10.87	10.73	9.15	9.24	8.21	11.04	10.66	10.12	9.13	9.93	13
14	8.67	10.71	10.35	9.81	8.67	11.53	11.42	9.65	9.72	8.61	11.72	11.34	10.73	9.60	10.52	14
15	9.04	11.26	10.86	10.29	9.04	12.16	12.07	10.12	10.18	8.99	12.36	11.99	11.30	10.05	11.07	15
16	9.39	11.79	11.35	10.75	9.39	12.75	12.70	10.57	10.61	9.34	12.97	12.60	11.85	10.47	11.59	16
17	9.71	12.28	11.82	11.19	9.71	13.31	13.28	10.99	11.02	9.67	13.55	13.18	12.36	10.87	12.09	17
18	10.02	12.75	12.25	11.59	10.02	13.84	13.84	11.39	11.41	9.98	14.09	13.73	12.85	11.25	12.56	18
19	10.31	13.19	12.66	11.98	10.30	14.34	14.36	11.77	11.78	10.27	14.60	14.24	13.31	11.61	13.00	19
20	10.57	13.60	13.05	12.34	10.57	14.80	14.84	12.12	12.13	10.54	15.08	14.73	13.74	11.95	13.41	20
21	10.82	13.99	13.41	12.68	10.82	15.24	15.31	12.45	12.46	10.79	15.53	15.18	14.14	12.27	13.80	21
22	11.06	14.36	13.76	13.00	11.06	15.66	15.74	12.76	12.77	11.03	15.96	15.61	14.52	12.58	14.17	22
23	11.28	14.70	14.08	13.30	11.28	16.05	16.15	13.06	13.07	11.25	16.36	16.01	14.88	12.87	14.51	23
24	11.49	15.02	14.38	13.58	11.49	16.41	16.53	13.33	13.35	11.46	16.74	16.39	15.22	13.14	14.83	24
25	11.68	15.33	14.66	13.85	11.68	16.76	16.89	13.59	13.62	11.66	17.09	16.75	15.53	13.40	15.14	25

[1]ELEC = Electricity; DIST = Distillate; LPG = Liquefied Petroleum Gas; NATGAS = Natural Gas; RESID = Residual; COAL = Steam Coal; and GASLNE = Gasoline.

Source: Lippiatt and Ruegg (1988). Notation is changed from UPW* to UPV** for consistency with other notation used in this book.

Table G-23 UPV** Discount Factors Adjusted for Average Fuel Price Escalation By End-Use Sector and Major Fuel[1]
Discount Rate = 7 Percent
Census Region 2 (Ohio, Indiana, Illinois, Michigan, Wisconsin, Minnesota, Iowa, Missouri, North Dakota, South Dakota, Nebraska, Kansas)

	RESIDENTIAL				COMMERCIAL					INDUSTRIAL					TRANSPORTATION	
N	ELEC	DIST	LPG	NATGAS	ELEC	DIST	RESID	NATGAS	COAL	ELEC	DIST	RESID	NATGAS	COAL	GASLNE	N
1	0.93	0.99	1.00	0.95	0.93	1.01	0.97	0.95	0.95	0.93	1.01	0.98	0.95	0.95	0.95	1
2	1.79	1.92	1.95	1.84	1.79	1.98	1.88	1.84	1.86	1.78	1.98	1.89	1.86	1.86	1.86	2
3	2.60	2.82	2.84	2.69	2.59	2.90	2.73	2.68	2.72	2.58	2.91	2.75	2.73	2.72	2.71	3
4	3.34	3.67	3.68	3.49	3.34	3.79	3.53	3.47	3.54	3.31	3.80	3.56	3.57	3.55	3.51	4
5	4.03	4.49	4.49	4.27	4.03	4.66	4.32	4.22	4.33	3.99	4.66	4.35	4.39	4.34	4.28	5
6	4.67	5.28	5.27	5.01	4.67	5.51	5.09	4.95	5.07	4.62	5.51	5.15	5.18	5.08	5.04	6
7	5.27	6.05	6.03	5.74	5.26	6.35	5.87	5.65	5.79	5.20	6.35	5.96	5.98	5.80	5.78	7
8	5.83	6.80	6.79	6.43	5.82	7.18	6.67	6.32	6.46	5.74	7.18	6.79	6.75	6.47	6.52	8
9	6.35	7.54	7.52	7.11	6.33	8.00	7.47	6.98	7.10	6.24	8.01	7.64	7.52	7.12	7.25	9
10	6.83	8.26	8.24	7.77	6.82	8.81	8.26	7.62	7.71	6.72	8.82	8.48	8.28	7.73	7.96	10
11	7.29	8.95	8.93	8.40	7.27	9.60	9.03	8.23	8.29	7.16	9.61	9.30	9.02	8.31	8.64	11
12	7.72	9.62	9.58	9.01	7.70	10.36	9.76	8.81	8.84	7.58	10.37	10.09	9.74	8.87	9.28	12
13	8.12	10.25	10.20	9.58	8.10	11.09	10.46	9.37	9.36	7.98	11.09	10.84	10.42	9.39	9.90	13
14	8.50	10.84	10.79	10.13	8.48	11.77	11.12	9.89	9.86	8.35	11.78	11.54	11.06	9.89	10.48	14
15	8.85	11.41	11.34	10.64	8.83	12.42	11.74	10.39	10.33	8.69	12.42	12.21	11.67	10.36	11.03	15
16	9.19	11.94	11.87	11.13	9.16	13.03	12.33	10.86	10.77	9.02	13.04	12.84	12.25	10.81	11.55	16
17	9.50	12.45	12.37	11.59	9.47	13.62	12.89	11.30	11.19	9.32	13.62	13.44	12.79	11.23	12.04	17
18	9.79	12.92	12.84	12.02	9.76	14.16	13.42	11.72	11.59	9.61	14.17	14.01	13.31	11.64	12.51	18
19	10.06	13.37	13.28	12.43	10.04	14.68	13.92	12.12	11.97	9.88	14.68	14.53	13.79	12.02	12.94	19
20	10.32	13.79	13.70	12.82	10.29	15.16	14.38	12.49	12.33	10.13	15.16	15.03	14.24	12.38	13.35	20
21	10.56	14.19	14.09	13.18	10.53	15.62	14.82	12.84	12.67	10.36	15.62	15.50	14.67	12.72	13.74	21
22	10.78	14.56	14.46	13.52	10.75	16.05	15.23	13.16	12.99	10.58	16.05	15.94	15.08	13.05	14.11	22
23	10.99	14.91	14.80	13.84	10.96	16.45	15.62	13.47	13.29	10.79	16.45	16.36	15.45	13.35	14.45	23
24	11.19	15.24	15.13	14.14	11.16	16.83	15.98	13.76	13.58	10.98	16.83	16.75	15.81	13.65	14.77	24
25	11.37	15.56	15.43	14.42	11.34	17.18	16.33	14.04	13.85	11.16	17.19	17.12	16.15	13.92	15.07	25

[1]ELEC = Electricity; DIST = Distillate; LPG = Liquefied Petroleum Gas; NATGAS = Natural Gas; RESID = Residual; COAL = Steam Coal; and GASLNE = Gasoline.

Source: Lippiatt and Ruegg (1988).

Table G-24 UPV Discount Factors Adjusted for Average Fuel Price Escalation By End-Use Sector and Major Fuel[1] Discount Rate = 7 Percent Census Region 3 (Delaware, Maryland, District of Columbia, Virginia, West Virginia, North Carolina, South Carolina, Georgia, Florida, Kentucky, Tennessee, Alabama, Mississippi, Arkansas, Louisiana, Oklahoma, Texas)**

	RESIDENTIAL				COMMERCIAL					INDUSTRIAL					TRANSPORTATION	
N	ELEC	DIST	LPG	NATGAS	ELEC	DIST	RESID	NATGAS	COAL	ELEC	DIST	RESID	NATGAS	COAL	GASLNE	N
1	0.93	0.98	0.99	0.95	0.93	1.01	0.98	0.95	0.95	0.92	1.01	0.98	0.96	0.95	0.95	1
2	1.79	1.92	1.93	1.84	1.78	1.97	1.90	1.84	1.85	1.77	1.97	1.89	1.88	1.85	1.86	2
3	2.58	2.80	2.81	2.69	2.58	2.89	2.76	2.68	2.71	2.56	2.90	2.75	2.76	2.70	2.71	3
4	3.32	3.65	3.65	3.49	3.31	3.78	3.57	3.46	3.53	3.28	3.79	3.55	3.62	3.50	3.51	4
5	4.00	4.46	4.44	4.26	3.99	4.64	4.37	4.22	4.31	3.95	4.65	4.34	4.47	4.27	4.28	5
6	4.64	5.25	5.21	5.01	4.63	5.49	5.17	4.94	5.05	4.57	5.50	5.13	5.31	4.99	5.04	6
7	5.24	6.01	5.96	5.73	5.23	6.32	6.00	5.64	5.75	5.15	6.33	5.93	6.16	5.68	5.79	7
8	5.80	6.75	6.69	6.42	5.79	7.14	6.85	6.32	6.42	5.70	7.16	6.75	6.99	6.33	6.53	8
9	6.33	7.48	7.40	7.09	6.31	7.96	7.71	6.97	7.05	6.22	7.97	7.58	7.84	6.95	7.26	9
10	6.83	8.18	8.09	7.75	6.81	8.76	8.58	7.61	7.64	6.72	8.77	8.40	8.68	7.53	7.97	10
11	7.30	8.86	8.76	8.38	7.27	9.54	9.42	8.22	8.21	7.18	9.56	9.21	9.51	8.08	8.65	11
12	7.74	9.51	9.39	8.98	7.72	10.30	10.24	8.80	8.75	7.62	10.31	9.98	10.31	8.60	9.31	12
13	8.15	10.13	9.98	9.55	8.13	11.01	11.00	9.35	9.26	8.03	11.03	10.70	11.08	9.09	9.92	13
14	8.54	10.71	10.55	10.09	8.52	11.69	11.73	9.87	9.75	8.42	11.70	11.40	11.80	9.56	10.51	14
15	8.91	11.26	11.08	10.60	8.88	12.33	12.42	10.37	10.21	8.78	12.34	12.05	12.48	10.01	11.06	15
16	9.25	11.79	11.59	11.09	9.22	12.93	13.07	10.84	10.64	9.12	12.95	12.67	13.13	10.43	11.58	16
17	9.57	12.28	12.07	11.55	9.54	13.51	13.69	11.28	11.05	9.44	13.53	13.25	13.75	10.82	12.08	17
18	9.87	12.75	12.52	11.98	9.84	14.05	14.28	11.70	11.44	9.74	14.07	13.81	14.33	11.20	12.55	18
19	10.15	13.19	12.95	12.38	10.12	14.56	14.82	12.09	11.81	10.02	14.58	14.33	14.87	11.56	12.99	19
20	10.41	13.60	13.35	12.76	10.39	15.04	15.33	12.46	12.16	10.28	15.06	14.81	15.38	11.90	13.40	20
21	10.66	13.99	13.72	13.12	10.63	15.49	15.82	12.81	12.49	10.53	15.51	15.27	15.86	12.22	13.79	21
22	10.89	14.35	14.08	13.46	10.86	15.91	16.28	13.14	12.81	10.76	15.93	15.70	16.31	12.52	14.15	22
23	11.11	14.70	14.41	13.78	11.08	16.31	16.70	13.44	13.11	10.98	16.33	16.11	16.74	12.81	14.50	23
24	11.31	15.02	14.72	14.08	11.28	16.68	17.11	13.73	13.39	11.18	16.71	16.49	17.14	13.08	14.82	24
25	11.50	15.33	15.02	14.36	11.47	17.04	17.49	14.01	13.66	11.37	17.06	16.85	17.52	13.34	15.13	25

[1]ELEC = Electricity; DIST = Distillate; LPG = Liquefied Petroleum Gas; NATGAS = Natural Gas; RESID = Residual; COAL = Steam Coal; and GASLNE = Gasoline.

Source: Lippiatt and Ruegg (1988).

Table G-25 UPV** Discount Factors Adjusted for Average Fuel Price Escalation By End-Use Sector and Major Fuel[1] Discount Rate = 7 Percent Census Region 4 (Montana, Idaho, Wyoming, Colorado, New Mexico, Arizona, Utah, Nevada, Washington, Oregon, California, Alaska, Hawaii)

	RESIDENTIAL				COMMERCIAL					INDUSTRIAL					TRANSPORTATION	
N	ELEC	DIST	LPG	NATGAS	ELEC	DIST	RESID	NATGAS	COAL	ELEC	DIST	RESID	NATGAS	COAL	GASLNE	N
1	0.94	0.99	0.99	0.95	0.94	1.01	0.97	0.95	0.96	0.94	1.01	0.98	0.95	0.96	0.96	1
2	1.81	1.93	1.93	1.84	1.81	1.97	1.88	1.84	1.89	1.81	1.98	1.89	1.87	1.89	1.86	2
3	2.62	2.82	2.81	2.69	2.62	2.89	2.74	2.67	2.79	2.62	2.90	2.75	2.74	2.77	2.71	3
4	3.38	3.67	3.65	3.50	3.38	3.78	3.54	3.46	3.65	3.37	3.79	3.56	3.59	3.62	3.52	4
5	4.09	4.49	4.44	4.27	4.09	4.64	4.32	4.20	4.48	4.08	4.66	4.35	4.41	4.44	4.29	5
6	4.76	5.29	5.21	5.02	4.76	5.48	5.10	4.92	5.28	4.75	5.51	5.14	5.22	5.23	5.04	6
7	5.39	6.06	5.95	5.74	5.38	6.31	5.89	5.61	6.05	5.38	6.35	5.95	6.03	5.99	5.79	7
8	5.98	6.82	6.69	6.44	5.97	7.14	6.69	6.27	6.78	5.98	7.19	6.78	6.83	6.70	6.53	8
9	6.54	7.55	7.40	7.12	6.53	7.95	7.49	6.92	7.49	6.54	8.01	7.62	7.62	7.39	7.26	9
10	7.07	8.27	8.09	7.78	7.07	8.75	8.29	7.54	8.17	7.09	8.82	8.46	8.40	8.05	7.98	10
11	7.58	8.97	8.75	8.41	7.57	9.53	9.07	8.13	8.83	7.61	9.61	9.28	9.17	8.69	8.66	11
12	8.06	9.63	9.38	9.02	8.05	10.29	9.82	8.70	9.46	8.10	10.38	10.06	9.91	9.30	9.31	12
13	8.50	10.26	9.98	9.60	8.50	11.00	10.52	9.23	10.05	8.56	11.11	10.80	10.61	9.88	9.93	13
14	8.92	10.86	10.54	10.14	8.92	11.67	11.19	9.74	10.62	8.99	11.79	11.50	11.28	10.43	10.51	14
15	9.32	11.42	11.07	10.66	9.31	12.31	11.82	10.22	11.16	9.40	12.44	12.17	11.91	10.95	11.06	15
16	9.69	11.96	11.58	11.15	9.68	12.92	12.42	10.68	11.67	9.78	13.06	12.80	12.51	11.44	11.59	16
17	10.03	12.47	12.06	11.61	10.02	13.49	12.99	11.11	12.15	10.13	13.64	13.40	13.08	11.91	12.08	17
18	10.36	12.95	12.51	12.05	10.34	14.03	13.52	11.51	12.61	10.47	14.19	13.96	13.61	12.35	12.55	18
19	10.66	13.39	12.94	12.46	10.65	14.54	14.03	11.89	13.04	10.78	14.71	14.49	14.11	12.77	12.99	19
20	10.94	13.82	13.34	12.84	10.93	15.02	14.50	12.25	13.45	11.07	15.19	14.98	14.58	13.17	13.40	20
21	11.21	14.21	13.71	13.21	11.20	15.47	14.94	12.59	13.84	11.34	15.65	15.45	15.02	13.55	13.79	21
22	11.46	14.59	14.06	13.55	11.44	15.89	15.36	12.91	14.21	11.60	16.08	15.89	15.44	13.90	14.16	22
23	11.69	14.94	14.40	13.87	11.68	16.29	15.75	13.21	14.56	11.84	16.48	16.30	15.83	14.24	14.50	23
24	11.91	15.27	14.71	14.17	11.90	16.66	16.12	13.49	14.89	12.06	16.86	16.69	16.20	14.56	14.82	24
25	12.12	15.58	15.00	14.46	12.10	17.01	16.47	13.75	15.20	12.28	17.22	17.06	16.55	14.86	15.13	25

[1]ELEC = Electricity; DIST = Distillate; LPG = Liquefied Petroleum Gas; NATGAS = Natural Gas; RESID = Residual; COAL = Steam Coal; and GASLNE = Gasoline.

Source: Lippiatt and Ruegg (1988).

Table G-26 UPV** Discount Factors Adjusted for Average Fuel Price Escalation By End-Use Sector and Major Fuel[1] Discount Rate = 7 Percent United States Average

	RESIDENTIAL				COMMERCIAL					INDUSTRIAL					TRANSPORTATION	
N	ELEC	DIST	LPG	NATGAS	ELEC	DIST	RESID	NATGAS	COAL	ELEC	DIST	RESID	NATGAS	COAL	GASLNE	N
1	0.93	0.99	0.99	0.94	0.93	1.01	0.97	0.95	0.95	0.93	1.01	0.98	0.96	0.95	0.95	1
2	1.80	1.92	1.93	1.84	1.79	1.97	1.88	1.83	1.86	1.79	1.98	1.89	1.87	1.86	1.86	2
3	2.60	2.81	2.82	2.69	2.59	2.89	2.74	2.67	2.72	2.58	2.90	2.75	2.75	2.72	2.71	3
4	3.34	3.66	3.66	3.49	3.34	3.77	3.54	3.45	3.54	3.31	3.79	3.55	3.59	3.54	3.51	4
5	4.03	4.47	4.45	4.26	4.02	4.63	4.33	4.20	4.33	3.99	4.66	4.34	4.42	4.32	4.28	5
6	4.67	5.26	5.23	5.00	4.67	5.47	5.12	4.92	5.07	4.62	5.51	5.13	5.25	5.06	5.04	6
7	5.27	6.03	5.98	5.72	5.27	6.30	5.92	5.61	5.78	5.22	6.34	5.93	6.06	5.77	5.79	7
8	5.84	6.78	6.72	6.41	5.83	7.12	6.74	6.28	6.46	5.77	7.17	6.75	6.87	6.44	6.53	8
9	6.37	7.50	7.43	7.09	6.36	7.93	7.58	6.93	7.10	6.29	7.99	7.59	7.68	7.07	7.26	9
10	6.87	8.21	8.13	7.74	6.86	8.73	8.41	7.55	7.71	6.79	8.80	8.41	8.48	7.68	7.97	10
11	7.34	8.90	8.80	8.36	7.33	9.50	9.22	8.15	8.29	7.26	9.58	9.22	9.27	8.25	8.65	11
12	7.79	9.55	9.44	8.96	7.78	10.25	10.00	8.73	8.84	7.71	10.34	9.99	10.03	8.79	9.30	12
13	8.20	10.17	10.04	9.53	8.20	10.96	10.73	9.27	9.36	8.12	11.06	10.73	10.76	9.31	9.92	13
14	8.59	10.76	10.61	10.07	8.59	11.63	11.43	9.78	9.86	8.51	11.74	11.42	11.44	9.80	10.50	14
15	8.96	11.32	11.15	10.58	8.95	12.26	12.09	10.27	10.33	8.88	12.39	12.07	12.09	10.27	11.05	15
16	9.30	11.85	11.66	11.06	9.30	12.87	12.71	10.73	10.77	9.22	13.00	12.70	12.70	10.70	11.58	16
17	9.63	12.34	12.15	11.52	9.62	13.44	13.30	11.16	11.19	9.54	13.58	13.29	13.28	11.12	12.07	17
18	9.93	12.81	12.61	11.95	9.92	13.97	13.86	11.57	11.59	9.85	14.12	13.84	13.83	11.52	12.54	18
19	10.21	13.26	13.04	12.35	10.20	14.48	14.38	11.96	11.97	10.13	14.63	14.36	14.34	11.89	12.98	19
20	10.47	13.67	13.44	12.73	10.47	14.95	14.87	12.32	12.33	10.39	15.11	14.85	14.83	12.24	13.39	20
21	10.72	14.06	13.82	13.09	10.71	15.40	15.34	12.67	12.66	10.64	15.57	15.31	15.28	12.58	13.78	21
22	10.95	14.43	14.18	13.43	10.95	15.82	15.77	12.99	12.99	10.87	15.99	15.74	15.71	12.90	14.14	22
23	11.17	14.78	14.51	13.75	11.16	16.21	16.18	13.29	13.29	11.09	16.39	16.15	16.11	13.20	14.49	23
24	11.37	15.10	14.83	14.04	11.37	16.59	16.57	13.57	13.58	11.29	16.77	16.53	16.49	13.48	14.81	24
25	11.56	15.41	15.13	14.33	11.56	16.94	16.93	13.84	13.85	11.48	17.13	16.90	16.85	13.75	15.11	25

[1] ELEC = Electricity; DIST = Distillate; LPG = Liquefied Petroleum Gas; NATGAS = Natural Gas; RESID = Residual; COAL = Steam Coal; and GASLNE = Gasoline.

Source: Lippiatt and Ruegg (1988).

Index

Index

MIX
Papier aus verantwortungsvollen Quellen
Paper from responsible sources
FSC® C105338

If you have any concerns about our products,
you can contact us on
ProductSafety@springernature.com

In case Publisher is established outside the EU,
the EU authorized representative is:
Springer Nature Customer Service Center GmbH
Europaplatz 3, 69115 Heidelberg, Germany

Printed by Libri Plureos GmbH
in Hamburg, Germany